The Ecology and Conservation of Seasonally Dry Forests in Asia

The Ecology and Conservation of Seasonally Dry Forests in Asia

Edited by

William J. McShea,
Stuart J. Davies,
and Naris Bhumpakphan

A SMITHSONIAN CONTRIBUTION TO KNOWLEDGE

Published in Cooperation with
ROWMAN & LITTLEFIELD PUBLISHERS, INC.

Smithsonian Institution
Scholarly Press

Washington, D.C.
2011

Published by SMITHSONIAN INSTITUTION SCHOLARLY PRESS

P.O. Box 37012, MRC 957
Washington, D.C. 20013-7012
www.scholarlypress.si.edu

In cooperation with
ROWMAN & LITTLEFIELD PUBLISHERS, INC.

Published in the United States of America
by Rowman & Littlefield Publishers, Inc.
A wholly owned subsidary of The Rowman & Littlefield Publishing Group, Inc.
4501 Forbes Boulevard, Suite 200, Lanham, Maryland 20706
www.rowmanlittlefield.com

Estover Road
Plymouth PL6 7PY
United Kingdom

British Library Cataloguing in Publication Information Available

Library of Congress Cataloging-in-Publication Data:
    The ecology and conservation of seasonally dry forests in Asia / edited by
William J. McShea, Stuart J. Davies, and Naris Bhumpakphan.
        p. cm. "Published in cooperation with Rowman & Littlefield Publishers, Inc."
    ISBN 978-1-935623-02-1 (cloth : alk. paper)
    1. Forest ecology—Asia. 2. Forest conservation—Asia. 3. Forest management—Asia.
I. McShea, William J. II. Davies, Stuart J. III. Bhumpakphan, Naris.
QH179.E264 2011
634.9095'0913—dc22   2010040900

Printed in the United States of America

∞ ™ The paper used in this publication meets the minimum requirements of American
National Standard for Information Sciences—Permanence of Paper for Printed Library
Materials, ANSI/NISO Z39.48-1992.

Contents

Introduction:
Seasonally Dry Forests of Tropical Asia
An Ecosystem Adapted to Seasonal Drought, Frequent Fire, and Human Activity

William J. McShea and Stuart J. Davies

Seasonally dry tropical forests (SDTFs) extend across the world within the band of tropical latitudes where high temperatures and seasonal rainfall result in extended periods of predictable water stress (Mooney et al. 1995; Miles et al. 2006). This seasonal water stress is reflected in the composition of the plant communities and the animal communities reliant on those plants, and influences the way forest-dependent people interact with their environment. SDTFs occur in all three tropical regions (America, Africa, and Asia; Miles et al. 2006) and are among the most extensive ecosystems within these regions, but are being degraded and eliminated by increasing human activity (Murphy and Lugo 1986; Janzen 1988; Stott 1990; Sanchez-Azofeifa et al. 2005). Previous studies of the dry forests of America (Bullock et al. 1995; Frankie et al. 2004; Pennington et al. 2009) and Africa (Mistry and Berardi 2006) have emphasized the interaction between dry forests and human activity. The interaction between humans and the environment in seasonally dry tropical Asia is particularly strong; high human densities are not a recent phenomenon in this area, and the historical record indicates that fluctuations in the intensity of the monsoon have significantly impacted human populations (Buckley et al. 2010; Cook et al. 2010).

On a global scale, SDTFs are under severe threat (Janzen 1988; Bullock et al. 1995; Pennington et al. 2006; Prance 2006). The situation in tropical Asia is no exception. Rapid economic growth and population expansion with their consequent pressure on forests led to rapid forest loss and conversion across the area during the twentieth century (Food and Agriculture Organization of the United Nations 2007). However, there remain widely divergent estimates of the spatial distribution of SDTF and the degree of forest degradation (Miles et al. 2006; Leimgruber et al., this volume). Considering SDTF as only those forests with a five- to six-month dry season, Miles et al. (2006) estimated that in 2001 only about 16 percent of continental Southeast Asia's original dry forest remained. Considering the entire region, including all degrees of seasonality, Myers et al. (2000) estimated that only 5 percent of the area remained in relatively

I

pristine condition. However, given the long history of human activity, it is unlikely that any of these forests have been unaffected by humans. Better geographical information on the extent and conservation status of the remaining SDTF in Asia will be an important part of future effective forest management (Leimgruber et al., this volume).

The degree of threat to SDTF ecosystems varies considerably among countries and forest types in the region. Although protection of the remaining forests has improved in recent years, they are still under threat from illegal extraction of timber and non-timber products, expansion of commercial agriculture, fragmentation, hunting and bushmeat extraction, and invasion of exotic species (Wikramanayake et al. 2002; Sodhi et al. 2004). Montane forests are naturally fairly well protected by geographic remoteness (but see Santisuk 1988), however, the expansion of rubber plantations in the hills of southern Yunnan is a potentially serious threat (Qiu 2009). Less than 10 percent of the remaining deciduous dipterocarp forest, a key dry forest type, is protected in Southeast Asia (McShea et al. 2005). This forest loss and degradation has important consequences for the region's fauna. In an analysis of the Indian subcontinent, the extent of dry deciduous forest was a strong predictor of the presence of fifteen of the twenty large mammal species examined (Karanth et al. 2009); only protected status of the land was a stronger predictor of large mammal species richness.

Priority areas of SDTF for conservation have been proposed (Wikramanayake et al. 2002), including the globally outstanding regions of dry and dry evergreen forests in Indochina and the eastern highlands of India. In addition to expanding protected areas designed for the conservation of threatened animal and plant species, there is an urgent need for SDTF ecosystems in Asia to be more sustainably managed (Ashton 1990). These management systems need to be carefully designed to incorporate sustainable agriculture and forest production, enable economic development of rural populations, and protect the region's remaining forests. In areas where substantial rural–urban migration and associated land abandonment are occurring, opportunities to restore SDTF exist, and techniques for doing this need to be developed (Blakesley et al. 2002; Sayer et al. 2004; Forest Restoration Research Unit 2008). In some areas, large-scale conservation efforts have been initiated, for example, within the Mekong River watershed (see Maxwell and Cox, this volume), but most other areas, such as the Irrawaddy River watershed in Myanmar, lack effective sustainable management (Rao et al. 2002).

Three features of SDTF are important to understanding its ecology and conservation: rainfall patterns, fire frequency, and interactions with humans. These factors influence seasonally dry forests throughout the tropics, but the nature of their influence is unique to Asian forests. Here we provide an overview of some of the key issues addressed in the following chapters.

RAINFALL DISTRIBUTION

SDTF in Asia, as elsewhere in the tropics, is characterized by having less than 1,500–1,800 mm of annual rainfall and a predictable annual dry season (Bullock et al. 1995; Pennington et al. 2009). The weather patterns of continental Southeast Asia are largely controlled by the movement of a low pressure trough, the Intertropical Convergence

Zone, that brings heavy "monsoonal" rain to the region from April/May to October/November during the Northern Hemisphere summer (Lau and Yang 1997). The climate experienced by Asian dry forests is seasonal, with a strong dry season of three to six months alternating with a season of consistently high rainfall in which all months receive at least 100 mm of rain and often considerably more. All forests are subject to at least some degree of predictable annual water deficit during the dry season, although the intensity of both the monsoon rains and the annual dry season vary considerably (Cook et al. 2010). SDTF in Asia includes forest types that range from evergreen to deciduous. Six of the major SDTF types are described in detail here (see Bunyavejchewin et al., this volume). Our definition of SDTF includes deciduous dipterocarp forest and pine-dominated forest, where an open canopy and frequent fire produces an abundant grass cover supporting a diverse community of large mammal species. Pennington et al. (2006, 2009), in review of dry forests in the neotropics, defines forest with a grass-dominated understory as *savanna*. We consider all seasonally dry, tree-dominated ecosystems in Asia as SDTF.

Along much of the regional gradient of seasonal water stress, evergreen and deciduous forests occur in close proximity, suggesting that water stress is not the only factor controlling forest-type distribution in Asian SDTFs. Complex topography, soil heterogeneity, perennial and seasonal stream beds, and complex disturbance histories result in a mosaic of vegetation types that create a diverse landscape for the flora and fauna of the region (Bunyavejchewin et al., this volume). High β-diversity, which is a characteristic of dry forests in the neotropics (Pennington et al. 2009), also characterizes Asian SDTF. In contrast to the high β-diversity of SDTFs in Asia, individual stands are dominated by relatively few species, resulting in lower α-diversity than aseasonal, wet forests in equatorial Asia (see Bunyavejchewin et al., this volume).

FIRE

A second characteristic feature of SDTF in Asia is the incidence of fire. Unlike many SDTFs in the neotropics (Pennington et al. 2009), lowland Asian dry forests burn periodically. Miles et al. (2006) estimated that 20 percent of dry forests in Southeast Asia burn each year. At any single location, fires may occur on annual to decadal frequencies and vary considerably in intensity (Kobsak Wanthongchai and Goldammer, this volume). Fire-sensitive succulents that are found within neotropical dry forests (Pennington et al. 2009) are almost absent from Asian dry forests. Grasses are the fuel for dry season fires in deciduous dipterocarp forests, and in moister deciduous forest types, where grasses are less abundant, dried bamboo clumps and the herbaceous layer support fire during particularly dry years (Baker et al. 2008). Fires remove downed woody debris and dense shrubs, and open the forests for movement of large mammals (Stott 1988; Kobsak Wanthongchai and Goldammer, this volume).

Little is known about long-term fire regimes in the region and fire's role in maintaining observed landscape patterns and dynamics (Miles et al. 2006; Baker and Bunyavejchewin 2009). However, it is clear that fire in these forests has both man-made and natural origins (Stott 1988, 1990; Lavorel et al. 2007). The last months of the dry season produce frequent lightning storms that can readily ignite dead and dried leaves from deciduous trees and senescent ground cover (Goldammer and Price 1998). Tree species found in Asian SDTF are adapted to ground fires with low average flame heights that occur primarily in the dry season (Stott 1990; Baker et al. 2008). Both deciduous and evergreen SDTFs experience dry season fires, and the limited evidence available indicates that both forest types, while suffering enhanced tree mortality among the smaller size classes, are adapted to periodic fire (Baker and Bunyavejchewin 2009). How the different SDTF types and their dominant tree species respond to different frequencies and intensities of fire remains poorly understood. New technologies that enable remote tracking of the extent of fires across Asia (Dwyer et al. 2000; van der Werf et al. 2008) provide an opportunity for better understanding of the relationships between fire regimes and the distribution of SDTFs of Asia. Coupled with this remote sensing work is the need for more-detailed field studies of the interactions between fire regimes and forest types with their constituent plant and animal species. These studies are urgently needed to provide the means for more-effective landscape management.

In the extensively studied savanna ecosystems of Africa and North America, fire plays a critical role in maintaining habitats that support abundant large mammal communities (Archibald et al. 2005; Fuhlendorf et al. 2008). Fuhlendorf et al. (2008) consider the restoration of large mammals to North American savannas to be possible only using a management system that includes prescribed fire. The authors describe a system where a combination of fire and herbivory restores and maintains grasslands. In Asian seasonal dry forests with a grass-dominated understory, large mammals and fire may play similar roles. Populations of large mammals, including elephant (*Elephas maximus*), banteng (*Bos jananicus*), gaur (*Bos gaurus*), and Eld's deer (*Rucervus eldii*), require succulent grass shoots and forbs for forage, and these can only be maintained by regular fire, whether natural or prescribed. SDTFs, however, have a landscape heterogeneity that is absent from many savanna systems, which makes direct comparisons difficult, and adds a complexity to Asian forest-community responses to fire.

HUMAN-FOREST INTERACTIONS

The third key feature of SDTF in Asia concerns the interactions of human communities with the forested environment. The Indochinese and Indian subcontinents have been occupied by people for tens of thousands of years with relatively high densities of humans being present in the area for millennia (Stott 1988). The array of forest products used by villagers (Songer et al., this volume), and the attitudes of rural people toward their natural environment (Allendorf, this volume), illustrate the dependence of people on the forest for sustenance, and the adaptations human culture has made to the cycles of the forest.

SDTFs provide valuable resources to local communities, including wood, livestock fodder, food, and non-timber products. Murphy and Lugo (1986) estimated that 80 percent of wood removed from tropical dry forests is used for the purpose of cooking and heating households rather than timber extraction. Songer et al. (this volume) studied natural resource use by villages surrounding a SDTF reserve in Myanmar. Over 90 percent of the households interviewed collected fuelwood from inside the sanctuary, including both the core area and buffer zone. This finding is consistent with reports of forest use from throughout the region (V. Singh and J. Singh 1989; Ganesan 1993; Wikramanayake et al. 2002; Davidar et al. 2007). Beyond fuelwood, other important products derived from dry forests by local villagers include medicines, rubber, resin, cooking oils, wild game, fruit, nuts, and thatch (Stott 1990; Murali et al. 1996). In the Songer et al. study (this volume), 75 percent of households also relied on the forests to provide food to supplement their harvest. Fodder for livestock is also an important component of SDTF use in the region. The grasses and forbs that are so beneficial for wildlife are a primary source of fodder for livestock pastured around and in the forests. Despite the importance to rural communities of fuelwood and fodder extraction, recent estimates of extraction in SDTFs in India indicate that current rates are unsustainable (Pandey and J. Singh 1984; V. Singh and J. Singh 1989; Ganesan 1993; Bahuguna 2000). Further work in this area is urgently needed.

An important component of the human-forest interaction in SDTFs in Asia involves the active use of fire for habitat control. Most fires in SDTFs in Asia are currently initiated by humans. Villagers living in dry forests have a complex code of conduct surrounding responsible fire management (MacInnes, this volume). Local villagers start fires to clear agricultural waste from fields, clear new fields, enhance new growth for livestock, and encourage the growth of non-timber forest products (e.g., thatch grass and mushrooms; MacInnes, this volume). These practices are not new (Maxwell 2004), but they may be increasing or having an increasing impact on the remaining SDTFs as populations increase and the area of rural land available to forest-dependent people declines with urban-area expansion (Stott 1990; Lavorel et al. 2007; Kobsak Wanthongchai and Goldammer, this volume).

A combination of reduced forest area, increased fragmentation of SDTFs, and changing intensity of the impact of rural communities on the remaining forests requires that more-active forest management systems be developed for SDTFs in Asia. A key part of this is the need to develop a better understanding of the impact of human-induced and natural fire regimes on each SDTF type, including an assessment of the consequences of different fire regimes for the different services required from these forests. Climate change–related shifts in environmental conditions are also likely to make the management of SDTFs in Asia more difficult. Natural fire regimes are likely to change with impending climate change (Goldammer and Price 1998; Lavorel et al. 2007; van der Werf 2008). Lightning strikes are projected to increase throughout the tropics, with a 20–50 percent increase in annual area burned (Goldammer and Price 1998). Increases in conditions favorable to fire may occur (Lavorel et al. 2007), and beyond changes in the fire regime itself, changes in the degree of annual water stress may further affect these forests through changes in fire susceptibility. Landscape managers need to engage local community knowledge and assistance in developing sustainable forest and fire management systems. Banning human-set fires within protected dry forests disrupts the human settlements that rely on the forests (Maxwell and Cox, this volume), and will potentially alter wildlife and plant communities within these areas. Increasing the use of prescribed fire in SDTF in Asia to set fire intervals at optimal levels for forest maintenance may be required; however, much greater understanding of how different fire regimes affect the forests and the interactions between people and forests is required before these management systems can be recommended.

This volume describes the ecology of seasonally dry forests in tropical Asia. Since none of these forests are beyond the influence of human activities, understanding the ecology of these forests requires a thorough investigation of the interactions between humans and the landscape. The volume is organized into three sections. The first section describes SDTF types in Asia, assesses the extent of this forest type throughout the region, and examines the role of disturbance and fire in creating the forest mosaic across the region. The second section provides a detailed description of the animal species and communities that depend on SDTF. The third and final section considers how humans interact with dry forest, including people's attitudes, the products they use, and how they manage the forests by controlling fire regimes. We hope this volume will be of use to the people charged with managing the biologically diverse but threatened SDTFs of Asia. We also hope it will stimulate ideas for future work by students and researchers interested in this dynamic region.

REFERENCES

Archibald, S., W. J. Bond, W. D. Stock, and D. H. K. Fairbanks. 2005. Shaping the Landscape: Fire-Grazer Interactions in an African Savanna. *Ecological Applications* 15:96–109.

Ashton, P. S. 1990. Thailand: Biodiversity Center for the Tropics of Indo-Burma. *Journal of Science and Society, Thailand* 16:107–16.

Bahuguna, V. 2000. Forests in the Economy of the Rural Poor: An Estimation of the Dependency Level. *Ambio* 29:126–29.

Baker, P. J., and S. Bunyavejchewin. 2009. Fire Behavior and Fire Effects across the Forest Landscape Mosaics of Continental Southeast Asia. In *Tropical Fire Ecology: Climate Change, Land Use and Ecosystem Dynamics*, ed. M. A. Cochrane, 311–34. Heidelberg, Germany: Springer-Praxis.

Baker, P. J., S. Bunyavejchewin, and A. P. Robinson. 2008. The Impacts of Large-Scale, Low-Intensity Fires on the Forests of Continental South-East Asia. *International Journal of Wildland Fire* 17:782–92.

Blakesley, D., S. Elliott, C. Kuarak, P. Navakitbumrung, S. Zangkum, and V. Anusarnsunthorn. 2002. Propagating Framework Tree Species to Restore Seasonally Dry Tropical Forest: Implications of Seasonal Seed Dispersal and Dormancy. *Forest Ecology and Management* 164:31–38.

Buckley, B. M., K. J. Anchukaitis, D. Penny, R. Fletcher, E. R. Cook, M. Sano, L. C. Nam, A. Wichienkeeo, T. T. Minhe, and T. M. Hong. 2010. Climate as a Contributing Factor in the Demise of Angkor, Cambodia. *Proceedings of the National Academy of Sciences, USA* 107:6748–52.

Bullock, S. H., H. A. Mooney, and E. Medina, eds. 1995. *Seasonally Dry Tropical Forests.* Cambridge: Cambridge University Press.

Cook, E. R., K. J. Anchukaitis, B. M. Buckley, R. D. D'Arrigo, G. C. Jacoby, and W. E. Wright. 2010. Asian Monsoon Failure and Megadrought during the Last Millennium. *Science* 328:486–89.

Davidar, P., M. Arjunan, P. Mammen, J. Garrigues, I. Puyravaud, and K. Roessingh. 2007. Forest Degradation in the Western Ghats Biodiversity Hotspot: Resource Collection, Livelihood Concerns and Sustainability. *Current Science* 93:1573–78.

Dwyer, E., J. M. C. Pereira, J. Gregoire, and C. C. DaCamara. 2000. Characterization of the Spatio-Temporal Patterns of Global Fire Activity Using Satellite Imagery for the Period April 1992 to March 1993. *Journal of Biogeography* 27:57–69.

Food and Agriculture Organization of the United Nations. 2007. *State of the World's Forest*, 7th ed. Rome: Food and Agriculture Organization of the United Nations.

Forest Restoration Research Unit. 2008. *Research for Restoring Tropical Forest Ecosystems: A Practical Guide*. Forest Restoration Research Unit, Biology Department, Science Faculty, Chiang Mai University, Thailand.

Frankie, G. W., A. Mata, and S. B. Vinson, eds. 2004. *Biodiversity Lessons in Costa Rica: Learning Lessons in a Seasonally Dry Forest*. Berkeley: University of California Press.

Fuhlendorf, S. D., D. M. Engle, J. Kerby, and R. Hamilton. 2008. Pyric Herbivory: Rewilding Landscapes through the Recoupling of Fire and Grazing. *Conservation Biology* 23:588–98.

Ganesan, B. 1993. Extraction of Non-timber Forest Products, including Fodder and Fuelwood, in Mudumalai, India. *Economic Botany* 47:268–74.

Goldammer, J. G., and C. Price. 1998. Potential Impacts of Climate Change on Fire Regimes in the Tropics Based on MAGICC and a GISS GCM-derived Lightning Model. *Climate Change* 39:273–96.

Janzen, D. 1988. *Tropical Dry Forests: The Most Endangered Major Tropical Ecosystem*, ed. E. O. Wilson, 130–37. Washington, DC: National Academy of Sciences / Smithsonian Institution.

Karanth, K., J. D. Nichols, J. E. Hines, K. U. Karanth, and N. L. Christensen. 2009. Patterns and Determinants of Mammal Species Occurrence in India. *Journal of Applied Ecology* doi:1 0.1111/j.1365-2664.2009.01710.

Lau, K. M., and S. Yang. 1997. Climatology and Interannual Variability of the Southeast Asian Summer Monsoon. *Advances in Atmospheric Sciences* 14:141–62.

Lavorel, S., M. D. Flannigan, E. F. Lambin, and M. C. Scholes. 2007. Vulnerability of Land Systems to Fire: Interactions among Humans, Climates, the Atmosphere, and Ecosystems. *Mitigation and Adaptation Strategies for Global Change* 12:33–53.

Maxwell, A. L. 2004. Fire Regimes in North-Eastern Cambodian Monsoonal Forests, with a 9300-year Sediment Charcoal Record. *Journal of Biogeography* 31:225–39.

McShea, W. J., K. Koy, T. Clements, A. Johnson, C. Vongkhamheng, and Myint Aung. 2005. Finding a Needle in the Haystack: Regional Analysis of Suitable Eld's Deer (*Cervus eldi*) Forest in Southeast Asia. *Biological Conservation* 125:101–11.

Miles, L., A. C. Newton, R. S. DeFries, C. Ravilious, I. May, S. Blyth, V. Kapos, and J. E. Gordon. 2006. A Global Overview of the Conservation of Tropical Dry Forests. *Journal of Biogeography* 33:491–505.

Mistry, J., and A. Berardi, eds. 2006. *Savanna and Dry Forests: Linking People with Nature*. Burlington, VT: Ashgate Publishing.

Murali, K., U. Shankar, R. Shaanker, K. Ganeshaiah, and K. Bawa. 1996. Extraction of Non-timber Forest Products in the Forests of Biligiri Rangan Hills, India. 2. Impact of NTFP Extraction on Regeneration, Population Structure, and Species Composition. *Economic Botany* 50:252–69.

Murphy, P. G., and A. E. Lugo. 1986. Ecology of Tropical Dry Forest. *Annual Review of Ecology and Systematics* 17:67–88.

Myers, N., R. A. Mittermeier, C. G. Mittermeier, G. A. B. de Fonseca, and J. Kent. 2000. Biodiversity Hotspots for Conservation Priorities. *Nature* 403:853–58.

Pandey, U., and J. S. Singh. 1984. Energy-Flow Relationships between Agrosystem and Forest Ecosystems in Central Himalaya. *Environmental Conservation* 11:45–53.

Pennington, R. T., M. Lavin, and A. Oliveira-Filho. 2009. Woody Plant Diversity, Evolution and Ecology in the Tropics: Perspectives from Seasonally Dry Tropical Forests. *Annual Review of Ecology and Systematics* 40:437–57.

Pennington, R. T., G. P. Lewis, and J. A. Ratter, eds. 2006. *Neotropical Savannas and Seasonally Dry Forests: Plant Diversity, Biogeography and Conservation*. Boca Raton, FL: CRC Press.

Prance, G. T. 2006. Tropical Savannas and Seasonally Dry Forests: An Introduction. *Journal of Biogeography* 33:385–86.

Qiu, J. 2009. Where the Rubber Meets the Garden. *Nature* 457:246–47.

Rao, M., A. Rabinowitz, and S. T. Khaing. 2002. Status Review of the Protected-Area System in Myanmar, with Recommendations for Conservation Planning. *Conservation Biology* 16:360–68.

Sanchez-Azofeifa, G. A., M. Kalacska, M. Quesada, J. C. Calvo-Alvarado, J. Nassar, and J. P. Rodriguez. 2005. Need for Integrated Research for a Sustainable Future in Tropical Dry Forests. *Conservation Biology* 19:1–2.

Santisuk, T. 1988. *An Account of the Vegetation of Northern Thailand*. Stuttgart, Germany: Franz Steiner Verlag Wiesbaden GmbH.

Sayer, J., U. Chokkalingam, and J. Poulsen. 2004. The Restoration of Forest Biodiversity and Ecological Values. *Forest Ecology and Management* 201:3–11.

Singh, V. P., and J. S. Singh. 1989. Man and Forests: A Case-Study from the Dry Tropics of India. *Environmental Conservation* 16:129–36.

Sodhi, N. S., L. P. Koh, B. W. Brooks, and P. K. L. Ng. 2004. Southeast Asian Biodiversity: An Impending Disaster. *Trends in Ecology and Evolution* 19:654–60.

Stott, P. 1988. The Forest as Phoenix: Towards a Biogeography of Fire in Mainland South-East Asia. *Geographical Journal* 154:337–50.

———. 1990. Stability and Stress in the Savanna Forests of Mainland South-East Asia. *Journal of Biogeography* 17:373–83.

van der Werf, G. R., J. T. Randerson, L. Giglio, N. Gobron, and A. J. Dolman. 2008. Climate Controls on the Variability of Fires in the Tropics and Subtropics, *Global Biogeochemical Cycles* 22, GB3028, doi:10.1029/2007GB003122.

Wikramanayake, E., E. Dinerstein, C. J. Loucks, D. M. Olson, J. Morrison, J. Lamoreux, M. McKnight, and P. Hedao. 2002. *Terrestrial Ecoregions of the Indo-Pacific: A Conservation Assessment*. Washington, DC: Island Press.

1

Seasonally Dry Tropical Forests in Continental Southeast Asia

Structure, Composition, and Dynamics

Sarayudh Bunyavejchewin, Patrick J. Baker, and Stuart J. Davies

The forests of continental Southeast Asia make up a significant portion of the Indo-Burma biodiversity hotspot (see McShea and Davies, this volume). These diverse forests are under severe threat from both land use and climate change, and urgently need to be more sustainably managed. This management of seasonally dry forests in Southeast Asia needs to be based on a sound understanding of the ecology of natural forests, the habitat requirements of their constituent species, and the response of these systems to natural disturbance dynamics. Natural forest management strategies will help protect habitat for wildlife, will limit the impact of nonnatural disturbances, and will lead to opportunities for restoring degraded lands to functional forests. An important first step in this process is to develop a much more refined description of the forests with a clear understanding of what controls their spatial variation in structure and composition across the region. The vegetation types and ecoregions described in recent mapping and conservation assessments in continental Southeast Asia—for example, Blasco et al. (1996) and Wikramanayake et al. (2002)—are useful for broad-scale threat assessment and priority setting; however, for the active forest management that is required in much of the region now, we need detailed understanding of the ecology and dynamics of specific forest types.

Our research aims to understand the controls on spatial distributions and temporal dynamics of seasonally dry tropical forest (SDTF) formations across the region. This work is principally based in Thailand, but since most of the forest types are widespread, with similar species often occurring across the seasonally dry areas of Indo-Burma (Champion and Seth 1968), this research will be broadly relevant in continental Southeast Asia.

Many authors (e.g., Miles et al. 2006) have defined dry tropical forests as those with 5–6 dry months (less than 100 mm rain per month) and average annual rainfall of less than 1,600 mm. In this chapter, we include all areas that experience a predictable, regular dry season in which the long-term mean monthly rainfall falls below 100 mm per month. This includes areas with 2–3 dry months and over 2,000 mm of rain per year in the far south and southeast of Thailand, as well as areas with as much as a 6-month dry season and 1,200 mm of rain in parts of the north and east of the region (Figure 1). We use this broader definition for our assessment of the ecology of SDTFs in Southeast Asia for the following reasons (see also Murphy and Lugo 1986; and Pennington et al. 2006). First, all forests in the seasonal areas of Southeast Asia share dynamics that are synchronized with the periodicity of seasonal water stress, including flowering and fruiting, leaf phenology (Ishida et al. 2006; Williams et al. 2008), carbon uptake (Yoshifuji et al. 2006), and growth and mortality (Baker and Bunyavejchewin 2006b). Farther south in aseasonal mixed dipterocarp forests, reproductive phenological patterns are dominated by supra-annual general flowering and mast-fruiting events (Appanah 1985), whereas to the north, phenological cycles are more strongly influenced by intra-annual temperature fluctuations, and particularly the impact of killing frosts. Second, continental Asian forests cannot be divided into wet and dry types based on deciduousness, as has been done elsewhere, because many parts of strongly seasonal Southeast Asia have evergreen forests (Champion and Seth 1968; Ashton 1990; Rundel and Boonpragob 1995). Evergreen and deciduous forests often occur interdigitated within landscapes. For example, at the Huai Kha Khaeng Wildlife Sanctuary in central western Thailand, deciduous dipterocarp, mixed deciduous, and evergreen forests all occur within hundreds of meters of each other. Third, biogeographers have long recognized the region of Indo-Burma as distinct from neighboring areas. To the south the region is bounded by the Kangar-Pattani Line, a relatively sudden division between seasonal and aseasonal forests, across which there is a dramatic

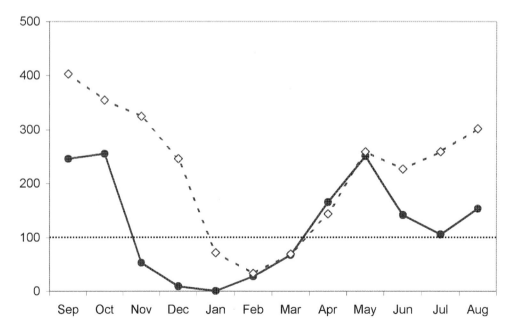

Figure 1. Rainfall distribution (mm per month) in Huai Kha Khaeng (solid line) and Khao Chong (dashed line) illustrating continental Southeast Asian climates with different dry season lengths.

floristic turnover (van Steenis 1950; Whitmore 1984; Baltzer et al. 2007). To the north the region is bounded by the foothills of the Himalayas and the Nan-Ling Mountains of southern China close to the Tropic of Cancer, and to the west the area is bounded by the delta of the great Brahmaputra River (Ashton 1990, 1997).

In continental Southeast Asia there is a range of seasonal forest types that vary widely in diversity, composition, and structural features. In this chapter we describe six of the main vegetation types that occur in seasonal climates in this area. We base this description on the published work of others as well as our own work. We also present preliminary, unpublished findings from an analysis of thirty-eight 1 ha plots that the senior author has established across Thailand over the past decade (Figure 2). In all thirty-eight plots, trees ≥ 10 cm dbh (diameter at breast height, "breast height" taken as 1.3 m) were mapped, measured, and tagged for long-term monitoring. The plots include over nine hundred species of trees with dbh ≥ 10 cm. Species were identified using regional floras and collections at the Department of National Parks, Wildlife and Plant Conservation (DNP) Forest Herbarium (BKF) in Bangkok, Thailand. Nomenclature follows Forest Herbarium (2001). Additional results from these plots will be presented in upcoming publications. Other important, though less-extensive, vegetation types occur in the region, including upper montane forests, mangroves, grasslands, swamp forests, and forests over limestone. These vegetation types are also important reservoirs of the region's unique biodiversity and require more detailed ecological study, but will not be further considered in this chapter.

Following the descriptions of the vegetation types, we discuss factors driving the dynamics of the different forest types, considering the impact of both natural and intrinsic disturbances, and the effects of human-induced changes in these systems. We use this analysis to provide the necessary context for discussing long-term, sustainable management of SDTFs in Southeast Asia.

DECIDUOUS DIPTEROCARP FOREST

Deciduous dipterocarp forest (DDF) occurs in areas with a relatively severe 5–6 month dry season and a total annual rainfall of about 1,000–1,500 mm (Bunyavejchewin 1983a). This forest type extends across continental Asia from northeast India, through Thailand, to Laos and Vietnam. In Thailand, DDF has been historically the most extensive forest type, covering as much as 45 percent of the total forest area (Neal 1967, in Bunyavejchewin 1983b; Rundel and Boonpragob 1995). The different names by which DDF has been described are summarized in Santisuk (1988).

DDF occurs mainly in the lowlands up to 900 m above sea level, but extends to higher elevations in some areas where it intergrades with pine forest. The soils over which DDF grows are acidic, shallow, sandy, and often lateritic (Santisuk 1988), and are generally considered poor and unable to sustain long-term agriculture (Cooling 1968).

DDF has a relatively low, open canopy composed of species with typically thick, leathery leaves. The canopy does not usually exceed 20 m in height and in xeric conditions can be 5–10 m tall. However, in very favorable sites, the main canopy dipterocarps and some *Terminalia* species can reach 25–30 m in height (Bunyavejchewin

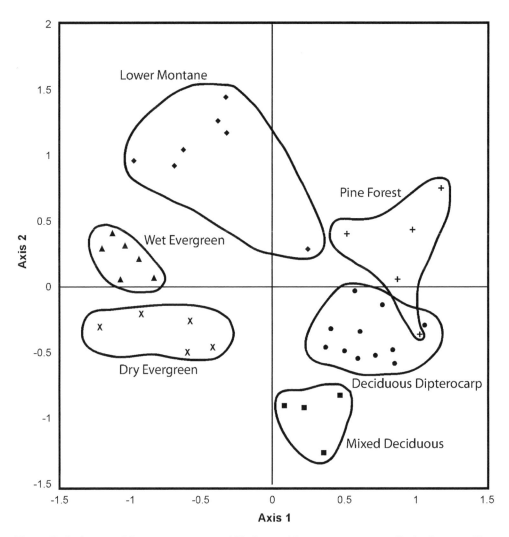

Figure 2. Ordination of floristic composition of 38 plots in 6 forest types in seasonally dry forests in Thailand. Non-metric multidimensional scaling analysis included total basal area of trees ≥10 cm dbh for the most important 300 species.

1983a; Santisuk 1988). Like most strongly seasonal forests, the abundance of stems in smaller size classes is low in DDF compared with evergreen rainforests. However, the mean stem density of trees ≥ 10 cm dbh overlaps with stem densities of other seasonal forests (Figure 3). Among the eleven 1 ha plots of DDF considered here, mean stem density was 440 stems per hectare (range: 269–646 stems/ha). Most species of the main canopy lose their leaves at the start of the cool, dry season in December or January, and regain new leaves before the start of the wet season in April or May (Bunyavejchewin 1983a; Williams et al. 2008).

DDF is composed of a very distinctive set of dominant species. One or more of the four deciduous dipterocarp species, *Shorea siamensis*, *S. obtusa*, *Dipterocarpus*

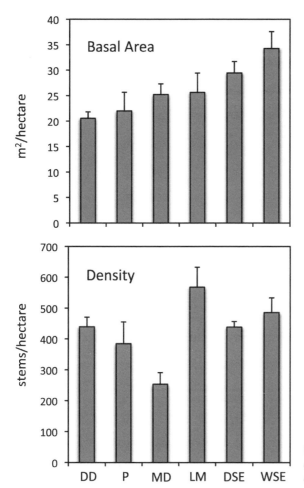

Figure 3. Mean basal area and stem density for the six forest types, based on 38 1-ha plots.

obtusifolius, and *D. tuberculatus*, usually form the core of DDF stands. In the eleven DDF plots studied, dipterocarps account for an average 51 percent of all stems ≥ 10 cm dbh, and 67 percent of the total basal area (Table 1). Other important species in the canopy of DDF include *Pterocarpus macrocarpus*, *Xylia xylocarpa*, *Gluta usitata*, and several species of *Terminalia*. The mid-story is often characterized by species such as *Aporosa villosa* and *Strychnos nux-blanda*. See Santisuk (1988) for a more-detailed list of species found in DDF for north Thailand. The most important species in our 1 ha plots are shown in Table 2. The understory of DDF is typically grass-dominated, with *Imperata cylindrica* and the dwarf bamboo, *Vietnamosasa pusilla*, as common species. The cycad, *Cycas siamensis*, and the stemless palm, *Phoenix acaulis*, are widespread in DDF.

Of the six forest types considered in this chapter, only pine forest is less diverse than DDF. For trees ≥ 10 cm dbh, species richness for the eleven 1 ha plots in DDF averaged 39.9 species/ha, and ranged from 24 to 64 species/ha (Table 1). Fisher's alpha averaged 10.9 across the eleven plots. Each stand tended to be dominated by one or a

Table 1. Comparison of species diversity in 38 one-hectare (ha) plots in six forest types in Thailand. Forest types follow the descriptions in the text as follows: DDF—deciduous dipterocarp forest, PF—pine forest, MDF—mixed deciduous forest, LMF—lower montane forest, DSEF—dry seasonal evergreen forest, and WSEF—wet seasonal evergreen forest. The percentage dominance is the percentage of all stems accounted for by the commonest species. The percentage of dipterocarps is the percentage of all stems (n) and basal area (ba) in each plot that are dipterocarps. Value ranges are given in parentheses.

Forest Type	DDF	PF	MDF	LMF	DSEF	WSEF
Number of plots	11	5	4	7	5	6
Total species	149	81	120	284	295	428
Species richness	39.9	23.4	46.0	63.0	83.0	128.7
(range per ha)	(24–64)	(6–39)	(33–64)	(28–82)	(68–105)	(99–151)
Fisher's alpha	10.9	5.7	16.7	18.4	31.1	60.5
(range per ha)	(6.4–20.2)	(1.1–9.1)	(10.5–22.8)	(7.9–26.1)	(21.4–45.9)	(34.9–80.8)
% dominance	35.4	44.3	35.4	17.7	12.9	9.3
(range)	(12.7–63.2)	(27.1–66.9)	(14.7–56.7)	(12.6–25.5)	(8.9–22.4)	(5.4–15.1)
% dipterocarp	51.0	39.5	0.1	0.5	9.4	11.8
(n, range)	(29.1–74.2)	(0.2–73.9)	(0.0–0.3)	(0.0–2.7)	(1.2–17.0)	(3.7–22.2)
% dipterocarps	67.4	33.6	0.1	0.5	23.5	23.2
(ba, range)	(50.2–85.7)	(0.1–66.0)	(0.0–0.2)	(0.0–2.3)	(10.7–33.6)	(5.5–39.8)

few species. Among the eleven plots, the most common species on average accounted for 35 percent of all stems.

Dry season fires occur relatively frequently in DDF. Many of the important species in these stands have thick, fire-resistant bark, and resprout vigorously following fire. Regeneration dynamics in DDF are not well understood. However, Troup (1921) noted both that successful regeneration required a combination of exposed mineral soil, ample sunlight, and rainfall within days of seedfall and that the appropriate combination of these conditions occurred less often than might be expected. Fire during the dry season may facilitate successful establishment of seedlings by removing leaf litter, which impedes penetration by the seed's radicle. Ogawa et al. (1961) suggested that in the absence of fire, the soil's organic matter might build up, improving soil fertility, and eventually leading to changes in forest composition toward other forest types such as mixed deciduous forest.

The stand structure and floristic composition of DDF varies significantly with elevation and soil chemical and physical properties. Several authors have described subtypes or dominance types within DDF based on structural features of the canopy and stand basal area and the relative dominance of the main canopy dipterocarps (Ogawa et al. 1961; Kutintara 1975; Bunyavejchewin 1983a). For example, based on fifty-two study plots Bunyavejchewin (1983a) recognized four DDF types plus a pine-dipterocarp type (see following section). The four main DDF types differed in composition, and this was linked to differences in elevation, topography, soil chemistry, and soil physical features that clearly are related to variation in the water-holding capacity

Table 2. Important and characteristic species in 38 one-hectare (ha) plots in seasonally dry forest in Thailand. Species were included based on either a high contribution to stem basal area or stem abundance in the plots, or on field observations that indicated that the species are characteristic of the respective forest types.

Deciduous Dipterocarp Forest (11 plots)

Dipterocarpaceae	*Shorea siamensis* Miq.
Dipterocarpaceae	*Shorea obtuse* Wall.
Dipterocarpaceae	*Dipterocarpus tuberculatus* Roxb.
Dipterocarpaceae	*Dipterocarpus obtusifolius* Teijsm. ex Miq.
Combretaceae	*Terminalia alata* Heyne ex. Roth
Lamiaceae	*Vitex peduncularis* Wall.
Combretaceae	*Terminalia mucronata* Craib and Hutch.
Fagaceae	*Quercus kerrii* Craib
Fabaceae	*Dalbergia cultrate* Graham
Fagaceae	*Lithocarpus dealbatus* (Hook.f. and Thomson.) Rehder
Phyllanthaceae	*Aporosa villosa* Baill.
Phyllanthaceae	*Aporosa octandra* (Buch.-Ham. ex D.Don) A. R.Vickery

Pine Forest (5 plots)

Pinaceae	*Pinus merkusii* Jungh. and de Vriese
Pinaceae	*Pinus kesiya* Royle ex Gordon
Dipterocarpaceae	*Dipterocarpus obtusifolius* Teijsm. ex Miq.
Dipterocarpaceae	*Dipterocarpus tuberculatus* Roxb.
Fagaceae	*Quercus brandisiana* Kurz
Anacardiaceae	*Gluta elegans* (Wall.) Hook.f.
Dipterocarpaceae	*Shorea obtusa* Wall. Ex Blume
Fagaceae	*Lithocarpus dealbatus* (Hook.f. and Thomson.) Rehder
Fagaceae	*Castanopsis tribuloides* A. DC.
Fagaceae	*Quercus truncate* King ex Hook.f.
Proteaceae	*Helicia nilagirica* Bedd.

Mixed Deciduous Forest (4 plots)

Lamiaceae	*Tectona grandis* L.f.
Lythraceae	*Lagerstroemia balansae* Koehne
Sapindaceae	*Schleichera oleosa* (Lour.) Oken
Moraceae	*Ficus geniculata* Kurz
Fabaceae	*Pterocarpus macrocarpus* Kurz
Phyllantaceae	*Antidesma acidum* Retz.
Fabaceae	*Xylia xylocarpa* (Roxb.) Taub.
Combretaceae	*Terminalia alata* Heyne ex Roth
Lamiaceae	*Vitex limonifolia* Wall.
Burseraceae	*Garuga pinnata* Roxb.
Lythraceae	*Lagerstroemia calyculata* Kurz.
Fabaceae	*Dalbergia cultrate* Graham
Fabaceae	*Afzelia xylocarpa* Craib

(Continued)

Table 2. Important and characteristic species in 38 one-hectare (ha) plots in seasonally dry forest in Thailand. Species were included based on either a high contribution to stem basal area or stem abundance in the plots, or on field observations that indicated that the species are characteristic of the respective forest types. *(Continued)*

Lower Montane (7 plots)

Fagaceae	*Castanopsis acuminatissima* (Blume) A. DC.
Hamamelidaceae	*Distylium indicum* Benth. ex C.B.Clarke
Podocarpaceae	*Dacrydium elatum* Wall.
Fagaceae	*Quercus semiserrata* Roxb.
Theaceae	*Gordonia axilleris* (Roxb. ex Ker Gawl.) D.Dietr
Myrtaceae	*Syzygium cacuminis* (Craib) Chantar. and J.Parn.
Apocynaceae	*Alstonia rostrata* C.E.C.Fisch.
Myrsinaceae	*Rapanea yunannensis* Mez
Fagaceae	*Castanopsis hystrix* A. DC.
Xanthophyllaceae	*Xanthophyllum affine* Korth. ex Miq.
Ulmaceae	*Gironniera subaequalis* Planch.
Fagaceae	*Castanopsis diversifolia* King ex Hook.f.

Dry Seasonal Evergreen Forest (5 plots)

Dipterocarpaceae	*Dipterocarpus turbinatus* C.F.Gaertn.
Dipterocarpaceae	*Dipterocarpus costatus* C.F.Gaertn.
Annonaceae	*Saccopetalum lineatum* Craib
Irvingiaceae	*Irvingia malayana* Oliver ex A.Benn.
Ebenaceae	*Diospyros pendula* Hasselt ex Hassk.
Dipterocarpaceae	*Hopea odorata* Roxb.
Fabaceae	*Dialium cochinchinense* Pierre
Dipterocarpaceae	*Shorea henryana* Pierre
Dipterocarpaceae	*Shorea thorelii* Pierre ex Laness.
Lythraceae	*Lagerstroemia calyculata* Kurz
Clusiaceae	*Garcinia speciosa* Wall.
Sapindaceae	*Dimocarpus longan* Lour.

Wet Seasonal Evergreen Forest (6 plots)

Dipterocarpaceae	*Parashorea stellata* Kurz
Euphorbiaceae	*Elateriospermum tapos* Blume
Dipterocarpaceae	*Dipterocarpus kerrii* King
Myrtaceae	*Syzygium attenuatum* (Miq.) Merr. and L.M.Perry
Dipterocarpaceae	*Dipterocarpus grandiflorus* Blanco
Sterculiaceae	*Scaphium scaphigerum* (G.Don) Guib. and Planch.
Fabaceae	*Millettia atropurpurea* Benth.
Fabaceae	*Cynometra malaccensis* Meeuwen
Dipterocarpaceae	*Shorea gratissima* Dyer
Dipterocarpaceae	*Shorea hypochra* Hance
Fagaceae	*Lithocarpus sundaicus* (Blume) Rehder
Dipterocarpaceae	*Anisoptera costata* Korth.

of the different soils. Further experimental approaches are required to understand the various habitat requirements of the dominant deciduous dipterocarps in these forests.

PINE FOREST

Pine forest in continental Southeast Asia occurs from as low as 400 m above sea level to over 1,600 m, and in insular Southeast Asia, pines can be found up to almost 2,500 m above sea level in the Philippines and Sumatra (Whitmore 1984). At the lower elevations, forest dominated by *Pinus* species has floristic similarities to DDF. The lower-elevation forests dominated by pine (particularly *Pinus merkusii*) have in the past been referred to as a subtype of DDF (Bunyavejchewin 1983a). Toward the higher elevations in Thailand, the *Pinus*-dominated forests (particularly those of *Pinus kesiya*) include more elements characteristic of lower montane forest, for example, species of Fagaceae (Santisuk 1988).

At lower elevations, *P. merkusii* associates strongly with several deciduous tree species typical of DDF, including *Shorea obtusa*, *Dipterocarpus obtusifolius*, and *D. tuberculatus* (Bunyavejchewin 1983a). At higher elevations, *P. kesiya* becomes more important (often sympatric with *P. merkusii*), deciduous dipterocarps are less important, and several evergreen genera of the Fagaceae and Lauraceae become important components of the community (Bunyavejchewin 1983a). Bunyavejchewin identified four distinct subtypes of pine forest differing in the degree of dominance by the two pine species and the identity of the codominants. The subtypes occur at overlapping but different altitudinal ranges. The range of variation of the codominants with the pine species across the elevation gradient is reflected in the ordination result for our five 1 ha pine forest plots (Figure 2). The lower elevation plot, with *P. merkusii* and many deciduous dipterocarps, was placed within the cluster of DDF sites in the ordination. The higher elevation plot, with more *P. kesiya* and oak species, was more similar to the lower montane forest plots.

Pine forest structure is similar to DDF structure, especially at lower elevations. At higher elevations, the pines often form an emergent canopy (reaching 40 m tall) well above the surrounding forest, and the upper branches of the pines are commonly adorned with lichens, mosses, and abundant epiphytes. The five plots included in this analysis had mean stem densities and basal areas similar to the DDF plots (Figure 3).

Soils under pine forest tend to be deep sandy to sandy-loams, with higher calcium concentrations and higher pH than DDF forests (Bunyavejchewin 1983a). Soil acidity decreases with elevation across the different pine forest subtypes.

Pine forest is the most species-poor of the seasonally dry forests in Southeast Asia considered in this chapter. For trees ≥ 10 cm dbh, species richness for the five 1 ha plots in pine forest averaged 23.4 species/ha, with a range of 6–39 species/ha (Table 1). Fisher's alpha averaged 5.7 among the plots. The degree of species-level dominance was high; the most common species on average accounted for 44 percent of all stems. Apart from the two pine species, *Dipterocarpus obtusifolius*, *D. tuberculatus*, and *Quercus brandisiana* were among the more-important species present in the five plots surveyed (Table 2).

As in DDF, fire is a relatively common occurrence in pine forest, particularly at lower elevations due to the grassy understory. Santisuk (1988) mentions that *P. merku-sii* is more fire tolerant than *P. kesiya*. The role of fire in controlling the distributions of the two species needs further field study.

MIXED DECIDUOUS FOREST

Mixed deciduous forest (MDF) covered an estimated 20 percent of Thailand (Neal 1967, in Bunyavejchewin 1983b). It is common in Cambodia (Tani et al. 2007), and is wide-spread across the Indo-Burmese region (Ashton 1990). Due to an abundance of valuable timber species, notably teak (*Tectona grandis*) and several rosewood (*Dalbergia*) species, and the generally favorable soil conditions on which this forest type develops, MDF has been heavily modified by human activity. MDF occurs from around 100–800 m above sea level in areas with a strong 5–6 month dry season and 1,000–1,800 mm of rain (Bunyavejchewin 1983b; Santisuk 1988).

MDF is characterized by a virtual absence of dipterocarps (Table 2); a canopy dominated by a diversity of deciduous species, particularly in the families Lamiaceae, Lythraceae, Fabaceae, and Combretaceae; the frequent presence of bamboo in the understory and in gaps; and a grass- and ginger-rich ground cover (Bunyavejchewin 1983b, 1985; Santisuk 1988). Although MDF occurs in the same climatic conditions and elevations as DDF, MDF occurs on less-acidic and more-fertile soils. The soils under MDF are typically moderately fertile sandy to clay loams of pH 5–6, varying considerably in depth (Bunyavejchewin 1983b, 1985). The best forests occur on the deeper and better-drained soils.

Teak occurs exclusively in MDF, although it does not occur in all MDF stands (Bunyavejchewin 1983b; Teejuntuk et al. 2002). Other important canopy species include several species of *Lagerstroemia* (including *L. calyculata* and *L. balansae*), *Xylia xylocarpa*, *Pterocarpus macrocarpus*, *Bombax ceiba*, *Anogeissus acuminata*, *Dalbergia oliveri*, and several species of *Terminalia* (including *T. mucronata*, *T. tomentosa*, and *T. alata*; Bunyavejchewin 1983b; Marod et al. 1999). The subcanopy usually includes a number of evergreen species including the widespread *Irvingia malayana*, and *Syzygium cumini*. *Phyllanthus emblica* is relatively common in the understory. See Santisuk (1988) for a full list of species.

MDF is a tall-statured forest, typically from 25 to 30 m tall, although in favorable conditions the canopy can be above 40 m (Ogawa et al. 1961; Bunyavejchewin 1983b; Tani et al. 2007). Among the four MDF plots described in this chapter, basal area averaged 25.6 m²/ha, although stem densities were relatively low in these stands (Figure 3). In more mature stands, basal area typically ranges from 30 to 45 m²/ha (Bunyavejchewin 1983b).

Species diversity of MDF is intermediate between the less-diverse pine and deciduous dipterocarp forests and the more-diverse evergreen forests (Table 1). The four 1 ha plots surveyed in this chapter included a total of 120 species ≥ 10 cm dbh. Individual plots averaged 46 species/ha, ranging from 33 to 64 species/ha among plots. Fisher's alpha ranged from 10.5 to 22.8 (mean: 16.7).

Forest structure and floristics of MDF vary with altitude and soil quality. Bunyave-jchewin (1985) describes several subtypes of MDF. MDF dominated by teak occurs on soils with higher calcium and phosphorus concentrations. Forests dominated by *Lager-stroemia calyculata* occupy less-fertile soils. In more-xeric conditions, forest stature is much reduced and the canopy more open (Santisuk 1988).

As in other forest types with a long dry season, forest fires occur regularly in MDF. When disturbed by logging or when fragmented due to agriculture, MDF also seems to be particularly vulnerable to invasion by weedy exotics such as *Chromolaena odorata*.

LOWER MONTANE FOREST

Montane forest in continental Southeast Asia occurs on hills and mountains higher than 800–1,000 m above sea level (Ohsawa et al. 1985; Santisuk 1988; Ashton 2003). Lower montane forest (LMF), occurring approximately 800–2,000 m above sea level, is distinguished from upper montane forest (UMF), occurring above 1,700 m, by dif-ferences in forest structure and species composition. LMF differs from UMF in having a greater canopy height (sometimes with emergent crowns), larger average leaf sizes, a distinct subcanopy layer, and typically fewer mosses and other non-vascular epi-phytes (Richards 1996; Ashton 2003). The altitudes at which the various vegetation transitions on tropical mountains occur (including lowland forest to LMF, and LMF to UMF) differ considerably from place to place depending in part on the size of the mountain massif and its proximity to the coast (Richards 1996; Ashton 2003; Kha-myong et al. 2004). The environmental factors that control the structural and floristic changes from lowland forest to LMF are not well understood. In addition to changes in precipitation, temperature, and insolation, all of which may influence the transition (Ohsawa et al. 1985; Ohsawa and Nitta 2002), systematic changes in soils at the low-land–LMF transition (800–1,200 m) may also have an important effect (Ashton 2003).

LMF in seasonally dry areas of continental Southeast Asia receives about 1,300–2,000 mm of rain per year (Santisuk 1988; Kanzaki et al. 2004). The soils under LMF are generally moderately fertile and moist, and have accumulations of organic matter on the surface (Santisuk 1988). In part due to the relatively high fertility of the soils, LMFs have been seriously degraded by swidden and industrial agriculture (Santisuk 1988; Ash-ton 2003). LMFs are evergreen and differ from lower elevation evergreen forests in both floristic composition and forest structure. The Dipterocarpaceae, dominants of lowland evergreen forests, are largely absent from LMF in continental Southeast Asia (Table 1). LMFs are usually dominated by species in the Fagaceae, Lauraceae, Theaceae, and Mag-noliaceae, and a range of less-diverse representatives of primarily northern temperate families and genera (e.g., Hamamelidaceae and Cornaceae; Table 2).

LMF is typically more species-rich than any of the deciduous forest types in continen-tal Southeast Asia, but is not as diverse as the lowland evergreen forests (Table 1). Among the seven 1 ha plots sampled here, we recorded a total of 284 tree species with dbh ≥ 10 cm dbh, with an average of 63 species/ha. Average Fisher's alpha among the seven plots was 18.4 (range: 7.9–26.1). The Lauraceae and Fagaceae were the most diverse families, often with numerous sympatric species in any one LMF stand. No individual species dominates

LMF, as can occur in seasonally dry lowland forests. Species specificity to microhabitats within LMF is similar to that found in lowland forests (Noguchi et al. 2007).

The canopy of mature LMF is typically about 20–35 m tall. In more-mesic sites, emergent trees can exceed 1.5 m in diameter and reach 50 m tall (see Kanzaki et al. 2004). Standing basal area averaged 25.6 m²/ha in the seven plots surveyed for this chapter (Figure 3). On one of the 1 ha plots, basal area was 40.4 m²/ha, and on Doi Inthanon, Thailand, basal area was 36.1 m²/ha in a 16 ha plot near the upper limit of LMF (Kanzaki et al. 2004). Stem densities are generally high in LMF; the average density of stems ≥ 10 cm dbh in the seven 1 ha plots was 567 stems/ha, higher than all other forest types in the survey (Figure 3). Stem densities in smaller size classes (less than 10 cm dbh) are also typically high in LMF; in the Doi Inthanon 16 ha plot, there were on average 4,391 stems 1–10 cm dbh per hectare (Kanzaki et al. 2004), a value more typical of forests in climates with a short or no dry season (Figure 4). Indeed, the values of both basal area and stem densities of undisturbed LMF are similar to lowland evergreen forests of continental Southeast Asia (Figure 3).

There is considerable variation in the composition of LMF across continental Indo-Burma. Santisuk (1988) recognized three types of LMF: a more-diverse type dominated by Lauraceae and Fagaceae, a moderately diverse type dominated by Fagaceae, and a less-diverse and more-open type dominated by *Pinus kesiya*, Fagaceae, and Theaceae. Santisuk attributed differences among these forest subtypes to the degree of degradation from human activities. However, even in relatively undisturbed LMF, there are strong patterns of variation in species distributions across the landscape (see Noguchi et al. 2007). Such patterns may be driven by variation in soil properties, the history of human and natural disturbance, or other factors. Among the seven 1 ha plots in this study, one of the LMF plots included several individuals of *Pinus kesiya*, and it clustered more closely with the pine forest plots (Figure 2). Long-term monitoring and experimental studies are required to better understand what controls the distribution of LMF and its constituent species.

DRY SEASONAL EVERGREEN FOREST

Dry seasonal evergreen forest (DSEF), with a predominantly evergreen canopy and understory, occurs in areas with 4–6 dry months and rainfall of 1,200–2,000 mm per year. The forest occurs across Indo-Burma, often in moist valleys and ravines among low- to medium-sized hills, and in low, wide river valleys and gallery forests (Santisuk 1988). DSEF and MDF often occur in a mosaic across low hills of continental Southeast Asia. We observed this in the Huai Kha Khaeng landscape, and Blanc et al. (2000) report a similar pattern in southern Vietnam (see also De Cauwer and De Wulf 1994, in Blanc et al. 2000). The soils under DSEF tend to be moderately fertile, with good to very good moisture retention capacity. The forest type is economically important due to the high value of several of the timber species, particularly *Hopea odorata*.

Evergreen species in the Dipterocarpaceae often form a major part of the main canopy in DSEF. Among the five 1 ha plots described here, dipterocarps accounted for on average 24 percent of the stand basal area, and 9.4 percent of the stems (Table 1).

Stem Density (per ha)

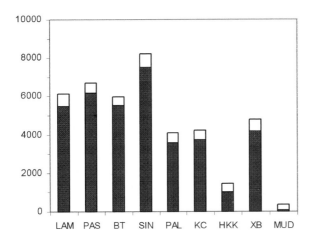

Basal Area (m²/ha)

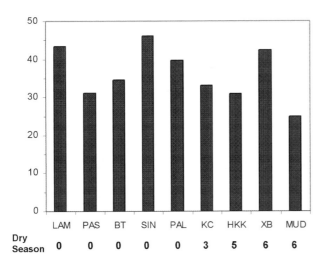

Figure 4. Mean basal area and stem density for all trees ≥1 cm dbh for 9 large CTFS plots in Asia (www.ctfs.si.edu). Stem densities are shown for trees 1-10 cm (black) and ≥10 cm in white. The length of the dry season is indicated in months. Plots are: LAM - Lambir, Malaysia; PAS - Pasoh, Malaysia; BT - Bukit Timah, Singapore; SIN – Sinharaja, Sri Lanka; PAL – Palanan, Philippines; KC – Khao Chong, Thailand; HKK – Huai Kha Khaeng, Thailand, XB – Xishuangbanna, China, and MUD – Mudumalai, India.

Up to seven dipterocarp species may coexist within one DSEF (Bunyavejchewin et al. 2003). However, in many stands that have been studied, one or two dipterocarp species dominate, for example, *Hopea odorata* (Bunyavejchewin et al. 2001, 2002), *H. ferrea* (Bunyavejchewin 1986, 1999), *Shorea henryana* (Bunyavejchewin 1986, 1999), *Dipterocarpus gracilis* (Kitamura et al. 2005), *D. alatus* (Blanc 2000), and *D. turbinatus* (Blanc 2000). These dipterocarps generally have smaller leaf sizes than the dipterocarp species that dominate DDF. The dipterocarp flora is largely endemic to this forest type within the area of Indo-Burma (Ashton 1990).

Other important trees in the canopy and subcanopy of DSEF include species in the families Meliaceae, Annonaceae, Lauraceae, and Sapindaceae. The relative importance of individual species within the canopy and subcanopy varies considerably from forest to forest. DSEF is often characterized by the presence of very large, fast-growing, light-demanding species including *Tetrameles nudiflora*, *Duabanga grandifolia*, and *Pterocymbium javanicum* (Bunyavejchewin et al. 2001; Santisuk 1988). Very large hemi-epiphytic "strangler" figs are also quite common in DSEF. The understory is mostly dominated by saplings of canopy and subcanopy species, and by shrubs in the Euphorbiaceae and Rubiaceae. The ground flora includes numerous seasonally leafless gingers and aroids (Santisuk 1988). Lianas are abundant and include many species that can reach the canopy. Bamboos are mostly absent, although they can be quite common when the forest is disturbed (Santisuk 1988).

Despite the name of this forest type, the foliar phenology of the canopy is relatively complex. The dominant dipterocarps in the DSEF are all evergreen; however, about 30 percent of the canopy of DSEF is deciduous (Williams et al. 2008). Among the deciduous species, there is a wide range in the duration of leaflessness; some species lose their leaves for 1–2 weeks, while others lose their leaves for six months. Curiously, many of the tree species that drop some or all of their leaves flush new leaves at the height of the dry season, well before the onset of the monsoon rains. The timing, intensity, and duration of leaflessness varies strongly both within a species and from year to year (Williams et al. 2008).

The canopy of DSEF is usually about 30–40 m tall (Bunyavejchewin 1986; Blanc et al. 2000; Tani et al. 2007). In exceptional conditions, emergent individual dipterocarps can reach 55–60 m tall, as in the 50 ha plot at Huai Kha Khaeng Wildlife Sanctuary (Bunyavejchewin et al. 2003). Stem densities in DSEF are similar to MDF and wet seasonal evergreen forest (Figure 3). Standing basal area for trees $\geq$ 10 cm dbh in the five plots averaged 29.5 m²/ha (range: 23.2–36.7 m²/ha; Figure 3). This is similar to most values for other DSEF forests (see Bunyavejchewin 1986; Bunyavejchewin et al. 2001; Blanc et al. 2000).

DSEF is among the most diverse seasonal forest types in continental Indo-Burma. Of the six forest types discussed here, only wet seasonal evergreen forest has more tree species. We recorded 295 species $\geq$ 10 cm dbh among the five 1 ha plots surveyed here. For individual plots, species richness averaged 83 species/ha, and ranged from 68 to 105 species/ha $\geq$ 10 cm dbh (Table 1). Average Fisher's alpha among the five plots was 31.1 (range: 21.4–45.9).

There is considerable variation in the composition of DSEF across Indo-Burma. The factors that define the ecological distributions of the fairly gregarious and locally dominant dipterocarp species (e.g., *Hopea odorata* and *H. ferrea*) need to be approached experimentally. Bunyavejchewin (1986) suggests that some of this variation may be due to soil chemical preferences, and Baker et al. (2005) describes the role of stand-changing disturbance events in creating within-stand variability in structure and composition.

In northern parts of continental Southeast Asia, seasonal evergreen forest is sometimes distinguished from DSEF as described here. Evergreen forests of this type occur in northern Myanmar (Ashton 1991), in parts of northeast India (Champion and Seth 1968), and in southern Yunnan (Zhu 1997; Cao et al. 2006, 2008). These forest types share floristic and structural similarities with the DSEF described here. Further work is required to understand the similarities and differences among the continental seasonal evergreen forest types.

WET SEASONAL EVERGREEN FOREST

At the southern limits of continental Asia there are relatively small areas in which the climate is shortly seasonal, with a dry period of only 2–3 months and total rainfall typically in excess of 2,000 mm. There are two main areas where this climate occurs in continental Southeast Asia: peninsular Thailand south of Chumphon, and southeast Thailand extending into Cambodia, an area sometimes referred to as the "Chantaburi pocket" (Ashton 1990; Whitmore 1984). The forest in these areas is a distinct form of evergreen forest, here referred to as wet seasonal evergreen forest (WSEF; cf. Ashton 1991). This forest type, despite being the most diverse in continental Southeast Asia, has been the subject of relatively limited study (Ogawa et al. 1961, 1965a, 1965b; Woodruff 2003; Baltzer et al. 2007).

In southern peninsular Thailand, this forest type abuts the northern side of the distinct climatic and floristic transition, the Kangar-Pattani Line (van Steenis 1950; Whitmore 1984). This line separates two of Asia's distinct biological hot spots: Sunda-land and Indo-Burma (Myers et al. 2000). To the south of the Kangar-Pattani Line is an everwet climate, and to the north a seasonally dry climate (having an approximately 2–3 month dry season) with little or no change in total annual rainfall (Ashton 1997). Forests to the north are WSEF and those to the south are aseasonal evergreen mixed dipterocarp forests. As noted above, a dramatic floristic transition occurs across the line. Ashton (1997) reports that there are 157 species of Dipterocarpaceae south of the line, only 27 of which cross the line into WSEF. An additional 19 dipterocarp species not present in the aseasonal forests are found north of the line. Van Steenis (1950) reports that 375 plant genera occurring south of the line reach their northern limits at the line, and 200 genera from the north reach their southern limits at the line. Strong patterns of faunistic turnover are also found in peninsular Thailand, although the exact position of the greatest species turnover may be somewhat further north for animals than for plants (Hughes et al. 2003; Woodruff and Turner 2009).

WSEF has a tall, heterogeneous canopy 30–40 m tall, with emergents considerably taller. The canopy is dominated by species of Dipterocarpaceae, Fabaceae, Sapotaceae, Myrtaceae, and Meliaceae. Although the total basal area contributed by dipterocarps can be high, the number of species of dipterocarps is considerably lower than in the aseasonal mixed dipterocarp forests occurring further south. Of 590 species in a recently established 24 ha plot in WSEF at Khao Chong, peninsular Thailand, only 10 species were dipterocarps (S. Bunyavejchewin and S. J. Davies, unpublished data). Despite the very high diversity and low species-level dominance in WSEF (Table 1), there are a number of fairly characteristic canopy species, including *Parashorea stellata*, *Shorea hypochra*, *S. gratissima*, *Dipterocarpus costatus*, *Millettia atropurpurea*, *Cynometra malaccensis*, *Intsia palembanica*, *Nephelium lappaceum*, and *Michelia champaca* (Table 2).

The subcanopy is rich in Myristicaceae, Sapindaceae, Annonaceae, Euphorbiaceae, and Ebenaceae. There are typically many subcanopy and understory palm species, and climbing "rattan" palms are an important part of the diverse liana community. There is also a much greater incidence of cauliflory than in more-seasonal forest types (Ogawa et al. 1965a). The ground cover is dominated by tree and liana recruitment. Bamboos are absent from undisturbed WSEF.

Species diversity in the six 1 ha plots of WSEF was highest of the forest types examined for this chapter (Table 1). We recorded 428 species ≥ 10 cm dbh among these six WSEF 1 ha plots. For individual plots, species richness averaged 129 species/ha, and ranged from 99 to 151 species/ha ≥ 10 cm dbh (Table 1). Average Fisher's alpha among the six plots was 60.5 (range: 34.9–80.8). In addition to the low species-level dominance in this forest, another key feature of the diversity was the large number of genera with many sympatric species (e.g., *Syzygium*, *Diospyros*, *Aglaia*, and *Ficus*). This is a feature usually associated with the exceptional local diversity of lowland everwet tropical rain forests.

Structurally, WSEF looks similar to the lowland forests in aseasonal areas further south. The average total basal area of the six 1 ha plots surveyed here was 34.3 m^2/ha (Figure 3). This is similar to the aseasonal forests further south (Figure 4). However, WSEF has generally lower stem densities than the everwet forests (Figure 4), a difference entirely due to the lower density of stems in the smaller size classes (1–5 cm dbh; S. Bunyavejchewin and S. J. Davies, unpublished data; and see also Bunyavejchewin et al. 2001 for the same effect in DSEF).

Despite its short, predictable dry season, WSEF is almost completely evergreen. Less than 5 percent of the trees in the WSEF at Khao Chong are deciduous, and then only briefly (S. Bunyavejchewin and S. J. Davies, unpublished data). The short, predictable dry season does, however, seem to significantly impact other aspects of forest function. The majority of canopy species appear to flower annually, although flowering intensity varies from year to year (Ashton 1991; S. Bunyavejchewin and S. J. Davies, unpublished data). Recent studies have shown that WSEF species have systematic differences in a range of physiological and demographic traits that allow them to tolerate drought more effectively than species of the southern aseasonal forests (Baltzer et al. 2007, 2008).

DISTURBANCE AND DISTURBANCE REGIMES IN THE SEASONALLY DRY FORESTS OF CONTINENTAL SOUTHEAST ASIA

The SDTFs of Southeast Asia are subject to a range of natural disturbances. These disturbances, which include fires, windstorms, and perhaps extreme droughts, vary in intensity, extent, and frequency. The nature of this variation over decades and centuries may have a profound influence in shaping the structure and sorting the composition of the region's forests. In this section, we review the current knowledge of disturbance regimes in the SDTFs of Southeast Asia and their role in the forests' dynamics across a range of spatial and temporal scales. We begin by discussing specific disturbance types and the (often limited) information available to describe their impacts on the forests.

Finally, we consider for a well-studied DSEF how different disturbances overlap at different scales to generate considerable structural and compositional complexity within the forest.

Types of Natural Disturbance in Seasonally Dry Forests of Southeast Asia

Fire

No other disturbance is more closely associated with SDTF than fire. The strong seasonality in precipitation that characterizes SDTFs across the region also predisposes them to fire. Depending on the latitude of the site relative to the annual migration of the subtropical high pressure zones, the length of the dry season may range from two months in weakly seasonal areas to as long as eight months in strongly seasonal areas. The absence of rainfall coincides with a period of high temperatures and strong insolation due to low cloud cover, which together produce a prolonged reduction in relative humidity. These climatic conditions are sufficient to dry out fine fuels such as grasses, leaves, and small twigs, and in areas with more-pronounced dry seasons, even larger fuels such as branches and small pieces of wood. When combined with an ignition source, the result is a system that is readily, and routinely, burned.

Fire has attracted considerable interest and sparked heated debate in South and Southeast Asia over the past 150 years (Troup 1921; see also Kobsak Wanthongchai and Goldammer, this volume). Early foresters trained in European forestry schools and focused on protecting timber resources advocated a program of fire suppression. More recently, with the economic development of many Southeast Asian countries, urban populations have been calling for fire suppression. In contrast, local communities living on the forest margins have long promoted cool season burning to control the accumulation of combustible biomass in the understory of SDTF (see Kobsak Wanthongchai and Goldammer, this volume). Wanthongchai and Goldammer argue that controlled burning provides green shoots for grazing livestock, reduces the risk of uncontrollable fires, and returns limited nutrients to the soil for agriculture. In many cases, fire has been a fundamental tool for converting SDTF to agricultural land.

Despite historical and contemporary concerns about the effects of fire in SDTFs, there is remarkably little empirical data on forest fires in continental Southeast Asia. A series of recent papers has provided new data quantifying fire effects in a landscape overlaid by a mosaic of SDTF types (Baker and Bunyavejchewin 2006a, 2009; Baker et al. 2008). The El Niño Southern Oscillation (ENSO) event of 1997–98 led to a severe and prolonged drought across tropical Southeast Asia that was accompanied by widespread wildfires across insular and continental Southeast Asia. In western Thailand, fires associated with this ENSO event burned through nearly two-thirds of the Huai Kha Khaeng Wildlife Sanctuary, a World Heritage Site. The presence of a 50 ha forest dynamics plot, which was established in 1991, and research scientists on the ground at the time of the fires in January and February 1998, allowed us to assess the behavior and record the impacts of a widespread fire event across a range of the forest types typical of SDTF. The Huai Kha Khaeng fires burned through all three of the dominant forest types in the sanctuary (DSEF, MDF, and DDF). As discussed above, these forest types vary considerably in their structure and composition; however, these differences also influence their susceptibility to fire and the behavior of fires when they do occur in each forest type.

Forest fires influence both the physical structure of the forest and the floristic composition of the forest by killing trees. Surveys at Huai Kha Khaeng in the wake of the 1998 fires demonstrated that the fires generated more gaps of greater size in the MDF than in either the DDF or DSEF. In the DSEF most of the gaps that formed after the fire were less than 25 m², although the largest was greater than 500 m². In the MDF there was a similar range in gap sizes, but the mean gap size was considerably larger (104 m²). The DDF suffered very few gaps, and they were generally small (Baker and Bunyavejchewin 2009). At the community level, we were surprised to find little evidence of differences among the forest types in fire intensity and fire-induced mortality (Baker et al. 2008; Baker and Bunyavejchewin 2009). In each of the three forest types, the fires had low flames (10–30 cm) of low intensity (11–25 kW/m), and mortality was primarily limited to trees of less than 5 cm dbh (Baker and Bunyavejchewin 2009). Mean scorch height ranged from 1.2 to 1.5 m among the three forest types; MDF had the lowest mean scorch height, but it was not significantly different from mean scorch heights in DDF and DSEF. Post-fire censuses of mortality across size classes showed similar patterns among the forest types. Seedling mortality was greater than 90 percent in all forests; sapling mortality ranged from a low of approximately 30 percent in DSEF to a high of approximately 50 percent in MDF. Among poles and trees, mortality was limited to less than 10 and 5 percent, respectively, in all three forest types. The tree species in these forests, however, have evolved in an environment in which fire has been a regular occurrence. Consequently, many of the species have adaptations to fire such as thick bark or the ability to resprout vigorously following fire. We recensused the same sites within each of the forest types nine months after the fires. We found that in the DDF and MDF nearly half of the seedlings that had been considered killed immediately after the fire had resprouted. Nearly all the saplings had resprouted, irrespective of forest type. Among the seedlings and saplings, the only class that did not resprout abundantly were DSEF seedlings.

Seasonally dry tropical forests of Southeast Asia often support a mosaic of evergreen and deciduous forest types. Deciduous forest types are more prone to burning because the annual loss of leaves during the dry season lowers the relative humidity and increases the temperature on the forest floor, drying the fine fuels that carry low-intensity fires. Fires enter evergreen forests less frequently because the year-round canopy cover prevents prolonged exposure of the forest floor to direct sunlight. The cooler temperatures and higher relative humidity make fire incursions relatively rare. Indeed, under normal conditions, it is often impossible to light fires intentionally in DSEF (L. Johnson, personal communication). There is a widespread perception that tree species in evergreen forests are, therefore, more susceptible to degradation as a consequence of fire. Two recent studies have suggested that this may not be the case. Baker and Bunyavejchewin (2006a) showed that bark thickness, a common proxy measure of fire susceptibility, was not related to abundance of saplings of common canopy and mid-canopy DSEF tree species in an area at Huai Kha Khaeng that had burned three times in the past decade. The tree species with the most abundant saplings, *Baccaurea ramiflora*, had the second-thinnest bark, whereas the tree species with thickest bark,

Hopea odorata, had almost no saplings. A broader examination of all species in the Huai Kha Khaeng 50 ha plot compared changes in population abundance of evergreen and deciduous tree species in the censuses before (1994) and after (2004) the 1998 ENSO fires. Neither the deciduous tree species, which are commonly considered to be better adapted to fire, nor the evergreen tree species, which are considered to be more susceptible to fire-induced damage and mortality, showed any evidence of a directional change in population abundance as a result of the fires (Baker and Bunyavejchewin 2009). These results suggest that evergreen forest types, such as DSEF, that co-occur with deciduous forest types across the seasonally dry landscapes of continental Southeast Asia may be relatively resilient to the effects of fires.

Wind

Sudden, intense bursts of wind can have locally destructive effects on SDTF. Convective downdrafts associated with squall lines are known to cause forest blowdowns ranging from a few to several thousand hectares in the Amazon Basin (Nelson et al. 1994). The local and regional weather conditions that give rise to these and other intense windstorms occur across continental Southeast Asia, and the windstorms that they generate impact upon forests in the region. Assessing the relative importance of windstorms as a disturbance on Southeast Asia's SDTFs is difficult due to the lack of studies describing them. Most of the evidence for windstorms damaging forests is anecdotal. Colonial foresters working in southwestern Burma in the early 1900s describe areas of forest of several square kilometers blown down by wind. In one instance, a localized tornado flattened a swath of forest roughly 0.5 km wide and 6 km long (Anonymous 1929, 1932). At Huai Kha Khaeng, an area of approximately 30 ha of MDF was knocked over by windstorms in 1987 (T. Prayurasiddhi, personal communication). In peninsular Thailand, bent tree stems in WSEF suggest moderate-scale blow-downs during the past century. Particularly in the southern parts of the region, where WSEF forests are more abundant, cyclones are a recurring event. The extreme winds and heavy rainfall associated with cyclones are capable of causing considerable damage to forests.

Drought

In continental Southeast Asia, the cycle of rainfall seasonality generates dry conditions for several months each year. The flora and fauna of the region have evolved phenological, physiological, and behavioral adaptations to accommodate these seasonal fluctuations in the availability of food and water (see Baltzer et al. 2008). For trees in SDTF, deciduousness is a common response to the environmental conditions of the dry season. The extent, timing, and intensity of leaf loss varies widely among sympatric species (Williams et al. 2008). Some species, such as *Tetrameles nudiflora* and *Spondias pinnata*, may be leafless for 4–5 months; other species, such as some *Shorea* (e.g., *S. roxburghii*, *S. siamensis*, *S. obtusa*) may lose their leaves for only about 1 month; and yet other species, such as many *Diospyros* and *Garcinia*, may lose only a fraction of the leaves in their crowns and never be fully deciduous (Williams et al. 2008). Phenological adaptations to seasonal drought, as well as the associated physiological responses, mean that trees in SDTFs are well-adapted to water stress and, as such, are relatively resilient to drought as a disturbance. In contrast, trees in everwet tropical forests, which do not regularly experience severe water stress, are more sensitive to the irregular occurrence of drought events. In Southeast Asia, extreme drought events are typically associated with severe ENSO events, the most recent of which occurred in 1997–98. While the 1997–98 ENSO-driven drought led to fires across much of continental and insular Southeast Asia, the fires killed primarily small (less than 10 cm dbh) trees. The widespread mortality of large trees that was observed in the everwet forests of Borneo in the wake of the same ENSO event was largely a consequence of the drought, not the fires (Potts 2003; van Nieuwstadt and Sheil 2005). There is little evidence that the 1997–98 drought had any adverse effects (e.g., increased mortality) on tree populations in SDTFs.

While intense droughts of short duration may not have a significant direct impact on tree populations in SDTFs, more prolonged droughts may. Recent studies in northern Thailand and Vietnam using tree rings to reconstruct historical drought conditions have shown that continental Southeast Asia has been subjected to several intense, multidecade "mega-droughts" over the past 400–600 years (Buckley et al. 2007; Sano et al. 2008). The teak tree-ring chronology from northwestern Thailand records a 20–25 year drought in the early 1700s (Buckley et al. 2007), while the *Fokienia hodginsii* chronology from central Vietnam points to an intense 20-year drought that began in the 1890s (Sano et al. 2008). Although trees in SDTFs are adapted to annual periods of water stress, it is unknown what the impact of such prolonged intense drought conditions would be. A potential consequence of prolonged drought is an increase in fire frequency or intensity. The limited paleoecological evidence does not point to a greatly heightened period of fire during these droughts, although the coarse temporal resolution of the few sediment cores from the region may constrain such interpretation (Maxwell 2004).

Disturbance Regimes—Spatial and Temporal Complexity

Describing the disturbance regime of a given area requires historical information on the frequency, intensity, and extent of the major disturbances acting on that area. In most cases, this involves the collection of historical data on disturbance events—either directly through carefully dated samples associated with the disturbance, or indirectly through reconstructions of past recruitment and mortality. It has been possible to reconstruct disturbance regimes for many temperate zone forests due to the presence of annual growth rings in most or all of the resident tree species. Until recently, no such studies had been conducted in tropical forests because of the widespread belief that tropical tree species do not form annual growth rings. We now know that some tropical tree species, particularly in strongly seasonal environments, do form annual growth rings (Worbes 2002), and that dendrochronological analyses of these species, in combination with other data on forest structure, can yield new insights into the role of disturbances in structuring tropical forests, particularly those of the seasonally dry tropics (Baker et al. 2005; Baker and Bunyavejchewin 2006b). In the first dendroecological reconstruction of the disturbance history of a tropical forest, Baker et al. (2005) demonstrated the complex influences of past disturbance events on a species-rich DSEF in western Thailand. Analyses of tree rings, stem diameter distributions, and life-history traits showed that the structurally complex, floristically rich site had largely been initiated in the wake of a catastrophic event in the early- to mid-1800s that killed most of the preceding forest over an area of approximately 500 ha. The discrete boundary of the modern-day DSEF, the lack of any distinct soil changes associated with this boundary (Baker 1997), and documented windstorms in the region suggest that this stand-initiating disturbance was most likely a windstorm. There are, however, individual trees that predate the disturbance, suggesting that mortality was not complete and that there existed a heterogeneous, post-disturbance environment of living trees, standing dead trees, and blown-down trees. Interestingly, many of the trees that survived the disturbance and are still alive are dominant tree species in MDF or DDF, but not DSEF. This suggests the possibility that the catastrophic disturbance of the 1800s may have precipitated a shift from a deciduous forest type to an evergreen forest type.

The catastrophic disturbance had a distinctive effect on the present-day DSEF structure and composition. This is most notable in the relative dominance of the canopy tree species *Hopea odorata*. In the Huai Kha Khaeng 50 ha plot, *Hopea odorata* accounts for less than 0.5 percent of the stems but more than 10 percent of the total stand basal area (Bunyavejchewin et al. 2001, 2009). The size structure of the *Hopea* population at the Huai Kha Khaeng plot and its known preference for growth in high-light environments clearly point to the *Hopea* overstory as being an even-aged population of trees that established simultaneously in the wake of a catastrophic disturbance. Dendroecological study of the DSEF at Huai Kha Khaeng, however, revealed that the catastrophic disturbance was only one component of the disturbance history of the site. In the decades that followed, a number of widespread but low-intensity disturbances occurred, causing pulses of recruitment scattered across hundreds of hectares of DSEF. Tree-ring analyses reveal four of these disturbance events, in the 1890s, 1920s, 1950s, and 1970s. Based on contemporary observations of forest dynamics in the wake of landscape-scale fires, Baker et al. (2005) suggest that widespread fires were the most likely cause of these disturbances (although the 1950s event may have been associated with a known cyclone). In addition to these disturbance events, the tree-ring analyses also demonstrated that small canopy gaps have been forming in every decade since the catastrophic disturbance (Baker et al. 2005). Taken together, the different types, intensities, and frequencies of disturbance that occurred in the Huai Kha Khaeng DSEF demonstrate the complex role of disturbance in structuring SDTF communities.

The only other forest type in the SDTFs of Southeast Asia that has received similar study is pine forest (PF) in northern Thailand. These forests are much simpler than DSEF in terms of forest structure, species richness, and, presumably, disturbance history. Zimmer and Baker (2009) looked at a range of PF stands across north and northwest Thailand to determine whether pine recruitment was driven by climatic variability as suggested by early foresters in the region (Troup 1921). They found that most pine forests were multi-aged, but that recruitment typically occurred in discrete pulses, creating distinct age cohorts. In addition, there was a strong relationship between pine recruitment and the occurrence of multiyear periods during which the dry season was cooler and wetter than the long-term mean conditions. While the dynamics and disturbance history of the PF were substantially simpler than those of DSEF, there were distinct interactions among stand, site, and regional-scale dynamics that created a relatively complex history of disturbance among the PF studied.

A RESEARCH AND CONSERVATION (MANAGEMENT) AGENDA FOR THE SEASONALLY DRY TROPICAL FORESTS OF CONTINENTAL SOUTHEAST ASIA IN THE COMING DECADES

Despite the importance of SDTFs and the ongoing threat to their persistence, we are still only beginning to understand what these forests are (in terms of both structure and composition), how they change over time, and the role of disturbances in their long-term dynamics. Our limited understanding of these threatened ecosystems represents a major hurdle to their conservation. Currently, most conservation efforts in the region focus on preserving the seasonal forests in protected areas such as national parks and wildlife sanctuaries. These areas are often small, isolated forest fragments within a broader matrix of degraded forest, swidden agriculture, and rural communities. Within the region there are some important exceptions—protected areas supporting thousands of square kilometers of contiguous forest. Notable examples include the Western Forest Complex of Thailand, nearly 10,000 km² of protected area divided among seventeen different parks and sanctuaries, and the Northern Forest Complex in Myanmar, which includes nearly 22,000 km² of land extending from lowland seasonal tropical forests to the treeline in the eastern Himalayas. In protected areas of this size, management can be coordinated among individual parks and sanctuaries to ensure that the biodiversity advantages obtained at a larger scale are realized. The mosaic distribution of the main SDTF types across continental Southeast Asia makes this level of coordination fundamentally necessary. Animals that depend on resources that vary seasonally must be able to move among forest types to find food as it becomes available. Many of the smaller protected areas support only one or two forest types, limiting the fauna to species that are specialized for particular forest types.

Management of protected areas in continental Southeast Asia, as elsewhere, is primarily focused on perpetuating the existing conditions. Boundaries are enforced, and infrastructure, such as access roads and ranger stations, is maintained. Threats to the forests and associated fauna such as poaching, fire, and invasive species are, in some cases, managed. For example, in the Huai Kha Khaeng Wildlife Sanctuary, the Thai government greatly reduced the high levels of poaching in the 1980s and early 1990s by establishing an army base adjacent to the sanctuary and initiating anti-poaching patrols with well-armed soldiers and forest rangers. In many instances, however, management of SDTFs and the disturbances that affect them is conducted on an ad hoc basis. Management of the most widespread disturbance within the region, fire, is generally opportunistic and reactive. Fire-fighting crews are deployed into remote areas on foot with little equipment or support and often with limited information about the fires. In most years this may be adequate, but in extreme fire years, such as 1997–98, the scale of the fires may overwhelm the capacity of fire management agencies and protected area staff. However, even with the best information and unlimited resources to fight fires, it is unclear what the management staff of protected areas should do because answers are lacking to fundamental ecological questions:

- What is the "natural" fire regime for these landscapes?
- How often do fires return and how do they vary in intensity?
- Is variation in fire frequency and intensity forest-type specific?
- Should some or even all fires be left to burn unchecked?
- Should some or even all fires be suppressed upon first detection?
- Should fires be managed differently in different forest types?

Our ability to answer these questions and others like them is further complicated by uncertainties associated with future climate change. For example, if the distribution of forest types within landscapes is in part controlled by environmental conditions, will a drier climate lead to a rearrangement of forest types across continental Southeast Asia's remaining forested landscapes? If the forest fauna require a mix of forest types within a landscape, how will directional change toward drier forest types such as DDF impact animal populations? Our current understanding of the factors that control the relative abundance of species within forest types and the relative abundance of forest types within landscapes is insufficient to answer these questions. However, the patterns of species distribution and abundance within forest types described in our permanent forest inventory plots point to testable hypotheses. As an example, for forest types dominated by dipterocarp species (i.e., DDF and DSEF), common garden experiments with 5–10 of the major species growing across gradients of water and light availability would elucidate the role of local environmental conditions on establishment within each forest type.

The underlying assumption behind the management of protected areas is that in the face of long-term, region-wide forest loss and fragmentation, the existing network of protected areas must be sufficient to protect regional biological resources. For many plant and animal species, this may be sufficient; however, for some particularly large, long-lived animals such as tigers, elephants, gaur, and hornbills, this may not be the case. In recent years there has been growing interest in the conservation value of secondary forests (see Barlow et al. 2007; Gardner et al. 2009; Norden et al. 2009) and the potential to reestablish native forests on abandoned agricultural land or to accelerate recovery of secondary forests (Lamb et al. 2005). If native forests could be reestablished at a significant scale, then the possibility would exist of reversing twentieth-century trends of forest loss and fragmentation. This could have potentially profound implications for conservation in continental Southeast Asia, particularly regarding highly threatened and endangered animals that require large home ranges. However, recreating native forests successfully would require a detailed understanding of life-history traits of tree species, forest-type-specific development patterns, and species-specific responses to disturbance. Our studies describing the structure and composition of the dominant forest types within the study region (and how they vary from site to site) and the role of disturbances in shaping these communities provide the necessary foundations for such an undertaking.

REFERENCES

Anonymous. 1929, 1932. Annual Reports on Working Plans and Silviculture in Burma. *Superintendent of Government Printing and Stationary*. Rangoon, Burma.

Appanah, S. 1985. General Flowering in the Climax Rain Forests of South-East Asia. *Journal of Tropical Ecology* 1:225–40.

Ashton, P. S. 1990. Thailand: Biodiversity Center for the Tropics of Indo-Burma. *Journal of Science and Society, Thailand* 16:107–116.

———. 1991. Toward a Regional Classification of the Humid Tropics of Asia. *Tropics* 1:1–12.

———. 1997. South Asian Evergreen Forests: Some Thoughts towards Biogeographic Reevaluation. *Tropical Ecology* 38:171–80.

———. 2003. Floristic Zonation of Tree Communities on Wet Tropical Mountains Revisited. *Perspectives in Plant Ecology, Evolution and Systematics* 6:87–104.

Baker, P. J. 1997. Seedling Establishment and Growth across Forest Types in an Evergreen/ Deciduous Forest Mosaic in Western Thailand. *Natural History Bulletin of the Siam Society* 45:17–41.

Baker, P. J., and S. Bunyavejchewin. 2006a. Bark Thickness and the Influence of Forest Fire on Tree Population Structure in a Seasonal Evergreen Tropical Forest. *Natural History Bulletin of the Siam Society* 54:215–25.

———. 2006b. Suppression, Release and Canopy Recruitment in Five Tree Species from a Seasonal Tropical Forest in Western Thailand. *Journal of Tropical Ecology* 22:521–29.

———. 2009. Fire Behavior and Fire Effects across the Forest Landscape Mosaics of Continental Southeast Asia. In *Tropical Fire Ecology: Climate Change, Land Use and Ecosystem Dynamics*, ed. M. A. Cochrane. Heidelberg, Germany: Springer-Praxis.

Baker, P. J., S. Bunyavejchewin, C. D. Oliver, and P. S. Ashton. 2005. Disturbance History and Historical Stand Dynamics of a Seasonal Tropical Forest in Western Thailand. *Ecological Monographs* 75:317–43.

Baker, P. J., S. Bunyavejchewin, and A. R. Robinson. 2008. The Impacts of Large-Scale, Low-Intensity Fires on the Forests of Continental South-East Asia. *International Journal of Wildland Fire* 17:782–92.

Baltzer, J. L., S. J. Davies, S. Bunyavejchewin, and N. S. Noor. 2008. The Role of Desiccation Tolerance in Determining Tree Species Distributions along the Malay-Thai Peninsula. *Functional Ecology* 22:221–31.

Baltzer, J. L., S. J. Davies, A. R. Kassim, N. S. Noor, and J. V. LaFrankie. 2007. Geographic Distributions in Tropical Trees: Can Geographic Range Predict Performance and Habitat Association in Co-occurring Tree Species? *Journal of Biogeography* 34:1916–26.

Barlow, J., and T. A. Gardner et al. 2007. Quantifying the Biodiversity Value of Tropical Primary, Secondary and Plantation Forests. *Proceedings of the National Academy of Sciences, USA* 104:18555–60.

Blanc, L., and G. Maury-Lechon et al. 2000. Structure, Floristic Composition and Natural Regeneration in the Forests of Cat Tien National Park, Vietnam: An Analysis of the Successional Trends. *Journal of Biogeography* 27:141–17.

Blasco, F., M. F. Bellan, and M. Aizpuru. 1996. A Vegetation Map of Tropical Continental Asia at Scale 1:5 Million. *Journal of Vegetation Science* 7:623–34.

Buckley, B. M., K. Palakit, K. Duangsathaporn, P. Sanguantham, and P. Prasomsin. 2007. Decadal Scale Droughts over Northwestern Thailand over the Past 448 Years: Links to the Tropical Pacific and Indian Ocean Sectors. *Climate Dynamics* 29:63–71.

Bunyavejchewin, S. 1983a. Canopy Structure of the Dry Dipterocarp Forest of Thailand. *Thai Forest Bulletin* 14:1–132.

———. 1983b. Analysis of the Tropical Dry Deciduous Forest of Thailand. I. Characteristics of the Dominance-Types. *Natural History Bulletin of the Siam Society* 31 (2): 109–22.

———. 1985. Analysis of the Tropical Dry Deciduous Forest of Thailand. II. Vegetation in Relation to Topographic and Soil Gradients. *Natural History Bulletin of the Siam Society* 33 (1): 3–20.

———. 1986. Ecological Studies of Tropical Semi-evergreen Rain Forest at Sakaerat, Nakhon Ratchasima, Northeast Thailand. I. Vegetation Patterns. *Natural History Bulletin of the Siam Society* 34:35–57.

———. 1999. Structure and Dynamics in Seasonal Dry Evergreen Forest in Northeastern Thailand. *Journal of Vegetation Science* 10:787–72.

Bunyavejchewin, S., P. J. Baker, J. V. LaFrankie, and P. S. Ashton. 2001. Stand Structure of a Seasonal Dry Evergreen Forest at Huai Kha Khaeng Wildlife Sanctuary, Western Thailand. *Natural History Bulletin of the Siam Society* 49:89–106.

———. 2002. Floristic Composition of a Seasonal Dry Evergreen Forest at Huai Kha Khaeng Wildlife Sanctuary, Western Thailand. *Natural History Bulletin of the Siam Society* 50:125–34.

Bunyavejchewin, S., J. V. LaFrankie, P. J. Baker, S. J. Davies, and P. S. Ashton. 2009. *Forest Trees of Huai Kha Khaeng Wildlife Sanctuary, Thailand: Data from the 50-hectare Forest Dynamics Plot*, 342. Bangkok, Thailand: National Parks, Wildlife and Plant Conservation Department.

Bunyavejchewin, S., J. V. LaFrankie, P. J. Baker, M. Kanzaki, P. S. Ashton, and T. Yamakura. 2003. Spatial Distribution Patterns of the Dominant Canopy Dipterocarp Species in a Seasonal Dry Evergreen Forest in Western Thailand. *Forest Ecology and Management* 175:87–101.

Cao, M., H. Zhu, H. Wang, G. Lan, Y. Hu, S. Zhou, X. Deng, and J. Cui. 2008. *Xishuangbanna Tropical Seasonal Rainforest Dynamics Plot: Tree Distribution Maps, Diameter Tables and Species Documentation*. Kunming, China: Yunnan Science and Technology Press.

Cao, M., X. Zou, M. Warren, and H. Zhu. 2006. Tropical Forests of Xishuangbanna, China. *Biotropica* 38:306–9.

Champion, H. G., and S. K. Seth. 1968. *A Revised Survey of the Forest Types of India*. Dehli, India: Government of India Press.

Cooling, E. N. G. 1968. *Fast Growing Timber Trees of the Lowland Topics*: Pinus merkusii, 312. Commonwealth Forestry Institute, Oxford University.

Forest Herbarium, Royal Forest Department. 2001. *Thai Plant Names, Tem Smitinand*, rev. ed. Bangkok: Royal Forest Department.

Gardner, T. A., J. Barlow, R. Chazdon, R. M. Ewers, C. A. Harvey, C. A. Peres, and N. S. Sodhi. 2009. Prospects for Tropical Biodiversity in a Human-Modified World. *Ecology Letters* 12:561–52.

Hughes, J. B., P. D. Round, and D. S. Woodruff. 2003. The Indochinese-Sundaic Faunal Transition at the Isthmus of Kra: An Analysis of Resident Forest Bird Species Distributions. *Journal of Biogeography* 30:569–80.

Ishida, A., S. Diloksumpun, P. Ladpala, D. Staporn, S. Panuthai, M. Gamo, K. Yazaki, M. Ishizuka, and L. Puangchit. 2006. Contrasting Seasonal Leaf Habits of Canopy Trees between Tropical Dry-Deciduous and Evergreen Forests in Thailand. *Tree Physiology* 26:643–56.

Kanzaki, M., M. Hara, T. Yamakura, T. Ohkubo, M. N. Tamura, K. Sri-Ngernyuang, P. Sahunalu, S. Teejuntuk, and S. Bunyavejchewin. 2004. Doi Inthanon Forest Dynamics Plot, Thailand. In *Forest Diversity and Dynamism: Findings from a Large-Scale Plot Network*, ed. E. C. Losos and E. G. Leigh, 474–81. Chicago: University of Chicago Press.

Khamyong, S., A. M. Lykke, D. Seramethakun, and A. S. Barfod. 2004. Species Composition and Vegetation Structure of an Upper Montane Forest at the Summit of Mt. Doi Inthanon, Thailand. *Nordic Journal of Botany* 23:83–97.

Kitamura, S., S. Suzuki, T. Yumoto, P. Chuailua, K. Plongmai, P. Poonswad, N. Noma, T. Maruhashi, and C. Suckasam. 2005. A Botanical Inventory of a Tropical Seasonal Forest in Khao Yai National Park, Thailand: Implications for Fruit-Frugivore Interactions. *Biodiversity and Conservation* 14:1241–62.

Kutintara, U. 1975. Structure of the Dry Dipterocarp Forest. PhD diss., Colorado State University.

Lamb, D., P. D. Erskine, and J. A. Parrotta. 2005. Restoration of Degraded Tropical Forest Landscapes. *Science* 310:1628–32.

Marod, D., U. Kutintara, C. Yarwudhi, H. Tanaka, and T. Nakashizuka. 1999. Structural Dynamics of a Natural Mixed Deciduous Forest in Western Thailand. *Journal of Vegetation Science* 10:777–86.

Maxwell, A. 2004. Fire Regimes in North-Eastern Cambodian Monsoonal Forests, with a 9300-year Sediment Charcoal Record. *Journal of Biogeography* 31:225–39.

Miles, L., A. C. Newton, R. S. DeFries, C. Ravilious, I. May, S. Blyth, V. Kapos, and J. E. Gordon. 2006. A Global Overview of the Conservation Status of Tropical Dry Forests. *Journal of Biogeography* 33:491–505.

Murphy, P. G., and A. E. Lugo. 1986. Ecology of Tropical Dry Forest. *Annual Review of Ecology and Systematics* 17:67–88.

Myers, N., R. A. Mittermeier, C. G. Mittermeier, G. A. B. da Fonseca, and J. Kent. 2000. Biodiversity Hotspots for Conservation Priorities. *Nature* 403:853–58.

Nelson, B. W., V. Kapos, J. B. Adams, W. J. Oliveira, O. P. G. Braun, and I. L. do Anural. 1994. Forest Disturbance by Large Blowdowns in the Brazilian Amazon. *Ecology* 75:853–58.

Noguchi, H., A. Itoh, T. Mizuno, K. Sri-Ngernyuang, M. Kanzaki, S. Teejuntuk, W. Sungpalee, M. Hara, T. Ohkubo, P. Sahunalu, and T. Yamakura. 2007. Habitat Divergence in Sympatric Fagaceae Tree Species of a Tropical Montane Forest in Northern Thailand. *Journal of Tropical Ecology* 23:549–58.

Norden, N., R. L. Chazdon, A. Chao, Y.-H. Jiang, and B. Vilchez-Alvarado. 2009. Resilience of Tropical Rain Forests: Tree Community Reassembly in Secondary Forests. *Ecology Letters* 12:385–94.

Ogawa, H., K. Yoda, and T. Kira. 1961. A Preliminary Survey on the Vegetation of Thailand. *Nature and Life in Southeast Asia* 1:20–158.

Ogawa, H., K. Yoda, T. Kira, K. Ogino, T. Shidei, D. Ratanawongse, and C. Apasutaya. 1965a. Comparative Ecological Study on Three Main Types of Forest Vegetation in Thailand. I. Structure and Floristic Composition. *Nature and Life in Southeast Asia* 4:13–48.

Ogawa, H., K. Yoda, K. Ogino, and T. Kira. 1965b. Comparative Ecological Study on Three Main Types of Forest Vegetation in Thailand. II. Plant Biomass. *Nature and Life in Southeast Asia* 4:49–80.

Ohsawa, M., P. H. J. Nainggolan, N. Tanaka, and C. Anwar. 1985. Altitudinal Zonation of Forest Vegetation on Mount Kerinci, Sumatra: With Comparisons to Zonation in the Temperate Regions of East Asia. *Journal of Tropical Ecology* 1:193–216.

Ohsawa, M., and I. Nitta. 2002. Forest Zonation and Morphological Tree-Traits along Latitudinal and Altitudinal Environmental Gradients in Humid Monsoon Asia. *Global Environmental Research* 6:41–52.

Pennington, R. T., G. P. Lewis, and J. A. Ratter. 2006. An Overview of the Plant Diversity, Biogeography and Conservation of Neotropical Savannas and Seasonally Dry Forests. In *Neotropical Savannas and Seasonally Dry Forests: Plant Diversity, Biogeography, and Conservation*, ed. R. T. Pennington, G. P. Lewis, and J. A. Ratter, 1–2. Boca Raton, FL: CRC Press.

Potts, M. D. 2003. Drought in a Bornean Everwet Forest. *Journal of Ecology* 91:467–74.

Richards, P. W. 1996. *The Tropical Rain Forest: An Ecological Study*. Cambridge: Cambridge University Press.

Rundel, P. W., and K. Boonpragob. 1995. Dry Forest Ecosystems of Thailand. In *Seasonally Dry Tropical Forests*, ed. S. H. Bullock, H. A. Mooney, and E. Medina, 93–123. Cambridge: Cambridge University Press.

Sano, M., B. M. Buckley, and T. Sweda. 2008. Tree-Ring Based Hydroclimate Reconstruction over Northern Vietnam from *Fokienia hodginsii*: Eighteenth-century Mega-drought and Tropical Pacific Influence. *Climate Dynamics* 30:153–62.

Santisuk, T. 1988. *An Account of the Vegetation of Northern Thailand*. Stuttgart, Germany: Franz Steiner Verlag Wiesbaden GmbH.

Tani, A., E. Ito, M. Kanzaki, S. Ohta, S. Khorn, P. Pith, B. Tith, S. Pol, and S. Lim. 2007. Principal Forest Types of Three Regions of Cambodia: Kampong Thom, Kratie, and Mondulkiri. In *Forest Environments in the Mekong River Basin*, ed. H. Sawada, M. Araki, N. A. Chappell, J. V. LaFrankie, and A. Shimizu, 201–13. Tokyo: Springer.

Teejuntuk, S., P. Sahunalu, K. Sakurai, and W. Sungpalee. 2002. Forest Structure and Tree Species Diversity along an Altitudinal Gradient in Doi Inthanon National Park, Northern Thailand. *Tropics* 12:85–102.

Troup, R. S. 1921. *The Silviculture of Indian Trees*. Oxford: Clarendon Press.

van Nieuwstadt, M. G. L., and D. Sheil. 2005. Drought, Fire and Tree Survival in a Borneo Rain Forest, East Kalimantan, Indonesia. *Journal of Ecology* 93:191–201.

van Steenis, C. G. G. J. 1950. The Delimitation of Malaysia and Its Main Geographical Divisions. *Flora Malesiana* 1:lxx–lxxv.

Whitmore, T. C. 1984. *Tropical Rain Forests of the Far East*. Oxford: Oxford University Press.

Wikramanayake, E., E. Dinerstein, C. J. Loucks, D. M. Olson, J. Morrison, J. Lamoreux, M. McKnight, and P. Hedao. 2002. *Terrestrial Ecoregions of the Indo-Pacific: A Conservation Assessment*. Washington, DC: Island Press.

Williams, L. J., S. Bunyavejchewin, and P. J. Baker. 2008. Deciduousness in a Seasonal Tropical Forest in Western Thailand: Interannual and Intraspecific Variation in Timing, Duration and Environmental Cues. *Oecologia* 155:571–82.

Woodruff, D. S. 2003. Neogene Marine Transgressions, Palaeogeography and Biogeographic Transitions on the Thai-Malay Peninsula. *Journal of Biogeography* 30:551–67.

Woodruff, D. S., and L. M. Turner. 2009. The Indochinese-Sundaic Zoogeographic Transition: A Description and Analysis of Terrestrial Mammal Species Distributions. *Journal of Biogeography* 36:803–21.

Worbes, M. 2002. One Hundred Years of Tree-Ring Research in the Tropics—A Brief History and an Outlook to Future Challenges. *Dendrochronologia* 20:217–31.

Yoshifuji, N., T. Kumagai, K. Tanaka, N. Tanaka, H. Komatsu, M. Suzuki, and C. Tantasirin. 2006. Inter-annual Variation in Growing Season Length of a Tropical Seasonal Forest in Northern Thailand. *Forest Ecology and Management* 229:333–39.

Zhu, H. 1997. Ecological and Biogeographical Studies on the Tropical Rain Forest of South Yunnan, SW China with a Special Reference to Its Relation with Rain Forests of Tropical Asia. *Journal of Biogeography* 24:647–62.

Zimmer, H., and P. J. Baker. 2009. The Role of Regional Climate and Local Disturbance on Pine Recruitment in Southeast Asia. *Forest Ecology and Management* 257:190–98.

2

Seasonally Dry Tropical Forests in Southern India

An Analysis of Floristic Composition, Structure, and Dynamics in Mudumalai Wildlife Sanctuary

Hebbalalu Satyanarayana Suresh, Handanakere Shivaramaiah Dattaraja, Nandita Mondal, and Raman Sukumar

India is a mega-diverse country (Myers et al. 2000) with 67.7 million hectares (20.6 percent of its total area) under forest (Forest Survey of India 2005). Due to its complex physiography and climatic regimes, India has a diverse spectrum of vegetation types, including seasonal tropical evergreen forest, deciduous forest, extensive areas of semi-arid desert, temperate vegetation at mid- to high elevations of the Himalayas, and mangroves along the coast (Subramanyam and Nayar 1974). About 86 percent of the total forest area of the country is tropical; of this, about 90 percent is seasonally dry (K. Singh and Kushwaha 2005). Seven different forest formations were recognized by Champion and Seth (1968), including wet evergreen, semi-evergreen, moist deciduous, dry deciduous, littoral and swamp, dry thorn, and dry evergreen types. Champion and Seth divided the dry deciduous forests of India into two broad categories: (1) southern tropical dry deciduous forest, or teak- (*Tectona grandis*, Lamiaceae) dominated forest; and (2) northern tropical dry deciduous forest, or sal- (*Shorea robusta*, Dipterocarpaceae) dominated forest. We henceforth refer to these forests as seasonally dry tropical forest (SDTF).

SDTF is characterized by marked seasonality and interannual variability in precipitation (Read and Lawrence 2006; Murphy and Lugo 1986), high densities of wildlife and domestic livestock (Sinclair and Norton-Griffiths 1979; Skarpe 1991), and natural and human-caused fire (Swaine 1992; Goldammer 1993). SDTFs in southern India are also influenced by high climatic variability, herbivory by wild mammals and livestock (Sukumar et al. 1998; Silori and Mishra 2001), and dry season fires (Kodandapani et al. 2004, 2008). SDTF is less rich in tree species than tropical moist forests (Gentry 1988; Suresh and Sukumar 2005). Phenological events such as leaf flushing and flowering occur during the dry season despite the moisture stress (Bullock and

Solis-Magallanes 1990; Borchert 1994; Murali and Sukumar 1993, 1994). Vegetative coppicing is an important strategy of recovery following disturbances such as fire (Hoffmann 1998; Sukumar et al. 1998, 2005). SDTF trees also invest in below-ground biomass to a greater extent than do trees of moist forest types (Murphy and Lugo 1986; Finkeldey and Hattemer 2007).

In India, the historical extent of SDTF is difficult to ascertain due to widespread changes resulting from human activity. Large tracts of SDTF have been modified for agricultural and other purposes (Gadgil and Guha 1993). Current trends in land-use change and landscape modification coupled with natural processes will determine the future distribution of SDTF. Models of potential vegetation response to climate change for India predict that as much as three-fourths of the present forest area, including the SDTF, will change from one vegetation type to another (Ravindranath et al. 2006). It is therefore important to understand the ecological characteristics and dynamics of SDTF in order to promote adaptive strategies for its conservation.

In this chapter we present an analysis of the woody vegetation of Mudumalai, an SDTF landscape in southern India. The analysis is carried out at several spatial scales representing rainfall gradient across the landscape, the vegetation characteristics of a particular forest type, and the fine-scale structure of a tree community over fifty hectares. As part of these analyses, we present information on the ecological setting of Mudumalai, describe the vegetation structure and composition, and present a brief review of forest community dynamics.

STUDY AREA

Mudumalai Wildlife Sanctuary (Mudumalai) is part of the Western Ghats, a mountain range that runs parallel to the west coast of India and is listed as a global "hot spot" of biodiversity (Myers et al. 2000). The sanctuary is located north of the Nilgiri massif, between 11°30′ and 11°39′ N and 76°27′ and 76°43′ E (Figure 1). Mudumalai is part of a more-extensive tract of tropical dry forest that lies to the east and in the rain shadow of the Western Ghats. It is incorporated into the Nilgiri Biosphere Reserve, a Project Elephant reserve, and a national tiger reserve. Spread over an area of 321 km^2 along a distinct east-west rainfall gradient, Mudumalai includes tropical dry thorn, dry deciduous, moist deciduous, and semi-evergreen forest types. Grassy swamps are also characteristic of the moister forests to the west, though most of these swamps are now under cultivation (Figure 2). Champion and Seth (1968) and Puri (1960) identified three phyto-associations along the moisture gradient in the Western Ghats, and these have been applied to Mudumalai as well: (1) dry thorn forest characterized by *Acacia* and *Ziziphus*; (2) dry deciduous forest characterized by an association of *Anogeissus*, *Tectona*, and *Terminalia*; and (3) moist deciduous forest characterized by *Tectona*, *Terminalia*, *Lagerstroemia*, and *Dalbergia* (Champion and Seth 1968; Puri 1960).

The forests of Mudumalai are administered by the Tamilnadu Forest Department of the Tamilnadu state government. To the north of Mudumalai are the forests of Bandipur National Park (Karnataka State), and to the west are the moist deciduous forests of Wyanad Wildlife Sanctuary (Kerala State). Mudumalai has been the site of a variety

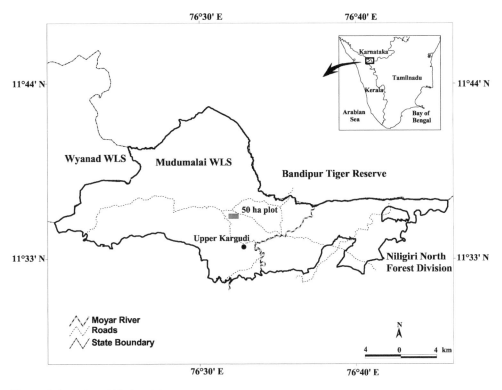

Figure 1. Location of Mudumalai.

of ecological investigations covering floristics (Sharma et al. 1977; Suresh et al. 1996, 2006); long-term forest dynamics (Sukumar et al. 1992, 1998, 2005); tree phenology and seed dispersal (Murali and Sukumar 1993, 1994; Suresh and Sukumar 2009; Prasad and Sukumar 2009); fire ecology (Kodandapani et al. 2008); non-timber forest product extraction and human impacts on forests (Ganesan 1993; Narendran et al. 2001; Silori and Mishra 2001); and wildlife ecology (Desai 1991; Baskaran et al. 1995; Varman and Sukumar 1995; Arivazhagan and Sukumar 2005; Vidya et al. 2005).

Geology and Soils

The terrain of Mudumalai is undulating with an elevation ranging from 350 m above sea level in the Moyar gorge in the northeast to 1,300 m above sea level in the west. The typical elevation over most of the area is 900–1,000 m above sea level. Geologically, the region belongs to the Archean continental landmass of the Indian peninsula composed of Pre-Cambrian volcanic rocks (Krishnan 1974). There are both igneous and metamorphic rocks dominated by biotite and granite gneisses composed of quartz- and feldspar-based rocks in combination with hornblende, pyroxene, and muscovite micas (George et al. 1988). The soils of Mudumalai have been classified into four orders, namely, inceptisols, alfisols, mollisols, and entosols, and nearly twenty-five

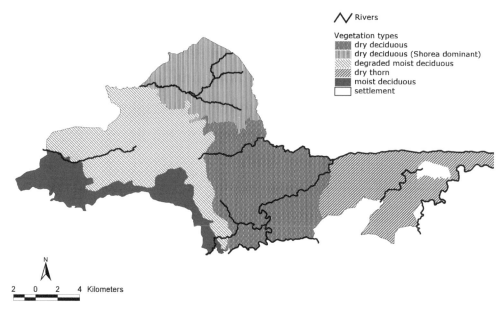

Rivers

Vegetation types
dry deciduous
dry deciduous (Shorea dominant)
degraded moist deciduous
dry thorn
moist deciduous
settlement

N

2   0   2   4 Kilometers

Figure 2. Map of the vegetation of Mudumalai with major rivers indicated. Data based on satellite imagery (IRSP6 LIS3) and ground verification.

soil series have been identified (George et al. 1988). The soils are generally loamy and red-black in color with an organic carbon content (dry weight) in the top layer (0–15 cm) varying from 1.5 percent in the drier east to 3.2 percent in the moister west (unpublished data).

Climate

The Koppen-Geiger global climate classification (Kottek et al. 2006) places this region under the equatorial winter dry type. Mean maximum temperature varies from 24°C in August to 30°C in March, and mean minimum temperature from 13.1°C in December to 17.9°C in May. A distinct precipitation gradient exists along an east-west axis, with the drier east receiving 600 mm and the moister west receiving 1,800 mm of rain on average annually (Figure 3). Under the monsoonal climate of the South Asian region, Mudumalai receives about 50 percent of its total rainfall from the southwest (summer) monsoon, which is active from June to September. The retreating, or winter (northeast), monsoon brings significant rainfall (over 30 percent) to the eastern side of Mudumalai during October and November. The dry season (less than 50 mm of rain per month) extends from December to March in the central part of Mudumalai. The Center for Ecological Sciences maintains weather stations at Kargudi (11°57'01" N and 76°64'61" E) and Masinagudi (11°57'42" N and 76°64'61" E). During 1990–2008, mean annual rainfall was 1,265 ± 288 mm at Kargudi, and 852 ± 203 mm at Masinagudi (Figure 4). Over these two decades the region witnessed an intense drought between 2000 and 2003 when rainfall was 28 percent below normal. For the

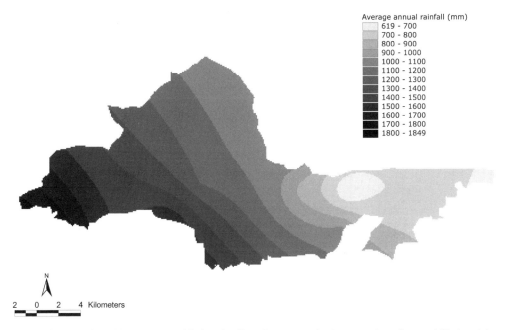

Figure 3. Map of rainfall contours in Mudumalai. Data from several rain gauges in and around Mudumalai.

landscape analysis, we also used rainfall estimates from thirty rain gauges set up by the Tamilnadu Electricity Board across the reserve. A more-detailed account of the climate of the Nilgiris is given by von Lengerke (1977).

Hydrology

The Moyar River, a tributary of the Cauvery, is the only perennial river running through Mudumalai. It is an important source of water for wildlife during the dry season. Important seasonal streams include Kekkanalla, Bidarahalla, and Benne Hole. Apart from the seasonal streams, there are two man-made, perennial water holes, namely, Hambetta Pond and Game-Hut Pond. The Maravakandi Dam, that stores water for a hydropower scheme, and a flume channel that carries water to the power plant at the foot of the gorge at Moyar, are also important water sources in the eastern part of the reserve (Figure 2).

Fauna

Mudumalai supports an intact fauna characteristic of peninsular India. The densities of the larger mammals at Mudumalai are among the highest recorded for similar forests anywhere in Asia (Varman and Sukumar 1995). Common herbivores present in Mudumalai include chital, or spotted deer (*Axis axis*), sambar (*Cervus unicolor*), gaur (*Bos gaurus*), and elephant (*Elephas maximus*). Other less-frequent herbivores include barking deer, or muntjac (*Muntiacus muntjak*), four-horned antelope (*Tetracerus*

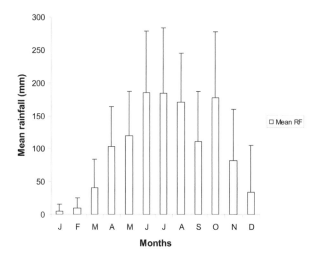

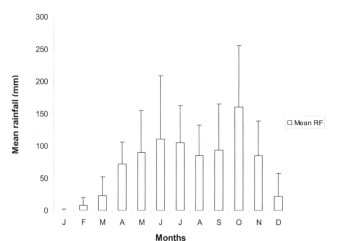

Figure 4. Average monthly rainfall at Kargudi and Masinagudi. Data from CES weather station for the years 1990–2008.

quadricornis), and black buck (*Antilope cervicapra*). Primates include bonnet macaque (*Macaca radiata*), leaf-eating monkeys, the hanuman langur (*Semnopithecus entellus*), and the rarely seen slender loris (*Loris tardigradus malabaricus*). Carnivore species include Asiatic wild dog (*Cuon alpinus*), leopard (*Panthera pardus*), striped hyena (*Hyaena hyaena*), and tiger (*Panthera tigris*). Omnivores include sloth bear (*Melursus ursinus*) and wild boar (*Sus scrofa*). There are several species of lesser cats, civets, and mongooses inhabiting Mudumalai. There are about 200 species of birds, 17 species of amphibians, and 42 species of reptiles recorded in Mudumalai.

History of Human Influence

Mudumalai has an ancient history of human activity, with remains of megalithic culture dating back to the early centuries of the common era (Nilgiri District Gazetteer

1995). Hunter-gatherer societies such as Kurubas, Irulas, and Paniyas have inhabited this region for several centuries with disease and wars regulating their populations (Hockings 1989; Prabhakar 1994). The forests of Mudumalai have also been exploited for timber since at least the early nineteenth century though systematic extraction began with the organization of the Forest Department by the British during the second half of the century (Ranganathan 1941). Species such as *Tectona grandis*, *Terminalia crenulata*, *Lagerstroemia microcarpa*, and *Dalbergia latifolia* were extracted for railway sleepers, shipbuilding, and construction in the Upper Nilgiri Plateau. Logging also enabled fires to penetrate the forests. Elephants were captured in these forests and trained for timber transportation (Krishnamurthy and Wemmer 1995). Coffee plantations were developed along the southeast, and tea plantations to the west of Mudumalai during the colonial period. Villagers also began cultivating the *vyals*, or swamps, in the moister western region. The antimalarial campaign during the 1950s, followed by development projects, facilitated an influx of people into this region from the plains. Villages such as Masinagudi, Moyar, and Singara now abut the reserve to the east, with the populous settlements of Thorapalli-Gudalur to the south. These villages have brought considerable pressure on the sanctuary through woodcutting, livestock grazing, and illegal hunting (Silori and Mishra 2001).

METHODS FOR ANALYZING VEGETATION STRUCTURE, COMPOSITION, AND DYNAMICS

Landscape-Level Analysis

Mudumalai was gridded into 1.5 × 1.5 km blocks, with plots of 400 m² (20 × 20 m) laid at the intersections of each gridline. All woody stems ≥ 1 cm dbh (diameter at breast height, "breast height" taken as 1.3 m) in each plot were identified and measured. An ocular estimate of height was made for each stem and reported in 5 m class intervals. A map depicting the rainfall gradient across the sanctuary (Figure 3), based on the thirty rain gauges, was derived using the "kriging with linear drift" interpolation procedure in a geographic information system (GIS; Burrough and McDonnell 1998). The average annual rainfall estimate for each plot was extracted from this map and used for further analysis.

A survey of fire occurrences in the sanctuary was undertaken each year between 1989 and 2003. Field surveys were conducted at the end of the dry season during April and May, and the extent of burning mapped at 1:50,000 scale on topographic sheets. The resulting fire maps were converted to GIS format and combined to produce a single map of fire frequency within the sanctuary, depicting the number of times the sampled plots had burned over the fifteen-year period. Fire frequency ranged from zero (not burned) to a maximum of eleven burns during the fifteen-year period.

Community and Population Analysis Using I ha Plots and a 50 ha Plot

Nineteen 1.04 ha (260 × 40 m) plots were established across Mudumalai during 1993–94 to study variation in forest structure and dynamics across the east-west rainfall gradient and among the major forest types. In this chapter we use these data to describe the vegetation characteristics of the forest types.

A 50 ha permanent forest dynamics plot was established in the central part of Mudumalai (in dry deciduous forest, close to the transition to moist deciduous forest) to describe forest structure and dynamics in relation to factors such as climate variability, fire, and herbivory by large mammals (Sukumar et al. 1992). The Mudumalai Forest Dynamics Plot (MFDP), a part of a global network of plots coordinated by the Center for Tropical Forest Science (CTFS), was established during 1988 and 1989 following standard CTFS protocols (Condit 1995, 1998). Detailed field methods employed for establishing the plot and initial stand composition are described in Sukumar et al. (1992, 2004, 2005).

RESULTS

Based on published literature and our own field knowledge, we recognize four main forest types within Mudumalai: dry thorn, dry deciduous, moist deciduous, and semi-evergreen forest. Dry deciduous forest covers about 47 percent of the total area of Mudumalai, moist deciduous forest 31 percent, dry thorn forest 15 percent, and semi-evergreen forest 6 percent (Figure 2; 1 percent of the total area represents habitations within the reserve). Some of these primary forest types, especially the moister types, are degraded due to logging and fire; thus extensive areas of moist deciduous forest that are recognized by their species composition resemble dry deciduous forest in structure.

FLORISTICS, FOREST STRUCTURE, AND DIVERSITY: RESULTS AT THE SCALE OF THE MUDUMALAI LANDSCAPE

Mean species richness among the eighty-nine 0.04 ha plots that were sampled across Mudumalai was 6.74 ± 3.3 (range: 1–17) species per 0.04 ha. There was distinct variation in species richness across the sanctuary, with high species richness both in the moist southwestern part of Mudumalai and the dry northeastern part of Mudumalai along the Moyar gorge, which has topographic heterogeneity (Figure 5). Species richness was lowest in the dry deciduous forest occupying the central part of Mudumalai.

Mean woody species diversity (Shannon-Weiner diversity index) of the sanctuary was 1.44 ± 0.45 (range: 0–2.5). The diversity pattern was similar to that of species richness, being low in the central portion in dry deciduous forest and higher in the moister southwestern and the drier northeastern parts of Mudumalai. Woody species diversity was weakly negatively correlated with fire frequency across the sanctuary ($r = 0.25$, $p = 0.06$).

Mean stem density ≥ 1 cm dbh was 558 ± 358 stems/ha (range: 25–1,925 stems/ha). Zones of high stem density were found mainly along the southwestern and northern boundaries. Stem density was also weakly negatively correlated with fire frequency across the sanctuary ($r = 0.23$, $p = 0.09$). Stand density was not influenced by rainfall.

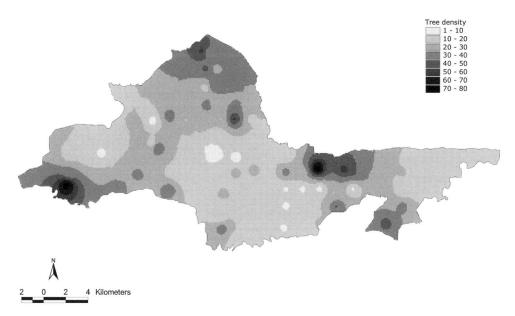

Figure 5. Species richness contour map of Mudumalai. Contours were generated using the number of species recorded from 0.04 ha sample plots.

Three canopy species, *Anogeissus latifolia* (Combretaceae), *Tectona grandis* (Lamiaceae), and *Terminalia crenulata* (Combretaceae), dominated the floristics at the landscape scale. They accounted for 42 percent of all stems (Table 1). There were 110 species recorded in the landscape-level enumeration. The only dipterocarp species of dry forests in peninsular India, *Shorea roxburghii*, accounted for 5 percent of all stems. The understory species *Catunaregam spinosa* (Rubiaceae) and *Cassia fistula* (Fabaceae) accounted for 4.3 percent and 4.1 percent of stems respectively. Thirty-one species accounted for a total of 88 percent of the stems, with each species having at least ten individuals ≥ 1 cm dbh. There were 34 species with one individual. A summary of tree diversity in the forest types of Mudumalai is presented in Table 1.

Size-class distribution differed among the three main forest types in Mudumalai (Figure 6). Moist deciduous forest had more stems per unit area compared to dry deciduous and dry thorn forests. The size-class distribution of individuals in moist deciduous forest was, however, not significantly different from that of dry deciduous forest (Kolmogorov-Smirnov [KS] test, $p > 0.05$), but was different from dry thorn forest (KS test, $p < 0.05$). The size-class distribution of stems in dry deciduous forest was not significantly different from that in dry thorn forest (KS test, $p > 0.05$).

Canopy height in both moist and dry deciduous forest was greater than 25 m while in dry thorn it was less than 15 m. Moist deciduous forest had the greatest mean tree height (11.9 ± 8.2 m), followed by dry deciduous (10.0 ± 6.5 m) and dry thorn forest (4.8 ± 3.6 m). The variability in tree heights among different forest types was significant ($F = 127.6$, $p < 0.01$).

Table 1. Summary of plant diversity in different forest types of Mudumalai Reserve. Numbers are total numbers for all plots in each forest type. Plots are 0.04 ha in area; n is the number of plots in each forest type.

Attributes	Moist Deciduous Forest (n = 27)	Dry Deciduous Forest (n = 49)	Dry Thorn Forest (n = 13)
Number of species	61	56	39
Number of individuals	660	979	303
Shannon-Weiner (H') diversity	3.26	2.67	2.94
Simpson's diversity	0.94	0.85	0.91
Evenness	0.430	0.258	0.486
Fisher's alpha	16.4	12.89	11.91

The four common canopy species (stems greater than 10 cm dbh) show differences in their spatial distributions across Mudumalai and we illustrate the distributions for two species in Figure 7. *Anogeissus latifolia* has a wider distribution across Mudumalai than the other species. *Lagerstroemia microcarpa* is more restricted to the western part where mean rainfall is over 1,300 mm per year. Although two other abundant species, *Terminalia crenulata* and *Tectona grandis*, extend into the high rainfall zone, most trees of these species are in the central part, which receives intermediate rainfall levels and has a higher frequency of dry season fire. *Tectona grandis* is abundant in the southern part of Mudumalai because of a Forest Department plantation.

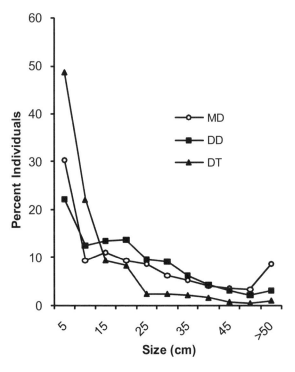

Figure 6. Size class distribution of individuals in different forest types. Percent of individuals in different size classes is plotted for each forest type.

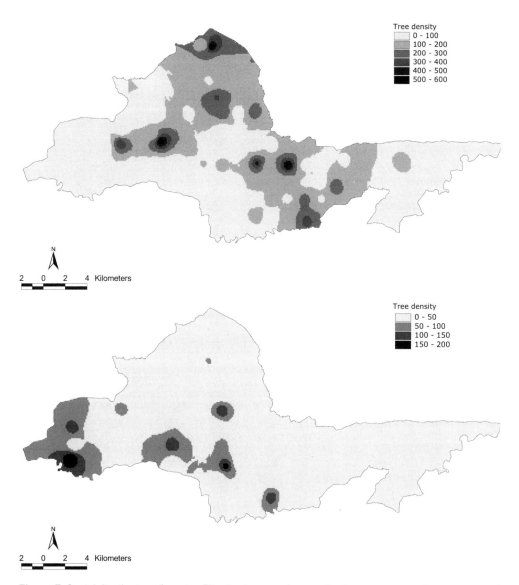

Figure 7. Spatial distribution of species. Distribution maps for two dominant canopy species are generated for individuals >10 cm dbh from 0.04 ha sample plots. The top figure shows the distribution of *Anogeissus latifolia* and the bottom figure shows *Lagerstroemia microcarpa*.

Floristic Composition

Fifty-five families of woody angiosperms, made up of a total of 221 species, were recorded in the various plots established in Mudumalai (Table 2). Fabaceae is the most species-rich family in Mudumalai with 25 species (11.3 percent of all species), including genera such as *Pterocarpus*, *Dalbergia*, *Ougeinia*, and *Butea*. One of the dominant tree genera in the dry thorn forest is *Acacia*, with four tree and two climbing species, including *A. concinna*, an important minor forest product, and *A. pennata*, a

Table 2. Diversity of plant families in Mudumalai Reserve.

Rank	Family	Number of Species	Relative Contribution (%)	Cumulative Contribution (%)
1	Fabaceae	25	11.46	11.46
2	Euphorbiaceae	23	10.5	22.01
3	Rubiaceae	15	6.88	28.89
4	Rutaceae	12	5.50	34.40
5	Lauraceae	9	4.12	38.52
6	Moraceae	9	4.12	42.66
7	Verbenaceae	9	4.12	46.78
8	Tiliaceae	7	3.21	50.00
9	Meliaceae	6	2.75	52.75
10	Sapindaceae	6	2.75	55.50
11	Anacardiaceae	5	2.29	57.79
12	Apocyanaceae	5	2.29	60.09
13	Bignoniaceae	5	2.29	62.38
14	Combretaceae	5	2.29	64.67
15	Flacourtiaceae	5	2.29	66.97
16	Oleaceae	5	2.29	69.26
17	Sterculiaceae	5	2.29	71.55
18	Rhamnaceae	4	1.83	73.39
19	Boraginaceae	3	1.37	74.77
20	Burseraceae	3	1.37	76.14
21	Ebenaceae	3	1.37	77.52
22	Melastomaceae	3	1.37	78.89
23	Myrtaceae	3	1.37	80.27
24	Sapotaceae	3	1.37	81.65
25	Annonaceae	2	0.91	82.56
26	Celastraceae	2	0.91	83.48
27	Clusiaceae	2	0.91	84.40
28	Elaeocarpaceae	2	0.91	85.32
29	Lythraceae	2	0.91	86.23
30	Myrsinaceae	2	0.91	87.15
31	Solanaceae	2	0.91	88.07
32	Violaceae	2	0.91	88.99
33	Araliaceae	1	0.45	89.44
34	Asclepiadaceae	1	0.45	89.90
35	Bomacaceae	1	0.45	90.36
36	Capparaceae	1	0.45	90.82
37	Caprifoliaceae	1	0.45	91.28
38	Convolvulaceae	1	0.45	91.74
39	Dipterocarpaceae	1	0.45	92.20
40	Gnetaceae	1	0.45	92.66
41	Lamiaceae	1	0.45	93.11
42	Lecythidaceae	1	0.45	93.57
43	Leeaceae	1	0.45	94.03
44	Linaceae	1	0.45	94.49
45	Loganiaceae	1	0.45	94.95
46	Malvaceae	1	0.45	95.41
47	Palmae	1	0.45	95.87
48	Pandanaceae	1	0.45	96.33
49	Poaceae	1	0.45	96.78
50	Protaceae	1	0.45	97.70
51	Rosaceae	1	0.45	98.16
52	Sabiaceae	1	0.45	98.62
53	Salicaceae	1	0.45	99.08
54	Santalaceae	1	0.45	99.54
55	Ulmaceae	1	0.45	99.99

browse species for elephants. Euphorbiaceae (including Phyllanthaceae) with 23 species (10.4 percent of all species) is the second most important family, including canopy trees such as *Bridelia*, *Bischofia*, and *Croton*, understory genera such as *Glochidion* and *Phyllanthus*, and shrubs such as *Breynia* and *Flueggea*. *Flueggea leucopyros* is one of the dominant shrubs in the dry thorn forest in Mudumalai. Rubiaceae with 15 species is the third most speciose family in Mudumalai, represented by species such as *Mitragyna parvifolia*, *Anthocephalus chinensis*, and *Hymenodictyon orixense* in the canopy; *Canthium diococcum* and *Catunaregam spinosa* in the understory; and *Tarenna asiatica*, *Gardenia turgida*, and *Canthium parviflorum* among the shrubs. Spinescent shrubs of the genera Gardenia and Canthium characterize the dry thorn forests. Rutaceae, the fourth most speciose family, had no species in the canopy of the dry deciduous forest, but in the moist forest it is represented by *Zanthoxylum rhetsa*, and in the dry thorn forest by *Naringi crenulata*. This family has many shrubs in all forest types, including species such as *Glycosmis pentaphylla*, *Clausena indica*, and *Murraya paniculata*. Other common families are Lauraceae, represented in wetter areas by genera such as *Cinnamomum* (*C. verum*), *Actinodaphne* (*A. malabarica*), and *Litsea* (nine species); and Moraceae, represented by nine species of *Ficus*, including *F. benghalensis*, *F. religiosa*, *F. roxburghii*, and *F. drupeaceae*. Lamiaceae includes one of the most common canopy trees, *Tectona grandis*, in addition to *Gmelina*, *Vitex*, and nine species of *Clerodendrum*. The timber-rich family Combretaceae is represented by two genera and five species: *Anogeissus latifolia* and four species of *Terminalia* (*T. crenulata*, *T. chebula*—an important medicinal plant—*T. bellerica*, and *T. paniculata*). Another timber family is Lythraceae, represented by *Lagerstroemia microcarpa* and *L. parviflora*.

There were twenty-four families with one species. They include the Dipterocarpaceae, a characteristic family of moist lowland tropical forests in Asia, but here represented only by *Shorea roxburghii*, which is common in the northern areas of the dry deciduous forest. The family Malvaceae is represented by the understory tree *Kydia calycina*, an important browse species for elephants. Sandalwood (*Santalum alba*), an economically important tree species, represents Santalaceae. *Bombax ceiba* (Bombacaceae), *Careya arborea* (Lecythidaceae), *Salix tetrasperma* (Salicaceae), *Strychnos potatorum* (Loganiaceae), and *Celtis tetrandra* (Ulmaceae) are some of the other notable species. There were three monocot families in Mudumalai: the palms (Arecaceae), represented by *Caryota urens*; the grasses (Poaceae), represented by the bamboo *Bambusa arundinacea*; and the Pandanaceae (Screw-pine family), represented by *Pandanus thwaitsii*.

STRUCTURE AND DIVERSITY OF THE MAJOR FOREST TYPES: RESULTS FROM 1 HA PLOTS

Dry Thorn Forest

This forest type is confined to the eastern part of Mudumalai, with mean annual rainfall of 850 mm (range: 450–1,230 mm) and a four-month dry season (rainfall less than 50 mm per month). The four 1 ha plots included a total of 3,842 individual stems ≥ 1 cm dbh belonging to 66 species and 24 families. Species richness was 32 ± 9.6 species/ha with a mean Shannon-Weiner diversity of 2.4 ± 0.7. Champion and Seth (1968) describe this forest as "southern tropical dry thorn forest" with *Acacia* and *Ziziphus* association. However, when trees (greater than 10 cm dbh) are considered, the most abundant species were *Anogeissus latifolia* (Combretaceae), *Acacia chundra* (Fabaceae), *Erthroxylon monogynum* (Erythroxylaceae), and *Ziziphus xylopyros* (Rhamnaceae). The understory (stems 1–10 cm dbh) was dominated by *Gardenia turgida* (Rubiaceae), *Flueggea leucopyros* (Phyllanthaceae), and *Canthium parviflorum* (Rubiaceae). The most abundant species were *Gardenia turgida* (Rubiaceae) and *Flueggea leucopyros* (Phyllanthaceae), accounting for 30 percent of all stems greater than 1 cm dbh. The grasses that grow in this vegetation type are mainly short grasses, such as *Digitaria bicornis*, *Cenchorus biflorus*, and *Aristida cetacea*. Many of these are annual grasses that grow only during the wet months from June to December. Basic vegetation parameters of dry thorn forest are given in Table 3.

Dry Deciduous Forest

Dry deciduous forest covers the central part of Mudumalai with a mean annual rainfall of about 1,200 mm (range: 750–1,900 mm) and a four-month dry season. Two distinct subtypes can be recognized within the dry deciduous forests: a relatively open, dry dipterocarp forest to the north characterized by *Shorea roxburghii* and experiencing the highest fire frequencies of all forest types/subtypes in Mudumalai (Kodandapani et al. 2008); and a typical *Anogeissus-Tectona-Terminalia* forest to the south. These areas are typical savanna-woodlands with a dense undergrowth of perennial grasses such as *Themeda* and *Cymbopogon*.

Six 1 ha permanent plots were established in this forest type. They included a total of 5,330 individuals (greater than 1 cm dbh) belonging to 71 species and 27 families. Species richness was 25.5 ± 7.7 species/ha with a mean Shannon-Weiner diversity of 2.7 ± 0.2. *Anogeissus latifolia* (30 percent of all trees) and *Shorea roxburghii* (20.4 percent) dominated the floristics. Other important species included *Tectona grandis*

Table 3. Commonness and rarity among trees of 1 cm or more diameter at breast height (dbh) for the three growth forms shrubs, understory trees, and canopy trees. The number of species making up 50%, 90%, and the bottom 1% of the trees or shrubs in each growth form are listed. The numbers in parentheses are the percentages of the species in each growth form. The "rare" species are defined here as those with no more than 25 individuals in the plot. The percentage of the top-ranking species in each growth form is also listed.

Growth Form	Number of Species	Median Abundance	% Rank-1	No. of Species 50%	90%	Bottom 1%	No. of "Rare" sps.
Shrub	8	15	96.6	1 (12.5)	1 (12.5)	4 (50.0)	7 (87.5)
Understory tree	17	32	58.0	1 (??)	4 (23.5)	8 (47.1)	7 (41.2)
Canopy	44	33	28.5	2 (4.5)	8 (18.2)	18 (40.9)	19 (43.2)

(11.3 percent) and *Terminalia crenulata* (9.1 percent). However, when only trees (greater than 10 cm dbh) are considered, *Anogeissus latifolia* (Combretaceae), *Tectona grandis* (Lamiaceae), and *Terminalia crenulata* (Combretaceae) dominated the species composition. When the two subtypes are considered separately, and for all stems > 1 cm dbh, a single 1 ha plot in the dry dipterocarp forest was dominated by *Shorea roxburghii* (54 percent), *Terminalia crenulata* (15 percent), and *Anogeissus latifolia* (13 percent); while for the five other plots in the dry deciduous forest, *Anogeissus latifolia* (38 percent), *Tectona grandis* (13 percent), and *Terminalia crenulata* (7 percent) dominated the forest. Grass species *Heteropogon contortus*, *Themeda triandra*, *Aristida cetacea*, and *Digitaria granularis* dominate the ground vegetation. Floristic attributes of dry deciduous forest are given in the Table 3.

Moist Deciduous Forest

This forest type is common in the western part of Mudumalai. An association of *Lagerstroemia*, *Tectona*, *Terminalia*, and *Dalbergia* characterize this community. Mean annual rainfall is about 1,580 mm (range: 760–2,230 mm) with a four-month dry season. Four 1 ha permanent plots were established in this forest type. They included 3,769 individuals ≥ 1 cm dbh belonging to 90 species and 51 families. Mean species richness was 46.2 ± 9.2 species/ha with a Shannon-Weiner diversity index of 2.85 ± 0.45. An understory shrub, *Helicteres isora* (17.1 percent of stems), and an understory tree, *Kydia calycina* (10.7 percent of stems), were the most common woody species. Trees such as *Lagerstroemia microcarpa* (7.7 percent) and *Tectona grandis* (3 percent) dominated the canopy. However when trees greater than 10 cm dbh are considered, *Lagerstroemia microcarpa* (Lythraceae), *Terminalia crenulata* (Combretaceae), *Anogeissus latifolia* (Combretaceae), and *Kydia calycina* (Malvaceae) dominated the stand. The understory (1–10 cm dbh) was dominated by species such as *Helicteres isora* (malvaceae), *Catunaregam spinosa* (Rubiaceae), and *Kydia calycina* (Malvaceae). The ground layer is dominated by grasses such as *Sacciolepis indica*, *Themeda cymbaria*, *Panicum notatum*, and *Axonopus compressus*; ferns; and herbs belonging to the Asteraceae and Zingiberaceae.

Semi-evergreen Forest

This forest community is found in patches in a narrow belt along the southwestern border of Mudumalai. Mean annual precipitation in this tract is about 1,750 mm. In five permanent 1 ha plots we enumerated 10,251 woody stems ≥ 1 cm dbh belonging to 135 species and 30 families. Species richness of all stems > 1 cm dbh was 66 ± 10.6 species/ha with a Shannon-Weiner diversity index of 2.95 ± 0.35. The most abundant tree was *Olea dioica*, a canopy tree belonging to the Oleaceae (12.2 percent of stems), followed by an understory tree, *Casearia ovoides* (Salicaceae, 11.7 percent of stems), and the shrub *Ardisia solanacea* (Myrsinaceae, 10.9 percent of stems). The ground vegetation cover is dominated by ferns and grasses such as *Oplismenus composites*, *Oryza granulata*, and *Centotheca lappacea*. Among the herbs, species belonging to Amaranthaceae, Asteraceae, and Zingiberaceae are important.

STRUCTURE AND DYNAMICS OF DRY DECIDUOUS FOREST: RESULTS FROM THE 50 HA PLOT

When it was established in 1988–89, the 50 ha Forest Dynamics Plot included 25,929 individuals ≥ 1 cm dbh belonging to 71 woody species and 1 bamboo species, *Bambusa arundinacea*. The most common species in the plot was an understory tree, *Kydia calycina*, with 5,175 individuals ≥ 1 cm dbh and representing 20.3 percent of all individuals recorded in the first census. The second most common species, a canopy tree, *Lagerstroemia microcarpa*, had 3,982 individuals (15.6 percent). The seven most abundant species, each with more than 1,500 individuals ≥ 1 cm dbh, accounted for about 81 percent of all individuals in the plot. Of these seven species, four were canopy trees, two were understory trees, and one was a shrub. The four dominant canopy species, *Lagerstroemia microcarpa*, *Anogeissus latifolia*, *Tectona grandis*, and *Terminalia crenulata*, accounted for more than 40 percent of the stems in the plot. At the other end of the abundance spectrum, eight species were represented by just one individual in the plot. A total of 25 species, or more than a third of all species, had ten or fewer individuals in the plot. A little over half the species, 39 species, accounted for 99 percent of all individuals ≥ 1 cm dbh, while the remaining 33 species accounted for only 1 percent of individuals. The mean abundance of species in the plot was 370 individuals (±959 SD), while the median abundance was over ten times lower at just 31. The median abundance corresponds to a density of less than one individual per hectare.

The species-area curve asymptotes by about 30 ha in this dry deciduous forest community (Figure 8). Although species accumulate rapidly with increasing area, the

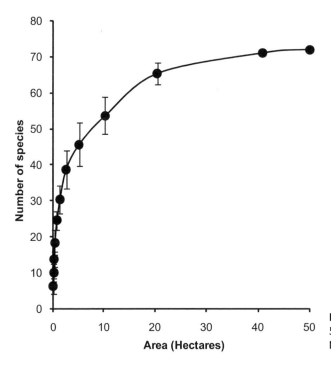

Figure 8. Species-area relations in a 50-ha permanent vegetation plot in Mudumalai.

abundances of different species are very unequal at the 50 ha scale. The dominance-diversity curve for all species in the 50 ha plot indicates the dominance of a few species in the community (Figure 9a). Different growth forms vary in their dominance-diversity patterns (Figure 9b). The curve for the understory trees and shrubs falls sharply, indicating higher levels of dominance for these life forms when compared to trees. The most common understory tree, *Kydia calycina*, was 2.7 times more abundant than the second most common understory tree, *Catunaregam spinosa*, and 6.7 times greater in abundance than *Phyllanthus emblica*, the third most common understory tree. *K. calycina* accounted for about 58 percent of all understory trees, and *C. spinosa* for 21 percent. In the 17 species of understory trees recorded, the five most common species accounted for 90 percent of all understory trees. The shrubs were almost entirely of one species, *Helicteres isora*, which had 2,569 individuals, 107 times more than the

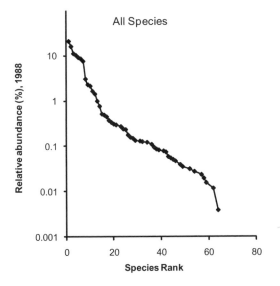

Figure 9a. Dominance–rank relations of individuals (>1.0 cm dbh) in 50-ha permanent plot in Mudumalai.

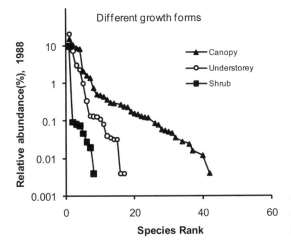

Figure 9b. Dominance–rank relations of individuals (>1.0 cm dbh) in different growth forms of 50-ha permanent plot in Mudumalai.

second-ranking shrub, *Allophylus cobbe*, with 24 individuals. Of the 8 species of shrubs recorded, *H. isora* alone accounted for over 96 percent of all individuals.

The forest canopy was also dominated by few species. Of the 44 species of canopy trees recorded, the top four species (*Anogeissus latifolia*, *Lagerstroemia microcarpa*, *Terminalia crenulata*, and *Tectona grandis*), each with more than 2,000 individuals, accounted for about 80 percent of all canopy trees. In all, 25 canopy species had more than 30 individuals each, and together they accounted for about 99 percent of all canopy trees ≥ 1 cm dbh.

Over a twenty-year period of monitoring (1988–2008), the 50 ha plot has shown considerable dynamism in ecological features such as population size, rank abundance of species, size distribution of stems, and biomass (Sukumar et al. 2005; Suresh et al. 2010). Herbivory by elephants on stems of *Kydia calycina*, *Helciteres isora*, *Eriolaena quinquilocularis*, and *Grewia tilaefolia* contributed to substantial declines in their populations. Extensive dry season ground fires during five years (1989, 1991, 1992, 1996, and 2002) resulted in high mortality rates of the smaller-sized stems, causing declines in overall population size. In periods without fire there has also been strong recruitment, especially since 2004. The population of nearly 26,000 individuals ≥ 1 cm dbh (72 species) in 1988–89 declined to its lowest level of just over 15,000 individuals (66 species) in 1996, but then increased to over 37,000 individuals (about 80 species) in 2008. The understory tree *Kydia calycina* was first replaced by a canopy tree, *Lagerstroemia microcarpa*, and then by a shrub, *Helicteres isora*, as the most abundant species. More than 90 percent of recruits (≥ 1 cm dbh) have been through vegetative means (coppicing of stems and root stock), with *Helicteres isora* (Malvaceae), *Cassia fistula* (Fabaceae), and *Kydia calycina* (Malvaceae) contributing to 72 percent of total recruits, though 40 species have produced at least ten recruits or more during this period. The larger-sized stems (greater than 30 cm dbh) increased from nearly 4,900 stems in 1988 to over 5,400 stems in 2008; basal area correspondingly increased from 24.4 m²/ha to 25.8 m²/ha; and above-ground biomass of woody stems increased from 173.5 tons/ha in 1988 to 188.4 tons/ha in 2008.

CONCLUSIONS

As with global patterns of tropical forest diversity (Murphy and Lugo 1995; Gentry 1995), the SDTFs of India are generally species poor and less diverse than moist tropical forests of the Western Ghats (Ayyappan and Parthasarathy 2001; Ganesh et al. 1996; Pascal and Pélissier 1996; Suresh and Sukumar 2005; Anitha et al. 2007). The sal- (*Shorea robusta*) dominated dry deciduous forests in north India also have low levels of diversity (S. Singh et al. 1994; Gautam et al. 2008). Deciduous forests have fewer species and individuals in comparison to forest types elsewhere in Asia such as in Thailand (Lamotte et al. 1998).

Our analyses of tree species diversity in seasonal forests in Mudumalai, spanning a significant rainfall gradient over a distance of just 50 km, while broadly consistent with previously observed patterns, also brings out an interesting aspect of diversity patterns. The number of tree species and measures of heterogeneity are lowest in the intermediate rainfall regime of the dry deciduous forest, and higher at either end of the rainfall spectrum, though the moister end of the gradient shows the highest diversity. This is because fire also influences species richness, and the high fire frequencies in the dry deciduous forests contribute to their lower diversity.

Seasonally dry tropical forests across the globe are among the most heavily modified forest ecosystems. Though dry forests account for 42 percent of the total global vegetation cover (Brown and Lugo 1982) the conservation status of many of these forests remains seriously threatened. In addition to the possible direct impacts of future climate change, indirect impacts of dry season fire and future exploitation of these forests have to be considered in planning for their management and conservation.

ACKNOWLEDGMENTS

We thank the Ministry of Environment and Forests of the Government of India for funding our research in Mudumalai. We thank the Forest Department of Tamilnadu for research permission. We also thank Mr. Sandeep for GIS help; our colleagues C. M. Bharanaiah, R. P. Harisha, R. Mohan, and several others for help during the field-work; and our field assistants Mr. Krishna, Mr. Bomman, Mr. Siddan, Mr. Kunmari, and Mr. Maran for assistance during the field-work. We thank the Smithsonian Institution, especially William J. McShea and Stuart J. Davies, for giving us the opportunity to present this analysis. We thank the Center for Tropical Forest Science for its long-term support of the Mudumalai program. Finally, Hebbalalu Satyanarayana Suresh would like to thank the Center for Tropical Forest Science for its support in his attendance of the FORTROP–II meeting in Thailand.

REFERENCES

Anitha, K., P. Balasubramanian, and S. N. Prasad. 2007. Tree Community Structure and Regeneration in Anaikatty Hills, Western Ghats. *Indian Journal of Forestry* 30:315–24.

Arivazhagan, C., and R. Sukumar. 2005. Comparative Demography of Asian Elephants (*Elephas maximus*) in Southern India. CES Technical Report 106. Bangalore: Centre for Ecological Sciences, Indian Institute of Science.

Ayyappan, N., and N. Parthasarathy. 2001. Patterns of Tree Diversity within a Large Scale Permanent Plot of Tropical Evergreen Forest, Western Ghats, India. *Ecotropica* 7:61–76.

Baskaran, N., S. Balasubramaniam, S. Swaminathan, and A. A. Desai. 1995. Home Range of Elephants in the Nilgiri Biosphere Reserve, South India. In *A Week with Elephants*, ed. J. C. Daniel and H. S. Datye, 296–313. Bombay: Bombay Natural History Society; New Delhi: Oxford University Press.

Borchert, R. 1994. Soil and Stem Water Storage Determine Phenology and Distribution of Dry Forest Trees. *Ecology* 75:1437–49.

Brown, S., and A. E. Lugo. 1982. The Storage and Production of Organic Matter in Tropical Forests and Their Role in the Global Carbon Cycle. *Biotropica* 14:161–87.

Bullock, S. H., and J. Solis-Magallanes. 1990. Phenology of Canopy Trees of a Deciduous Forest in Mexico. *Biotropica* 22:22–35.

Burrough, P. A., and R. A. McDonnell. 1998. *Principles of Geographical Information Systems*. Oxford: Oxford University Press.

Champion, H. G., and S. K. Seth. 1968. *A Revised Survey of the Forest Types of India*. Delhi: Publication Division, Government of India.

Condit, R. 1995. Research in Large, Long-Term Tropical Forest Plots. *Trends in Ecology and Evolution* 10:18–22.

———. 1998. *Tropical Forest Census Plots: Methods and Results from Barro Colorado Island, Panama and a Comparison with other Plots*. Berlin: Springer-Verlag.

Desai, A. A. 1991. The Home Range of Elephants and Its Implications for Management of the Mudumalai Wildlife Sanctuary, Tamil Nadu. *Journal of the Bombay Natural History Society* 88:145–56.

Finkeldey, R., and H. H. Hattemer. 2007. Sexual and Asexual Reproduction in Tropical Forests. Chap. 4 in *Tropical Forest Genetics*. Berlin: Springer-Verlag.

Forest Survey of India. 2005. *The State of Forest Report 2003*. Dehradun: Forest Survey of India, Ministry of Environment and Forests, Government of India.

Gadgil, M., and R. Guha. 1993. *This Fissured Land: An Ecological History of India*. New Delhi: Oxford University Press.

Ganesan, B. 1993. Extraction of Non-timber Forest Products including Fodder and Fuelwood in Mudumalai, India. *Economic Botany* 47:268–74.

Ganesh, T., R. Ganesan, M. Soubhdra Devy, P. Davidar, and K. S. Bawa. 1996. Assessment of Plant Biodiversity at a Mid-elevation Evergreen Forest of Kalakad-Mundanthurai Tiger Reserve, Western Ghats, India. *Current Science* 71:379–92.

Gautam, M. K., A. K. Tripathi, and R. K. Manhas. 2008. Plant Diversity and Structure of Sub-tropical *Shorea robusta* Gaertn f. (Sal) Forests of Doon Valley, India. *Indian Journal of Forestry* 31:127–36.

Gentry, A. H. 1988. Changes in Plant Community Diversity and Floristic Composition on Environmental and Geographic Gradients. *Annals of the Missouri Botanical Garden* 75:1–34.

———. 1995. Diversity and Floristic Composition of Neotropical Dry Forests. In *Seasonally Dry Tropical Forests*, ed. S. H. Bullock, H. A. Mooney, and E. Medina, 146–94. Cambridge: Cambridge University Press.

George, M., M. Gupta, and J. Singh. 1988. *Forest Soil Vegetation Survey: Report on Mudumalai Forest Division, Tamilnadu*. Coimbatore, India: Soil-Vegetation Survey, Southern Division.

Goldammer, J. G. 1993. Fire Management. In *Tropical Forestry Handbook*, ed. L. Pancel, 2:1221–68. Berlin: Springer-Verlag.

Hockings, P., ed. 1989. Blue Mountains: The Ethnography and Biogeography of a South Indian Region. New York: Oxford University Press.

Hoffmann, W. A. 1998. Post-burn Reproduction of Woody Plants in a Neotropical Savanna: The Relative Importance of Sexual and Vegetative Reproduction. *Journal of Applied Ecology* 35:422–33.

Kodandapani, N., M. A. Cochrane, and R. Sukumar. 2004. Conservation Threat of Increasing Fire Frequencies in the Western Ghats, India. *Conservation Biology* 18:1553–61.

———. 2008. A Comparative Analysis of Spatial, Temporal and Ecological Characteristics of Forest Fires in Seasonally Dry Tropical Ecosystems in the Western Ghats, India. *Forest Ecology and Management* 296:607–17.

Kottek, M., J. Grieser, C. Beck, B. Rudolf, and F. Rubel. 2006. World Map of the Köppen-Geiger Climate Classification Updated. *Meteorologische Zeitschrift* 15:259–63.

Krishnamurthy, V., and C. Wemmer. 1995. Timber Elephant Management in the Madras Presidency of India (1844–1947). In *A Week with Elephants*, ed. J. C. Daniel and H. S. Datye, 456–72. Bombay: Bombay Natural History Society; New Delhi: Oxford University Press.

Krishnan, M. S. 1974. Geology. In *Ecology and Biogeography in India*, ed. M. S. Mani, 60–98. The Hague: Dr. W. Junk b.v. Publishers.

Lamotte, S., J. Gajaeni, and F. Malaisse. 1998. Structure Diversity in Three Forest Types of North-Eastern Thailand (Sakaerat Reserve, Pak Tong Chai). *Biotechnology, Agronomy, Society and Environment* 2:192–202.

Murali, K. S., and R. Sukumar. 1993. Leaf Flushing and Herbivory in a Tropical Deciduous Forest, Southern India. *Oecologia* 94:114–19.

———. 1994. Reproductive Phenology of Tropical Dry Forest in Mudumalai, Southern India. *Journal of Ecology* 82:759–67.

Murphy, P. G., and A. E. Lugo. 1986. Ecology of Tropical Dry Forest. *Annual Review of Ecology and Systematics* 17:67–88.

———. 1995. Dry Forests of Central America and the Caribbean. In *Seasonally Dry Tropical Forests*, ed. S. H. Bullock, H. A. Mooney, and E. Medina, 9–34. Cambridge: Cambridge University Press.

Myers, N., R. A. Mittermeier, C. G. Mittermeier, G. A. B. da Fonseca, and J. Kent. 2000. Biodiversity Hotspots for Conservation Priorities. *Nature* 403:853–58.

Narendran, K., I. K. Murthy, H. S. Suresh, H. S. Dattaraja, N. H. Ravindranath, and R. Sukumar. 2001. Non-Timber Forest Product Extraction: A Case Study from the Nilgiri Biosphere Reserve, Southern India. *Economic Botany* 55:528–38.

Nilgiri District Gazetteer. 1995. Nilgiri District Gazetteer, Tamilnadu. Chennai, India: Government of Tamilnadu.

Pascal, J.-P., and R. Pélissier. 1996. Structure and Floristic Composition of a Tropical Evergreen Forest in Southwest India. *Journal of Tropical Ecology* 12:191–214.

Prasad, S., and R. Sukumar. 2009. Context-Dependency of a Complex Fruit-Frugivore Mutualism: Temporal Variation in Crop Size and Neighborhood Effects. *Oikos* (in press).

Prabhakar, R. 1994. Resource Use, Culture and Ecological Change: A Case Study of the Nilgiri Hills of Southern India. PhD diss., Indian Institute of Science, Bangalore.

Puri, G. S. 1960. *Indian Forest Ecology: A Comprehensive Survey of Vegetation and Its Environment in the Indian Subcontinent*, vols. 1, 2. New Delhi: Oxford University Press.

Ranganathan, C. R. 1941. *Working Plan for the Nilgiris Division*. Madras, India: Government Press.

Ravindranath, N. H., N. V. Joshi, R. Sukumar, and A. Saxena. 2006. Impact of Climate Change on Forests in India. *Current Science* 90:354–61.

Read, L., and D. Lawrence. 2006. Interactions between Water Availability and Nutrient Cycling in Dry Tropical Forests. In *Dryland Ecohydrology*, ed. P. D. Odorica and A. Porporato, 217–32. The Netherlands: Springer.

Sharma, B. D., B. V. Shetty, K. Vivekanandan, and N. C. Rathakrishnan. 1977. Flora of Mudumalai Wildlife Sanctuary, Tamil Nadu. *Journal of the Bombay Natural History Society* 75:13–42.

Silori, C. S., and B. K. Mishra. 2001. Assessment of Livestock Grazing Pressure in and around the Elephant Corridors in Mudumalai Wildlife Sanctuary, South India. *Biodiversity and Conservation* 10:2181–95.

Sinclair, A. R. E., and M. Norton-Griffiths, eds. 1979. *Serengeti—Dynamics of an Ecosystem*. Chicago: University of Chicago Press.

Singh, K. P., and C. P. Kushwaha. 2005. Emerging Paradigms of Tree Phenology in Dry Tropics. *Current Science* 89:964–75.

Singh, S. P., B. S. Adhikari, and D. B. Zobel. 1994. Biomass, Productivity, Leaf Longevity and Forest Structure in the Central Himalaya. *Ecological Monographs* 64:401–21.

Skarpe, C., 1991. Impact of Grazing in Savannah Ecosystems. *Ambio* 20:351–56.

Subramanyam, K., and M. P. Nayar. 1974. Vegetation and Phytogeography of the Western Ghats. In *Ecology and Biogeography in India*, ed. M. S. Mani, 187–96. The Hague: Dr. W. Junk b.v. Publishers.

Sukumar, R., H. S. Dattaraja, H. S. Suresh, J. Radhakrisnan, R. Vasudeva, S. Nirmala, and N. V. Joshi. 1992. Long-Term Monitoring of Vegetation in Tropical Deciduous Forest in Mudumalai, Southern India. *Current Science* 62:608–16.

Sukumar, R., R. Ramesh, R. K. Pant, and G. Rajagopalan. 1993. A δ13 Record of Late Quaternary Climate Change from Tropical Peats in Southern India. *Nature* 364:703–6.

Sukumar, R., H. S. Suresh, H. S. Dattaraja, and N. V. Joshi. 1998. Dynamics of a Tropical Deciduous Forest: Population Changes (1988 through 1993) in a 50-hectare Plot at Mudumalai,
Southern India. In *Forest Biodiversity Research, Monitoring and Modelling: Conceptual Background and Old World Case Studies*, ed. F. Dallmeier and J. A. Comiskey, 495–506. Man in
the Biosphere 20. Paris: UNESCO; Carnforth, UK: Parthenon Publishing Group.

Sukumar, R., H. S. Suresh, H. S. Dattaraja, N. V. Joshi, and R. John. 2004. The Mudumalai
Forest Dynamics Plot. In *The Global Network of Large Forest Plots*, ed. E. Losos and E. G.
Leigh Jr. Chicago: University of Chicago Press.

Sukumar, R., H. S. Suresh, H. S. Dattaraja, S. Srinidhi, and C. Nath. 2005. Dynamics of a Tropical Dry Forest at Mudumalai (India): Climate, Fire, Elephants and the Evolution of Life History Strategies. In *Biotic Interactions in the Tropics*, ed. D. Burslem. Cambridge: Cambridge
University Press.

Suresh, H. S., H. S. Dattaraja, Harish R. Bhat, and R. Sukumar. 2006. Revised Flora of Mudumalai. *Journal of Economic Botany and Taxonomy* 30 (1).

Suresh, H. S., H. S. Dattaraja, and R. Sukumar. 1996. Tree Flora of Mudumalai Sanctuary,
Southern India. *Indian Forester* 122:507–19.

———. 2010. Relationship between Annual Rainfall and Tree Mortality in a Tropical Dry Forest: Results of a 19-year Study at Mudumalai, Southern India. *Forest Ecology and Management* 259:762–69.

Suresh, H. S., and R. Sukumar. 2005. Trends in Diversity and Floristics of Forest Vegetation in
Nilgiri Biosphere Reserve: A Comparison with Other Tropical Forests. In *Wildlife Conserva-
tion, Research and Management*, ed. Y. V. Jhala, R. Chellam, and Q. Qureshi. Dehra Dun:
Wildlife Institute of India.

———. 2009. Influence of Climatic Variability on Tree Phenology in the Tropical Dry Forests
of Mudumalai, Southern India. Submitted to proceedings of FORTROP II Thailand.

Swaine, M. D. 1992. Characteristics of Dry Forest in West Africa and the Influence of Fire.
Journal of Vegetation Science 3:365–74.

Varman, K. S., and R. Sukumar. 1995. The Line Transect Method for Estimating Densities of
Large Mammals in a Tropical Deciduous Forest: An Evaluation of Models and Field Experiment. *Journal of Biosciences* 20:273–27.

Vidya, T. N. C., P. Fernando, D. J. Melnick, and R. Sukumar. 2005. Population Differentiation
within and among Asian Elephant (*Elephas maximus*) Populations in Southern India. *Hered-
ity* 94:71–80.

von Lengerke, J. H. 1977. *The Nilgiris: Weather and Climate of a Mountain Area in South
India*. Wiesbaden, Germany: Franz Steiner Verlag.

3

The Uncertainty in Mapping Seasonally Dry Tropical Forests in Asia

Peter Leimgruber, Melanie Delion, and Melissa Songer

Seasonally dry tropical forests (SDTF) are some of the most threatened, but least protected, tropical ecosystems globally (Janzen 1988; Wikramanayake et al. 2002; Miles et al. 2006). Despite this recognition, there are few reliable maps or established monitoring systems for SDTF, making it difficult to develop strategies for conserving these forests. The lack of mapping and monitoring probably can be attributed to the difficulty in consistently defining SDTF, which includes a broad range of ecosystems with vastly different forest structure and species composition (see Bunyavejchewin et al., this volume, for a definition of SDTF). SDTF mapping is further hampered because established remote sensing techniques tend not to perform well in delineating open-canopy forests (Jensen 1996). Also, some SDTFs are difficult to separate from other mixed and open habitats, including shrubland and even degraded forest (Grainger 1999).

Although availability of satellite maps for SDTF is limited, researchers have produced regional and global land cover maps, as well as data sets quantifying canopy cover (DeFries et al. 1998, 2000; Hansen et al. 2000, 2003; Loveland et al. 2000; Bartholome and Belward 2005). These are regularly used for studies on the extent, location, and condition of different ecosystem types in ecological research (e.g., Sanderson et al. 2002; Leimgruber et al. 2003; Hubener et al. 2005; Dinerstein et al. 2007; Miles et al. 2006) and may represent a first step for analyzing the remaining SDTFs. Results from these data, however, can be contradictory and conflicting when assessing location and extent of SDTF. For example, estimates for the late 1980s, provided in the *Conservation Atlas for Tropical Forests*, report approximately 452,000 km^2 of SDTF in Southeast Asia (Collins et al. 1991). World Wildlife Fund (WWF) ecoregion assessment in the late 1990s placed SDTF at close to 600,000 km^2 (Wikramanayake et al. 2002), which would suggest a 25 percent increase over the intervening decade, an unlikely occurrence.

The most recent global assessment of SDTF by Miles et al. (2006) relied on a combination of remotely sensed data, regression models to estimate canopy cover, and coarse ancillary information on ecoregions and aridity zones. Though not ground-truthed, the estimates of SDTF cover provided by the authors were spatially explicit and on a global scale. Based on their threat assessment, the authors estimated that only 16.7 percent (approx. 172,000 km²) of the world's 1,048,700 km² of SDTF remained in the Eurasian tropics (i.e., South Asia and mainland Southeast Asia). This would represent a 71 percent decline in SDTF from WWF's estimate (Wikramanayake et al. 2002), an improbable loss that is not obvious on the ground.

These and other examples demonstrate troubling discrepancies in existing estimates, drawing further into question how much we really know about SDTF distribution and conservation status. Vast overestimation of losses or underestimation of what remains will seriously reduce the credibility of conservation scientists trying to argue for conservation of SDTF. Careful assessment of existing maps is needed to determine the extent of our current knowledge and to understand better the uncertainties and limitation in mapping SDTF. Using such an assessment, it should be possible to determine future ways to delineate and monitor SDTF changes at country and landscape levels.

The original intent of our research was to provide an accurate picture of the extent and condition of remaining SDTF in Asia. Currently this cannot be accomplished using existing maps and data. Here we compile existing land cover, forest cover, and SDTF maps for South and Southeast Asia to

1. Compare existing global, continental, and regional land cover and SDTF maps to assess how much they agree on extent, amount, and location of SDTF;
2. Assess the accuracy of these data by focusing on one region and one SDTF type, the deciduous dipterocarp forests of Myanmar;
3. Determine threat levels for SDTF by assessing differences in past and current human population densities; and
4. Develop and propose a framework for future, spatially explicit mapping of assessments of SDTF distribution, decline, and threats.

METHODS

Study Area

SDTF, found throughout Asia, are characterized by wide variation in vegetation form, degree of deciduousness, canopy closure, height, and structure. Regional variation makes SDTF delineation itself difficult, and mapping problems increase when attempting to produce a single SDTF map for a wide geographic region. We restricted analyses to South Asia (Nepal, India, Bangladesh, and Sri Lanka), mainland Southeast Asia (Myanmar, Thailand, and peninsular Malaysia), and Indochina (Cambodia, Laos, and Vietnam). We included dry forests in Nepal, though strictly speaking they lie outside the tropics. The sal (*Shorea robusta*) forests of Nepal's Terai Arc are similar to many of the SDTFs found in tropical South Asia and represent a conservation hot spot, especially for tigers (*Panthera tigris*) and Indian rhinos (*Rhinoceros unicornis*).

To assess and compare accuracy in SDTF delineation among maps, we defined an evaluation area inside the study area. For this, we focused on upper Myanmar (upper left: lat. 26°14′3″ N, long. 92°36′36″ E; lower right: lat. 19°45′21″ N, long. 98°6′20″ E), an area where the authors have extensive experience in mapping forest cover and deforestation patterns (McShea et al. 1999; Koy et al. 2005; Leimgruber et al. 2005; Myint Aung et al. 2004; Songer 2006; Songer et al. 2009). This region is dominated by a special type of SDTF, the deciduous dipterocarp forest, which is characterized by strong seasonality, with a long and severe dry season, and relatively low canopy cover (Koy et al. 2005; Songer et al. 2009; see Bunyavejchewin et al., this volume, for a definition of deciduous dipterocarp forest relative to SDTF).

Seasonally Dry Tropical Forests in Existing Regional Maps

We used four available maps to approximate and compare SDTF area, extent, and distribution in South and Southeast Asia (Table 1). These were the Global International Geosphere-Biosphere Programme (IGBP) DISCover (Belward et al. 1999; Loveland et al. 2000); the MODIS/Terra Land Cover Classification (MLCC; Strahler et al. 1999; Friedl et al. 2002); the Global Land Cover 2000 (GLC2000; Bartholome and Belward 2005; Stibig et al. 2007); and the Global Distribution of Tropical Dry Forest (GDTDF; Miles et al. 2006).

All maps are based on analysis of medium- to coarse-resolution satellite data (500–1,000 m). Only the GDTDF was created specifically to delineate SDTF and was based on a combination of satellite information with WWF's terrestrial ecoregions (Global 200; Olson et al. 2001) and the *World Atlas of Desertification* (Middleton and Thomas 1997). The Vegetation Continuous Field (VCF) map (Hansen et al. 2003; http://glcf.umiacs.umd.edu) produced from Moderate Resolution Imaging Spectroradiometer (MODIS) data provided the starting point for creation of the GDTDF. The VCF approximates percentage tree cover based on regression tree analysis, linking finer resolution training data from Landsat satellites to MODIS imagery. For the GDTDF, SDTF was defined as all tropical areas with ≥ 40 percent tree cover that fall inside four biomes contained within WWF ecoregions (for details see Miles et al. 2006).

All other data sets in our analyses were intended as general land cover maps and did not include well-defined SDTF categories. For each, we selected land cover categories that best characterize SDTF (Table 1). For additional details on land cover classification methods, refer to the original sources of the maps as noted above.

For each map we calculated total SDTF area and percentage in protected status within the study region as well as for each country. Protection was determined by overlaying a protected areas layer including all IUCN level I–V areas (World Conservation Monitoring Centre 2007). We used cross-tabulation operations in a geographic information system (GIS) to calculate pairwise overlap for SDTF among the maps. To determine similarity in distribution patterns among countries, we used Spearman rank correlation coefficients.

Developing a Hierarchical SDTF Map

We developed two hierarchical maps of SDTF areas by first delineating climatic areas for SDTF and then integrating these with forest cover from the four land cover data sets. Though details vary by author, SDTFs are commonly defined by their seasonality, the length of the dry season, the amount of rainfall during dry and wet seasons, and elevation (Stamp 1925; Ruangpanit 1995; Blasco et al. 1996, 2000; World Conservation

Table 1. Existing regional and global land cover classifications evaluated for use as seasonally dry tropical forest (SDTF) maps in South Asia and mainland Southeast Asia.

Map / Classification[a]	Source	Spatial Resolution Data (m)	Imagery / Data Used[b]	Data Acquisition	Potential SDTF Classes	Overall Accuracy Assessed
GDTDF	World Conservation Monitoring Centre	500	VCF, Global 2000, WAD	Oct. 2000– Dec. 2001	One class	Not assessed
MLCC	Boston University (http://edcims www.cr.usgs. gov/pub/ imswelcome/)	1,000	MODIS	Oct. 2000– Oct. 2001	Deciduous broadleaf forest, open and closed shrubland, savanna, woody savanna	70.7% (Strahler et al. 1999)
GLC2000	European Commission's Joint Research Centre (www-gvm.jrc.it/ glc2000/)	1,000	SPOT 4 VEGET- ATION	Nov. 1999– Dec. 2000	Tropical mixed and dry deciduous forest, deciduous shrubland/mosaic of deciduous shrubcover and cropping, deciduous thorny shrubland	72% for forest, 65% overall (Stibig et al. 2007)
IGBP DISCover	U.S. Geological Survey EROS DAAC (http://edc. usgs.gov/products/ landcover/glcc. html)	1,000	AVHRR	April 1992– March 1993	Deciduous broadleaf forest, open and closed shrubland, savanna, woody savanna	73.5%–78.7% (Scepan 1999)

[a] GDTDF = global distribution of tropical dry forest (Miles et al. 2006); MLCC = MODIS/Terra Land Cover Classification (Strahler et al. 1999; Friedl et al. 2002); GLC2000 = Global Land Cover 2000 (Bartholome and Belward 2005; Stibig et al. 2007); IGBP DIS-Cover = International Geosphere-Biosphere Programme (IGBP) DISCover (Belward et al. 1999; Loveland et al. 2000).
[b] VCF = Vegetation Continuous Field; Global 200 = WWF terrestrial ecoregions (Olson et al. 2001); WAD = *World Atlas of Desertification* (Middleton and Thomas 1997); MODIS = Moderate Resolution Imaging Spectroradiometer; SPOT = Satellite Pour l'Observation; AVHRR = Advanced Very High Resolution Radiometer.

Monitoring Centre 1997). SDTFs experience a distinct dry period of four to six months with rainfall of less than 100 mm per month. Average annual rainfall is relatively high (approx. 1,600–2,000 mm) due to monsoon rains. Most SDTFs in Asia are found in the rainshadow of mountain ranges such as India's Western Ghats and Myanmar's Rakhine Yoma, and tend to be restricted to elevations below 1,200 m. We used BIOCLIM and DIVA-GIS (Hijmans et al. 2005) to model the ecological and climate niche of SDTF based on the climate thresholds defined above. For our first SDTF map (SDTF_A), we included all SDTF and other forest areas shared by at least two of the other maps and inside our climate envelope. For our second SDTF map (SDTF_B), we included all forest areas shared by at least two of the other maps and inside our climate envelope. To reduce noise in the resulting SDTF data set, we excluded all patches whose extent was less than 200 km².

Accuracy Assessments

Focusing on our evaluation area in upper Myanmar, we compared presence or absence of SDTF for 0.01 degree blocks among all maps, using a GIS. This block-based approach allowed us to integrate validation data collected from finer resolution imagery (30 m Landsat ETM+ and 15 m ASTER imagery) and provided an appropriate scale for evaluating our map. We randomly selected 500 grid cells and used a subset of 188 cells for which seasonal pairs of ASTER or Landsat ETM+ were available. Within each cell we determined presence or absence of SDTF based on imagery. We found 110 cells contained SDTF, while 78 did not. We estimated accuracy by comparing SDTF presence or absence in an error matrix.

Human Population Density as a Threat Factor

To assess the role of human populations in reducing the extent of SDTF, we used the History Database of the Global Environment (HYDE, www.mnp.nl/en/themasites/hyde/index.html; Goldewijk 2001). Historical data on human demography and distribution are crucial for developing historical and future models of human impacts on the environment. HYDE integrates historical and current human population density and distribution data into a spatially explicit model of human population development, on a 0.5 by 0.5 degree grid. We used HYDE data for 1200 AD to present, and compared mean human population densities inside and outside SDTF climate areas.

RESULTS

Country-level estimates of remaining SDTF vary considerably among existing land and forest cover maps (Table 2; Figure 1), with greatest discrepancies found between GDTDF and all other maps. GDTDF provides a very low estimate for total remaining SDTF and indicates a very high level of protection (Table 2). With the exception of Cambodia, Malaysia, and Sri Lanka, GDTDF estimates of SDTF are markedly lower than other maps, sometimes by more than tenfold, with no SDTF in Bangladesh, Malaysia, and Nepal. Discrepancies are most pronounced for India and Myanmar, with area differences greater than 600,000 km^2 and 150,000 km^2, respectively. This would indicate that previous estimates are 96–99 percent higher when compared to GDTDF, or that there has been a near total loss of SDTF in these two countries.

Comparisons among IGBP, MLCC, and GLC2000 show slightly greater agreement in area estimates and protection status (Table 2; Figure 2). However, there remain considerable differences for India (a 21 percent difference between IGBP and MLCC), and for parts of Southeast Asia and Indochina, especially Myanmar, Cambodia, and Vietnam.

While area estimates are correlated among all data sets, there is relatively little spatial overlap, ranging only from 3 to 27 percent (Table 3). GDTDF consistently has the lowest spatial overlap and shows no correlation with the other data sets in SDTF distribution across countries. IGBP and MLCC have the greatest spatial overlap, whereas in terms of SDTF distribution across countries, GLC2000 and MLCC are best correlated.

Table 2. Total area seasonally dry tropical forest (SDTF; km^2) and percentage protected seasonally dry tropical forest calculated from different regional and global land cover/forest cover data sets.

Country	IGBP[a]	GLC2000[b]	MLCC[c]	GDTDF[d]
Bangladesh	4,997	2,365	15,608	0
	(0%)	(1%)	(1%)	
Cambodia	19,049	48,659	51,070	55,876
	(16%)	(25%)	(22%)	(28%)
India	631,954	774,436	796,122	22,931
	(7%)	(5%)	(5%)	(4%)
Laos	47,455	29,680	37,237	21,082
	(0%)	(0%)	(0%)	(0%)
Malaysia	3,876	0	3,441	0
	(1%)		(2%)	
Myanmar	183,101	155,913	191,430	1,616
	(2%)	(1%)	(1%)	(2%)
Nepal	35,926	24,467	45,583	0
	(12%)	(10%)	(8%)	
Sri Lanka	1,037	14,556	12,358	13,248
	(3%)	(26%)	(12%)	(28%)
Thailand	85,683	110,863	118,359	20,998
	(29%)	(23%)	(18%)	(62%)
Vietnam	47,718	36,560	72,422	33,493
	(4%)	(3%)	(3%)	(10%)
Total	1,060,796	1,197,499	1,343,630	169,244
	(8%)	(7%)	(6%)	(22%)

[a]International Geosphere-Biosphere Programme (IGBP) DISCover (Belward et al. 1999; Loveland et al. 2000).
[b]Global Land Cover 2000 (Bartholome and Belward 2005; Stibig et al. 2007).
[c]MODIS/Terra Land Cover Classification (Strahler et al. 1999; Friedl et al. 2002).
[d]GDTDF = global distribution of tropical dry forest (Miles et al. 2006).

When comparing the SDTF maps to known sites in northern Myanmar, overall and SDTF-classification accuracy is highest for our maps SDTF_A and SDTF_B, both derived from combinations of the other four SDTF maps (Table 4). SDTF_A, combining all areas previously identified as SDTF by at least two maps, provides the highest overall accuracy, second-best accuracy (along with IGBP) for the "other" category, and second-best accuracy for classifying SDTF. SDTF_B, combining all forested areas previously identified by at least two maps, has the best accuracy in identifying SDTF, but the lowest accuracy for classifying the "other" category, and only the second-best overall accuracy.

Of the remaining data sets, MLCC had the best SDTF accuracy, and GLC2000 had the best accuracy for the "other" category (Table 4). Both achieved an overall accuracy of 73 percent, only slightly lower than SDTF_B.

Based on SDTF_A and SDTF_B, forested areas that may include SDTFs cover approximately 600,000–1,000,000 km^2, with about 9–12 percent protection (Table 5;

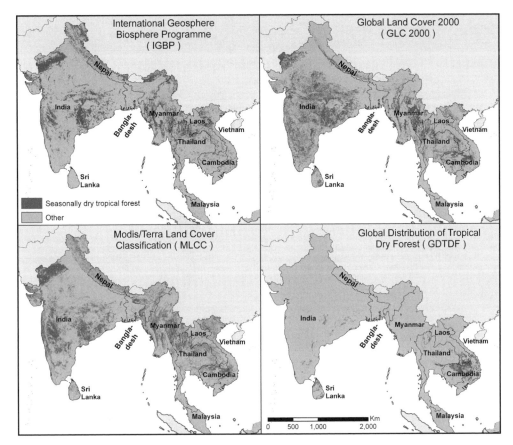

Figure 1. Spatial distribution (extent and location) of seasonally dry tropical forest (SDTF) for different existing land cover / land use data sets.

Figure 2). The largest potential SDTF areas are found in India, Myanmar, and Thailand. The highest percentage of protected SDTFs are found in Sri Lanka and Thailand.

Climate zones that favor SDTF have experienced dramatically higher and faster population growth than other climate areas combined (Figure 3). Mean population densities are more than three times higher in SDTF climate zones. Within the SDTF climate zones, SDTFs persist in areas with low population densities.

DISCUSSION

Uncertainty in Existing SDTF Maps

Our research indicates that currently existing regional and global maps do not accurately delineate SDTF in Asia. All previously produced maps analyzed in our study (IGBP, GLC2000, MLCC, GDTDF) demonstrate a very low accuracy in delineating SDTFs, show little overlap, and differ drastically in the amount of SDTF predicted for different

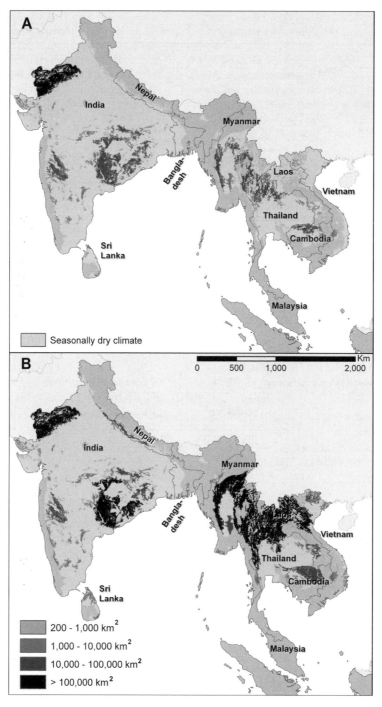

Figure 2. Combination of climate and seasonally dry tropical forest (SDTF) data sets delineating potential SDTF areas throughout South Asia and Indochina. (*A*) All areas with SDTF delineated in at least two previous maps and located within dry forest climates. (*B*) All forested areas delineated in at least two previous maps and located within dry forest climates.

Table 3. Percentage overlap in seasonally dry tropical forest (SDTF) and Spearman rank correlation of total SDTF area for existing regional and global land cover / forest cover data sets.

	IGBP[a]		GLC2000[b]		MLCC[c]	
	Overlap (%)	r	Overlap (%)	r	Overlap (%)	r
GLC2000	20	0.891**				
MLCC	27	0.939**	26	0.964**		
GDTDF[d]	3	0.386[n.s.]	4	0.607[n.s.]	3	0.497[n.s.]

**< 0.01
[n.s.]not significant
[a]International Geosphere-Biosphere Programme (IGBP) DISCover (Belward et al. 1999; Loveland et al. 2000).
[b]Global Land Cover 2000 (Bartholome and Belward 2005; Stibig et al. 2007).
[c]MODIS/Terra Land Cover Classification (Strahler et al. 1999; Friedl et al. 2002).
[d]Global distribution of tropical dry forest (Miles et al. 2006).

Table 4. Percentage accuracy of global / regional land cover / forest cover maps in predicting seasonally dry tropical forest (SDTF) in upper Myanmar based on 188 samples.

Map	SDTF	Other Land Cover	Overall
IGBP[a]	44	74	65
GLC2000[b]	61	91	73
MLCC[c]	78	66	73
GDTDF[d]	2	50	41
SDTF_A[e]	83	74	79
SDTF_B[f]	94	47	74

[a]International Geosphere-Biosphere Programme (IGBP) DISCover (Belward et al. 1999; Loveland et al. 2000).
[b]Global Land Cover 2000 (Bartholome and Belward 2005; Stibig et al. 2007).
[c]MODIS/Terra Land Cover Classification (Strahler et al. 1999; Friedl et al. 2002).
[d]Global distribution of tropical dry forest (Miles et al. 2006).
[e]SDTF_A includes all SDTF areas that were found in at least two previous maps (IGBP, GLC2000, MLCC, and GDTDF) and were located within the dry forest climate zone.
[f]SDTF_B: includes all forested areas found in at least two previous maps (IGBP, GLC2000, MLCC, and GDTDF) and were located within the dry forest climate zone.

Table 5. Potential area and percentage protected of seasonally dry tropical forest (SDTF; km^2) calculated from combined seasonally dry tropical forest maps.

Country	SDTF_A[a]	SDTF_B[b]
Bangladesh	0 (0%)	0 (0%)
Cambodia	26,150 (24%)	54,139 (22%)
India	397,110 (7%)	454,269 (7%)
Laos	11,708 (0%)	97,376 (0%)
Malaysia	0 (0%)	0 (0%)
Myanmar	84,314 (1%)	209,984 (2%)
Nepal	1,113 (1%)	13,432 (17%)
Sri Lanka	2,475 (59%)	11,543 (30%)
Thailand	58,102 (30%)	123,508 (46%)
Vietnam	9,910 (7%)	72,831 (9%)
Total	590,882 (9%)	1,037,082 (12%)

[a]SDTF_A includes all SDTF areas that were found in at least two previous maps (IGBP, GLC2000, MLCC, or GDTDF) and were located within the dry forest climate zone.
[b]SDTF_B includes all forested areas that were found in at least two previous maps (IGBP, GLC2000, MLCC, or GDTDF) and were located within the dry forest climate zone.

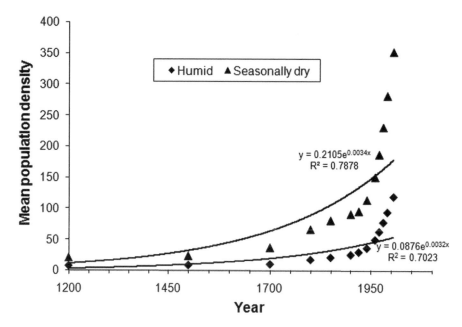

Figure 3. Human population development compared between seasonally dry tropical forest areas and tropical humid forest areas.

geographic regions. At least for Asia, the only map developed specifically for delineating SDTFs and assessing their conservation status, GDTDF, performed worst of all the Asia maps, with area estimates that are likely much too low (Tables 3 and 4; Figure 1).

None of the data sets provide clear and detailed error or accuracy assessments. This lack of evaluation is surprising considering how often global and regional data are used for analyzing other biological or geophysical properties, or for delineating ecosystem properties and conservation priorities (Sanderson et al. 2002; Leimgruber et al. 2003; Hubener et al. 2005; Dinerstein et al. 2007).

Perhaps most important, our results illustrate that the uncertainty in locating SDTF based on regional and global maps is so high that it might be best to avoid using broad-scale data for region-wide SDTF conservation assessments. Threat assessments are also hampered by this spatial uncertainty since one cannot accurately assess how much SDTF may be conserved inside protected areas. We consider it fruitless to compare old and new SDTF maps because the error in classifying these ecosystems is likely higher than the amount of change that may have occurred. Our analysis highlights the urgent need for developing better data sets for this ecosystem.

Using Existing Land Cover Data Sets

There are several possible explanations for the low accuracy and agreement in regional and global land cover data, including differences in SDTF definitions, canopy densities mapped, characteristics of the satellite imagery used, availability of ground reference, classification techniques, and scale and resolution.

Differences in the definition of what constitutes SDTF are among the most important reasons for divergent delineations. Definitions of SDTF can include a wide range of deciduous or evergreen forest, shrub, savanna, and open woodland ecosystems (Blasco et al. 1996, 2000; Bunyavejchewin et al., this volume). In the broad sense used by these authors, it includes vegetation formations with nearly closed canopies such as the *Shorea robusta* forests of central India, open woodlands such as the dry dipterocarp forests of Indochina, and dense, low-canopy evergreen forests such as the dry forests of Sri Lanka.

Previous studies have shown that thematic accuracy and overlap between land cover classes can be low for the data sets used in this study (Giri et al. 2005; McCallum et al. 2006; Heiskanen 2008). Differences in the spatial arrangement of land cover classes (i.e., overlap) is most problematic for transition zones and transitional land cover types (Hansen and Reed 2000; Heiskanen 2008). Heiskanen (2008) has demonstrated this for the tundra-taiga transition zone in Finland. SDTF is also a transitional land cover type, and this poses a major problem for delineating it from existing land cover maps. Additionally, land cover categories differ between the maps and frequently cannot easily be matched to conditions on the ground, a condition that likely increases difficulties in SDTF mapping because of the wide range of vegetation forms in this ecosystem type. For these reasons, it is difficult to determine how much of the disagreement in SDTF spatial arrangement between maps is the result of misclassification, or which maps are more accurate (Heiskanen 2008).

Our restricted accuracy assessment should not be extrapolated across the entire region because regional differences in SDTF structure and form may affect which maps better predict SDTF extent and location. However, the differences in accuracy among the selected maps for upper Myanmar raise our concerns about accuracy of the data in other regions.

The main reason current land cover maps (IGBP, GLC2000, MLCC) do not perform well in mapping SDTF is that the ecosystem is transitional in space and time. In space, the transition has to do with a gradient of canopy cover from dry deciduous forest types to mixed and evergreen forest types. In time, the transition stems from a pronounced seasonal change in tree phenology. During the dry season, many SDTFs have little canopy cover, making them spectrally akin to deforested or agricultural areas. To complicate matters, seasonal phenology also varies across space, depending on geographic differences in the onset of the monsoon rains. The spatial and temporal factors increase the risk of classifying SDTF either as open and agricultural areas during the dry season, or as moist tropical forests during the rainy season, and of doing so inconsistently across space because the seasons in different regions transition over different time periods. Conventional broad-scale land cover mapping may not provide sufficient detail to separate SDTF accurately on either side of the canopy cover spectrum.

Using the Global Distribution of Tropical Dry Forest Maps

Intuitively, GDTDF should perform better at SDTF delineation than regular land cover maps because Miles et al. (2006) relied strongly on the VCF data set. VCF was created by linking tree density estimates from finer resolution imagery to mid-resolution MODIS data via regression trees (Hansen et al. 2002). Instead of assigning broad spectral signatures to sometimes ambiguous land cover categories, VCF attempts to link spectral information directly to percentage tree cover. However, there may be four reasons for poor performance of GDTDF:

1. In their definition of SDTF, Miles et al. (2006) concluded that climatic approaches to delineating SDTF (e.g., definitions and frameworks provided by Blasco et al. 1996, 2000) are ambiguous, so they did not include climatic data in their definition. We consider climatic conditions, especially pronounced seasonality in rainfall, essential to the existence of SDTF (Bunyavejchewin et al., this volume).
2. Koy et al. (2005) found the average for canopy cover for SDTF in upper Myanmar was 34.2 percent ± 2.46 percent, while Miles et al. (2006) considered in their analysis only areas with ≥ 40 percent canopy cover. SDTF may have more open canopies than Miles et al. (2006) allowed for in their definition, possibly resulting in an underestimation of SDTF areas.
3. Although the base data was at 500 m spatial resolution, Miles et al. (2006) used coarse-scale ecoregional delineations (i.e., WWF ecoregions, and the *World Atlas of Desertification*) to "refine" their estimate. Using these coarse data sets resulted in the exclusion of major SDTF areas throughout Asia.
4. As with many global data sets, VCF has not yet been thoroughly ground-truthed (but see Hansen et al. 2002). Studies in other transitional ecosystems have demonstrated the limited accuracy with which VCF portrays actual percentage tree cover (White et al. 2005; Heiskanen 2008).

IDENTIFYING FORESTS FOR FUTURE FINE-SCALE MAPPING OF SDTF

Hierarchical approaches to identifying SDTF areas may be most promising for future accurate delineation and monitoring of SDTF. Our results demonstrated that broad-scale data on forest cover, combined with climatic data, may provide a first-level approximation of areas that may support SDTF (Figure 2). Areas delineated in these maps should be priority areas for fine-scale mapping using higher resolution imagery or improved sensors (e.g., hyperspectral and lidar). This hierarchical approach will also be helpful in efforts to initiate monitoring of SDTF changes and losses over time. Dramatic declines in forest cover in areas that may support SDTF should raise alarm for conservation scientists concerned about SDTF status. Based on our accuracy assessments in upper Myanmar, SDTF_A may provide a more realistic picture of the current locations and extent of possible SDTF areas because it has a much higher overall classification accuracy and the map performs better in separating other areas from SDTF areas. SDTF_B provides a more optimistic outlook on the current conditions of SDTF in Asia, but likely overestimates the current extent of SDTF, which is expressed in the low accuracy for other areas and for the map overall. However, if we want to take a precautionary approach in identifying as much of the remaining SDTF as possible, SDTF_B might be the best starting point.

To assess better the current extent, distribution, and recent declines of SDTF, more-detailed country- or landscape-level mapping is needed. Examples of such finer-scale mapping that includes accuracy assessments exist for Myanmar, where researchers have delineated SDTFs because they represent critical habitat to an endangered species, Eld's deer (*Rucervus eldii*; McShea et al. 1999; Koy et al. 2005; Songer 2006). All these studies have relied on extensive ground data collection, calibrating classification models using these ground data, and evaluating accuracy of the final map with an independent test sample. McShea et al. (1999) showed that this landscape-level mapping produced SDTF cover estimates that deviated widely from previous broad-scale mapping.

MAPPING AND CONSERVATION

Through advances in remote sensing and computer technology, we are now able to monitor and map the globe in ways not possible just twenty years ago. It is easy to use these free, readily available global data for multiple objectives. However, before employing them, it is important to consider how these data sets were created, their purposes, and their specifications. Producing data sets at a global scale presents many challenges in dealing with variation in phenology and habitat types across regional scales. Working at 1 km resolution makes it difficult to distinguish subtle habitat changes, and analysts rarely have detailed on-the-ground experience around the globe. When selecting an existing land cover classification, it is important to consider the methodology and assumptions that have gone into that classification within the context of specific objectives.

Despite major advances in remote sensing and the satellite mapping of ecosystems, we still lack usable quantitative and spatially explicit information on the current extent and status of SDTF. These data are urgently needed because SDTFs, and especially deciduous dipterocarp forests, are among the most threatened and least protected forests, and they are found in areas with rapid human population growth. Broad-scale remote sensing and land cover mapping not specifically targeted at SDTF is not likely to produce the needed assessments and maps, because SDTFs are complex in phenology, structure, species composition, and spatial configuration. Rather, a hierarchical approach is needed, with broad-scale maps used to identify areas requiring finer-scale assessments.

Our comparisons of human population growth between SDTF and humid tropical climate zones demonstrate the immense development pressures SDTF must withstand in Asia. Despite our lack of accurate and spatially explicit data, current information is sufficient to indicate SDTFs are in serious trouble primarily through replacement by agriculture or other human land uses.

REFERENCES

Bartholome, E., and A. S. Belward. 2005. GLC2000: A New Approach to Global Land Cover Mapping from Earth Observation Data. *International Journal of Remote Sensing* 26:1959–77.
Belward, A. S., J. E. Estes, and K. D. Kline. 1999. The IGBP-DIS Global 1-km Land Cover Data set DISCover: A Project Overview. *Photogrammetric Engineering and Remote Sensing* 65:1013–20.

Blasco, F., M. F. Bellan, and M. Aizpuru. 1996. A Vegetation Map of Tropical Continental Asia at Scale 1:5 Million. *Journal of Vegetation Science* 7:623–34.

Blasco, F., T. C. Whitmore, and C. Gers. 2000. A Framework for the Worldwide Comparison of Tropical Woody Vegetation Types. *Biological Conservation* 95:175–89.

Collins, N. M., J. Sayer, and T. C. Whitmore, eds. 1991. *The Conservation Atlas of Tropical Forests*. London: Macmillian.

DeFries, R. S., M. Hansen, J. R. G. Townshend, A. Janetos, and T. Loveland. 2000. A New Global Data Set of Percent Tree Cover Derived from Remote Sensing. *Global Change Biology* 6:247–54.

DeFries, R. S., M. Hansen, J. R. G. Townshend, and R. Sohlberg. 1998. Global Land Cover Classifications at 8 km Spatial Resolution: The Use of Training Data Derived from Landsat Imagery in Decision Tree Classifiers. *International Journal of Remote Sensing* 19:3141–68.

Dinerstein, E., C. Loucks, E. Wikramanayake, J. Ginsberg, E. Sanderson, J. Seidensticker, J. Forrest, G. Bryja, A. Heydlauff, S. Klenzendorf, P. Leimgruber, J. Mills, T. G. O'Brien, M. Shrestha, R. Simons, and M. Songer. 2007. The Fate of Wild Tigers. *Bioscience* 57:508–14.

Friedl, M. A., D. K. McIver, J. C. F. Hodges, S. Y. Zhang, D. Muchoney, A. H. Stahler, C. E. Woodcock, S. Gopal, A. Schneider, A. Cooper, A. Baccini, F. Gao, and C. Schaaf. 2002. Global Land Cover Mapping from MODIS: Algorithms and Early Results. *Remote Sensing of Environment* 83:287–302.

Giri, C., Z. Zhu, and B. Reed. 2005. A Comparative Analysis of the Global Land Cover 2000 and MODIS Land Cover Datasets. *Remote Sensing of Environment* 94:123–32.

Goldewijk, K. K. 2001. Estimating Global Land Use Change over the Past 300 Years: The HYDE Database Source. *Global Biogeochemical Cycles* 15:417–33.

Grainger, A. 1999. Constraints on Modeling the Deforestation and Degradation of Tropical Open Woodlands. *Global Ecology and Biogeography* 8:179–90.

Hansen, M. C., R. S. DeFries, J. R. G. Townshend, M. Carroll, C. Dimiceli, and R. A. Sohlberg. 2003. Global Percent Tree Cover at a Spatial Resolution of 500 Meters: First Results of the MODIS Vegetation Continuous Fields Algorithm. *Earth Interactions* 7:1–15.

Hansen, M. C., R. S. DeFries, J. R. G. Townshend, L. Marufu, and R. Sohlberg. 2002. Development of a MODIS Tree Cover Validation Data Set for Western Province, Zambia. *Remote Sensing of the Environment* 83:320–35.

Hansen, M., R. DeFries, J. R. G. Townshend, and R. Sohlberg. 2000. Global Land Cover Classification at 1 km Resolution Using a Decision Tree Classifier. *International Journal of Remote Sensing* 21:1331–65.

Hansen, M. C., and B. Reed. 2000. A Comparison of the IGBP DISCover and University of Maryland 1 km Global Land Cover Products. *International Journal of Remote Sensing* 21:1365–73.

Heiskanen, J. 2008. Evaluation of Global Land Cover Data Sets over the Tundra-Taiga Transition Zone of Northernmost Finland. *International Journal of Remote Sensing* 13:3727–51.

Hijmans, R. J., L. Guarino, A. Jarvis, R. O'Brien, P. Mathur, C. Bussink, M. Cruz, I. Barrantes, and E. Rojas. 2005. *DIVA-GIS*, version 5.4, www.diva-gis.org (accessed January 2009).

Hubener, H., M. Schmidt, M. Sogalla, and M. Kerschgens. 2005. Simulating Evapotranspiration in a Semi-arid Environment. *Theoretical and Applied Climatology* 80:153–67.

Janzen, D. H. 1988. Tropical Dry Forests: The Most Endangered Major Tropical Ecosystem. In *Biodiversity*, ed. E. O. Wilson and F. M. Peters, 130–44. Washington, DC: National Academy Press.

Jensen, J. R. 1996. *Introductory Digital Image Processing: A Remote Sensing Perspective*, 2nd ed. Upper Saddle River, NJ: Prentice Hall.

Koy, K., W. J. McShea, P. Leimgruber, B. N. Haack, and Myint Aung. 2005. Percentage Canopy Cover—Using Landsat Imagery to Delineate Habitat for Myanmar's Endangered Eld's Deer (*Cervus eldi*). *Animal Conservation* 8:289–96.

Leimgruber, P., J. B. Gagnon, C. Wemmer, D. S. Kelly, M. A. Songer, and E. R. Selig. 2003. Fragmentation of Asia's Remaining Wildlands: Implications for Asian Elephant Conservation. *Animal Conservation* 6:347–59.

Leimgruber, P., D. S. Kelly, M. Steininger, J. Brunner, T. Mueller, and M. Songer. 2005. Forest Cover Change Patterns in Myanmar (Burma) 1990–2000. *Environmental Conservation* 32:356–64.

Loveland, T. R., B. C. Reed, J. F. Brown, D. O. Ohlen, Z. Zhu, L. Yang, and J. W. Merchant. 2000. Development of a Global Land Cover Characteristics Database and IGBP DISCover from 1 km AVHRR Data. *International Journal of Remote Sensing* 21:1303–30.

McCallum, I., M. Obersteiner, S. Nilson, and A. Shvidenko. 2006. A Spatial Comparison of Four Satellite Derived 1 km Global Land Cover Datasets. *International Journal of Applied Earth Observation and Geoinformation* 8:246–55.

McShea, W. J., P. Leimgruber, Myint Aung, S. L. Monfort, and C. Wemmer. 1999. Range Collapse of a Tropical Cervid (*Cervus eldi*) and the Extent of Remaining Habitat in Central Myanmar. *Animal Conservation* 2:173–83.

Middleton, N., and D. Thomas, eds. 1997. *World Atlas of Desertification*. London: Edward Arnold Publishers.

Miles, L., A. C. Newton, R. S. DeFries, C. Ravilious, I. May, S. Blyth, V. Kapos, and J. E. Gordon. 2006. A Global Overview of the Conservation Status of Tropical Dry Forests. *Journal of Biogeography* 33:491–505.

Myint Aung, Khaing Khaing Swe, Thida Oo, Kyaw Kyaw Moe, P. Leimgruber, T. Allendorf, C. Duncan, and C. Wemmer. 2004. The Environmental History of Chatthin Wildlife Sanctuary, a Protected Area in Myanmar (Burma). *Journal of Environmental Management* 72 (4): 205–16.

Olson, E. M., E. Dinerstein, E. D. Wikramanayake, N. D. Burgess, G. V. N. Powell, E. C. Underwood, J. A. D'Amico, I. Itoua, H. E. Strand, J. C. Morrison, C. J. Loucks, T. F. Allnutt, T. H. Ricketts, Y. Kura, J. F. Lamoreux, W. W. Wettengel, P. Hedao, and K. R. Kassem. 2001. Terrestrial Ecoregions of the World: A New Map of Life on Earth. *BioScience* 51:933–38.

Ruangpanit, N. 1995. Tropical Seasonal Forests in Monsoon Asia: With Emphasis on Continental Southeast Asia. *Vegetation* 121:31–40.

Sanderson, E. W., M. Jaiteh, M. A. Levy, K. H. Redford, A. V. Wannebo, and G. Woolmer. 2002. The Human Footprint and the Last of the Wild. *BioScience* 52:891–904.

Scepan, J. 1990. Thematic Validation of High Resolution Global Land Cover Data Sets. *Photogrammetric Engineering and Remote Sensing* 65:1051–60.

Songer, M. 2006. *Endangered Dry Deciduous Forests of Upper Myanmar (Burma): A Multiscale Approach for Research and Conservation*. PhD diss., University of Maryland.

Songer, M., Myint Aung, B. Senior, R. DeFries, and P. Leimgruber. 2009. Spatial and Temporal Deforestation Dynamics in Protected and Unprotected Dry Forests: A Case Study from Myanmar (Burma). *Biodiversity and Conservation* 18:1001–18.

Stamp, L. D. 1925. *The Vegetation of Burma from an Ecological Standpoint*. Calcutta: Thacker, Spink and Company.

Stibig, H.-J., A. S. Belward, P. S. Roy, U. Rosalina-Wasrin, S. Agrawal, P. K. Joshi, Hildanus, R. Beuchle, S. Fritz, S. Mubareka, and C. Giri. 2007. A Land-Cover Map for South and Southeast Asia Derived from SPOT-VEGETATION Data. *Journal of Biogeography* 34:625–37.

Strahler, A., D. Muchoney, J. Borak, M. Friedl, S. Gopal, E. Lambin, and A. Moody. 1999. *MODIS Land Cover Product: Algorithm Theoretical Basis Document (ATBD)*, version 5.0. Boston: Boston University.

White, M. A., J. D. Shaw, and R. D. Ramsey. 2005. Accuracy Assessment of the Vegetation Continuous Field Tree Cover Product Using 3954 Ground Plots in the South-Western USA. *International Journal of Remote Sensing* 12:2699–2704.

Wikramanayake, E., E. Dinerstein, C. J. Loucks, D. M. Olson, J. Morrison, J. Lamoreux, M. McKnight, and P. Hedao. 2002. *Terrestrial Ecoregions of the Indo-Pacific: A Conservation Assessment*. Washington, DC: Island Press.
World Conservation Monitoring Centre. 1997. *A Global Overview of Forest Conservation*. CD-ROM, www.unep-wcmc.org/forest/fp_background.htm#. (accessed January 2009).
———. 2007. *World Database on Protected Areas*, version 2007, http://glcf.umiacs.umd.edu/data/wdpa/ (accessed January 2009).

4

The Role of Disturbance in Seasonally Dry Tropical Forest Landscapes

Virginia H. Dale

*D*isturbance can be defined as "any relatively discrete event in time that disrupts ecosystem, community, or population structure and changes resources, substrate availability, or the physical environment" (White and Pickett 1985). This definition requires that the spatial and temporal scales of the system and disturbance be determined. Disturbances are typically characterized by their size, spatial distribution, frequency or return time, predictability, and magnitude (which includes both intensity and severity; White and Pickett 1985). These disturbance attributes set the parameters for the suite of species, both plant and animal, that can persist within a given system. As such, an understanding of seasonally dry tropical forests in Asia requires an understanding of disturbance within the region. However, disturbances are relatively poorly understood in dry topical forests, partly because of the weak seasonality in temperature and high tree species diversity of these forests relative to most forest systems of the world (Huete et al. 2008).

Miles et al. (2006) report that there are about 1,048,700 km² of dry tropical forest worldwide and that only 3 percent of this land is in conservation status. In other words, 97 percent of the world's seasonally dry tropical forest is at risk of human disturbance. About half of this forest occurs in South America, where most of the conservation lands are located. Satellite imagery based on MODIS (Moderate Resolution Imaging Spectroradiometer) data shows that only about 3.8 percent of the world's dry tropical forests are in Australia and Southeast Asia (Miles et al. 2006). The susceptibility of these forests to human disturbances is of great concern and is largely unstudied. Because natural disturbance regimes shape the ecosystem structure and are in many ways integral to these forest systems, it is critical to know how natural disturbance affects dry forest in order to understand the effects of human activities on these forests. Even basic information about disturbances in dry tropical forests is only recently available. Therefore this chapter brings together much of the available information from dry tropical forest throughout the world with the goal of developing an understanding of the role of disturbance in Asian dry forests.

Most ecologists now recognize that disturbances, rather than being catastrophic agents of destruction, are a normal, perhaps even an integral, part of long-term system dynamics (Weatherhead 1986; Lugo et al. 2006; Imbert and Portecop 2008). The composition, structure, organization, and developmental and trophic dynamics of most forest systems are the products of disturbances (Denslow 1980; Leach and Givnish 1996; Lugo and Scatena 1996; Wootton et al. 1996; Werneck and Franceschinelli 2004; Kennard and Putz 2005; Otterstrom et al. 2006; Chazdon et al. 2007; Urquiza-Haas et al. 2007; Alvarez-Yepiz et al. 2008; Arredondo-Leon et al. 2008; Sahu et al. 2008; Mehta et al. 2008a, 2008b; Poorter et al. 2008; Anitha et al. 2009; Sapkota et al. 2009; Zimmer and P. Baker 2009). For example, Anitha et al. (2009) compared forest composition for two disturbances in the Anaikatty Hills of Western Ghats, where the low disturbance was from past logging followed by cutting and illicit felling and grazing and the high disturbance was due to human presence, past logging, and fuelwood collection. They found higher species richness and Shannon-Wiener diversity index for the low-disturbance forest (98 and 3.9, respectively) compared to the high-disturbance stand (45 and 2.71, respectively) as well as significant differences in mean basal area of trees, density of seedlings, number of species, density and diversity of shrubs, and number of species and diversity of herbs.

Some ecological systems contain species that have evolved in response to disturbances. Adaptations typical of dry tropical forest plants are drought tolerance, seed dispersal mechanisms, and the ability to sprout subsequent to disturbance (see Kozlowski and Pallardy 2002; Otterstrom et al. 2006). In contrast, Farwig et al. (2008) found evidence that human disturbance in Kakamega Forest of western Kenya has significantly reduced allelic richness and heterozygosity, increased inbreeding, and slightly reduced gene flow in *Prunus africana* in the past century.

DISTURBANCE SCALE, DISTRIBUTION, FREQUENCY, AND MAGNITUDE

The scale of disturbance can vary from those disturbances larger than typical management activities (e.g., large-scale fires or hurricanes) to those at the microscale (e.g., collection of firewood in a locale). Martini et al. (2008) studied patterns of tree mortality and damage in an old-growth, semi-deciduous seasonal forest at the Caetetus Ecological Station, southeastern Brazil. Of the 743 disturbance events recorded, 650 involved the demise of entire trees (33 percent uprooted, 28 percent dead-standing, 26.5 percent snapped, 10.5 percent inclined, and 2 percent complex events) while 93 involved only parts of trees (60.2 percent branch-falls and 39.8 percent stem-fall from multi-stemmed trees). The most abundant trees in the area experienced much resprouting, suggesting that this ability increases the competitive potential of these species, especially in forests with high disturbance rates.

Sapkota et al. (2009) found that the spatial distribution, advanced regeneration, and stand structure of *Shorea robusta*–dominated forests in five seasonally dry deciduous sal forests in the Nawalparansi district of Nepal were influenced by disturbance intensity. The most disturbed forest plots had less tree species richness in the more-disturbed plots, greater density of saplings (trees with less than 1.5 cm diameter at breast height [dbh]), and no significant difference in stem basal area. The overall stand density changed quadratically across the disturbance gradient, with moderate disturbance intensity not only ensuring high stand density but also enhancing the advanced regeneration of socioeconomically important tree species and affecting their dispersion patterns. The study recorded 67 tree species in the forest plots, with 41 species in the least disturbed forests and only 10 species in the most heavily disturbed forests. Ten species varied in their dispersion patterns across the disturbance gradient, and yet most of the socioeconomically important tree species considered had little or no regeneration in both the least and most heavily disturbed forests. In addition, select species had unique responses to disturbance ranging from "tolerant" (*Shorea robusta*, *Lagerstroemia parviflora*, and *Symplocos* spp.) to "sensitive" (*Trewia nudiflora*, *Adina cardifolia*, and *Terminalia alata*).

Sahu et al. (2008) also documented that tree density, diversity, and structure at forty-two sites in the Achanakmar-Amarkantak Biosphere Reserve of central India responded to anthropogenic disturbance. The number of species and indices of species diversity were positively associated with copicing and also with total disturbance (which included foot trails and dung piles as evidence of livestock grazing).

Even though epiphytic assemblages are a notable part of these ecosystems, existing information on how epiphytes are affected by different degrees of human disturbance, and what drives their recovery, is limited or region-specific. Yet a recent study by Werner and Gradstein (2009) in northern Ecuador focused on how sensitive these plants, especially bryophytes, are to changes in relative humidity, which are often associated with disturbance events. They found that the density of bryophytes species was significantly lower in edge habitat and on isolated trees than in closed forest, in contrast to species of vascular epiphytes, which did not change between habitat types.

The intensity of disturbances can be large enough to set the system back to an earlier stage in its developmental sequence or to initiate a different seral pathway. However, very intense disturbances are rare (Turner et al. 1997). Other disturbances may be so small in intensity, or common in time or space (or both), that some managers may consider them to be a part of the natural functioning of the system (see Werner and Gradstein 2009).

Disturbance magnitude has two attributes: intensity, which refers to the physical force of the event on a per area basis, and severity, which refers to the impact on an organism, community, or ecosystem. Severity measures the extent to which the predisturbance vegetation is damaged and ecosystem properties are disrupted. Vegetation at a site will often reestablish more slowly after a severe disturbance than after a mild disturbance (Turner et al. 1998). Human activities, the landscape context, and site fertility typically greatly influence vegetation reestablishment after a disturbance to dry topical forests (as Velazquez and Gomez-Sal [2008] documented subsequent to a landslide in Nicaragua). However, severity can also be measured as an effect on ecosystem function. A disturbance may affect soil properties and long-term nutrient dynamics. In one such case, Bormann and Likens (1979) found that a severe disturbance resulted in such substantial erosion and nutrient losses that the ecosystem might require decades to recover its key properties.

Both spatial and temporal regimes of disturbances can vary greatly (Tilman 1989), and direct measurement of the effects of broad-scale disturbances on these forests are quite difficult and rarely done (but see P. Baker et al., 2005). Yet Vieira and de Carvalho (2008) found that high genetic diversity of *Protium spruceanum* correlates with the diversity of the forest fragments that occurs in the vegetation corridors that connect small remnants of undisturbed primary forest in the Lavras landscape in Brazil. Furthermore, evidence of recent bottlenecks by anthropogenic disturbance was also detected in these fragments. Thus Vieira and de Carvalho found high levels of gene diversity in the vegetation corridors and absence of any inbreeding. Given that they also found bottlenecks in landscape patterns, these corridors provide a high potential for in situ genetic conservation and for seed collection destined to restore degraded areas. These results suggest that landscape management strategies should both create new vegetation corridors and manage the protection of existing ones.

TYPES OF DISTURBANCE

Fire

The frequency, size, intensity, seasonality, and type of fires depend on weather and climate in addition to forest structure and composition. Fires are such an integral part of many seasonally dry tropical forest systems that some managers do not consider them a disturbance at all. Fire initiation and spread depend on the amount and frequency of precipitation, the presence of ignition agents, and conditions (e.g., lightning, fuel availability, forest distribution, topography, temperature, relative humidity, and wind velocity; Janssens 2000).

Fire effects on dry tropical forests include acceleration of nutrient cycling, mortality of individual trees, shifts in successional direction, induced seed germination, loss of soil seed bank, increased landscape heterogeneity, alteration of surface-soil organic layers, changes in underground plant-root and reproductive tissues, and volatilization of soil nutrients (Whelan 1995). Erosion can occur where soil disturbance accompanies fire (e.g., during fire fighting or timber salvage operations). Fire affects forest value for wildlife habitat, timber, and recreation; and through smoke, it affects human health (Moreno 2006).

The spatial and temporal heterogeneity created by fire is important to ungulates that occupy these forests (McShea and Baker, this volume). Although fires are not the prime cause of local extirpation of any plant species, the spatial heterogeneity created by fires in dry tropical forests over time is critical to providing habitat for the diverse animals that occur in these dry tropical forests (e.g., elephants and tigers; Fernando and Leimgruber, this volume; Smith et al., this volume).

Fire was found to have a strong effect on bamboo distribution in the tropical seasonal mixed deciduous forest of the Mae Klong Watershed Research Station in Kanchanaburi Province of western Thailand (Marod et al. 1999). Mortality was size dependent, with middle-sized trees (30–50 cm) having the lowest mortality and smallest trees (5–10 cm) having the highest mortality. Tree recruitment was mostly in the area where dieback of undergrowth bamboo occurred. Hence "both the fire disturbance regime and bamboo life-cycle greatly influence the structure and dynamics of this seasonal tropical forest" (Marod et al. 1999).

The rapid response of fire regimes to changes in climate (Flannigan et al. 1998, 2000; Stocks et al. 1998) can potentially overshadow the direct effects of climate change on species distribution and migration. Modeling results predict great variation in future fire-weather patterns. Some fire-history studies suggest that the frequency of fire might decrease despite warmer temperatures because of increased precipitation (see Bergeron and Archambault 1993). In a study of northeastern Cambodia, Maxwell (2004) determined that the greatest fire activity ended about 8,000 years ago, and the subsequent period experienced low levels of fire and a strong summer monsoon. There was then an increase in forest disturbance, and fire activity increased, commencing about 5,500 years ago. A small change in the record occurred about 3,500 years ago, and more marked change occurred about 2,500 years ago when fire frequency, and maybe human control over fire, became more important. It is difficult to extrapolate for all of Asia using this one study or to determine the long-term human impacts from burning. Nevertheless, Maxwell (2004) found that current charcoal input from fire activity is one of the lowest of the past 9,300 years. He suggests that "anthropogenic fire is an adaptation to the monsoonal environment."

Because fire is a frequent disturbance in the seasonally dry tropical forests of Central America, Otterstrom et al. (2006) conducted an experimental burn in a tropical dry forest of western Nicaragua to evaluate plant survivorship and recruitment. They found that the most common dry forest species are fire tolerant, for these species had high survivor rates or used seeding or sprouting as an alternative for persistence in the forest system. This phenomenon also occurs in Jamaica (McLaren et al. 2005). In addition, mechanisms for seed dispersal, particularly via wind, were important for recruitment success after the fire.

Aguilar-Fernandez et al. (2009) report that rates of land conversion from forest to cultivated land by slash-and-burn practices are higher in tropical dry forest than in any other neotropical forest type. The short-term consequences of the slash-and-burn process on arbuscular mycorrhizal fungi were that infective propagules were reduced in pasture plots during the first year after slash-and-burn, whereas spore abundance and intraradical colonization remained similar in forest and pasture plots over two years (Aguilar-Fernandez et al. 2009). These results suggest that forest conversion via slash-and-burn, followed by cultivation, resulted in few immediate changes in the arbuscular mycorrhizal fungi communities, likely because of the low heat conductivity of the soil and rapid combustion of plant residues (Aguilar-Fernandez et al. 2009).

Kennard and Putz (2005) monitored tree establishment and growth response to canopy opening, aboveground biomass removal, and experimental burns of low and high intensities in a lowland dry forest in Bolivia. Shade-intolerant species reestablished mostly from seed and had the highest survival and growth rates following high-intensity burns. Shade-tolerant species were abundant in gap control and plant removal treatments and showed little difference between the treatments in height growth. Trees with the ability to sprout roots were most abundant following plant removal and low-intensity burn treatments. This variety of regeneration strategies likely facilitates recovery from the diversity of disturbances to which these forests can be exposed.

Drought

Droughts occur in nearly all forest ecosystems. Extreme drought effects are influenced by soil texture and depth, exposure, species present, life stage, and the frequency, duration, and severity of the drought. Seasonally dry tropical forests are largely water limited; and in areas of forest disturbance, carbon fluxes respond more to seasonal water availability than to availability of sunlight (Huete et al. 2008). The primary immediate response of forests to drought is to reduce net primary production (NPP) and water use; both reductions are driven by reduced soil moisture and stomatal conductance. Under severe conditions, plants die. Small plants, such as seedlings and saplings, are usually the first to die and can succumb under moderate conditions. Deep rooting and stored carbohydrates and nutrients make large trees susceptible only to severe droughts. Secondary effects also occur. When reductions in NPP are extreme or sustained over multiple growing seasons, increased susceptibility to insects or disease is possible, especially in dense stands (see Negron 1998). Drought can also reduce decomposition processes and thus slow down the nutrient cycling process. However, reduced decomposition can also lead to a buildup of organic matter on the forest floor that may increase fire frequency or intensity, which thereby enhances nutrient cycling. Brandeis and Woodall (2008) found that forest litter decreased from wetter to drier forest life zones in tropical forests of Puerto Rico and the U.S. Virgin Islands. In addition, when compared to continental tropical forests, the island forests had fewer coarse woody fuels (2.91 Mg/ha) and relatively greater quantities of smaller-sized fine woody fuels (10.18 Mg/ha and 10.82 Mg/ha for duff and litter, respectively). These data suggest that continued decomposition of coarse woody debris from a previous hurricane is balanced by increasing inputs of fine woody debris from recovering and maturing secondary forest (Brandeis and Woodall 2008).

The consequences of drought depend on annual and seasonal climate changes and whether the current drought adaptations offer resistance and resilience to new conditions (Hanson and Weltzin 2000). Forests tend to grow to a level of maximum leaf area that almost fully uses soil water during the growing season (Neilson and Drapek 1998). A small increase in growing-season temperature could increase evaporative demand, triggering moisture stress.

Diversity in the lowland mixed dipterocarp forest at Lambir Hills National Park in Sarawak, East Malaysia, may have been enhanced by the 1998 drought (Potts 2003). The Lambir Hills forest is likely one of the richest forests in terms of tree species diversity left in the Old World. This is an area that experiences droughts almost every year, but the severity of the 1998 drought was on the order of once in every century (Nakagawa et al. 2000). Stem mortality was higher during this severe drought when compared to the pre-drought period, regardless of tree size or soil type. Smaller trees and large common trees experienced higher mortality, whereas the mortality of large rare trees did not differ significantly from other years (because of the small sample size; Potts 2003). Some species are more drought-mortality prone at maturity than others. In the middle of the Lambir forest, frequent slight droughts on deep sandy soils are associated with a diffuse but highly structured canopy supporting abundant but clumped emergent trees less than 70 m tall (Ashton and Hall 1992). Potts (2003) postulates that extreme droughts enhance diversity via greater mortality of large trees and taxonomic differences in rates of mortality.

Windstorms

Scale and frequency of disturbance have profound impacts on the dynamics and composition of flora and fauna in wet tropical forests (Ashton and Hall 1992), but their role in seasonally dry forests is unknown outside of the Caribbean and Central America. In a study subsequent to Hurricane George's striking Puerto Rico, Van Bloem et al. (2005) found that hurricane-induced mortality of large trees was significant enough that the low, dense structure of Caribbean dry forest could be maintained by hurricane damage to larger stems. Furthermore, the dry forest trees of the Caribbean sprout near the base following hurricane disturbance even if the trees have not incurred structural damage (Van Bloem et al. 2006). Even though the density and proportion of surviving stems decreased over the ensuing seven-year period, the sprouting rate was still three times higher and the proportion of sprouting stems five times higher than before the hurricane (Van Bloem et al. 2007). Mortality of sprouting and non-sprouting stems was similar (13.9 percent and 15.4 percent) after seven years. These results demonstrate that, despite some thinning, hurricane-induced sprouts survive to influence the structural characteristics of Caribbean dry forests long after the hurricane disturbance (Van Bloem et al. 2007).

Forest structure itself is often a result of wind disturbance in these seasonally dry tropical systems. Lugo et al. (2006) report that the low floristic diversity and stature and high density of small- and medium-sized trees of Antillean dry forests are a result of periodic hurricanes, anthropogenic disturbance, and the fact that most of these forests occur on limestone substrate, which induces water stress and nutrient limitations. A great diversity of life forms, high resistance to wind, a high proportion of root biomass, high soil carbon and nutrient accumulation below ground, the ability of most tree species to resprout, and their high nutrient-use efficiency allow the plants to survive the stressful environment (Lugo et al. 2006). These attributes also make them resilient to disturbance (Lugo et al. 2006). On the other hand, the forests become less resilient and species invasion occurs subsequent to opening the canopy, eroding the soil, and removing root biomass (Lugo et al. 2006).

EFFECTS OF FOREST DISTURBANCES ON ANIMALS

Studies are limited (but see McShea and Baker, this volume; Bhumpakphan and McShea, this volume), but there is compelling evidence that animals respond strongly to disturbances in seasonally dry tropical forests. Sanchez et al. (2007) report a low abundance of bats in Patia, Columbia, compared to nearby dry tropical forest, which the authors relate to human disturbance. Also, two ground-dwelling sympatric couas species that occupy the dry forest of Madagascar are insectivores that feed at ground level but differ in size, and logging affects their foraging via changing their ability to climb in the vegetation (Chouteau 2009). Furthermore, the two species responded differentially to burning, likely because of changes in microhabitat variables relating to the foraging sites of the two species as related to stem density, understory vegetation, cover, and prey detectability (Choutaeu 2007).

Andresen (2005) documented the effects of both rainfall seasonality and forest structure on the dung beetle community of a Mexican tropical dry forest. This study supports the concept idea that dung beetle communities are useful bioindicators of habitat disturbance for dry tropical forest, as already has been observed in tropical rain forests.

Zelikova and Breed (2008) measured how ant communities affect seed removals and seed dispersal distances for *Acacia collinsii* and papaya seeds in tropical dry forests of Guanacaste Province, Costa Rica, to determine if habitat disturbance affects ant community composition and associated interactions between ants and seeds. The forests differed with regard to land-use and disturbance history. Habitat disturbance affected the ant community and the relative abundance of key seed-dispersing ant species, with consequential effects on seed removals and seed dispersal distances.

Furthermore, Achury et al. (2008) found that disturbance increases the probability of dominance by the little fire ant, *Wasmannia auropunctata* (Formicidae: Myrmicinae), an indicator species of low-diversity ant communities in tropical dry forest fragments of the Cauca River Valley, Colombia. At each of four locations, three biotopes were distinguished: forest interior, forest margin, and the surrounding matrices (sugar cane or pasture). Competition ability indices were calculated for the 66 ant species in 30 genera that were attracted to bait traps. Ant species composition was found to be primarily influenced by disturbance conditions.

Suaz-Ortuno et al. (2008) documented impacts of forest conversion to agricultural mosaic on 18 anuran, 18 lizard, 23 snake, and 3 turtle species assemblages of neotropical dry forests. The assemblages each responded differently to disturbance. The species richness, diversity, and abundance of lizards were higher in disturbed forests. Anuran diversity and species richness were lower in disturbed forest, and abundance was similar in both disturbance types. The diversity, richness, and abundance of turtles were lower in disturbed forests, whereas the structure and composition of snake assemblages did not differ between forest conditions.

Finally, Suaz-Ortuno et al. (2007) examined the diet of the marbled toad (*Bufo marmoreus*) in conserved and disturbed tropical dry forest on the coast of Jalisco, Mexico. During 2000 and 2001, toads in the conserved area consumed greater proportions of ants (36.7 percent by volume), whereas toads in disturbed forest consumed greater proportions of beetles (53.1 percent by volume). Furthermore, the diversity of diet was significantly lower in the disturbed area. However, prey availability was not affected by disturbance because the abundance, size, and weight of the toads were similar in both areas.

EFFECTS OF DISTURBANCE ON ECOSYSTEMS

Disturbances, both human-induced and natural, shape forest systems by influencing their composition, structure, and functional processes. Indeed, forests throughout the world are molded by their land-use and disturbance history. Natural disturbances having the greatest effects on forests include fire, drought, introduced species, insect and pathogen outbreaks, hurricanes and monsoons and other wind storms, landslides, and volcanoes (Dale et al. 1998). Each disturbance affects forests differently. Large disturbances such as hurricanes can result in massive tree mortality (see Van Bloem et al. 2005; Imbert and Portecop 2008), and most disturbances to dry tropical forests affect tree-size distribution (see Poorter et al. 2008; Chazdon et al. 2007).

Forest disturbance in seasonally dry tropical forests can lead to changes in ecosystem properties. In areas of forest disturbance of monsoon Asia, carbon fluxes closely track seasonal water availability (Huete et al. 2008). In addition, dipterocarp rain forest in southwest China, which has a species composition similar to that of forests in tropical Asia, has higher seed storage subsequent to intense disturbance and forest fragmentation (Tang et al. 2006). As another example, Sagar and Singh (2005) studied the impact of disturbance on the diversity patterns, forest structure, and regeneration of tree species in the Vindhyan dry tropical forests of India based on more than 1,500 quadrats spread over five sites that varied in disturbance level. Alpha diversity increased with decreasing disturbance intensity. Both the level of disturbance and the nature of the species strongly affect regeneration (as revealed in the regression analysis).

Forest disturbances influence how much carbon is stored in trees, dead wood, and soils. Abril et al. (2005) examined soil balance (CO_2 production as affected by soil and litter organic carbon) in areas under varying degrees of disturbance by overgrazing and fire in the Dry Chaco woodland, Argentina. Under undisturbed conditions, soil carbon was relatively constant throughout the year (range: 23–24 g/kg), whereas both burned and grazed sites experienced carbon values fluctuating markedly (range: 21–31 g/kg). In non-grazed, burned areas, soil carbon content increased 16 percent, whereas at burned and overgrazed sites, it decreased 38 percent. Hence overgrazing has a more-significant, adverse effect on soil carbon balance than does fire when both factors act separately. The apparent reason for this pattern is that burned but non-grazed areas appear to slowly recover their initial carbon balance, whereas in chronically overgrazed sites there is a tendency to permanent carbon loss.

Human disturbance of dry tropical forests can be intense. Between 1880 and 1980, the area of cultivated land increased by 106 M ha in South and Southeast Asia (Richards and Flint 1994). Human population density and livestock density in these forest systems have also tripled in the past century (Table 1). Yet because of the uncertainty of estimating carbon storage and forest density both in 1880 and for recent times, the exact magnitude of these changes is not entirely apparent (Table 2). Even so, there have been strong declines in carbon storage over thirteen countries in South and Southeast Asia (Flint and Richards 1994) during the past century. Combined with natural disturbances, these human-induced changes have greatly altered the ecosystem services provided by seasonally dry tropical forests in Asia. Furthermore, the changes in human population over recent decades have been even greater than in the past (Table 1), but their effects on livestock density, forest area, and carbon content are not recorded in a way that allows comparisons to estimates over the past century. Even so, the decline in forest carbon content and forest area is undoubtedly increasing—likely at a steep rate.

A critical effect of disturbances in dry tropical forest is change in water flow. Bruijnzeel (2004) found that in Southeast Asia the effects of forest disturbance and

Table 1. Estimated human and livestock densities and total populations for 13 South Asian nations (India, Bangladesh, Sri Lanka, Myanmar, Thailand, Cambodia, Laos, Vietnam, Malaysia, Brunei, Singapore, Indonesia, and the Philippines). Countries have a total area of 791.3 million ha.

	Year				
Attribute	1880	1920	1950	1980	2008
Human population density (millions/ha)	0.39	0.50	0.74	1.44	4.93
Livestock density in natural vegetation (millions/ha)	0.37	0.47	0.66	1.04	2.56
Total human population (millions)	310.4	397.3	589.3	1,140.5	3,908.2
Total livestock population (in natural vegetation, millions)	221.1	271.1	360.4	507.9	731.9

Source: Richards and Flint (1994) for 1880–1980; United Nations estimates for 2008.

Table 2. Carbon in the vegetation and soils of seasonally dry tropical forests.

Attribute	Value
Area of forest[a] (ha; in millions)	31
Carbon in vegetation[b] (tC/ha)	
High estimate	60
Low estimate	40
Carbon in soils[b] (tC/ha)	50
Estimated biomass density for undisturbed seasonally dry forests of tropical Asia[c] (Mg/ha)	
Ecologically based estimates from global land-use models:	
Potential biomass	120
Model projections for 1880	97
Model projection for 1980	84
Geographic information system–based estimates:	
Potential biomass	215
Model projections for 1880	150
Model projection for 1980	80

Note: Based on information in Houghton and Hackler (1994).
[a]Houghton et al. 1985, 1987; Palm et al. 1986.
[b]Food and Agriculture Organization / United Nations Environment Programme 1981.
[c]Dale et al. 1994.

conversion from undisturbed to disturbed forest on rainfall infiltration and subsequent water flow are less than those for conversion from forest to grassland because, after disturbance, the radiative properties of the secondary forest are similar to those of the original forest. Actual amounts of annual water yield differ between sites and years because of differences in rainfall and degree of surface disturbance, with the greatest yield associated with complete forest clearing. However, with limited surface disturbance, most increases in water yield occurs at times of low flows, and if groundwater reserves are not replenished during the wet season, declines in dry-season flows may be extreme. Bruijnzeel's survey (2004) of more than sixty catchment sediment yield studies from Southeast Asia demonstrates the significant effects of common forest disturbances, such as selective logging and clearing for agriculture, plantation creation, urbanization, mining, and road construction. He finds that the "low-flow problem" is a critical watershed issue that requires further research to address the time delay between upland soil conservation measures and potential changes in downstream sediment yield, particularly at increasing distances from the disturbance. The net effect on forests of changes in water storage include the potential for vegetation dieback and species shifts, although the high diversity in these tropical dry forests means that disturbances are not a key factor in maintaining alpha diversity (Leigh et al. 2004).

In contrast, however, sal (*Shorea robusta* Gaertn. f.) forest occurs in Nepal as part of dry deciduous systems typical at elevations of less than 1,000 m. Webb and Sah (2003) found that the disturbance associated with forest management and regeneration of the sal forest in Nepal resulted in a severe decline in botanical diversity and robust regeneration of *Shorea*. Even though total tree density was high, all species except *S. robusta* declined in abundance in managed forest after twenty years of protected regeneration. Although successional sal forest recovered most of its species richness in twenty years, most tree species were in lower abundances. Thus Webb and Sah (2003) call for a full analysis of forest "value" to establish optimal land-use practices. It should be noted that rainfall and past disturbances (fire and anthropogenic use) are mainly responsible for different community types of sal forest in lowland Nepal.

Virtually all dry tropical forests, whether at stand or regional scale, are in some state of seral succession because of more or less occasional landscape-scale catastrophes (Ashton 1993). Sagar et al. (2003) found that the disturbance gradient explained both tree diversity and density in a dry tropical forest in India. They compared community composition and species diversity of the understory vegetation among five dry tropical forest sites in northern India using data from 1,500 quadrats distributed over 15 one-ha permanent plots differing in degree of disturbance. Both human disturbance intensity as well overall disturbance regimes combined with the soil's water-holding capacity to explain understory composition and organization. Understory species diversity decreased with increasing human disturbance intensity, reflecting utilization pressure and decreased soil fertility. The high correlation between overstory and understory tree diversity suggests that understory trees may soon replace the overstory.

Most landscapes are a mosaic of historical catastrophic events, which create a diversity of tree ages and species and follow the intermediate disturbance hypothesis that local species diversity is at a maximum at an intermediate level of disturbance (Connell 1979). Bongers et al. (2009) tested this hypothesis in tropical forests using data from 2,504 one-ha plots and found that it explains more variation in the diversity of seasonally dry rather than wet tropical forests because early-successional species diversity increases with disturbance, shade-tolerant species decrease in number, and intermediate species diversity does not change. In summary, Connell (1979) is right: intermediate disturbance (toward but not at the low end) favors tree species diversity and, therefore, insect, microorganism, and overall biodiversity.

CLIMATE INFLUENCES ON FOREST DISTURBANCES

Forest disturbances are influenced by climate and in some cases might be exacerbated by climate change. The effects on each type of disturbance are partly tempered by prior adaptations. The species present in a dry tropical forest often reflect past disturbances. Drought-prone sites typically support species that survive well under dry conditions with variable rates of annual rainfall. Thus, if climate change alters the distribution, extent, frequency, or intensity of any of these disturbances, large impacts (such as loss of species regeneration) could be expected. The effects on species or communities already at the margin of their range may be particularly severe. For example, a synergistic convergence of climatic, technological, and socioeconomic factors explains the loss of the dry tropical Chaco forests of central Argentina (Zak et al. 2008). During the past three decades of the twentieth century, about 80 percent of an area that was originally undisturbed forest became occupied by crops, pastures, and secondary scrub.

As another example, Poorter et al. (2008) compared size distribution of tree species in forty-four Ghanaian tropical forest communities based on data from 880 one-ha plots and more than 118,000 trees of 210 species. The variation in the size distribution of these forest species is continuous and log-normal and related to both rainfall and disturbance. Size distributions in dry forests were more skewed than those in wet forests, with more tall species and a wider height range.

Finally, analysis of palaeoecological records for the Amazon by Mayle and Power (2008) suggests the forests in most of Amazonia appear to have been resilient to climatic conditions that were significantly drier than they are today, even though there is widespread evidence of forest burning. The authors surmise that the great spatial and temporal variation in patterns of vegetation disturbance and fire in the Amazon Basin likely reflect complex heterogeneity in precipitation and seasonality patterns across the basin, as well as interactions among climate change, drought and fire susceptibility of the forests, and Palaeo-Indian land use.

INTERACTIONS AMONG DISTURBANCES

Many disturbances are cascading. Drought often weakens tree vigor, leading to insect infestations, disease, or fire. Insect infestations and disease promote future fires by increasing fuel loads, and fires promote future infestations by compromising tree defenses. Increased fire intensity or extent can enhance the potential for landslides. Also, changes in land use, forest management, and atmospheric chemistry can interact with these natural disturbances. Agriculture and road establishment in landslide-prone areas, coupled with increased wetness, can result in more landslides. In some cases, however, a combination of disturbances may ameliorate impacts. Under drought-prone conditions, stomata tend to close, reducing the effects of high ozone exposure.

Zimmer and P. Baker (2009) found statistical evidence of synchronous recruitment at stand and site scale, but not at a regional scale, for seasonal tropical pine forests of northern Thailand. Recruitment success was often related to favorable climatic conditions, such as multiyear periods of cool and wet dry seasons. Nevertheless, there are clearly complex interactions among local disturbance history, regional climate variability, and pine recruitment (Zimmer and P. Baker 2009).

When ecosystems experience more than one disturbance, the compounded effects can lead to new ecological states (Paine et al. 1998). A new ecological state is entered when the system has not recovered from the first disturbance before a second perturbation occurs, leading the system to a new long-term condition. For instance, a combination of climatically driven wildfires, fragmentation caused by agricultural settlement, and logging often results in significant and unprecedented changes in forest composition (see Weir 1996). Invasive nonnative species are sometimes able to modify existing disturbance regimes or introduce entirely new disturbances (Mack and D'Antonio 1998). Under climate change, these compounded interactions may be unprecedented and unpredictable. They are likely to appear slowly and be difficult to detect because trees live for so long.

Some disturbances can be a function of both natural and human conditions. On the slopes of the Javan volcanoes of Indonesia, where disturbances are endemic, people opportunistically grow crops in response to volcanic activity and El Niño–related forest fires or longer climatic anomalies, while land degradation from state-organized deforestation is often buried by new volcanic activity (Lavigne and Gunnell 2006). Such responses are also seen in Latin American dry forests that lie in the shadow of volcanoes.

At the current time, a human disturbance to dry tropical forests that appears to be on the rise is the establishment of bioenergy crops in areas that were previously forests (Lim 2001). It is not clear from remote-sensing studies how much forest degradation or clearing occurred prior to the establishment of the bioenergy crops. In some cases it may be that establishment of commercial perennial bioenergy crops on land that had previously been used for slash-and-burn agriculture may actually provide a sustainable lifestyle that effectively reduces pressure for deforestation (Kline and Dale 2008). Developing policies that reduce pressure for forest degradation or clearing requires a broad-scale perspective and consideration of the role of the diversity of human activities and forest-ecosystem benefits. As an example, Urquiza-Haas et al. (2007) assessed the effects of combined disturbances on forest basal area, stand-level wood specific gravity, and aboveground biomass of forests of the Yucatan, where the predominant land use is still slash-and-burn agriculture. Intensive logging and fire reduced aboveground biomass in late-successional plots (30–50 years) by 36 percent and 37 percent, respectively.

In addition, Imbert and Portecop (2008) studied forest recovery in old-growth dry forest subsequent to Hurricane Hugo on the calcareous island of Grande-Terre, Guadeloupe, which has experienced several centuries of anthropogenic disturbances, as well as natural, catastrophic disturbances such as hurricanes. The authors measured floristic composition, litterfall, stem density, basal area, and tree growth for trees ≥ 10 cm dbh in a set of 2,400 m^2 permanent inventory plots from 1987 to 1998. The hurricane disturbance in these dry forests affected the forest structure on a longer time-scale than the functioning of the forest ecosystem.

The effects of livestock grazing, fuelwood extraction, and burning on vegetation in Bandipur National Park in southern India were analyzed using data from two hundred sites in four watersheds spanning a degradation gradient (Mehta et al. 2008). Vegetation plot height, canopy cover, and tree diameter were negatively correlated with field disturbance, resulting in stunted forest stature in degraded sites. Vegetation composition in degraded watersheds was dominated by small-stature woody tree species and a greater diversity of shrub species.

Laurance et al. (2006) report that even moderate hunting pressure can markedly alter the structure of mammal communities in central Africa, based on the authors' studies in southern Gabon. Roads also reduced the abundance of some species, with avoidance of roads being stronger where hunting was permitted outside wildlife reserves.

MANAGEMENT NEEDS RELATED TO DISTURBANCES

Management plans need to include the potential for disturbances. Infrequent disturbances may not be included in planning because people's perceptions are most influenced by what has occurred during their lifetimes (Christensen et al. 1989). Management goals typically do not consider what aims are appropriate in the face of disturbance, but they should identify the components, time frame, and spatial resolution needed to achieve specific goals (Rogers 1997).

The realization of the integral nature of disturbance alters the way we consider management of disturbance-prone systems in that both disturbances and their management are often part of the overall management system rather than a result and response subsequent to a particular disturbance. Sagar and Singh (2006) documented the relationships between forest basal area and diversity components (number of species and evenness) for a disturbed dry tropical forest of northern India. Recurring disturbances concentrated biomass or stems in only a few strong competitors. The conservation activities the authors deemed to be important were management of fuelwood plantations near human settlements, deferred grazing, and canopy enrichment through multispecies plantations of desirable species within the forest patches of low basal area. As might be expected, small, frequent disturbances are easier to study, understand, and incorporate into planning and management than are large or infrequent disturbances. Aspects of silviculture are designed to replicate small disturbances in forests, yet large disturbances are seldom a part of forest management plans (W. Baker 1992).

When disturbances do occur, they may trigger large departures from average events, and these changes can result in an outcry from the media and the public for immediate and costly action to quickly return the situation to seemingly normal conditions (Schullery 1989). Yet the costs and benefits of management actions are assessed at relatively small temporal and spatial scales. Planning for response to disturbances typically does not consider the length of time or full cost required for repair to occur naturally (Dale et al. 1998).

For management purposes, the conditions that lead to or result from disturbances need to be understood so that alternatives for manipulation can be developed when possible. Our poor understanding of some disturbances emphasizes the need for simultaneous programs of system management and research. Ideally, active adaptive-management efforts (Walters and Hilborn 1978) could be focused on systems where a large disturbance is either imminent or recent. Such a focus would allow evaluation of the implications of management actions. Management plans may focus on the potential for managing the system prior to the disturbance, the disturbance itself, the system immediately after the disturbance, or the recovery process. Sabogol (1992) found that management strategies for natural forests in the tropical dry zones of Nicaragua could focus on encouragement of advanced growth of desirable tree species, inducement of natural regeneration, coppice management, and compensatory planting as a way to avoid the disturbance created by traditional utilization.

In managing the system prior to disturbance, actions should be designed so that the system responds to disturbances in ways that do not compromise management goals. Therefore, it is useful to think of potential management options in view of a conceptual framework for management of potential disturbance to a system (Figure 1). The seasonally dry tropical forest system can be manipulated to alter its vulnerability, its resistance, or its response to a disturbance. An example of changing a system's vulnerability is the alteration of fuel loads in a forest so that fire is more or less intense when it occurs. Resistance can be increased at the location in the system where the agent first exerts the disturbing force (e.g., through fire breaks, spraying for insects, or planting windbreaks). Then the disturbing force can still occur, but the system will not be altered to the extent it would have been without management intervention. Maxwell (2004) argues for consideration of the importance of indigenous land-use customs as a part of forest management and biological conservation.

The disturbance itself can be managed in one of two ways: no action, or prevention and thus some control of the disturbance. Action should not be taken when nothing can be done about preventing the disturbance (as in the case of many volcanoes and hurricanes), or when one can accept the consequences of the event. Prevention is usually motivated by the desire to satisfy or protect a human concern. Managers attempt to prevent disturbance when the costs of such efforts are perceived to be balanced or exceeded by the anticipated benefits. As examples, dams, levees, reservoirs, and coastal barriers are implemented to prevent disturbances caused by water action.

Managing the system after the disturbance requires recognition of the potential for natural processes of recovery. Sometimes management actions have been both costly

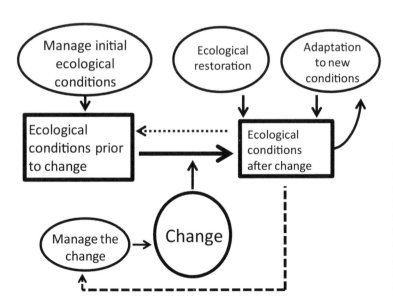

Figure 1. Diagram of how managers can deal with disturbances to seasonally dry tropical forests. As indicated by the dashed line, over time future changes as well as changing ecological conditions may alter these managed systems.

and detrimental to the natural recovery process. An example of such management is that, after Hurricane Hugo hit Puerto Rico in 1989, stream channels in the Luquillo Experimental Forest were filled with trees knocked down by the high winds. Some of the logjams were removed, but those left in place provided the environmental benefit of greater cover and food for stream organisms and retained valuable nutrients, sediments, and organic matter (Covich et al. 1991; Covich and McDowell 1996; Covich and Crowl 1990).

Because managing recovery can be thought of as managing succession, knowledge of natural succession processes increases the likelihood that the management plan will be achievable (see Luken 1990). The first step in managing recovery is evaluating the site potential, which includes the spatial pattern of the disturbance effects, the biological and physical residuals of the disturbance, the site's environmental conditions, the potential for propagule rain to aid in the recovery, and the influence of structural heterogeneity on the reestablishment of plants and animals. A common goal of recovery management is to shorten the process of succession or to maintain the process of succession at one particular state that is considered desirable for human purposes.

In developing recovery plans for seasonally dry tropical forests, managers should focus on critical stages of the successional process (MacMahon 1987; Luken 1990): migration, establishment, biotic interactions (such as competition, predation, and mutualism), and the reaction of the systems to the changes that occur over successional time. For each successional stage, a management prescription can be developed that offers three alterative interventions: add something to the system, kill or remove a biological component, or alter the structure of the system. Alternatively, one could physically remove existing propagules from the system, exclude animals that might eat the propagules, or remove the dispersal agent of undesirable seeds.

Successional trajectories after a disturbance event can follow a variety of directions (McCune and Allen 1985), but in many cases they appear to culminate in functional communities similar in structure and composition to the original ones. Alternatively, the successional trajectory can involve novel or unfamiliar seral communities when a disturbance sets succession back to early or uncommon stages. For example, after Hurricane Hugo in Puerto Rico, stands in the Luquillo Experimental Forest were invaded by vines and herbaceous species not previously observed at that site in more than fifty years of research (Scatena et al. 1996). Yet this early stage in the recovery of the forest quickly gave way to more-familiar successional trees, such as *Cecropia peltata*. Seven years after the hurricane, vines and herbaceous vegetation had assumed their previous low abundance, yet their dominance after the hurricane was critical in the immobilization of nutrients released by the disturbance and in the reestablishment of forest conditions (Scatena et al. 1996). Thus, while not always predictable, succession events after a disturbance often shape the reestablishment process.

CONCLUSIONS

The research on disturbances to seasonally dry tropical forests has been limited because of the difficulty of studying these long-lived systems and the prevailing cumulative human influences on these forests. A key research need identified by Bruijnzeel (2004) is including the role of underlying geological controls of catchment hydrological behavior when analyzing the effect of land-use change on (low) flows or sediment production. Baker et al. (2005) emphasize the importance of understanding the role of legacies from prior disturbances in changes to these dry tropical forests. Clearly, the interactions of human activities with forest disturbances and changes thereafter will continue to be a critical part of understanding and managing seasonally dry tropical forests. There is also a need to characterize disturbance rates, intensities, and extents. Much of the current theory on how forest systems respond to disturbances was developed in temperate systems; there needs to be more testing and data collection to examine how these theories play out in seasonally dry tropical forests.

ACKNOWLEDGMENTS

Peter Ashton provided useful comments on an earlier draft of these ideas. Frederick O'Hara did an excellent job of editing the manuscript. Marcia Davitt, William Mc-Shea, and anonymous reviewers provided many useful comments on the manuscript. This research was supported by the U.S. Department of Energy (DOE) under the Office of the Biomass Program. Oak Ridge National Laboratory is managed by UT-Battelle, LLC, for the U.S. Department of Energy under contract DE-AC05-00OR22725.

REFERENCES

Abril, A. P., and E. H. Bucher. 2005. The Effect of Fire and Overgrazing Disturbes on Soil Carbon Balance in the Dry Chaco Forest. *Forest Ecology and Management* 206:339–405.

Achury, R., P. Chacon de Ulloa, and Á. M. Arcila. 2008. Ant Composition and Competitive Interactions with *Wasmannia auropunctata* in Tropical Dry Forest Fragments. *Revista Colombiana De Entomologia* 34:209–16.

Aguilar-Fernandez, M., V. J. Jaramillo, L. Varela-Fregoso, and M. E. Gavito. 2009. Short-Term Consequences of Slash-and-Burn Practices on the Arbuscular Mycorrhizal Fungi of a Tropical Dry Forest. *Mycorrhiza* 19:179–86.

Alvarez-Yepiz, J. C., A. Martinez-Yrizar, A. Burquez, and C. Lindquist. 2008. Variation in Vegetation Structure and Soil Properties Related to Land Use History of Old-Growth and Secondary Tropical Dry Forests in Northwestern Mexico. *Forest Ecology and Management* 256:355–66.

Andresen, E. 2005. Effects of Season and Vegetation Type on Community Organization of Dung Beetles in a Tropical Dry Forest. *Biotropica* 37:291–300.

Anitha, K., S. Joseph, E. V. Ramasamy, and S. N. Prasad. 2009. Changes in Structural Attributes of Plant Communities along Disturbance Gradients in a Dry Deciduous Forest of Western Ghats, India. *Environmental Monitoring and Assessment* 155:393–405.

Arredondo-León, C., J. Muñoz-Jiménez, and A. García-Romero. 2008. Recent Changes in Landscape-Dynamics Trends in Tropical Highlands, Central Mexico. *Interciencia* 33:569–77.

Ashton, P. S. 1993. The Community Ecology of Asian Rain Forests, in Relation to Catastrophic Events. *Journal of Biosciences* 18:501–14.

Ashton, P. S., and P. Hall. 1992. Comparisons of Structure among Mixed Dipterocarp Forests of North-Western Borneo. *Journal of Ecology* 80:459–81.

Baker, P. J., S. Bunyavejchewin, C. D. Oliver, and P. S. Ashton. 2005. Disturbance History and Historical Stand Dynamics of a Seasonal Tropical Forest in Western Thailand. *Ecological Monographs* 75:317–43.

Baker, W. L. 1992. The Landscape Ecology of Large Disturbances in the Design and Management of Nature Reserves. *Landscape Ecology* 7:181–94.

Bergeron, Y., and S. Archambault. 1993. Decreasing Frequency of Forest Fires in the Southern Boreal Zone of Québec and its Relation to Global Warming since the End of the "Little Ice Age." *Holocene* 3:255–59.

Bongers, F., L. Poorter, W. D. Hawthorne, and D. Sheil. 2009. The Intermediate Disturbance Hypothesis Applies to Tropical Forests, but Disturbance Contributes Little to Tree Diversity. *Ecology Letters* 12:798–805.

Bormann, E. H., and G. E. Likens. 1979. *Pattern and Process in a Forested Ecosystem*. New York: Springer-Verlag.

Brandeis, T. J., and C. W. Woodall. 2008. Assessment of Forest Fuel Loadings in Puerto Rico and the U.S. Virgin Islands. *Ambio* 37:557–62.

Bruijnzeel, L. A. 2004. Hydrological Functions of Tropical Forests: Not Seeing the Soil for the Trees? *Agriculture Ecosystems & Environment* 104:185–228.

Chazdon, R. L., S. G. Letcher, M. van Breugel, M. Martínez-Ramos, F. Bongers, and B. Finegan. 2006. Rates of Change in Tree Communities of Secondary Neotropical Forests following Major Disturbances. *Philosophical Transactions of the Royal Society B—Biological Sciences* 362:273–89.

Chouteau, P. 2007. The Impact of Burning on the Microhabitat Used by Two Species of Couas in the Western Dry Forest of Madagascar. Ostrich 78:43–49.

———. 2009. Impact of Logging on the Foraging Behaviour of Two Sympatric Species of Couas (*Coua coquereli* and *Coua gigas*) in the Western Dry Forest of Madagascar. *Comptes Rendus Biologies* 322:567–78.

Christensen, N. L., J. K. Agee, P. F. Brussard, J. Hughes, D. H. Knight, G. W. Minshall, J. M. Peek, S. J. Pyne, F. J. Swanson, B. F. Thomas, S. Wells, S. E. Williams, and H. A. Wright. 1989. Interpreting the Yellowstone Fires of 1988. *BioScience* 39:678–85.

Connell, J. H. 1979. Intermediate-Disturbance Hypothesis. *Science* 204:1345.

Covich, A. P., and T. A. Crowl. 1990. Effects of Hurricane Storm Flow on Transport of Woody Debris in a Rain Forest Stream (Luquillo Experimental Forest, Puerto Rico). In *Tropical Hydrology and Caribbean Water Resources*, ed. J. H. Krishna, V. Quiñones-Aponte, and F. Gomez-Gomez, 197–205. San Juan, Puerto Rico: American Water Resources Association.

Covich, A. P., T. A. Crowl, S. L. Johnson, D. Varza, and D. L. Certain. 1991. Post-Hurricane Hugo Increases in Atyid Shrimp Abundance in a Puerto Rican Montane Stream. *Biotropica* 23:448–54.

Covich, A. P., and W. H. McDowell. 1996. The Stream Community. In *The Food Web of a Tropical Rain Forest*, ed. D. P. Reagan and R. B. Waide, 434–59. Chicago: University of Chicago Press.

Dale, V. H., S. Brown, E. P. Flint, C. A. S. Hall, L. R. Iverson, and J. Uhlig. 1994. Estimating CO_2 Flux from Tropical Forests. In *Effects of Land Use Change on Atmospheric CO_2 Concentrations: Southeast Asia as a Case Study*, ed. V. H. Dale, 365–74. New York: Springer-Verlag.

Dale, V. H., A. Lugo, J. MacMahon, and S. Pickett. 1998. Ecosystem Management in the Context of Large, Infrequent Disturbances. *Ecosystems* 1:546–57.

Denslow, J. S. 1980. Patterns of Plant Species Diversity during Succession under Different Disturbance Regimes. *Oecologia* 46:18–21.

Farwig, N., C. Braun, and K. Boehning-Gaese. 2008. Human Disturbance Reduces Genetic Diversity of an Endangered Tropical Tree, *Prunus africana* (*Rosaceae*). *Conservation Genetics* 9:317–26.

Flannigan, M. D., Y. Bergeron, O. Engelmark, and B. M. Wotton. 1998. Future Wildfire in Circumboreal Forests in Relation to Global Warming. *Journal of Vegetation Science* 4:469–76.

Flannigan, M. D., B. J. Stocks, and B. M. Wotton. 2000. Climate Change and Forest Fires. *Science of the Total Environment* 262:221–29.

Flint, E. P., and J. F. Richards. 1994. Trends in Carbon Content of Vegetation in South and Southeast Asia Associated with Changes in Land Use. In *Effects of Land Use Change on Atmospheric CO_2 Concentrations: Southeast Asia as a Case Study*, ed. V. H. Dale, 201–300. New York: Springer-Verlag.

Food and Agriculture Organization / United Nations Environment Programme. 1981. *Tropical Forest Resource Assessment Project: Forest Resources of Tropical Asia*. Rome: Food and Agriculture Organization.

Hanson, P. J., and J. F. Weltzin. 2000. Drought Disturbance from Climate Change: Response of United States Forests. *Science of the Total Environment* 262:205–20.

Houghton, R. A., R. D. Boone, J. R. Fruci, J. Hobbie, J. M. Melillo, C. A. Palm, B. J. Peterson, G. R. Shaver, G. M. Woodwell, and B. Moore. 1987. The Flux of Carbon from Terrestrial Ecosystems to the Atmosphere in 1980 Due to Changes in Land Use: Geographic Distribution of the Global Flux. *Tellus* 39B:122–39.

Houghton, R. A., R. D. Boone, J. M. Melillo, C. A. Palm, G. M. Woodwell, N. Myers, B. Moore, and D. Skole. 1985. Net Flux of CO_2 from Tropical Forests in 1980. *Nature* 316:617–20.

Houghton, R. A., and J. L. Hackler. 1994. The Net Flux of Carbon from Deforestation and Degradation in South and Southeast Asia. In *Effects of Land Use Change on Atmospheric CO_2 Concentrations: Southeast Asia as a Case Study*, ed. V. H. Dale, 301–28. New York: Springer-Verlag.

Huete, A. R., N. Restrepo-Coupe, P. Ratana, K. Didan, S. R. Saleska, K. Ichii, S. Panuthai, and M. Gamo. 2008. Multiple Site Tower Flux and Remote Sensing Comparisons of Tropical Forest Dynamics in Monsoon Asia. *Agricultural and Forest Meteorology* 148:748–60.

Imbert, D., and J. Portecop. 2008. Hurricane Disturbance and Forest Resilience: Assessing Structural vs. Functional Changes in a Caribbean Dry Forest. *Forest Ecology and Management* 255:3494–3501.

Janssens, M. L. 2000. *Introduction to Mathematical Fire Modeling*, 2nd ed. Boca Raton, FL: CRC Press.

Kennard, D. K., and F. E. Putz. 2005. Differential Responses of Bolivian Timber Species to Prescribed Fire and Other Gap Treatments. *New Forests* 30:1–20.

Kline, K. L., and V. H. Dale. 2008. Biofuels: Effects on Land and Fire. *Science* 321:199–200.

Kozlowski, T. T., and S. G. Pallardy. 2002. Acclimation and Adaptive Responses of Woody Plants to Environmental Stresses. *Botanical Review* 68:270–334.

Laurance, W. F., B. M. Croes, L. Tchignoumba, S. A. Lahm, A. Alonso, M. E. Lee, P. Campbell, and C. Ondzeano. 2006. Impacts of Roads and Hunting on Central African Rainforest Mammals. *Conservation Biology* 20:1251–61.

Lavigne, F., and Y. Gunnell. 2006. Land Cover Change and Abrupt Environmental Impacts on Javan Volcanoes, Indonesia: A Long-Term Perspective on Recent Events. *Regional Environmental Change* 6:86–100.

Leach, M. K., and T. J. Givnish. 1996. Ecological Determinants of Species Loss in Remnant Prairies. *Science* 273:1555–58.

Leigh, E. G., Jr., P. Davidar, C. W. Dick, J.-P. Puyravaud, J. Terborgh, H. ter Steege, and S. J. Wright. 2004. Why Do Some Tropical Forests Have So Many Species of Trees? *Biotropica* 36:447–73.

Lim, K. O. 2001. Biomass—A Fuel with a Bright Future, ed. K. Sopian, M. Y. H. Othman, S. A. Rahman, and S. Shaari. *Advances in Malaysian Energy Research* 2000:255–62.

Lugo, A. E., E. Medina, J. C. Trejo-Torres, and E. Helmer. 2006. Botanica and Ecologica Basis for the Resilience of Antillean Dry Forests. In *Neotropical Savannas and Seasonally Dry Forests: Plant Diversity, Biogeography, and Conservation*, ed. R. T. Pennington, G. P. Lewis, and J. A. Ratter, 359–81. Systematics Association Special Volume Series. Boca Raton FL: CRC Press.

Lugo, A. E., and F. N. Scatena. 1996. Background and Catastrophic Tree Mortality in Tropical Moist, Wet, and Rain Forests. *Biotropica* 28:585–99.

Luken, J. O. 1990. *Directing Ecological Succession.* London: Chapman and Hall.

Mack, M. C., and C. M. D'Antonio. 1998. Impacts of Biological Invasions on Disturbance Regimes. *Trends in Ecology and Evolution* 13:195–98.

MacMahon, J. A. 1987. Disturbed Lands and Ecological Theory: An Essay about a Mutualistic Association. In *Restoration Ecology*, ed. W. R. Jordan, M. E. Gilpin, and J. D. Aber, 221–37. Cambridge: Cambridge University Press.

Marod, D., U. Kutintara, C. Yarwudhi, H. Tanaka, and T. Nakashizuka. 1999. Structural Dynamics of a Natural Mixed Deciduous Forest in Western Thailand. *Journal of Vegetation Science* 10:777–86.

Martini, A. M. Z., R. A. F. Lima, G. A. D. C. Franco, and R. R. Rodrigues. 2008. The Need for Full Inventories of Tree Modes of Disturbance to Improve Forest Dynamics Comprehension: An Example from a Semideciduous Forest in Brazil. *Forest Ecology and Management* 255:1479–88.

Maxwell, A. L. 2004. Fire Regimes in North-Eastern Cambodian Monsoonal Forests, with a 9300-year Sediment Charcoal Record. *Journal of Biogeography* 31:225–39.

Mayle, F. E., and M. J. Power. 2008. Impact of a Drier Early-Mid-Holocene Climate upon Amazonian Forests. *Philosophical Transactions of the Royal Society B—Biological Sciences* 363:1829–38.

McCune, B., and T. F. H. Allen. 1985. Will Similar Forests Develop on Similar Sites? *Canadian Journal of Botany* 63:367–76.

McLaren, K. P., M. A. McDonald, J. B. Hall, and J. R. Healey. 2005. Predicting Species Response to Disturbance from Size Class Distributions of Adults and Saplings in a Jamaican Tropical Dry Forest. *Plant Ecology* 181:69–84.

Mehta, V. K., P. J. Sullivan, M. T. Walter, J. Krishnaswamy, and S. D. DeGloria. 2008. Impacts of Disturbance on Soil Properties in a Dry Tropical Forest in Southern India. *Ecohydrology* 1:161–75.

Miles, L., A. C. Newton, R. S. DeFries, C. Ravilious, I. May, S. Blyth, V. Kapos, and J. E. Gordon. 2006. A Global Overview of the Conservation Status of Tropical Dry Forests. *Journal of Biogeography* 33:491–505.

Moreno, A. R. 2006. Climate Change and Human Health in Latin America: Drivers, Effects, and Policies. *Regional Environmental Change* 6:157–64.

Nakagawa, M., K. Tanaka, T. Nakashizuka, T. Ohkubo, T. Kato, T. Maeda, K. Sato, H. Miguchi, H. Nagamasu, K. Ogino, S. Teo, A. A. Hamid, and L. H. Seng. 2000. Impact of Severe Drought Associated with the 1997–1998 El Niño in a Tropical Forest in Sarawak. *Journal of Tropical Ecology* 16 (3): 355–67.

Negron, J. F. 1998. Probability of Infestation and Extent of Mortality Associated with the Douglas-Fir Beetle in the Colorado Front Range. *Forest Ecology and Management* 107:71–85.

Neilson, R. P., and R. J. Drapek. 1998. Potentially Complex Biosphere Responses to Transient Global Warming. *Global Change Biology* 4:505–21.

Otterstrom, S. M., M. W. Schwartz, and I. Velazquez-Rocha. 2006. Responses to Fire in Selected Tropical Dry Forest Trees. *Biotropica* 38:592–98.

Paine, R. T., M. J. Tegner, and A. E. Johnson. 1998. Compounded Perturbations Yield Ecological Surprises: Everything Else Is Business as Usual. *Ecosystems* 1:535–45.

Palm, C. A., R. A. Houghton, J. M. Melillo, and D. L. Skole. 1986. Atmospheric Carbon Dioxide from Deforestation in Southeast Asia. *Biotropica* 18:177–88.

Poorter, L., W. Hawthorne, F. Bongers, and D. Sheil. 2008. Maximum Size Distributions in Tropical Forest Communities: Relationships with Rainfall and Disturbance. *Journal of Ecology* 96 (3): 495–504.

Potts, M. D. 2003. Drought in a Bornean Everwet Rain Forest. *Journal of Ecology* 91:467–74.

Richards, J. F., and E. P. Flint. 1994. A Century of Land-Use Change in South and Southeast Asia. In *Effects of Land Use Change on Atmospheric CO_2 Concentrations: Southeast Asia as a Case Study*, ed. V. H. Dale, 15–66. New York: Springer-Verlag.

Rogers, K. H. 1997. Operationalizing Ecology under a New Paradigm: An African Perspective. In *The Ecological Basis of Conservation: Heterogeneity, Ecosystems, and Biodiversity*, ed. S. T. A. Pickett, R. S. Ostfeld, M. Shachak, and G. E. Likens, 60–77. New York: Chapman and Hall.

Sabogol, C. 1992. Regeneration of Tropical Dry Forests in Central-America, Examples from Nicaragua. *Journal of Vegetation Science* 3:407–16.

Sagar, R., A. S. Raghubanshi, and J. S. Singh. 2003. Tree Species Composition, Dispersion and Diversity along a Disturbance Gradient in a Dry Tropical Forest Region of India. *Forest Ecology and Management* 186:61–71.

Sagar, R., and J. S. Singh. 2006. Tree Density, Basal Area and Species Diversity in a Disturbed Dry Tropical Forest of Northern India: Implications for Conservation. *Environmental Conservation* 33 (3): 256–62.

Sagar, R., and R. S. Singh. 2005. Structure, Diversity, and Regeneration of Tropical Dry Deciduous Forest of Northern India. *Biodiversity and Conservation* 14:935–59.

Sahu, P. K., R. Sagar, and J. S. Singh. 2008. Tropical Forest Structure and Diversity in Relation to Altitude and Disturbance in a Biosphere Reserve in Central India. *Applied Vegetation Science* 11:461–70.

Sánchez, F., J. Alvarez, C. Ariza, and A. Cadena. 2007. Bat Assemblage Structure in Two Dry Forests of Colombia: Composition, Species Richness, and Relative Abundance. *Mammalian Biology* 72:82–92.

Sapkota, I. P., M. Tigabu, and P. C. Oden. 2009. Spatial Distribution, Advanced Regeneration and Stand Structure of Nepalese Sal (*Shorea robusta*) Forests Subject to Disturbances of Different Intensities. *Forest Ecology and Management* 257:1966–75.

Scatena, F. N., S. Moya, C. Estrada, and J. D. Chinea. 1996. The First Five Years in the Reorganization of Aboveground Biomass and Nutrient Use following Hurricane Hugo in the Bisley Experimental Watersheds, Luquillo Experimental Forest, Puerto Rico. *Biotropica* 28:424–40.

Schullery, P. 1989. The Fires and Fire Policy. *BioScience* 39:686–94.

Stocks, B. J., M. A. Fosberg, T. J. Lynham, L. Mearns, B. M. Wotton, Q. Yang, J. Z. Jin, K. Lawrence, G. R. Hartley, J. A. Mason, and D. W. McKenney. 1998. Climate Change and Forest Fire Potential in Russian and Canadian Boreal Forests. *Climatic Change* 38:1–13.

Suaz-Ortuno, I., J. Alvarado-Diaz, and M. Martinez-Ramos. 2008. Effects of Conversion of Dry Tropical Forest to Agricultural Mosaic on Herpetofaunal Assemblages. *Conservation Biology* 22:362–74.

Suaz-Ortuno, I., J. Alvarado-Diaz, E. Raya-Lemus, and M. Martinez-Ramos. 2007. Diet of the Mexican Marbled Toad (*Bufo marmoreus*) in Conserved and Disturbed Tropical Dry Forest. *Southwestern Naturalist* 52:305–9.

Tang, Y., M. Cao, and X. F. Fu. 2006. Soil Seedbank in a Dipterocarp Rain Forest in Xishuangbanna, Southwest China. *Biotropica* 38:328–33.

Tilman, D. 1989. Ecological Experimentation: Strengths and Conceptual Problems. In *Long-Term Studies in Ecology: Approaches and Alternatives*, ed. G. E. Likens, 136–57. New York: Springer-Verlag.

Timilsina, N., M. S. Ross, and J. T. Heinen. 2007. A Community Analysis of Sal (*Shorea robusta*) Forests in the Western Terai of Nepal. *Forest Ecology and Management* 241:223–34.

Turner, M. G., W. L. Baker, C. J. Peterson, and R. K. Peet. 1998. Factors Influencing Succession: Lessons from Large, Infrequent Natural Disturbances. *Ecosystems* 1:511–23.

Turner, M. G., V. H. Dale, and E. H. Everham. 1997. Fires, Hurricanes, and Volcanoes: Comparing Large Disturbances. *BioScience* 47:758–68.

Urquiza-Haas, T., P. M. Dolman, and C. A. Peres. 2007. Regional Scale Variation in Forest Structure and Biomass in the Yucatan Peninsula, Mexico: Effects of Forest Disturbance. *Forest Ecology and Management* 247:80–90.

Van Bloem, S. J., A. E. Lugo, and P. G. Murphy. 2006. Structural Response of Caribbean Dry Forests to Hurricane Winds: A Case Study from Guanica Forest, Puerto Rico. *Journal of Biogeography* 33:517–23.

Van Bloem, S. J., P. G. Murphy, and A. E. Lugo. 2007. A Link between Hurricane-Induced Tree Sprouting, High Stem Density and Short Canopy in Tropical Dry Forest. *Tree Physiology* 27:475–80.

Van Bloem, S. J., P. G. Murphy, A. E. Lugo, R. Ostertag, M. R. Costa, I. R. Bernard, S. M. Colon, and M. C. Mora. 2005. The Influence of Hurricane Winds on Caribbean Dry Forest Structure and Nutrient Pools. *Biotropica* 37:571–83.

Velázquez, E., and A. Gómez-Sal. 2008. Landslide Early Succession in a Neotropical Dry Forest. *Plant Ecology* 199:295–308.

Vieira, F. D., and D. de Carvalho. 2008. Genetic Structure of an Insect-Pollinated and Bird-Dispersed Tropical Tree in Vegetation Fragments and Corridors: Implications for Conservation. *Biodiversity and Conservation* 17:2305–21.

Walters, C. J., and R. Hilborn. 1978. Ecological Optimization and Adaptive Management. *Annual Review of Ecology and Systematics* 9:157–88.

Weatherhead, P. J. 1986. How Unusual Are Unusual Events? *American Naturalist* 128:150–54.

Webb, E. L., and R. N. Sah. 2003. Structure and Diversity of Natural and Managed Sal (*Shorea robusta* Gaertn.f.) Forest in the Terai of Nepal. *Forest Ecology and Management* 176:337–53.

Weir, J. M. H. 1996. The Fire Frequency and Age Mosaic of a Mixed Wood Boreal Forest. Master's thesis, University of Calgary.

Werneck, M. D., and E. V. Franceschinelli. 2004. Dynamics of a Dry Forest Fragment after the Exclusion of Human Disturbance in Southeastern Brazil. *Plant Ecology* 174:337–46.

Werner, F. A., and S. R. Gradstein. 2009. Diversity of Dry Forest Epiphytes along a Gradient of Human Disturbance in the Tropical Andes. *Journal of Vegetation Science* 20:59–68.

Whelan, R. J. 1995. *The Ecology of Fire.* Cambridge: Cambridge University Press.

White, P. S., and S. T. A. Pickett. 1985. Natural Disturbance and Patch Dynamics: An Introduction. In *The Ecology of Natural Disturbance and Patch Dynamics*, ed. S. T. A. Pickett and P. W. White, 3–13. New York: Academic Press.

Wootton, J. T., M. S. Parker, and M. E. Power. 1996. Effects of Disturbance on River Food Webs. *Science* 273:1558–61.

Zak, M. R., M. Cabido, D. Cáceres, and S. Díaz. 2008. What Drives Accelerated Land Cover Change in Central Argentina? Synergistic Consequences of Climatic, Socioeconomic, and Technological Factors. *Environmental Management* 42:181–89.

Zelikova, T. J., and M. D. Breed. 2008. Effects of Habitat Disturbance on Ant Community Composition and Seed Dispersal by Ants in a Tropical Dry Forest in Costa Rica. *Journal of Tropical Ecology* 24:309–16.

Zimmer, H., and P. Baker. 2009. Climate and Historical Stand Dynamics in the Tropical Pine Forests of Northern Thailand. *Forest Ecology and Management* 257:190–98.

5

Fire Management in South and Southeast Asia's Seasonally Dry Forests

Colonial Approaches, Current Problems, and Perspectives

Kobsak Wanthongchai and Johann Georg Goldammer

Seasonally dry tropical forest ecosystems are prone to fire and have long been subjected to periodic, predominantly short fire-return interval surface fires (Rundel and Boonpragob 1995; Stott et al. 1990). Thus, tree species in these forests are well adapted to fire and exhibit many fire-resistant characteristics, for example, a thick bark, an ability to heal fire scars, a capacity to resprout through coppicing or by means of epicormic shoots from dormant buds and lignotubers, and special seed characteristics (Whelan 1995). Fire regimes, however, are shaped by human-induced fire and other cultural activities. Too-frequent (annual) burning impedes and retards natural regeneration and—in conjunction with overmaturing and collapsing overstory—will gradually lead to increasingly dry forest communities and eventually to grasslands dominated by *Imperata cylindrica* (Goldammer 2002). Conversely, attempts to eliminate fire from some forest ecosystems, which are in a steady-state equilibrium with recurrent fires, may also have consequences to the biodiversity of flora and fauna.

As elsewhere in tropical Asia, fire has long been used in South and Southeast Asia's seasonally dry forests to clear vegetation, or to maintain a forest structure to produce a specific range of non-timber forest products (NTFP) that are favored by fire (Goldammer 1993a; Stott et al. 1990). However, in most countries of the region it is now illegal to set forests on fire for any reason. There are conflicts between national fire prevention policies and the need of fire by local landholders in maintaining their land-use systems. In addition, it is now known that the pyrogenic release of greenhouse gases and particulates into the atmosphere is affecting the composition and functioning of the atmosphere. Forest degradation as a consequence of excessive burning practices or wildfires, through a net release of carbon to the atmosphere, is contributing to global warming and climate change (Andreae and Merlet 2001). Vice-versa, extreme climate variability resulting from climate change, notably extreme droughts, is impacting forest ecosystems and hence resulting in increasing risk of high-intensity and -severity wildfires. Despite the fact that dry forest ecosystems dominate mainland South and

Southeast Asia where fire has constituted a long-term evolutionary factor, information on fire history, fire regimes, fire effects, and fire management history is limited. The complexity and controversial issues related to the role of fire in deciduous dipterocarp forests (DDFs) is a challenge for the development of principles of fire management in the tropical dry forests of Asia. This chapter examines fire management history during the British colonial period in India and Myanmar, as well as that of some selected Southeast Asian countries. Thereafter, we elaborate on current fire problems and management in seasonally dry tropical forests of these regions. Finally, recommendations for fire management approaches for the next decade are provided.

FIRE MANAGEMENT HISTORY

Significance of Historical Fire Regimes in Tropical Asia

Prehistoric occurrence of natural vegetation fires has been documented since the evolution of land plants some 350 to 400 million years ago (see Clark and Robinson 1993; Scott and Glasspool 2006). Archaeological and paleoenvironmental research in mainland Southeast Asia has provided insights into the use of fire around early settlements. Early charcoal records show high fire activity during the late Pleistocene, through the period 38,000 BP to 12,000 BP (Penny 2001). Fire regimes, i.e., frequency, intensity, seasonality, fire-return intervals, and affected areas, have exerted influence on the development of vegetation structure and function through evolutionary time periods (Hope et al. 2005). However, in paleoecological records from insular Southeast Asia and New Guinea, it is extremely difficult to distinguish between climatic and anthropogenic factors that have influenced vegetation changes (Maxwell 2004; Goldammer and Seibert 1989).

Historical information on fire management in seasonally dry forests in South Asia is available for former British India and Burma (Myanmar), where teak (*Tectona grandis*) was a key species managed under sustainable forestry regimes during the colonial period (Goldammer 2007). Debates on the contribution of fire in regeneration of teak were widely held among foresters during the nineteenth and early twentieth centuries. By contrast, comparatively little is known about the history of fire management in Southeast Asian countries other than Myanmar. This is despite the importance of fire as an agent of land-use and land cover change in the region. Over millennia, fire has been applied widely in land cultivation all over insular and mainland South and Southeast Asia. Besides the traditional and still ongoing use of fire in shifting cultivation, there was systematic use of fire in converting native primary and secondary vegetation, including peat-swamp biomes, to tree and other crop plantations.

Myanmar

By 1896, the basic concepts of forest fire prevention had been initiated and concentrated in mixed deciduous forest (MDF), where teak occurs naturally. Since the late 1880s, colonial decrees had regulated that the Forest Department was obliged to liaise with the local administrative bodies to coordinate the issuance of forest fire prevention instructions for compliance by forest officers. The teak forest area under fire protection increased from 1,856 sq. miles to 8,153 sq. miles between 1897 and 1907 (Thinn 1999).

In May 1896 an article titled "Too Much Fire Protection in Burma" was published in the journal *Indian Forester* (Slade 1896). Slade wrote that there was no doubt of the benefit of the fire protection of plantations in their early years. However, he argued against a trend to advocate an enlargement of fire-protected areas and fire protection for all age classes of teak forests. Further, Slade argued the cost of protection could be reduced because the cost of subsequent weeding would be reduced. He concluded that fire protection in teak forests should be concentrated in areas of teak seedlings. Fire protection should be stopped when the seedlings reached the sapling stage at which the individual trees can tolerate a fire. Subsequent research and recommendations similar to those of Slade were generally confirmed by the chief conservator of forests, F. B. Bryant, in 1906, and by Sir George Hart, the inspector-general of forests, in 1911 (Bryant 1907; Thinn 1999). However, their articles triggered controversial discussions among forest officers in Burma, since they traditionally had given high priority to fire prevention. Reduction of fire protection was slow in spite of the opinion of many forest officers in favor of burning.

As a result of the global economic depression, and the development of the ideas related to fire management mentioned above, fire protection in the teak forests of Myanmar had fallen to only 142 sq. miles of coverage by 1923, and it came to an end in 1930; fire protection then was carried out only in plantations. However, fire prevention was continued in young regeneration areas, and measures were undertaken in the Reserved Forests and adjacent forest areas. Under the new policy, the use of prescribed fire in teak forest was allowed under the supervision and control of the forest officers, who were able to use or to exclude fire as the conditions of teak reproduction demanded.

India

The first forest fires in India occurred about 200 million years ago, coinciding with the evolution of early mammals on the earth. (Narendran 2001). Saha (2002) reported that anthropogenic fire in India can be dated to 50,000 years ago when hunter-gatherers colonized India. Fire was mostly used for agricultural purposes. In addition, people burned the hills "with almost religious fervor" in the hope that the ash would wash down to waiting fields (Mukhopadhyay 2007).

The policy on fires in Indian forests during the colonization period was historically one of strict suppression. This was first officially articulated in the Indian Forest Act of 1927, which considered the setting of fire a punishable offence. In addition, the act made it mandatory for all forest-dependent people to provide assistance in preventing and controlling fires (Hiremath 2007). However, whether complete fire suppression suited the forest ecosystems of India and the culture of the local people was critically debated among land managers, foresters, and local people. The indigenous people argued that they understood how fire supported agricultural activities. Further, fire kept tigers and cobras away from their villages. Some people in India believe that forest ecosystem conservation and development requires fire as one of its fundamental elements, and they trust in the spiritual belief and rational knowledge of the local people in the use of fire (Mukhopadhyay 2007). In addition, fire helps the regeneration of some valued timber species, in particular teak, since the thickness and hardness of the seed coat of this species requires high temperatures for cracking to enable germination. By 1914, it was generally recognized that the natural regeneration of sal (*Shorea robusta*) had ceased throughout the fire-protected moist forests of Assam and Bengal. Since then, controlled burning, ideally at intervals of four to five years, has been applied to sal forests in Assam (cf. Troup 1921). As a result, rather than abolish fire, as they had intended, the Europeans adopted and adjusted their fire management approaches from the complete exclusion of fire to the use of prescribed burning or wildfire in some forest ecosystems of India (Pyne 2001).

Thailand, Laos, Cambodia, and Vietnam

Fire history information for Thailand before 1970 is not available because forest fire had not been a serious concern to either government officials or the Thai people. However a visitor, J. C. Macleod, delegated by the government of Canada in 1971, proposed that forest fire control in Thailand must by all means be seriously taken into account. Consequently, the Forest Fire Control Section was established in 1976 to tackle forest fire problems. Even though many forest fire prevention campaigns were launched over the following decades throughout the country to protect forests from fire, it became clear that fire exclusion would not meet the challenges of reality because fires in some forest types such as DDF are almost constant throughout the dry season due both to regular fire use inside and around the forest and to the fact that forest litter constitutes abundant flammable fuel (Akaakara 2003).

In-depth study of the fire regimes of northeastern Cambodian monsoonal forest demonstrates that the Cambodian people started to use fire about 8,000 years ago. Fire frequency, however, has been increasing, possibly due to human control over fire, becoming more important about 2,500 years ago and continuing up to the present (Maxwell 2004). There is the suggestion that anthropogenic fire is an adaptation to the monsoonal environment and may act to conserve forest cover in the open forest formations of Cambodia.

The main fire regimes of Vietnam include (1) regularly occurring fires in seasonally flammable deciduous forests, (2) wildfires in pine forest ecosystems, (3) wildfires in other natural and degraded vegetation, (4) targeted land-use fires (shifting agriculture, forest conversion), and (5) fires used on intensively treated agricultural lands (Goldammer 1992). The peak of burning activities in Vietnam occurs during the mid- to late dry season (January to April). In many of the DDFs of Vietnam, wildfires occur almost annually, for example, in the central plateau areas near the border of Cambodia. The dominating dipterocarps (e.g., *Dipterocarpus intricatus*) resprout after these fires. As in the neighboring countries of Cambodia, Laos, Thailand, Myanmar, and India, these seasonal forests are well adapted to the regular occurrence of fire. Much of the lowlands and the high plateau of Vietnam that were formerly covered by seasonal or evergreen broadleaved forests is now degraded shrub, tree, and grass savanna. This vegetation is utilized extensively and frequently affected by wildfires. The fires are not set for specific purposes. They occur largely as a result of carelessness, or are set intentionally but without any land-treatment purpose (J. G. Goldammer, personal observation). Shulman (2002) recommended a deliberate strategy of fire use, in combination with an effective community fire prevention program—based on economic incentives extraneous to park resources—that would result in a decrease in unplanned ignitions. Improving the capacity of fire fighting forces to suppress unwanted fires would be the final step in enhancing this overall fire management program. Recently the concept of prescribed burning has entered forest management of the pine plantations of Vietnam (Van Huong 2007).

During the Vietnam War, large tracts of forests were destroyed by herbicides (e.g., "Agent Orange"), and some of these forests were purposely ignited after chemical treatment. The Military Assistance Command–Vietnam (MACV) ordered the Sherwood Forest and Pink Rose operations, which involved chemically defoliating the jungle to create dry fuel and then dropping incendiary weapons to start a firestorm (Lewis 2005). These operations were not successful in terms of destroying vegetation cover for military ground forces.

CURRENT FIRE PROBLEMS AND MANAGEMENT

Current Fire Status

Forest fire occurrence in mainland South and Southeast Asia varies spatially and temporally, depending on forest types, local weather conditions, and cultural practices.

The fire season in mainland Southeast Asia is usually from December to April. In South Asia, however, various regions of India have different normal and peak fire seasons, which occur variably from January to June. In the Himalayan region, fires are common in May and June (Roy 2007). The vast majority of wildfires in the seasonally dry forest ecosystems of the South and Southeast Asia regions are of anthropogenic origin. Lightning as a wildfire source plays a very minor role (Goldammer 1993a). Rural populations use fire as a traditional tool for clearing forest understory and managing pasture lands. Fire is also used to facilitate the gathering of NTFPs, and in hunting and herding. In many cases, fires that are set for clearing agricultural debris spread out of control into adjacent forests. Fortunately, many dry forests are located on less-fertile sites, in particular the DDFs, where slash-and-burn is not practiced. In these forests, people set fires only for gathering NTFPs and for hunting. In MDFs, however, where soils are more fertile, people also set fires for shifting cultivation, and these fires can escape, affecting the forest beyond the intended area.

In Thailand, extensive forest areas are regularly affected by human-caused surface fires (Figure 1). Both DDFs and MDFs are burned by surface fires once every year (Akaakara 2000). The peak fire period is between February and March (Figure 2), following leaf fall, when the availability of surface fuels is high. It has been estimated that, up to 1984, the annual forest area burned was about 18,772,000 ha. Since 1986, fire surveys have been conducted continuously using reports made by ground staff. Burned-area assessments based on interpretation of satellite images have been in place

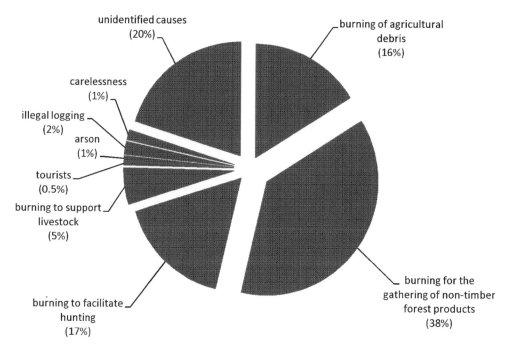

Figure 1. Causes of forest fire in Thailand (Forest Fire Control Office 2005).

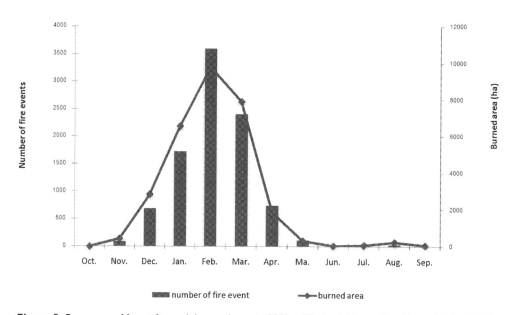

Figure 2. Frequency of forest fire and damaged area in 2005 in Thailand (Forest Fire Control Office 2005).

since 1992. The majority of these burned areas are concentrated in the northern and western regions, where most of the forests occur (Forest Fire Control Office 2002). In the years 2000 and 2002, fires occurring in these deciduous forests accounted for 91 percent and 72 percent of the total forest area burned, respectively (Ongsomwang et al. 2000, 2002). DDFs located in protected areas such as national parks and wildlife sanctuaries, by contrast, have been less affected by fires due to an attempt to suppress all fires. Owing to general environmental concerns, many fire prevention programs have been launched throughout all the forested areas of Thailand, including DDFs and MDFs. During extended periods of drought related to the El Niño Southern Oscillation, fires have been more widespread, extending to a certain extent into seasonally dry evergreen forests, lower montane evergreen forests, and even into some parts of moist evergreen forests (Goldammer 2002). In recent years, a notable number of crown fires have taken place in pine plantations in the northern part of the country, and there have also been ground fires in the swamp forests of southern Thailand (Akaakara 2000).

Forest Survey of India data indicate 50 percent of the country's forest areas are fire prone. With 63 million ha of forest in India, approximately 3.73 million ha can be presumed to be affected by fire annually. At this level, the annual losses from forest fires in India for the entire country can be conservatively estimated at US$107 million (Roy 2007). Fires in the forests of India are today almost entirely attributed to burning by people. People burn forests for a variety of reasons: to encourage a fresh flush of fodder for grazing livestock, to facilitate the collection of fuelwood and certain NTFPs, to clear the forest understory to improve access, and to fulfill religious beliefs or cultural practices (Goldammer 1993b). Fires are also used as a management practice to maintain wildlife habitat. In addition, fires are sometimes set as a form of protest against restrictive forest policies. And as noted above, many fires spread accidentally from agricultural burning-and-clearing (Hiremath 2007). India witnessed its most severe fires in recent years during the summer of 1995 across the hills of Uttar Pradesh and Himachal Pradesh in the Himalayas in the northern part of India. Approximately 677,700 ha of forest lands was affected by these fires, resulting in approximately US$45 million in losses of quantifiable timber. Lack of adequate manpower, poor communication, and reduced water availability in the hills allowed these fires to spread rapidly.

Influences of Human-Caused Fires on Burning Regimes

The seasonally dry forest ecosystems of South and Southeast Asia are characterized by diverse climate patterns, land-use systems, and socioeconomic and cultural settings, resulting in diverse fire regimes and vulnerabilities. As a consequence of the increase of human populations and land-use pressure, we assume that the character of the fire regimes in some of these ecosystems may have been altered in recent decades, possibly resulting in much shorter fire-return intervals and larger affected areas. Outside protected areas, fire-return intervals in natural forests in general are shorter. This is because unprotected forests have much higher human population densities and are more intensively used for various cultural activities, are more fragmented, and are located closer in proximity to agricultural lands that are burned annually. Kodandapani et al. (2004) reported that the fire-return interval for Mudumalai Wildlife Sanctuary in the

Western Ghats of India over a period of 14 years (1989–2002) was only 3.3 years. This report found a threefold increase in fire frequency over the past 80 years. The problem has been compounded by land cover transformations in the surrounding landscape. Large tracts of forests have been lost to agricultural areas (Subramanyam and Nayar 2001; Ramesh 2001), resulting in fragmented forest remnants. It is hypothesized that fragmentation of forests makes them more vulnerable to escaped agricultural fires along their extensive edges and that reduced patch size makes it more likely that entire fragments will burn during a single fire event.

A single fire will not result in the destruction or long-term degradation of most dry forest types. However, recurrent fires may lead to a loss of habitat, forest ecosystem biodiversity, and nutrients. Also, complete fire prevention or long-term fire suppression and the consequent build-up of fuel loads gives rise to the risk of high-intensity fires, which may cause high nutrient losses and have severe consequences for ecosystem stability (Figure 3). Unless burning frequency is appropriate, the ecosystem properties and functions within that forest will gradually decline. As the fire-return intervals decrease, it becomes more unlikely that a majority of tree species will be able to recruit new trees to a size resistant to mortality from the frequent fires. Therefore, apparently over-mature dry forests are found in large forest areas of mainland Southeast Asia and South Asia. Saha and Howe (2003) reported that annual low-intensity burning in dry deciduous forest of Mendha village, central India, resulted in reduced diversity of tree seedlings, and that more than 80 percent of tree diversity could be lost within a 100–200 year period if this process continued. Hence, burning frequency is crucial to the maintenance of forest composition and the ecological functions of dry forest ecosystems. The effects of short fire-return intervals have contributed to shape monodominant stands of *Shorea roxburghii* in India (Kodandapani 2001). Marod et al. (1999) suggested that if fires occurred less frequently, for example, at intervals of three to four years, seedlings in MDF in western Thailand could become large enough to tolerate fire. In addition, there are reports that too-frequent burning contributes to

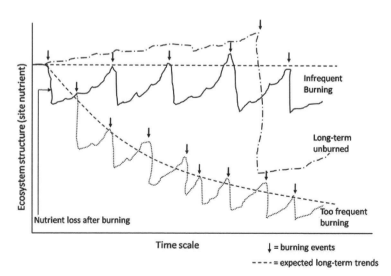

Figure 3. Relationship between available nutrient capital of a site and burning frequency. Dotted lines indicate expected long-term trends, and arrows represent forest burns.

rapid invasion by exotic, fire-adapted species. In India, *Lantana camara* and *Chromolaena odorata* have colonized regions subjected to repeated forest fires (Kodandapani 2001). Besides competing with native species for resources and space, these and other exotic invasive species may also alter the fire behavior in these forests by changing the fuel structure and hence potentially creating more-intense fires that could further accelerate the loss of native species and nutrients.

Recently, Wanthongchai (2008) reported a disparity in the proportions of dominant species in DDF (i.e., *Shorea siamensis* and *S. obtusa*) at annually burned sites in Huay Kha Khaeng Wildlife Sanctuary, Thailand. The number of these saplings increased with the length of fire-free intervals, suggesting that four to seven fire-free years may be required for seedlings to reach sapling stage (Figure 4). Furthermore, Wanthongchai also reported that soil fertility at frequently burned sites was generally lower than for less-frequently burned sites. In addition, the relative nutrient losses from aboveground biomass burning were significantly higher for sites that had frequently burned in the past, compared to other less-frequently burned sites (Table 1). Therefore, DDF at this sanctuary should be subjected to prescribed fire every six to seven years to maintain ecosystem structure and function (Wanthongchai 2008). Prescribed burning at this frequency can be safely managed as low-intensity fire according to Cheney (1994; Table 2).

The Influence of El Niño on Burning Regimes

In the tropical rain forest biome, the prolonged droughts caused by El Niño Southern Oscillation (ENSO) events drastically change the fuel complex and the flammability of the vegetation, and hence encourage the application of fire in land use and land-use change, resulting in altered burning regimes (Goldammer and Seibert 1990). El Niño–related droughts tend to occur every two to seven years, and last for a period of twelve to eighteen months at a time. However, analysis of data by the U.S. National Oceanic and Atmospheric Administration (NOAA) from the ten strongest El Niños of the past century has shown that El Niños are occurring more frequently, and are becoming progressively more intense. Since 1970, ENSO events occurred in 1972, 1976, 1982–83, 1987, 1991–92, 1994, and 1997–98. It has been suggested that rising temperatures, as a consequence of climate change, will result in more-frequent and

Table 1. Fire-related gross loss of selected nutrients in a 10-year period of different burning regimes: frequently burned (7 out of 10 years), infrequently burned (3 out of 10 years), and rarely burned (once in 10 years).

Burning Regime[a]	Gross Loss (kg/ha)			
	N	P	K	Ca
Frequently burned	−179.2	−6.3	−97.3	−132.3
Infrequently burned	−99.3	−9.6	−57.3	−71.7
Rarely burned	−52.7	−3.5	−11.3	−19.5

Source: Wanthongchai 2008 (modified).
[a]N = nitrogen; P = phosphorus; K = potassium; Ca = calcium.

Table 2. Quantitative fire behavior characteristics recorded for experimental fires applied to sites with different burning regimes: frequently burned (FB), infrequently burned (IB), and rarely burned (RB). Standard error is given in parentheses.

Fire Characteristics	Burning Regime		
	FB	IB	RB
Rate of spread (m/min)	2.7[a]	2.6[a]	1.3[a]
	(1.0)	(0.3)	(0.2)
Flame height (m)	1.2[a]	1.5[a]	1.2[a]
	(0.1)	(0.7)	(0.1)
Flame length (m)	1.51[a]	1.53[a]	1.27[a]
	(0.22)	(0.09)	(0.11)
Fireline intensity (kW/m)	361.1[a]	466.8[a]	291.2[a]
	(149.9)	(61.5)	(43.1)
Heat released (kJ/m^2)	132.8[a]	190.4[a]	229.5[a]
	(3.5)	(20.4)	(20.7)

Source: Wanthongchai 2008.
[a]Significant differences in the fire characteristics of the burning regimes (ANOVA, $p < 0.05$, followed by Duncan's multiple range test, or Kruskal-Wallis test followed by Mann-Whitney U-test).

more-intense ENSO occurrences. This may represent a positive feedback cycle in which climate change will be exacerbated by the net transfer of terrestrial carbon by fires that cause deforestation and degradation. This will contribute to a climate change–related increase in the frequency and severity of ENSO events, which in turn may create the conditions that lead to further excessive burning activities and wildfires. This will particularly affect equatorial ecosystem structures, processes, and dynamics in the rain forests of the region. Regarding the extent of burned areas during El Niño years, data from selected mainland South Asian countries reveal an increase in areas burned by wildfires (Table 3).

Current Smoke Pollution Problems

The impact of fire on forest ecosystem structure and function is not the only focal point for current fire management. A fire management strategy that concerns the negative impact of gas emission and smoke pollution on human health is also important and has been implemented through the forest fire control policies of each country.

Reports on smoke pollution resulting from fires and its affect on Southeast Asia have been available since at least the late 1800s (Potter 2001). The pyrogenic emissions constitute a bouquet of pollutants including aerosols, radiatively active trace gases, and toxic compounds that are affecting both human health in the region (Heil and Goldammer 2001; Goldammer et al. 2008) and the functioning of the global atmosphere. Peatland and rain forest conversion fires, in Indonesia in particular, are the major source of smoke-haze pollution in this region, whereas the pyrogenic pollution from other ecosystems is relatively less significant. In 2008, however, at the border between northern Thailand and Myanmar, slash-and-burn practices and wildfires in dry forest

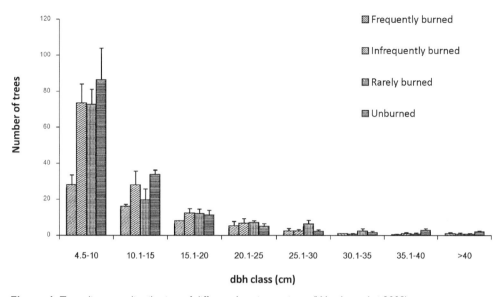

Figure 4. Tree diameter distribution of different burning regimes (Wanthongchai 2008).

ecosystems resulted in significant impacts on human health, the environment, and the transportation system of northern Thailand. Recently, transboundary haze pollution arising from forest fires and slash-and-burn agriculture has been a major political issue among the Southeast Asia countries (Goldammer 2007). Beginning in 1992, as a consequence of regional smog problems caused by land-use fires, member states of the Association of South East Asian Nations (ASEAN) agreed to cooperate in combating the problems arising from transboundary haze pollution. ASEAN workshops held in Balikpapan (1992) and Kuala Lumpur (1995) summarized the problems and urged

Table 3. Annual burned area from forest fire for selected countries.

Year	Area Burned (ha)			Remark
	Malaysia[a]	Thailand[b]	Vietnam[c]	
1992	418	2,030,160	n.a.	El Niño year
1993	56	1,459,617	n.a.	
1994	156	763,648	n.a.	El Niño year
1995	25	643,799	n.a.	
1996	18	490,303	n.a.	
1997	26	660,208	1,360	El Niño year
1998	1,646	1,145,452	15,088	El Niño year
1999	27	293,480	2,191	
2000	6	93,324	n.a.	

[a]Shields et al. 2006.
[b]Ganz 2003.
[c]Goldammer and Mutch 2001.

appropriate initiatives (Goldammer 1997). Most important in future ASEAN-wide cooperation in fire management will be the sharing of resources. Three of the main resource-sharing targets are (1) predicting fire hazard and fire effects on ecosystems and the atmosphere; (2) detecting, monitoring, and evaluating fires; and (3) sharing fire suppression resources and technologies. Although this cooperation has been accepted in principle, it should be noted that not all ASEAN member countries have ratified the agreement and given it their full endorsement.

A PERSPECTIVE FOR THE NEXT DECADE

Fire Management Based on Forest Ecosystem Classification

As a general rule for forest fire management, it is not sufficient to invest solely in fire prevention and suppression strategies. Likewise, uncontrolled anthropogenic burning as currently practiced should be revised. Therefore, it is necessary to consider seriously the potential use of planned, prescribed fire as a management tool for successfully achieving the goals of forest management. Top-down approaches to fire management, or indeed national forest fire policies applied uniformly and irrespective of the ecosystem, as are currently in operation in some South and Southeast Asian countries, should be revised. Fire management based on forest ecosystem classification (FEC) should be integrated into national forest fire policy (McRae 1996). The first order of FEC should be a fire-related classification of all ecosystems. Each category should then be assigned an adapted system of forest fire management. Fire prevention by any management means may be a necessary approach in ecosystems with species of little fire resistance and low resilience, such as moist evergreen forest (as postulated by Kowal [1966] and questioned by Baker et al. [2008]), dry evergreen forest, and montane evergreen forest. However, in the fire-dependent ecosystems of tropical Asia, such as DDF and MDF, fire management schemes should not concentrate solely on prevention and suppression, and annual low-intensity burning is not recommended either. Variation both within and between forest ecosystems should be subjected to different fire treatments. For example, DDFs in Thailand have been classified according to many subcommunities (see Bunyavejchewin 1983; Sahunalu and Dhanmanonda 1995; Kutintara 1975; Ogawa et al. 1961; Sukwong et al. 1977). Therefore, the decision as to which fire management strategies should be employed in these subcommunity forests should be explored by long-term monitoring studies.

Patch Burns and Landscape Diversity

Varying burning frequency as well as patch size may have beneficial effects for an ecosystem, just as regular fire frequency may have adverse ecological effects (Morrison 2002; Keith and Bradstock 1994; Tolhurst et al. 1992). Evidence suggests that a variable interval is more appropriate for biodiversity conservation than a constant one (McCarthy and Burgman 1995). Variation of the burning interval, intensity, and area may result in both a spatial and temporal mosaic across the landscape. For example,

Walker and Rabinowitz (1992) suggested that seasonal shifts in habitat use by the small mammal *Maxomys surifer* at Huay Kha Khaeng Wildlife Sanctuary indicate that patchiness might allow for higher population densities than would be possible in a more homogenous DDF. Wanthongchai (2008) suggested that to maintain the diversity of the landscape, a spatial and temporal variability in burning frequency can be applied throughout DDF.

Fire as a Tool for Reforestation

The destruction of South and Southeast Asia's seasonally dry tropical forests is a serious problem. When burning frequency is inappropriate, dry forests within the tropics are often degraded to secondary open forest land or converted to other land-use systems. In many cases, reforestation may be required to restore ecosystem goods and services. Hardwick et al. (2004) proposed an alternative restoration method, "accelerated natural regeneration," to encourage natural establishment of indigenous trees and shrubs in seasonal tropical forests in Thailand. This method requires a low input of labor, but a high input of ecological information, especially regarding fire disturbance. These authors proposed the necessity of burning for restoration; however, the potential use of controlled burning to accelerate natural regeneration should be further investigated. Very little research has explored how prescribed burning might be used most effectively. For example, should controlled burns be used to reduce fuel loads once trees have reached a certain size, and what about the fire susceptibility of each successional stage? In the case of fire-prone grasslands being restored to forests, it is particularly important to identify fast-growing native species that can regenerate following ground fires. We have not reached an information level where we can adequately restore dry forests in Asia.

Community-Based Fire Management

It is clear that most forest fires in tropical South and Southeast Asia are caused by humans and can be linked to daily human activities. Therefore, it is not surprising that local communities are often blamed as "the igniter." This view explains governments' perception of local communities as part of the problem. However, an underlying reason for local communities' failure to control fires is not a lack of awareness or carelessness, but rather the lack of incentives to protect forest resources (Food and Agriculture Organization of the United Nations 2003). Therefore, community-based fire management (CBFiM) has emerged as a new adaptive mechanism for working with and managing fire in many regions of the world. Small user communities, with their respective economic interests, may use managed fires to contribute to a patchwork of sustainable land-use practices that satisfy all stakeholder interests in forest land, fuel, fodder, and NTFP. Under this approach, however, there must be guidelines for the use of fire. Schmerbeck and Seeland (2007) concluded that land-use policy in post-colonial India, after more than fifty years of independence, should abandon colonial attitudes toward CBFiM. Forest and environmental laws and regulations should accept the fire practices of local forest users as contributions to sustainable forest management and recognize that they are more of a solution to natural resource use than a problem.

It is widely accepted that the seasonally dry forests of tropical Asia are an important "local food reserve." Mushrooms, ant eggs, the fresh new leaves of edible plants, and small firewood are the main NTFPs extracted from these forests. Specifically, many people believe that fire increases NTFPs productivity and encourages the development of tasty edible plants, a belief not rigorously tested. Preventing local people from lighting fires is admittedly a difficult task. However, dividing an experimental forest into sectors that can be burned for NTFP management, subject to the agreement of both the forest management authority and the villagers, may represent a suitable compromise. This strategy may reduce conflicts over the contrasting land-use objectives of the administration and the local villagers, and serve to maintain ecosystem productivity. Therefore, we recommend a system of participative fire management between local communities and local and regional authorities.

Fire and Global Warming

The degradation of forest ecosystems, or deforestation induced by excessive fire occurrence and land-clearing, is contributing to a net transfer of carbon to the atmosphere, thus to global climate change. However, the emission pulses of nutrients, trace gases, and particles from fires that occur periodically in fire-adapted ecosystems need to be considered as a phenomenon that has influenced biogeochemical cycles over historical and prehistoric periods, and thus cannot be attributed to recent human-induced climate change. Vice-versa the effects of climate change on fire regimes have been predicted. A number of recent studies for the temperate and boreal zone have used general circulation models (GCM) to simulate future climate. For example, it has been estimated there will be a 44 percent increase in lightning-caused fires, with an associated 78 percent increase in area burned, for a $2 \times CO_2$ climate scenario (Price and Rind 1994). Stocks et al. (1998) indicated significant increases in fire weather severity for Canada, Alaska, and Russia, which may be accompanied by an increase in area burned. The authors also indicated an earlier start to the fire season. Recent data for human-caused ignitions suggest increases of 18 percent and 50 percent by 2050 and 2100, respectively (Wotton et al. 2003). Flannigan et al. (2006) concluded that a future warmer climate is expected to result in more-severe fire weather, more area burned, more ignitions, and a longer fire season.

The implications of this research on temperate-boreal ecosystems may be useful for predicting future tropical dry forest conditions. If the El Niño phase of ENSO becomes stronger as global warming continues into the next century, then fire regimes in South and Southeast Asia may also change. The climatic conditions affected by El Niño, including reduced annual rainfall and prolonged drought periods, may affect fire dynamics in tropical dry forest ecosystems. In the absence of targeted fire management interventions, these changes are likely to lead to increased fire disturbance. Further, severe problems of land degradation such as soil erosion, loss of nutrients, and landslides are expected to occur more often as a consequence of high-severity fires in these seasonally dry forests.

The global warming crisis poses further problems for fire management in tropical dry forests. Forests can be both a significant carbon source, for example, as a consequence of fire-induced degradation or destruction (and replacement by less carbon-rich ecosystems), and a carbon sink—if sustainable management and protection results in higher stocks of biomass and thus in greater carbon storage.

CONCLUSIONS

Appropriate fire management in seasonally dry forests should develop strategies that reduce adverse impacts on human health and excessive emission of carbon to the atmosphere, consider the unavoidable recurrence of human-caused fires, and take advantage of the beneficial effects of fire on ecosystem biodiversity, stability, and productivity. Therefore, in order to define appropriate fire management approaches, more research is needed, particularly to address spatial and temporal aspects of burning. In addition, well-documented, long-term fire exclusion and long-term versus short-term interval burning research is critical since this information will support the development of fire management plans for these ecosystems. The impact of climate change on fire regimes is another area that requires additional research. However, the fundamental goal is to understand the ecological amplitude of seasonally dry tropical forests in their adjustment and adaptation to fire. It is recognized that tropical dry forests have evolved with fires, but we still need to ask how much fire they need, and how much fire they can tolerate while still providing ecological services and functionality. Thus, the establishment of future fire regimes is essential for informed decisions in forest and fire management.

REFERENCES

Akaakara, S. 2000. *Forest Fire Control in Thailand*. Bangkok: Royal Forest Department.
———. 2003. *Global Forest Fire Situation and Management*. Bangkok: Department of Natural Park, Wildlife and Plant Conservation.
Andreae, M. O., and P. Merlet. 2001. Emission of Trace Gases and Aerosols from Biomass Burning. *Global Biogeochemical Cycles* 15:955–66.
Baker, P. A., S. Bunyavejchewin, and A. P. Robinson. 2008. The Impacts of Large-Scale, Low-Intensity Fires on the Forests of Continental South-East Asia. *International Journal of Wildland Fire* 17:782-92.
Bryant, F. B. 1907. Fire Conservancy in Burma. *Indian Forester* 33:537–49.
Bunyavejchewin, S. 1983. Canopy Structure of Dry Dipterocarp Forest of Thailand. *Thai Forest Bulletin* 14:1–132.
Cheney, P. 1994. The Effectiveness of Fuel Reduction Burning for Fire Management. In *Fire and Biodiversity: The Effects and Effectiveness of Fire Management*, Biodiversity Series, paper no. 8, 9–16, www.environment.gov.au/biodiversity/publications/series/paper8/paper1.html (accessed October 2010).
Clark, J., and J. Robinson. 1993. Paleoecology of Fire. In *Fire in the Environment: The Ecological, Atmospheric, and Climatic Importance of Vegetation Fires*, ed. P. Crutzen and J. G. Goldammer, 193–214. Chichester, UK: John Wiley & Sons.
Flannigan, M., B. Amiro, K. Logan, B. Stocks, and B. Wotton. 2006. Forest Fires and Climate Change in the 21st Century. *Mitigation and Adaptation Strategies for Global Change* 11:847–59.
Food and Agriculture Organization of the United Nations. 2003. *Community-Based Fire Management: Case Studies from China, The Gambia, Honduras, India, the Lao People's Democratic Republic and Turkey*. Bangkok: Food and Agriculture Organization of the United Nations.

Forest Fire Control Office. 2002. *Forest Fire Control Yearly Report: 2002*. Bangkok: Royal Forest Department.

———. 2005. *Forest Fire Control Yearly Report: 2005*. Bangkok: Department of National Park, Wildlife and Plant Conservation.

Ganz, D. 2003. *Framing Fires: A Country-by-Country Analysis of Forest and Land Fires in the ASEAN nations*. Jakarta: Project FireFight South East Asia.

Goldammer, J. G. 1992. Viet Nam—A Fire Problem Analysis. *International Forest Fire News* 7:13–16.

———. 1993a. *Feuer in Waldökosystemen der Tropen und Subtropen*. Basel-Boston: Birkhäuser-Verlag.

———. 1993b. Fire Management. In *Tropical Forestry Handbook*, vols. 1–2, ed. L. Pancel, 1221–68. New York: Springer-Verlag.

———. 1997. Overview of Fire and Smoke Management Issues and Options in Tropical Vegetation. In *Transboundary Pollution and the Sustainability of Tropical Forests: Towards Wise Forest Fire Management—The Proceedings of the AIFM International Conference*, ed. H. A. Hassan, D. Taha, M. P. Dahalan, and A. Mahmud, 189–217. Kuala Lumpur: ASEAN Institute for Forest Management / Ampang Press.

———. 2002. *International Forest Fire News: Asia and Oceania Fire Special*. Geneva: United Nations Economic Commission for Europe.

———. 2007. History of Equatorial Vegetation Fires and Fire Research in Southeast Asia before the 1997–98 Episode: A Reconstruction of Creeping Environmental Changes. *Mitigation and Adaptation Strategies for Global Change* 12:13–32.

Goldammer, J. G., and R. W. Mutch. 2001. *Global Forest Fire Assessment 1990–2000*. Rome: Food and Agriculture Organization of the United Nations.

Goldammer, J. G., and B. Seibert. 1989. Natural Rain Forest Fires in Eastern Borneo during the Pleistocene and Holocene. *Naturwissenschaften* 76:518–20.

———. 1990. The Impact of Droughts and Forest Fires on Tropical Lowland Rain Forest of Eastern Borneo. In *Fire in the Tropical Biota: Ecosystem Processes and Global Challenges*, ed. J. G. Goldammer, 11–31. Ecological Studies 84. Berlin: Springer-Verlag.

Goldammer, J. G., M. Statheropoulos, and M. O. Andreae. 2008. Impacts of Vegetation Fire Emissions on the Environment, Human Health and Security: A Global Perspective. In *Wildland Fires and Air Pollution*, ed. A. Bytnerowicz, M. Arbaugh, A. Riebau, and C. Andersen, 3–36. Developments in Environmental Science 8. Amsterdam: Elsevier.

Hardwick, K., J. R. Healey, S. Elliott, and D. Blakesley. 2004. Research Needs for Restoring Seasonal Tropical Forests in Thailand: Accelerated Natural Regeneration. *New Forests* 27:285–302.

Heil, A., and J. G. Goldammer. 2001. Smoke-Haze Pollution: A Review of the 1997 Episode in Southeast Asia. *Regional Environmental Change* 2 (1): 24–37.

Hiremath, A., and J. Schmerbeck. 2007. Forest Fires in India: Extent, Justification and Policy. In *Forest Fires in India: Workshop Proceedings*, ed. J. Schmerbeck, A. Hiremath, and C. Ravichandran, 18–20. Bangalore, India: Ashoka Trust for Research in Ecology and Environment.

Hope, G., U. Chokkalingam, and S. Anwar. 2005. The Stratigraphy and Fire History of the Kutai Peatlands, Kalimantan, Indonesia. *Quaternary Research* 64:407–17.

Keith, D. A., and R. A. Bradstock. 1994. Fire and Competition in Australian Heat: A Conceptual Model and Field Investigations. *Journal of Vegetation Science* 5:347–54.

Kimmins, J. P. 1997. *Forest Ecology: A Foundation for Sustainable Management*, 2nd ed. Upper Saddle River, NJ: Prentice Hall.

Kodandapani, N. 2001. Forest Fires: Origins and Ecological Paradoxes. *Resonance* 6:34–41.

Kodandapani, N., M. A. Cochrane, and R. Sukumar. 2004. Conservation Threat of Increasing Fire Frequencies in the Western Ghats, India. *Conservation Biology* 18:1553–61.

Kowal, N. E. 1966. Shifting Cultivation, Fire and Pine Forest in the Cordillera Central, Luzón, Philippines. *Ecological Monograph* 36:389–419.

Kutintara, U. 1975. Structure of Dry Dipterocarp Forest. PhD diss., Colorado State University.

Lewis, J. G. 2005. *The Forest Service and the Greatest Good: A Centennial History*. Forest History Society, www.historycooperative.org/journals/eh/11.3/lewis.html (accessed March 2008).

Marod, D., U. Kutintara, C. Yarwudhi, H. Tanaka, and T. Nakashizuka. 1999. Structural Dynamics of a Natural Mixed Deciduous Forest in Western Thailand. *Journal of Vegetation Science* 10:777–86.

Maxwell, A. L. 2004. Fire Regimes in North-Eastern Cambodian Monsoonal Forests, with a 9300-year Sediment Charcoal Record. *Journal of Biogeography* 31:225–39.

McCarthy, M. A., and M. A. Burgman. 1995. Coping with Uncertainty in Forest Wildlife Planning. *Forest Ecology and Management* 74:23–36.

McRae, D. J. 1996. Use of Forest Ecosystem Classification Systems in Fire Management. *Environmental Monitoring Assessment* 39:559–70.

Morrison, D. A. 2002. Effects of Fire Intensity on Plant Species Composition of Sandstone Communities in the Sydney Region. *Austral Ecology* 27:433–41.

Mukhopadhyay, D. 2007. Culture of Fire in the Forest of India. In *Forest Fires in India: Workshop Proceedings*, ed. J. Schmerbeck, A. Hiremath, and C. Ravichandran, 27–29. Bangalore, India: Ashoka Trust for Research in Ecology and Environment.

Narendran, K. 2001. Forest Fire: Origin and Ecological Paradoxes, www.ias.ac.in/resonance/Nov2001/pdf/Nov2001p34-41.pdf (accessed November 2007).

Ogawa, H. K., K. Yoda, and T. Kira. 1961. A Preliminary Survey on the Vegetation of Thailand. *Nature and Life in Southeast Asia* 1:20–158.

Ongsomwang, S., S. Akaakara, C. Chantaraprapa, N. Nuchplian, A. Ransipanit, and A. Ratansuwan. 2000. *Assessment of Burned Forest Area in 2000 Using Remote Sensing and GIS*. Bangkok: Royal Forest Department.

Ongsomwang, S., S. Akaakara, A. Ransipanit, N. Nuchplian, A. Ratansuwan, and S. Pungkul. 2002. *Assessment of Burned Forest Area in 2002 Using Remote Sensing and GIS*. Bangkok: Royal Forest Department.

Penny, D. 2001. A 40,000 Year Palynological Record from North-East Thailand; Implications for Biogeography and Palaeo-Environmental Reconstruction. *Palaeogeography, Palaeoclimatology, Palaeoecology* 171:97–128.

Potter, L. 2001. Drought, Fire and Haze in the Historical Record of Malaysia. In *Forest Fires and Regional Haze in Southeast Asia*, ed. P. Eaton and M. Radojevic, 23–40. New York: Nova Science Publishers.

Price, C., and D. Rind. 1994. The Impact of a $2 \times CO_2$ Climate on Lightning-Caused Fires. *Journal of Climate* 7 (10): 1484–94.

Pyne, S. J. 2001. *Fire: A Brief History*. Seattle: University of Washington Press.

Ramesh, B. R. 2001. Patterns of Vegetation, Biodiversity and Endemism in the Western Ghats. *Memoir Geological Society of India* 47:973–81.

Roy, P. S. 2007. Forest Fire and Degradation Assessment Using Satellite Remote Sensing and Geographic Information System, www.wamis.org/agm/pubs/agm8/Paper-18.pdf (accessed 2007).

Rundel, P. W., and K. Boonpragob. 1995. Dry Forest Ecosystems of Thailand. In *Seasonally Dry Tropical Forests*, ed. S. H. Bulloock, H. Mooney, and E. Medina, 93–123. New York: Cambridge University Press.

Saha, S. 2002. Anthropogenic Fire Regime in a Deciduous Forest of Central India. *Current Science* 82:1144–47.

Saha, S., and H. F. Howe. 2003. Species Composition and Fire in a Dry Deciduous Forest. *Ecology* 84:3118–23.

Sahunalu, P., and P. Dhanmanonda. 1995. Structure and Dynamics of Dry Dipterocarp Forest, Sakaerat, Northeastern Thailand. In *Vegetation Science in Forestry*, ed. E. O. Box, 465–94. The Netherlands: Kluwer Academic Publishers.

Schmerbeck, J., and K. Seeland. 2007. Fire Supported Forest Utilization of a Degraded Dry Forest as a Means of Sustainable Local Forest Management in Tamil Nadu/South India. *Land Use Policy* 24:62–71.

Scott, A. C., and I. J. Glasspool. 2006. The Diversification of Paleozoic Fire Systems and Fluctuations in Atmospheric Oxygen Concentration. *Proceedings of the National Academy of Sciences, USA* 103 (29): 10861–65.

Shields, B. J., R. W. Smith, and D. Ganz. 2006. *Fire Management Working Papers: Global Forest Resources Assessment 2005 Report on Fires in the South East Asian (ASEAN) Region. Working Paper FM/10/E.* Rome: Food and Agriculture Organization of the United Nations.

Shulman, D. 2002. Fire Management Assessment, Tram Chim National Park, Dong Thap Province. *International Forest Fire News* 26:106–13.

Slade, H. 1896. Too Much Fire Protection in Burma. *Indian Forester* 22:172–76.

Stocks, B. J., M. A. Fosberg, T. J. Lynham, L. Mearns, B. M. Wotton, Q. Yang, J. Z. Jin, K. Lawrence, G. R. Hartley, J. A. Mason, and D. W. McKenney. 1998. Climate Change and Forest Fire Potential in Russian and Canadian Boreal Forests. *Climatic Change* 38:1–13.

Stott, P. A., J. G. Goldammer, and W. L. Werner. 1990. The Role of Fires in the Tropical Lowland Deciduous Forest of Asia. In *Fire in the Tropical Biota: Ecosystem Processes and Global Challenges*, ed. J. G. Goldammer, 32–44. Ecological Studies 84. Berlin: Springer-Verlag.

Subramanyam, K., and M. P. Nayar. 2001. Vegetation and Phytogeography of the Western Ghats. *Memoirs Geological Society of India* 47:945–59.

Sukwong, S., L. Chantanaparb, U. Kutintara, P. Sahunalu, S. Pongumphai, B. Thaiutsa, S. Thammincha, S. Siripatanadilok, and S. Kaitpraneet. 1977. *Quantitative Studies of the Seasonal Tropical Forest Vegetation in Thailand*, annual report no. 2. Bangkok: Faculty of Forestry, Kasetsart University.

Thinn, U. Myat. 1999. Forest Fire Prevention and Management in Myanmar. *International Forest Fire News* 20:21–28.

Tolhurst, K. G., D. W. Flinn, R. H. Loyn, A. A. G. Wilson, and A. Foletta. 1992. *Ecological Effects of Fuel Reduction Burning in a Dry Sclerophyll Forest: A Summary of Principal Research Findings and Their Management Implications.* Victoria, Australia: Forest Research, Department of Conservation and Environment.

Troup, R. S. 1921. The Silviculture of Indian Trees, vols. 1–3. London: Oxford University Press.

Van Huong, L. 2007. Fuel Assessment and Fire Prevention in Pine Plantations during the Tending Stage in Dalat, Lam Dong Province, Viet Nam. *International Forest Fire News* 36:76–86.

Walker, S., and A. Rabinowitz. 1992. The Small-Mammal Community of a Dry Tropical Forest in Central Thailand. *Journal of Tropical Ecology* 8:57–71.

Wanthongchai, K. 2008. Effects of Different Burning Frequencies on Fire Behaviour, Nutrient Dynamics, Soil Properties, and Vegetation Structure and Composition in Dry Dipterocarp Forest, Huay Kha Khaeng Wildlife Sanctuary, Thailand. PhD diss., University of Freiburg.

Whelan, Robert J. 1995. *The Ecology of Fire.* Melbourne: Cambridge University Press.

Wotton, B. M., D. L. Martell, and K. A. Logan. 2003. Climate Change and People-Caused Forest Fire Occurrence in Ontario. *Climatic Change* 60:275—95.

6

The Evolution and Ecology of Dry Tropical Forest Gingers (Zingiberaceae) in Southeast Asia

W. John Kress

One of the most conspicuous plant components of the understory of seasonally dry forests in Southeast Asia are gingers. These dry forest plants, which range in height from several centimeters to several meters, dominate the forest floor during the wet season, but go dormant and are completely hidden in the soil during the dry months of the year. A wealth of knowledge is now available about the taxonomy, classification, and phylogeny of the gingers and their relatives in the order Zingiberales as a result of investigations carried out during the past several decades. However, less is known about their ecology. Here I begin with a brief overview of the systematics and phylogeny of the tropical gingers and their closest relatives. This background then serves as the basis for an analysis of the ecology and adaptation of gingers to the seasonality of dry forests in Southeast Asia that reveals a complex pattern of evolution in response to these specialized habitats.

TAXONOMIC DIVERSITY AND CLASSIFICATION

With 53 genera and over 1,300 species the Zingiberaceae, or ginger family, is the largest of the eight families in the order Zingiberales. Recently the family has received significant taxonomic study, and many new species (see Poulsen et al. 1999; Williams et al. 2003) and even new genera (see Larsen and Mood 1998; Sakai and Nagamasu 2000; Kress and Larsen 2001; Larsen and Jenjittikul 2001) have been described in the past decade. Other members of the order Zingiberales include the bananas, the birds-of-paradise, the heliconias, the prayer plants, and the cannas (Table 1; Kress 1995). The Angiosperm Phylogeny Group accepts the Zingiberales as one of the ten primary orders in the monocotyledons (APG III 2009). Although less than 4 percent of extant monocot species are contained within this clade, the unique features of its members, including chemical, anatomical, and macromorphological features (Kress 1990, 1995), make it an easily recognizable order. The usually large leaves with long petioles, a central midrib, and transverse venation, and the colorful, bracteate inflorescences,

Table 1. The current classification of the order Zingiberales.

Order Zingiberales Nakai
Suborder Musineae W. J. Kress
Family Musaceae A. L. Jussieu (3 genera/42 species)
Suborder Strelitziineae W. J. Kress
Family Strelitziaceae Hutchinson (3 genera/7 species)
Family Lowiaceae Ridley (1 genus/15 species)
Suborder Heliconiineae W. J. Kress
Family Heliconiaceae Nakai (1 genus/225 species)
Suborder Zingiberineae W. J. Kress
Superfamily Zingiberariae W. J. Kress
Family Zingiberaceae Lindley (53 genera/1,300 species)
Family Costaceae Nakai (7 genera/130 species)
Superfamily Cannariae W. J. Kress
Family Cannaceae A. L. Jussieu (1 genus/10 species)
Family Marantaceae Petersen (30 genera/510 species)

Source: Kress et al. 2001.

while not unique among the monocotyledons, serve to identify members of the order in the field. Data from both morphological traits as well as DNA sequences have been utilized to establish the phylogenetic relationships of these eight families. The banana families (Musaceae, Strelitiziaceae, Lowiaceae, and Heliconiaceae) form a basal paraphyletic grade with the four ginger families (Zingiberaceae, Costaceae, Marantaceae, and Cannaceae) as a monophyletic terminal clade (Figure 1; Kress et al. 2001).

Early classifications of the family Zingiberaceae first proposed in 1889 and refined by others since that time recognized four tribes (Globbeae, Hedychieae, Alpinieae, and Zingibereae) based on morphological features such as number of locules and placentation in the ovary, development of staminodia, modifications of the fertile anther, and rhizome-shoot-leaf orientation. Recent phylogenetic analyses based on DNA sequences of the nuclear ITS and plastid *matK* regions (Kress et al. 2002) have suggested that some of these morphological traits are homoplasious and therefore three of the tribes in the earlier classification are paraphyletic and taxonomically unacceptable (Figure 2; Table 2). Two (Alpinieae and Hedychieae) of the four former tribes can be reformulated as monophyletic groupings by excluding several misplaced genera; the other two tribes, Globbeae and Zingibereae, must be included within the tribe Hedychieae. In order to accommodate these anomalies, the new phylogeny-based classification of the Zingiberaceae considerably revised the main categories. Four subfamilies and six tribes are now recognized (Table 2): subfamily Siphonochiloideae (with tribe Siphonochileae), subfamily Tamijioideae (with tribe Tamijieae), subfamily Alpinioideae (with tribes Alpinieae and Riedelieae), and subfamily Zingiberoideae (with tribes Zingibereae and Globbeae). A supporting suite of morphological features are congruent with this classification (Table 3).

The molecular phylogeny provided strong support for the existence of two major evolutionary groups of gingers: group one included most of the taxa included in the former tribes Hedychieae, Globbeae, and Zingibereae, while group two included most members of the traditional Alpinieae (Kress et al. 2002; Figure 2). Two genera from

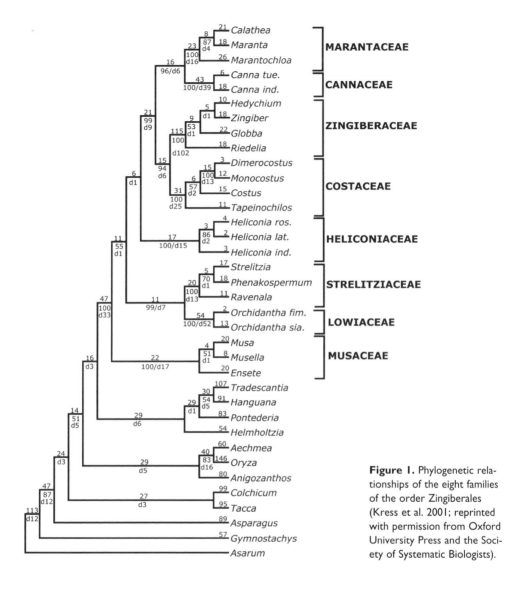

Figure 1. Phylogenetic relationships of the eight families of the order Zingiberales (Kress et al. 2001; reprinted with permission from Oxford University Press and the Society of Systematic Biologists).

the Hedychieae are clearly part of group two, and one member of the Alpinieae is part of group one. Two of the most interesting results of the analyses were that the genus *Tamijia*, with a single species endemic to Borneo, is sister to these two major groups, and that the African genera *Siphonochilus* and *Aulotandra* (see Harris et al. 2006) are sister to all other members of the family and therefore the most primitive genera among the gingers. Additionally, some of the largest genera in the Zingiberaceae, including *Alpinia*, *Amomum*, and *Curcuma*, are not monophyletic and therefore in need of substantial taxonomic revision (see Kress et al. 2005, 2007). This well-supported phylogeny based on DNA sequence data now provides the basis for exploring the patterns of evolution of various features of the gingers (see below).

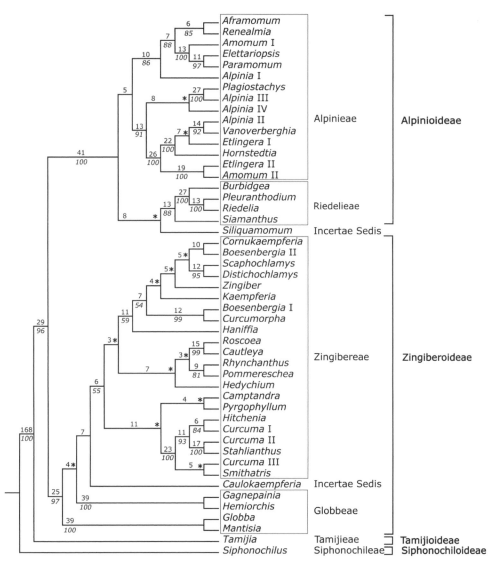

Figure 2. Phylogenetic relationships and classification of the family Zingiberaceae (Kress et al. 2002; reprinted with permission from the Botanical Society of America).

Based on the phylogeny of the Zingiberales derived from morphological and molecular characters (Figure 1), Kress and Specht (2005, 2006) developed a hypothesis for the spatial and temporal evolution of the Zingiberales using a dispersal-vicariance analysis combined with a local molecular clock technique. The divergence of the Zingiberales from their closest relatives in the monocots was found to have occurred around 124 million years ago (mya). Major family-level lineages became established in the late Cretaceous (80–110 mya) and crown lineages within each family began to diversify during the early to mid-Tertiary (29–64 mya). The broad ancestral distribution

Table 2. Current classification of the family Zingiberaceae.

Subfamily Siphonochiloideae W. J. Kress	Subfamily Tamijioideae W. J. Kress	Subfamily Alpinioideae Link	Subfamily Zingiberoideae Haask.
Tribe Siphonochileae W. J. Kress (2 genera/ 21 species)	Tribe Tamijieae W. J. Kress (1 genus/1 species)	Tribe Alpinieae A. Rich. (15 genera/686 species) Tribe Riedelieae W. J. Kress (4 genera/94 species	Tribe Zingibereae Meisn. (25 genera/ 382 species) Tribe Globbeae Meisn. (3 genera/110 species)
		Incertae Sedis (1 genus/1 species)	Incertae Sedis (1 genus/10 species)

Source: Kress et al. 2002.

of the common ancestors of the eight families in Gondwanaland followed by multiple secondary dispersal events within families during the Tertiary are sufficient to explain the main biogeographic events leading to the current pantropical distribution of members of this order (Figure 3).

The rapid radiation of the ancestral Zingiberales between 100 and 110 mya led to the divergence of a number of independent lineages early in the history of the group. The Musaceae was the first to diverge about 110 mya, having an ancestral distribution of America and Southeast Asia at that time. After the rapid diversification of the next three basal lineages of the Zingiberales (Strelitziaceae, Lowiaceae, and Heliconiaceae), the remaining crown group of the four "Ginger Families" began to further diversify about 106 mya. The ancestral distribution of these Ginger Families appears to have been in the Americas, which is an interesting result considering the pantropical nature of the extant members of the group. The current distribution appears to be the result of an early widespread occurrence coupled with short dispersal events starting at 106 mya when many of the southern continents were still in contact or in close enough proximity so as to permit short-distance dispersals. The Ginger Families as a clade bifurcated into two main lineages, each undergoing further cladogenesis to create a total of four extant lineages within the crown clade now considered to be families. The Costaceae-Zingiberaceae lineage quickly split into the two families recognized today. The ancestral range of the Costaceae/Zingiberaceae was Africa, America, Melanesia, and Southeast Asia, reflecting the current distribution of the members of these two families. The Zingiberaceae then began its major divergence at the very end of the Cretaceous about 65 mya.

HABITATS AND ECOLOGY

Today the gingers are geographically distributed in the tropical zones around the world with one genus (*Renealmia*) found in the neotropics, four genera (*Aframomum*,

Table 3. Selected characters of the main taxonomic groups of the family Zingiberaceae.

Character	Subfamilies and Tribes					
	Siphonochiloideae: Siphonochileae	Tamijioideae: Tamijieae	Alpinioideae: Riedelieae	Alpinioideae: Alpineae	Zingiberoideae: Zingibereae	Zingiberoideae: Globbeae
1) Seasonality	Dormancy period	Evergreen	Evergreen	Evergreen	Dormancy period	Dormancy period
2) Rhizomes	Fleshy	Fibrous	Fibrous	Fibrous	Fleshy	Fleshy
3) Plane of distichy of leaves	Perpendicular to rhizome	Perpendicular to rhizome	Perpendicular to rhizome	Perpendicular to rhizome	Parallel to rhizome	Parallel to rhizome
4) Extrafloral nectaries	Absent	Absent	Present on leaf blades	Absent	Absent	Absent
5) Lateral staminodes	Petaloid, fused to labellum	Petaloid, fused to labellum	Small or absent, never petaloid	Small or absent, never petaloid	Petaloid, free from or fused to labellum	Petaloid, free from labellum and sometimes connate to filament
6) Labellum	Not connate to filament	Not connate to filament	Not connate to filament	Not connate to filament	Not connate to filament	Connate to filament in slender tube
7) Filament	Short	Short	Medium	Medium, sometimes arching	Short to long	Short to long, sometimes arching
8) Anther crest	Petaloid	Petaloid	Petaloid or absent	Petaloid or absent	Absent, petaloid, or well-developed and wrapped around style	Spurred or absent
9) Ovary	3-locular (sometimes incompletely so)	1-locular	1- or 3-locular	3-locular	3-locular (sometimes incompletely so)	1-locular
10) Placentation	Axial	Parietal	Axial or parietal	Axial or free central	Axial, basal, or free columnar	Parietal
11) Capsule	Fleshy	Unknown	Silique-like, opening by longitudinal slits	Indehiscent or fleshy	Fleshy and dehiscent	Globbose and dehiscent

Source: Kress et al. 2002.

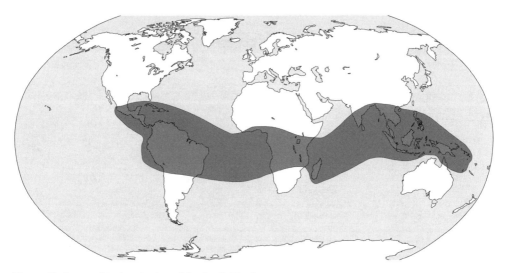

Figure 3. Geographic distribution of the family Zingiberaceae.

Aulotandra, *Siphonochilus*, and *Renealmia*) found in Africa, and the majority of genera distributed in east Asia and the Pacific islands (Figure 3). Although a few gingers are native to high-elevation zones in the Himalayas (i.e., *Cautleya* and *Roscoea*) and a few extend to the warm temperate regions of China and Japan (e.g., some species of *Alpinia*), members of the family are nearly exclusively tropical. Gingers vary in the habitats that they occupy, ranging from seasonally dry, monsoon forests and evergreen wet tropical forests to cooler alpine regions in the Himalayas. It would be difficult to find a tropical lowland or middle-elevation forest in which at least several members of the family are not prominent components of the understory flora. Members of the Zingiberaceae are mostly small- to medium-sized herbs, but a few (e.g., *Alpinia boia* Seem. and *Alpinia carolinensis* Koidz. from Oceania) have vegetative shoots that can attain heights surpassing ten meters. In addition to primary forest understory habitats, some gingers prefer brightly sunlit areas and occupy light gaps in forests, forest margins, and open secondary growth along rivers and streams.

Many members of the Zingiberaceae, mostly in the subfamily Zingiberoideae, are adapted to the monsoonal climates of Southeast Asia (Kress et al. 2002; Wood et al. 2000). These plants have evolved the capacity to go into dormancy during the dry season when almost no rain falls for a period of four to six months. During this time all aboveground photosynthetic and reproductive parts of the plant body are shed and the plants "over-winter" as thick, fleshy underground rhizomes (Figure 4). Some species also possess starch- and water-filled roots or tubers. In addition to these structural adaptations for water storage and retention, some gingers have also evolved a special suture zone at the base of each shoot that causes a rapid abscission of the photosynthetic organ when the rains cease and the arid season approaches. At the end of the dry period just prior to or at the earliest sign of the wet season, individuals will break dormancy by producing new vegetative and reproductive shoots, and complete their life cycle during the intense rainy months before the next dry period begins. It

Figure 4. Underground water storage organs such as fleshy rhizomes and tubers in dormant members of the Zingiberaceae. Left: *Kaempferia candida* Wall.; center: *Zingiber* sp.; right: *Curcuma arracanensis* W. J. Kress, ined. (all photos by W. J. Kress).

is during this period of rejuvenation that water and nutrients stored in the rhizomes, fleshy roots, and tubers are quickly assimilated by the plants for rapid growth and expansion. Some taxa in other families of the Zingiberales, such as the Marantaceae, Costaceae, and Musaceae, have also evolved this dormancy capability and are found in the same dry forest habitats with the Zingiberaceae. Despite the ability of these plants to withstand significant periods of extremely dry conditions, no species of the Zingiberales are naturally found in true desert habitats.

POLLINATORS

The great diversity in form and color of flowers of the Zingiberales in general and the Zingiberaceae in particular (Figure 5) has evolved in response to a diverse array of animal pollinators, including bats, birds, and insects (Kress and Specht 2005). Abiotic pollination by either wind or water has not been reported in the order. Six of the eight families of the Zingiberales contain taxa specialized for pollination by vertebrates, which is the primitive type of pollination in the order. Of these six families, two are exclusively vertebrate-pollinated (Strelitziaceae and Heliconiaceae), and the other four include both vertebrate and insect pollination. The Marantaceae and Lowiaceae are predominantly specialized for insect pollinators.

Figure 5. Representative gingers with examples from both dry and wet forest habitats. Top row (left to right): *Hitchenia glauca* Wall. (dry forest); *Globba wengeri* (C. E. C. Fischer) K. J. Williams (dry forest); *Curcuma attenuata* Wall. (dry forest). Center (left to right): *Kaempferia candida* Wall. (dry forest); *Zingiber kerrii* Craib. (wet forest). Bottom row (left to right): *Hemiorchis rhodorrahchis* K. Schum. (dry forest); *Smithatris myanmarensis* W. J. Kress (wet forest); *Alpinia officinarum* Hance (wet forest; all photos by W. J. Kress).

Within the Zingiberaceae, reports have documented pollination by birds, butterflies, moths, and bees. Although actual field observations are still lacking, floral characteristics, such as flower size and shape and flowering time, also suggest that some members of the Zingiberaceae (e.g., species of *Alpinia* and *Vanoverbergia* in the Pacific islands) may be primarily bat-pollinated (W. J. Kress, unpublished data). In Borneo, sunbird and spiderhunter pollination has been reported for a number of wet forest understory genera, including *Amomum*, *Etlingera*, *Hornstedtia*, *Plagiostachys*, and *Siamanthus* (Classen 1987; Kato et al. 1993; Kato 1996; Larsen and Mood 1998; Sakai et al. 1999). Honeyeaters have been documented as pollinators of the genus *Hornstedtia* in tropical Australia (Ippolito and Armstrong 1993). In the neotropics further fieldwork is needed to investigate the role that hummingbirds play as pollinators of members of the Zingiberaceae (e.g., *Renealmia*; Stiles 1975; Maas 1977; Kress and Beach 1994).

With regard to insect pollination, some genera in the Zingiberaceae, such as *Hedychium*, *Hitchenia*, and *Curcuma*, have flowers with long floral tubes that open in the evenings and produce a strong, sweet fragrance, which are all characteristics suggesting pollination by long-tongued hawkmoths (Mood and Larsen 2001; W. J. Kress, unpublished data). In Southeast Asia, pollination by both small halictid and medium-sized anthophorid (*Amegilla*) bees has been demonstrated (Kato et al. 1993; Kato 1996; Sakai et al. 1999) in a number of genera of the Zingiberaceae (*Alpinia, Amomum, Boesenbergia, Elettaria, Elettariopsis, Globba, Plagiostachys, Zingiber*). In the neotropics, pollination by bees is especially common in species of *Renealmia* (Zingiberaceae; Maas 1977; Kress and Beach 1994). Although reports are not extensive, most species of ginger appear to be pollinated by insects, whereas bird pollination (*Etlingera, Hornstedtia, Amomum*) and bat pollination (a few species of *Alpinia* and *Vanoverbergia*) have evolved independently in these insect-pollinated clades.

A very unusual floral mechanism, called flexistyly (Li et al. 2001, 2002), has evolved in species of *Alpinia* and *Amomum* in Southeast Asia that are pollinated by large bees (e.g., *Xylocopa*). In this mechanism, two specific floral phenotypes are present: cataflexistyle individuals possess flowers with a stigma that is held erect when the flower opens in the morning and turns downward in the late morning after the pollen of the flower has been shed; anaflexistyle individuals have stigmas that are turned downward and are receptive at flower opening then become reflexed backward out of the way of the dehiscing anther of that flower in the late morning. This mechanism promotes pollen movement between individuals of the two floral types in the population and has apparently evolved independently several times to ensure outcrossing and successful pollination (Kress et al. 2005). The great floral diversity in the family Zingiberaceae suggests that many interesting mating systems and floral phenomena, such as flexistyly, are yet to be discovered in these plants. In particular, very few investigations of the pollination ecology of gingers in dry forests have been published.

THE EVOLUTION OF DORMANCY IN GINGERS

The available evidence and data on the ecology, phylogeny, and diversification of the gingers provide a means to understand how seasonality and dormancy originated and evolved in the family. A number of patterns immediately become apparent by mapping the occurrence of seasonality versus evergreen habit onto the molecular phylogeny of the gingers (Figure 6). The first pattern is that two of the four main lineages of the family are characterized by seasonality (Siphonochiloideae and Zingiberoideae) and two lineages by the evergreen habit (Tamijioideae and Alpinioideae). The most basal lineage, the Siphonochiloideae with two genera *Siphonochilus* and *Aulotandra*, is made up of plants that inhabit the seasonal dry zones of Africa and exhibit a distinct period of dormancy. These African plants go dormant for part of the year in a very similar fashion to members of subfamily Zingiberoideae, which are all endemic to the Asian tropics. This basal phylogenetic position of Siphonochiloideae in the family Zingiberaceae suggests that dormancy first evolved in the common ancestor of all extant gingers probably at least 66 mya. However, the presence of an evergreen habit in the second-most primitive lineage, the Tamijioideae (with the single species *Tamijia flagellaris* S. Sakai and Nagam. native to the evergreen wet forests of Borneo), and the more-advanced subfamily Alpinioideae, suggests that the ability to go dormant was

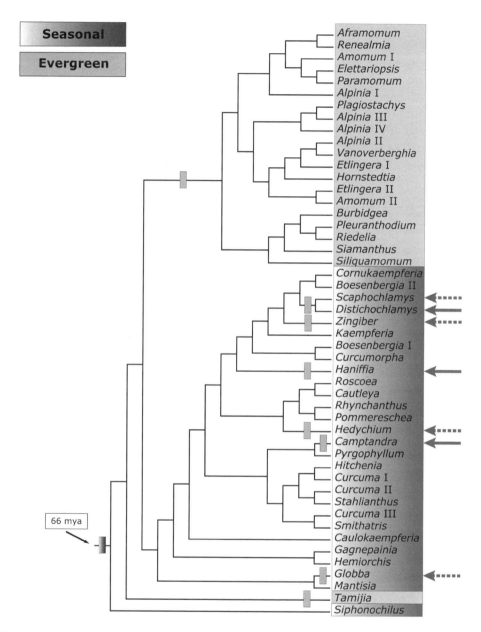

Figure 6. The evolution of seasonal (with a dormancy period) and evergreen habits in the family Zingibera-ceae. Solid gray shading indicates gingers with evergreen shoots and fibrous rhizomes that normally inhabit wet, non-deciduous tropical forests. Gradual gray shading indicates seasonal gingers that undergo a dormant period during the dry season and possess fleshy underground rhizomes, tubers, and/or an abscission zone at the base of each erect shoot. Arrows indicate genera in which the evergreen habit has evolved secondarily within the season clade of gingers (broken arrows are genera in which both evergreen and seasonal species have evolved; solid arrows are genera in which all species are evergreen). The oblique arrow indicates the position of the common ancestor of the family in which seasonality first evolved at least sixty-six million years ago.

independently lost in these two lineages and replaced with an evergreen life cycle by plants occupying wet tropical habitats.

In the advanced subfamily Zingiberoideae, seasonality and the ability to go dormant was maintained in the earliest ancestors of this lineage as they invaded the dry forests of Asia and rapidly diverged about 40 mya. Interestingly, although most species are seasonal within the subfamily Zingiberoideae, the evergreen habit re-evolved independently at least six times as some members (e.g., *Distichochlamys* and *Haniffia*) colonized evergreen forests in tropical Southeast Asia. In several genera (*Scaphochlamys*, *Zingiber*, *Hedychium*, and *Globba*), seasonality and the evergreen habit have repeatedly evolved among closely related species, suggesting that species within these genera have evolutionarily switched between these two modes. It is interesting that within the evergreen subfamily Alpinioideae the ability to become dormant has not evolved a single time, whereas the evergreen habit has repeatedly evolved in lineages of mainly dormant taxa in subfamily Zingiberoideae. These observations suggest that once dormancy is lost in a lineage it is difficult to re-evolve. This hypothesis is supported by the habitat distribution of contemporary species in which some normally dormant genera, such as *Curcuma*, may also occur in evergreen forests, but evergreen taxa, such as *Alpinia* and *Amomum*, are never found in dry forest habitats.

THE EVOLUTION OF DORMANCY IN OTHER ZINGIBERALES

Within the order Zingiberales, in addition to the family Zingiberaceae, dormancy and seasonality have evolved several times in three other families: the Musaceae, the Costaceae, and the Marantaceae (Figure 7). If dormancy is mapped onto the molecular phylogeny of the order, the evergreen habit is clearly ancestral and seasonality has independently evolved in a number of monsoonal forest taxa at least six to eight times. In the banana family, the Musaceae, which is basal in the order, species that respond by going dormant in dry forest habitats are found in all three genera. Some of the Asian species of the genus *Ensete* (e.g., *E. superbum* [Roxb.] Cheesman) and the Chinese *Musella* (*M. lasiocarpa* [Franch.] H. W. Li) exhibit a peculiar type of dormancy in which the pyramidal-shaped herbaceous shoots, which can be several meters in height, do not die back to their underground corms in the dry season, but instead lose the leaves on their stems and maintain the aboveground shoot. When the rains resume, the bulbous stems, which act as water-storage organs, sprout new leaves through their apical meristems. Although almost all members of the third genus of the banana family, *Musa*, are found in evergreen forests, at least one species, *M. laterita* Cheesman found in the dry forests of Myanmar and Thailand, has the ability to shed the entire aboveground stem and leaves during the dry period and resume growth after the rains arrive (W. J. Kress and M. Bordelon, personal observation).

At least one of the three genera in the family Strelitziaceae, *Strelitzia*, occupies dry forest habitats in South Africa, but the plants do not go dormant. Instead, they have evolved thick, leathery leaves that reduce transpiration and desiccation. The other two genera, *Ravenala* and *Phenakospermum*, occupy evergreen wet habitats. All species in the families Lowiaceae and Heliconiaceae inhabit evergreen landscapes in their native

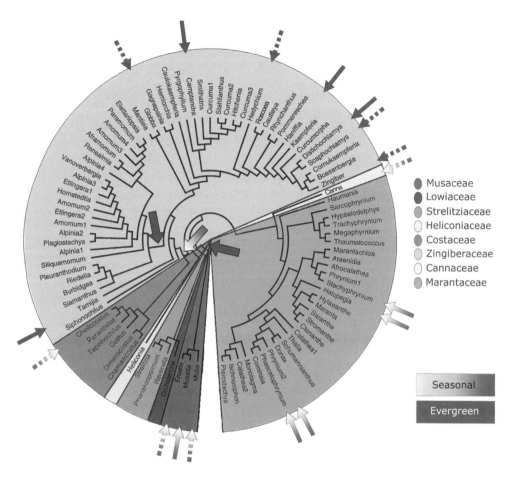

Figure 7. Phylogenetic distribution of seasonality in the order Zingiberales (after Kress and Specht 2005). The wide solid gray arrows in the center indicate the evergreen common ancestor of the order Zingiberales and the origin of the evergreen clade in the family Zingiberaceae. The narrow solid gray arrows around the periphery indicate the independent origins of the evergreen habitat within the dormant clade of seasonal gingers (see Figure 6). The wide gradual gray arrow toward the center indicates the origin of seasonality in the common ancestor of the family Zingiberaceae, while the narrow gradual gray arrows around the periphery indicate the independent evolution of dormancy in the families Musaceae, Costaceae, Cannaceae, and Marantaceae. The broken arrows are genera in which both seasonal and evergreen species occur.

regions of Southeast Asia and the neotropics, respectively. Most species in the family Cannaceae also inhabit wet, evergreen forests or swampy areas in the neotropics and do not go dormant. However, several species of *Canna* (e.g., *C. flaccida* Salisb. and *C. indica* L.), which can grow in subtropical zones, may also show some seasonality and dormancy during the cooler, winter months.

Finally, members of the families Costaceae and Marantaceae, which are found in the dry forests of Southeast Asia, have also evolved dormancy. In the Costaceae, several species of *Cheilocostus* (e.g., *C. speciosus* [J. Koenig] C. D. Specht; *C. lacerus* [Gagnep.] C. D. Specht; and *C. globosus* [Blume] C. D. Specht; formerly in the genus

Costus; Specht 2006) respond to the seasonal change in moisture regime by shedding their leaves or shoots during the dry months. Like the gingers, they quickly respond to the rains and sprout new leaves and shoots followed by flowers in the early part of the rainy season. The family Marantaceae, the second-largest family in the order Zingiberales, although most diverse in the American tropics, is represented by several genera in the Asian dry forests, including *Stachyphrynium*, *Halopegia*, *Phrynium*, and *Phacelophrynium* (P. Suksathan, PhD diss.). Members of these four genera have independently evolved the ability to go dormant in their seasonal habitats, sprouting new leaves and flowering in the wet season.

In addition to members of the order Zingiberales as described above, many other families and genera that inhabit the dry forests of Southeast Asia have evolved their own method of coping with the pronounced dry season. Monocot genera, such as *Stemona* (Stemonaceae), *Cyanotis* and *Murdannia* (Commelinaceae), *Dioscorea* (Dioscoreaceae), *Disporum* and *Tupistra* (Convallariaceae), *Amorphophalus* and *Arisaema* (Araceae), and many others have converged on the same solutions for surviving the water deficit period from November to April in order to complete their life cycle during the six months of the wet season.

CONSERVATION OF DRY FOREST GINGERS

Of the 29 genera and over 500 species in the subfamily Zingiberoideae, most are restricted to the seasonally dry forests of Southeast Asia. Many species of the larger genera, such as *Globba*, *Zingiber*, *Curcuma*, and *Kaempferia*, and several of the smaller genera (e.g., *Cornukaempferia*, *Hitchenia*, *Laosanthus*, *Pyrgophyllum*, and *Smithatris*), which are monotypic or contain only a few species, are endemic to only a single country or a region within a country. In addition to this high level of country endemicity, the very limited number of field collections that have been made by botanists for many of these species also suggests that their geographic distribution is quite narrow (Kress et al. 2003). As a result of this relative rarity, many species of gingers inhabiting the dry forests are most likely threatened and endangered with extinction as these specialized habitats are severely degraded due to unsustainable logging practices and conversion to agriculture. The fact that no species of dry forest ginger has yet been reported to be endangered or extinct is undoubtedly due to a lack of data on their distribution rather than the absence of any threat to their survival.

CONCLUSIONS

This brief review of our current knowledge of the ecology and evolution of the gingers of the understory of seasonally dry forests of Southeast Asia suggests that we have much to learn about the taxonomy and adaptation of these plants to the seasonal monsoon forests of this region. Current data indicate that the ability to respond to a prolonged dry season evolved in these plants at least 66 mya, which suggests that the debate over the possible anthropogenic origin of seasonally dry forests during the past several millennia may be misguided. Comparative investigations of genera that contain species found in both evergreen and dry forest should provide new insights into our understanding of the timing and patterns of adaptation to seasonality. In addition, detailed studies of the pollination biology of dry forest taxa are needed to document how these species focus their reproductive efforts during the contracted growth period of the wet season. Finally, it is clear from recent discoveries of new species and genera of gingers inhabiting the monsoonal forests of Southeast Asia that we are far from a complete inventory of these dry forest understory herbs, yet the rapid degradation of their habitats puts them under significant threat of extinction.

REFERENCES

APG III. 2009. An Update of the Angiosperm Phylogeny Group Classification for the Orders and Families of Flowering Plants: APGIII. *Botanical Journal of the Linnean Society* 161:105–21.

Classen, R. 1987. Morphological Adaptations for Bird Pollination in *Nicolaia elatior* (Jack) Horan (Zingiberaceae). *Gardens' Bulletin of Singapore* 40:37–43.

Harris, D. J., M. F. Newman, M. L. Hollingsworth, M. Möller, and A. Clark. 2006. The Phylogenetic Position of *Aulotandra* Gagnep. (Zingiberaceae). *Nordic Journal of Botany* 23:725–34.

Ippolito, A., and J. E. Armstrong. 1993. Floral Biology of *Hornstedtia scottiana* (Zingiberaceae) in a Lowland Rain Forest of Australia. *Biotropica* 25:281–89.

Kato, M. 1996. Plant-Pollinator Interaction in the Understory of a Lowland Mixed Dipterocarp Forest in Sarawak. *American Journal of Botany* 83:732–43.

Kato, M., T. Itino, and T. Nagamitsu. 1993. Melittophily and Ornithophily of Long-Tubed Flowers in Zingiberaceae and Gesneriaceae in West Sumatra. *Tropics* 2:129–42.

Kress, W. J. 1990. The Phylogeny and Classification of the Zingiberales. *Annals of the Missouri Botanical Garden* 77:698–721.

———. 1995. Phylogeny of the Zingiberanae: Morphology and Molecules. In *Monocotyledons: Systematics and Evolution*, ed. P. Rudall, P. J. Cribb, D. F. Cutler, and C. J. Humphries, 443–60. Kew, UK: Royal Botanic Gardens.

Kress, W. J., and J. H. Beach. 1994. Flowering Plant Reproductive Systems at La Selva Biological Station. In *La Selva: Ecology and Natural History of a Neotropical Rain Forest*, ed. L. A. McDade, K. S. Bawa, H. Hespenheide, and G. Hartshorn, 161–82. Chicago: University of Chicago Press.

Kress, W. J., R. A. DeFilipps, E. Farr, and Daw Yin Yin Kyi. 2003. A Checklist of the Trees, Shrubs, Herbs and Climbers of Myanmar (Revised from the Original Works by J. H. Lace and H. G. Hundley). *Contributions from the U.S. Natural Herbarium* 45:1–590.

Kress, W. J., and K. Larsen. 2001. *Smithatris*, a New Genus of Zingiberaceae from Southeast Asia. *Systematic Botany* 26:226–30.

Kress, W. J., A.-Z. Liu, M. Newman, and Q.-J. Li. 2005. The Molecular Phylogeny of *Alpinia* (Zingiberaceae): A Complex and Polyphyletic Genus of Gingers. *American Journal of Botany* 92:167–78.

Kress, W. J., M. Newman, A. Poulsen, and C. D. Specht. 2007. An Analysis of Generic Circumscriptions in Tribe Alpinieae (Alpinioideae: Zingiberaceae). *The Gardens' Bulletin Singapore* 59:113–28.

Kress, W. J., L. M. Prince, W. J. Hahn, and E. A. Zimmer. 2001. Unraveling the Evolutionary Radiation of the Families of the Zingiberales Using Morphological, Molecular, and Fossil Evidence. *Systematic Biology* 51:926–44.

Kress, W. J., L. M. Prince, and K. J. Williams. 2002. The Phylogeny and a New Classification of the Gingers (Zingiberaceae): Evidence from Molecular Data. *American Journal of Botany* 89:1682–96.

Kress, W. J., and C. D. Specht. 2005. Between Cancer and Capricorn: Phylogeny, Evolution, and Ecology of the Tropical Zingiberales. *Proceedings of a Symposium on Plant Diversity and Complexity Patterns—Local, Regional and Global Dimensions*, ed. I. Friis and H. Balslev, 459–78. Biologiske Skrifter. Copenhagen: Royal Danish Academy of Sciences and Letters.

————. 2006. The Evolutionary and Biogeographic Origin and Diversification of the Tropical Monocot Order Zingberales. In *Monocots: Comparative Biology and Evolution*, 2 vols., ed. J. T. Columbus, E. A. Friar, J. M. Porter, L. M. Prince, and M. G. Simpson, 621–32. Claremont, CA: Rancho Santa Ana Botanic Garden.

Larsen, K., and T. Jenjittikul. 2001. *Laosanthus*, a New Genus of Zingiberaceae from Laos. *Nordic Journal of Botany* 21:135–38.

Larsen, K., and J. Mood. 1998. *Siamanthus*, a New Genus of Zingiberaceae from Thailand. *Nordic Journal of Botany* 18:393–97.

Li, Q.-J., W. J. Kress, Z.-F. Xu, Y.-M. Xia, L. Zhang, X.-B. Deng, and J.-Y. Gao. 2002. Stigmatic Behavior and the Pollination Biology of *Alpinia kwangsiensis* (Zingiberaceae). *Plant Systematics and Evolution* 232:123–32.

Li, Q.-J., Z.-F. Xu, W. J. Kress, Y.-M. Xia, L. Zhang, X.-B. Deng, J.-Y. Gao, and Z.-L. Bai. 2001. Flexible Style that Encourages Outcrossing. *Nature* 410:432.

Maas, P. J. M. 1977. *Renealmia* (Singiberoideae) and Costoideae Additions (Zingiberaceae). *Flora Neotropica*, monograph no. 18. Bronx: New York Botanical Garden.

Mood, J., and K. Larsen. 2001. New Curcumas from South-East Asia. *New Plantsman* 8:207–17.

Poulsen, A. D., J. Mood, and H. Ibrahim. 1999. Three New Species of *Etlingera* (Zingiberaceae) from Borneo. *Nordic Journal of Botany* 19:139–47.

Sakai, S., M. Kato, and T. Inoue. 1999. Three Pollination Guilds and Variation in Floral Characteristics of Bornean Gingers (Zingiberaceae and Costaceae). *American Journal of Botany* 86:646–58.

Sakai, S., and H. Nagamasu. 2000. Systematic Studies of Bornean Zingiberaceae: III. *Tamijia*: A New Genus. *Edinburgh Journal of Botany* 57:245–55.

Specht, C. D. 2006. Systematics and Evolution of the Tropical Monocot Family Costaceae (Zingiberales): A Multiple Dataset Approach. *Systematic Botany* 31:89–106.

Stiles, F. G. 1975. Ecology, Flowering Phenology, and Hummingbird Pollination of Some Costa Rican *Heliconia* Species. *Ecology* 56:285–301.

Williams, K. J., W. J. Kress, and Thet Htun. 2003. A Striking New Epiphytic *Hedychium* (Zingiberaceae) from Myanmar with a Discussion of Several Anomalous Related Genera. *Edinburgh Journal of Botany* 60:43–48.

Wood, T. H., W. M. Whitten, and N. H. Williams. 2000. Phylogeny of *Hedychium* and Related Genera (Zingiberaceae) Based on ITS Sequence Data. *Edinburgh Journal of Botany* 57:261–70.

7

Seasonality in Avian Communities of a Dipterocarp Monsoon Forest and Related Habitats in Myanmar's Central Dry Zone

John H. Rappole, Nay Myo Shwe, Myint Aung, and William J. McShea

Myanmar is an extraordinarily diverse country, stretching over 2,400 km from the China border with Kachin State at 27° N to the Thai border with Tenasserim at 10° N. In addition to this diversity, most habitats in Myanmar are extremely seasonal (Peacock 1933; Smythies 1953). The classic banding studies by H. E. McClure (1974) in Southeast Asia indicate that many members of the avian community in the Old World tropics undergo complex seasonal movements over the course of an annual cycle. Our investigations of the avifauna at Chatthin Wildlife Refuge, which straddles the Tropic of Cancer, similarly have found sharp changes in the composition of the bird community in association with the changes in season (Rappole and Wemmer 1998; Rasmussen et al. 1995). No previous long-term investigation of avian abundance in seasonally dry forest has been performed in Myanmar or anywhere else in Southeast Asia to our knowledge, and there has been no systematic documentation of the seasonal changes in the avian community in this habitat type. As noted by Lekagul and Round (1991, 23), there is very little precise information on avian migration patterns for the region, although several general summaries are available (e.g., King et al. 1987; Lekagul and Round 1991; Robson 2000; Rasmussen and Anderton 2005).

In this study we address the question of seasonality in the avian community of the deciduous dipterocarp forest zone of the Southeast Asian tropics and subtropics, and relate our findings to information reported in the literature.

STUDY AREA

Chatthin Wildlife Sanctuary is located 160 km NNW of Mandalay at 23°43′ N, 95°31′ E in Myanmar's central dry zone. Established in 1941, this sanctuary covers 268.2 km². Elevations at the refuge range from 250 to 500 m. Climate is characterized by a hot rainy season (June–October), a cool dry season (November–February), and a hot dry season (March–May). Basic data on climate, habitats, birds, and other information relating to the site are given in Salter and Sayer (1983). Principal habitats occurring at the refuge and its immediate vicinity include the following:

Dipterocarp Forest—A subtropical dry deciduous forest made up of over 100 tree species, but dominated by several trees of the genus *Dipterocarpus*. This forest type, known as *Indaing* in Myanmar, has a relatively open understory of grasses and low shrubs maintained by regular, anthropogenic spring burns. Dominant trees in the forest lose their leaves in March during the height of the dry season. They leaf out again in June at the onset of the monsoon rains, at which time a lush, herbaceous understory develops.

Mixed Deciduous Forest—This forest type includes both evergreen and deciduous species with teak (*Tectona grandis*), *Xylia doabriformis*, and *Pterrocarpus macrocarpus* as dominant trees, and several species of bamboo (*Bambusa* sp.) in the understory. This habitat is found along streams and on hillsides at Chatthin.

Second Growth—Areas from which semi-evergreen or dipterocarp trees have been cut develop into thickets of woody vegetation within two to three years. Undergrowth becomes gradually thinner as trees mature and shade out the understory, but it takes fifteen to twenty years for young forest to replace dense second growth, during which time, the habitat has its own particular avifauna with significant differences from that found in forests or fields.

Hedgerows—Hedgerows are areas associated with second growth forest that is restricted to a narrow strip, a few meters in width, bordering agricultural fields, roads, and streams.

Rice Paddies, Cane Fields, and Other Agricultural Areas—These are the most common habitats bordering the east and south sides of the refuge. They are mostly dry stubble from November to May, and crop fields from June to October.

Lake, Lake Shore, and Marsh—Several seasonal streams flow through the refuge, one of which is dammed to produce a lake in the northwest portion. Most of these streams, as well as various swales and the rice fields, have some water in them from June to November, but are dry from December to May. Included within this habitat type are areas subject to prolonged flooding during the monsoon period, which develop into natural grasslands often surrounding central pools or marshes. The pools and marshes are dry generally by December or early January. This habitat type is known as *Lwin* in Myanmar.

The systematic, twice-monthly transects that form the main focus for this chapter were conducted entirely in dipterocarp forest.

METHODS

We have conducted field studies at Chatthin Refuge since December of 1994. Initially these studies involved general observations and collection of specimens. Banding studies, visual transect surveys, vegetation transects, and a daily log of ornithological observations were initiated in March of 1996 and were conducted at least biweekly from then until September 2001. A full-time field team of four biologists from the Myanmar Forestry Ministry mist-netted birds, ran visual surveys, and kept a daily log of observations of avian behaviors throughout this period. The data on which this paper focuses are based on 21 km of visual transects in dipterocarp forest that were run twice a month throughout the year from 27 March 1996 to 15 January 1999. Running of these transects involved two researchers: an observer and a recorder. Distance and direction from the observer were recorded for each individual bird seen within 25 m of the transect line. The observer reported all birds associated (i.e., within 10 m) with the first-sighted bird where appropriate, as in intra- or inter-specific flocks. Researchers also kept a daily log by species for ad hoc observations relevant to creating an understanding of the dynamics of the avian community (courtship, nest locations, feeding of young, foods used, predators, territorial encounters, song). The transect data were used to calculate monthly changes in bird numbers by species in dipterocarp forest. These data were also used to determine the seasonality and composition of same-species and mixed-species foraging flocks (King and Rappole 2001a, 2001b, 2002).

RESULTS

A total of 192 species was observed on the transects during the study period, of which 152 were recorded too irregularly or on too few occasions to present discernible patterns. Based on the information in Table 1, and supplemented by data from the literature as presented in the discussion below, the remaining 40 species recorded were assigned to one of the following movement categories: (1) Asian Subtropical Dry Forest Resident—no apparent seasonal movement; (2) Asian Palearctic Migrant; (3) Asian South Temperate/Himalayan Migrant; (4) Asian Subtropical/Intratropical Migrants; and (5) Local/Regional Migrant.

DISCUSSION

The occurrence patterns we have observed, when considered with data from the literature, are suggestive of five major movement types (Table 1). We consider these movement types as working hypotheses to be tested by further investigation involving banding and intensive surveys in neighboring habitats and other sites in Myanmar's central dry zone, and similar activities elsewhere in Myanmar and other countries of Southeast Asia and beyond. In our discussion of movement types, we define each type according to its breeding destination (e.g., species that breed in the Asian Palearctic and winter in the Asian subtropics and tropics are defined as Asian Palearctic Migrant).

Movement Types

Asian Subtropical Dry Forest Resident

This movement type is represented in Table 1 by 20 species, of which the Small Minivet is exemplary (Figure 1). Populations of these birds appear to be largely sedentary. Changes in numbers observed do occur over the course of the annual cycle, but these seem likely to be related to production of offspring or stochastic factors relating to

Table 1. Average number of birds detected on transects by month for three years (27 March 1996–15 January 1999) for species with sufficient sample size.

Species	Jan	Feb	Mar	Apr	May	Jun	Jul	Aug	Sep	Oct	Nov	Dec	Movement Type
Chinese Francolin *Francolinus pintadeanus*	0	1.5	1.0	1.0	1.7	2.7	2.0	0	0.3	0	0	0.7	Local/Regional Migrant (5)
Red-wattled Lapwing *Vanellus indicus*	0	5.0	1.0	0	2.0	0	0	0	1.3	0	0	0	Local/Regional Migrant (5)
Alexandrine Parakeet *Psittacula eupatria*	0.7	3.0	5.3	8.3	5.3	0.3	0	6.5	3.0	0.7	1.3	4.0	Asian Subtropical Dry Forest Resident (1)
Indian Roller *Coracias benghalensis*	0	2.5	8.3	6.3	10.3	5.0	2.0	1.5	2.0	1.3	0.3	0	Local/Regional Migrant (5)
Green Bee-eater *Merops orientalis*	0	7.5	9.3	11.0	7.7	1.3	1.7	0	5.7	3.0	2.0	0	Local/Regional Migrant (5)
Blue-tailed Bee-eater *Merops philippinus*	0	0	1.0	0	0	0	0	0	0	0	0	0	Asian South Temperate/Himalayan Migrant (3)
Eurasian Hoopoe *Upupa epops*	1.0	2.5	8.0	6.3	1.7	3.0	3.0	3.0	1.7	4.7	0.3	1.7	Asian Subtropical Dry Forest Resident (1)
Linneated Barbet *Megalaima lineata*	1.0	2.5	3.3	3.0	2.7	0.7	1.0	1.5	1.3	0.3	0.7	1.0	Asian Subtropical Dry Forest Resident (1)
Grey-capped Pygmy Woodpecker *Dendrocopos canicapillus*	6.3	2.5	5.3	6.7	2.3	4.3	2.0	1.0	7.0	5.3	6.3	4.7	Asian Subtropical Dry Forest Resident (1)
Yellow-crowned Woodpecker *Dendrocopos mahrattensis*	2.7	1.5	2.7	2.7	4.3	5.7	1.3	1.5	3.0	0.7	2.0	2.3	Asian Subtropical Dry Forest Resident (1)
White-bellied Woodpecker *Dryocopus javensis*	0.3	1.5	1.3	0	0.7	0.3	1.3	0.5	0.7	1.0	1.3	1.0	Asian Subtropical Dry Forest Resident (1)

Species													Status
Streak-throated Woodpecker *Picus xanthopygaeus*	0.7	0	0	3.0	1.0	2.0	2.7	1.5	0.3	0.7	1.7	1.3	Asian Subtropical Dry Forest Resident (1)
Common Goldenback *Dinopium javanense*	2.7	2.5	1.3	2.7	3.7	5.0	3.0	2.5	1.0	4.3	1.0	2.0	Asian Subtropical Dry Forest Resident (1)
Common Woodshrike *Tephrodornis pondicerianus*	26.0	25.0	37.0	36.0	25.7	37.7	33.7	28.0	136.0	35.3	29.0	19.7	Asian Subtropical Dry Forest Resident (1)
Common Iora *Aegithina tiphia*	8.7	0.5	3.7	4.3	2.3	12.0	4.3	2.5	10.0	5.7	10.3	5.3	Asian Subtropical Dry Forest Resident (1)
Ashy Minivet *Pericrocotus divaricatus*	0	0	2.7	0	0	0	0	0	0	0	0	0	Asian Palearctic Migrant (2)
Small Minivet *Pericrocotus cinnamomeus*	87.7	52.5	116.0	54.0	68.0	85.3	45.7	50.5	83.7	89.3	55.3	81.3	Asian Subtropical Dry Forest Resident (1)
Black-naped Oriole *Oriolus chinensis*	0	0	10.7	7.7	12.0	7.0	2.33	2.0	4.0	0.7	0.7	0	Asian Subtropical/Intratropical Migrant (4)
White-browed Fantail *Rhipidura aureola*	11.7	26.0	14.0	13.3	13.7	25.7	17.0	23.0	26.3	26.3	15.3	15.7	Asian Subtropical Dry Forest Resident (1)
Black Drongo *Dicrurus macrocercus*	4.7	7.5	27.3	53.0	22.7	29.3	10.7	4.0	2.0	4.0	4.3	3.3	Asian Subtropical/Intratropical Migrant (4)
Ashy Drongo *Dicrurus leucophaeus*	0.3	0.5	1.7	2.7	4.3	5.3	0.3	0.5	0.7	0	0	0.3	Asian Subtropical/Intratropical Migrant (4)
Asian Paradise Flycatcher *Terpsiphone paradisi*	0	0	0	1	1	2.7	0	0	0	0.3	0	0	Asian Subtropical/Intratropical Migrant (4)
Red-billed Blue Magpie *Urocissa erythrorhyncha*	2.7	0.5	0.3	4.3	0	0	0	0	0	0	1.7	0.3	Asian South Temperate/Himalayan Migrant (3)

(Continued)

Table 1. Average number of birds detected on transects by month for three years (27 March 1996–15 January 1999) for species with sufficient sample size. (Continued)

Species	Jan	Feb	Mar	Apr	May	Jun	Jul	Aug	Sep	Oct	Nov	Dec	Movement Type
Rufous Treepie *Dendrocitta vagabunda*	1.7	3.5	7.7	1.7	4.7	5.3	6.3	5.5	0.3	1.7	1.0	1.0	Asian Subtropical Dry Forest Resident (1)
Great Tit *Parus major*	6.0	6.5	7.0	5.0	6.3	5.7	3.0	7.5	4.7	3.0	4.3	4.0	Asian Subtropical Dry Forest Resident (1)
Singing Bush Lark *Mirafra cantillans*	1.7	1.5	1.0	0	0	0	0.7	0	0	3.3	2.0	0	Asian South Temperate/Himalayan Migrant (3)
Common Tailorbird *Orthotomus sutorius*	12.7	3.5	0.7	2.0	2.3	8.3	10.3	10.0	20.7	22.7	10.7	12.33	Asian Subtropical Dry Forest Resident (1)
Red-vented Bulbul *Pycnonotus cafer*	3.3	31.0	24.3	58.0	29.0	35.7	22.3	39.5	13.7	10.3	9.0	3.0	Local/Regional Migrant (5)
Streak-eared Bulbul *Pycnonotus blanfordi*	0.7	2.0	1.0	5.0	2.7	1.0	2.7	6.0	0.7	1.0	0.3	0.7	Asian Subtropical Dry Forest Resident (1)
Greater-necklaced Laughingthrush *Garrulax pectoralis*	17.7	5.5	7.7	11.3	9.0	5.0	7.7	23.0	6.0	5.3	15.3	7.7	Asian Subtropical Dry Forest Resident (1)
Chestnut-bellied Nuthatch *Sitta castanea*	15.7	13.5	13.3	11.3	21.7	27.7	18.0	21.5	28.3	15.0	18.0	13.7	Asian Subtropical Dry Forest Resident (1)
Vinous-breasted Starling *Sturnus burmannicus*	0	9.0	2.7	10.7	2.0	12.0	3.7	0	0	0	0	0	Asian Subtropical/Intratropical Migrant (4)
Oriental Magpie Robin *Copsychus saularis*	0.7	1.0	2.0	2.0	1.3	2.7	2.7	0.5	0.7	0.3	2.3	0.3	Asian Subtropical Dry Forest Resident (1)
Grey Bush Chat *Saxicola ferreus*	0	0.5	0	0	0	0	0	0	0	0	0.7	1.3	Asian South Temperate/Himalayan Migrant (3)
Red-breasted Flycatcher *Ficedula parva*	10.3	4.0	2.0	1.3	0.7	0	0.3	0	3.3	23.0	13.3	9.3	Asian Palearctic Migrant (2)

Species													Category
Blue-throated Blue Flycatcher *Cyornis rubeculoides*	1.0	0	0	0	0	0	0	0	0.3	0.3	1.3	0.3	Asian South Temperate/Himalayan Migrant (3)
Golden-fronted Leafbird *Chloropsis aurifrons*	3.0	4.5	2.7	2.7	1.7	1.7	1.3	0.5	2.0	2.7	4.0	5.3	Asian Subtropical Dry Forest Resident (1)
Plain-backed Sparrow *Passer flaveolus*	0	9.5	17.7	21.0	10.0	1.7	0.7	0	0	0	0	0	Local/Regional Migrant (5)
Grey Wagtail *Motacilla cinerea*	0	0.5	4.7	0.7	0.7	0	0	0	1.7	0	0	0.3	Asian Palearctic Migrant (2)
Olive-backed Pipit *Anthus hodgsoni*	15.3	7.0	5.3	0.3	2.7	0.7	0	3.0	2.0	9.7	24.7	11.0	Asian South Temperate/Himalayan Migrant (3)

Note: Nomenclature and taxonomic order from Gill and Wright (2006).

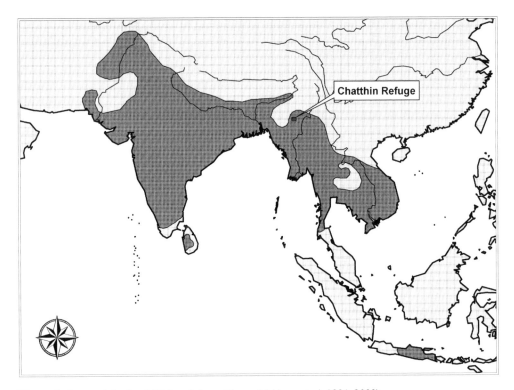

Figure 1. Range of the Small Minivet (adapted from del Hoyo et al. 1986–2008).

probability of flock encounters on the transect by observers. Individuals, pairs, or family groups of many of these species are common members of single-species or mixed-species flocks in dipterocarp forest (King and Rappole 2001a).

Perhaps unexpectedly, year-round residency is a difficult movement type to confirm because it is often hard to establish when large portions of the population are absent, so long as a few individuals remain. For instance, the avian community in subtropical, deciduous thorn scrub and second growth in the semi-arid, inland areas of south Texas is reported to consist largely of resident species (Oberholser 1974), but year-round, intensive sampling has revealed that, at least in one Texas county, there are no species in which the majority of the population is resident throughout the annual cycle (Vega and Rappole 1994). Therefore, good samples are required from throughout the year, as in the case of the minivet, and even then, it is hard to be certain without long-term banding data that no migratory movements are taking place. Based on our data, about 50 percent of the dipterocarp forest community (20 of 40 species with sufficient sample size) is composed of resident species.

Asian Palearctic Migrant

Members of this group breed in boreal and temperate regions of the Asian Palearctic and winter in the Asian subtropics and tropics. They are recorded at Chatthin only during one or more periods corresponding to Palearctic fall (September–October), winter (November–February), or spring (March–May), presumably as transients and/ or winter residents. Nay Myo Shwe et al. (1999) reported that 44 of the 236 species recorded for the Chatthin Refuge fall into this category. However, most of these occur in open or wetland habitats. Only three Asian Palearctic Migrants (8 percent of the 40 species in Table 1) occur regularly in dipterocarp forest: two as winter residents, the Red-breasted Flycatcher (Figure 2) and the Grey Wagtail; and one, the Ashy Minivet, as a spring transient (Table 1). The pattern of occurrence seen in the Red-breasted Fly-catcher illustrates two notable aspects of this group. The first is that birds begin arrival from Palearctic breeding habitats in September, but they remain as winter residents only until February when the majority seem to leave Chatthin dipterocarp forest. We do not know where these birds go, but it is obviously not to their breeding grounds in temperate and boreal Eurasia, where they are not recorded until April (Taylor 2006). Rappole and Jones (2003, 527) note that "movement during the wintering season (Nov–Mar), is typical for many migrants exploiting Afrotropical savanna habitats, a group that constitutes as much as 20 percent of the savanna avian community (Jones 1998)." The authors go on to state that the majority of these species are insectivores (like the Red-breasted Flycatcher) for which available resources in extremely seasonal African habitats change over the course of the wintering period. It is possible that a

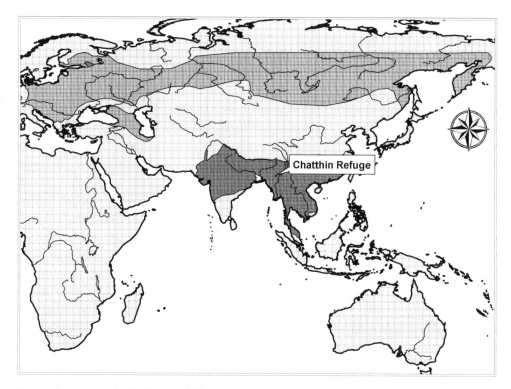

Figure 2. Range of the Red-breasted Flycatcher. Light gray = breeding range; dark gray = winter range (adapted from Taylor 2006).

similar explanation applies to the Red-breasted Flycatcher at Chatthin. The Grey Wagtail demonstrated a similar pattern, arriving at Chatthin as a winter resident in September but remaining only until December or early January (J. Rappole, field notes), by which time muddy borders to intermittent streams and depressions, the principal habitat for this species in dipterocarp forest, were mostly dry.

A second noteworthy aspect of Red-breasted Flycatcher data for Chatthin is the presence of a few individuals through the Palearctic summer (May–August). The tendency for some individuals of Nearctic or Palearctic migrant species to remain at transient or wintering sites through the period corresponding to the northern summer has been well-documented in shorebirds (Rappole and Blacklock 1985; Morrison and Myers 1989). However, this phenomenon, although not unrecorded, is poorly documented for forest-related species (Rappole 1995). P. D. Round, in comments on an earlier draft of this chapter, points out that the few summer records for Red-breasted Flycatchers at Chatthin could represent misidentifications. This explanation is, of course, quite possible. Nevertheless, the fact that recent theories concerning the development of migratory behavior in populations posit rapid change in migration patterns based on success or failure of individual choices in movement (Rappole and Jones 2003; Rappole et al. 2003) make the presence of these individuals of considerable interest in serving to emphasize the need to document such occurrences more thoroughly.

Asian South Temperate/Himalayan Migrant

Members of this group breed in South Asian mountains and east Asian temperate regions and occur at Chatthin as transients and/or winter residents. Six of 40 species in Table 1 (15 percent) showed this pattern, and information from other sources indicate an additional 17 species that occur at Chatthin may fit this category (Table 2). The distinction between members of this and the previous group (Asian Palearctic Migrant) is not always clear. Both groups have a similar occurrence pattern at Chatthin, i.e., they are found during one or more of the months from September to May. However in general, members of the Asian South Temperate/Himalayan Migrant grouping have breeding areas located much farther south on average than those of most of the Asian Palearctic migrants. In addition, many members of the group have breeding populations in both the subtropics and tropics, often broadly overlapping with wintering areas for individuals from their Asian Palearctic populations. Several of these species are altitudinal migrants, breeding in the temperate habitats of the Himalayas or other mountainous regions, and wintering in neighboring or more-southern lowlands. The Red-billed Blue Magpie (Figure 3), which is a member of this group, appears to be principally a seasonal transient in Chatthin's dipterocarp forest, with highest numbers in April. What separates members of this group from the following is the timing of their apparent movements. South Temperate/Himalayan Migrant movements appear to coincide with temperate seasonal change (i.e., the birds occur as transients at Chatthin in fall or spring and/or as winter residents in winter). Subtropical/Intratropical Migrant movements seem to coincide with monsoon climatic patterns.

Table 2. List of species hypothesized as South Temperate/Himalayan migrants based on Chatthin occurrence data, specimen data, and/or information in the literature as cited.

English Name	Scientific Name	Evidence
Fulvous Treeduck	Dendrocygna bicolor	del Hoyo et al. 1986–2008
Spot-billed Duck	Anas poecilorhyncha	King et al. 1987
Black Stork	Ciconia nigra	Rasmussen and Anderton 2005
Spot-billed Pelican	Pelecanus philippensis	Rasmussen and Anderton 2005
Indian Shag	Phalacrocorax fuscicollis	Rasmussen and Anderton 2005
Black Kite	Milvus migrans	King et al. 1987
Red Turtle-Dove	Streptopelia tranquebarica	Smythies 1953
Large Hawk-Cuckoo	Cuculus sparverioides	Banding data (McClure 1974); specimen collected at Chatthin with heavy fat, 7 Dec 1994; Robson 2000.
Brown Hawk-Owl	Ninox scutulata	Banding data (McClure 1974); specimen collected at Chatthin with heavy fat, 3 Dec 1994.
Long-tailed Nightjar	Caprimulgus macrurus	Specimen collected at Chatthin with heavy fat, 3 Dec 1994.
Brown Needletail	Hirundapus giganteus	Rasmussen and Anderton 2005
Blue-tailed Bee-Eater	Merops philippinus	Chatthin transect data (Table 1); Robson 2000
Grey-backed Shrike	Lanius tephronotus	Smythies 1953
Red-billed Blue Magpie	Urocissa erythrorhyncha	Chatthin transect data (Table 1)
Singing Bushlark	Mirafa javanica	Chatthin transect data (Table 1); Smythies 1953
Blue Rock-Thrush	Monticola solitarius	Smythies 1953
Grey Bushchat	Saxicola ferrea	Chatthin transect data (Table 1); Smythies 1953
Little Pied Flycatcher	Ficedula westermanni	Smythies 1953
Pale Blue Flycatcher	Cyornis unicolor	Rasmussen and Anderton 2005
Blue-throated Blue Flycatcher	Cyornis rubeculoides	Chatthin transect data (Table 1); Smythies 1953
Grey-headed Canary-Flycatcher	Culicicapa ceylonensis	Smythies 1953
Richard's Pipit	Anthus richardi	Smythies 1953
Olive-backed Pipit	Anthus hodgsoni	Chatthin transect data (Table 1); Smythies 1953

Note: Nomenclature and taxonomic order from Gill and Wright (2006).

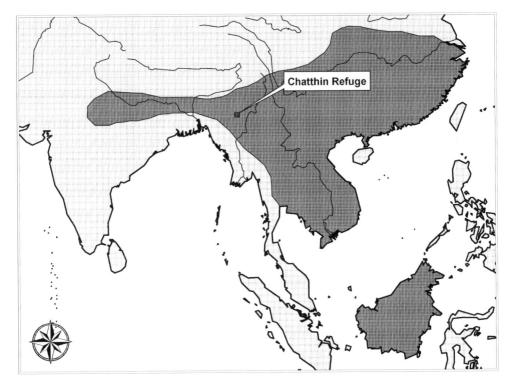

Figure 3. Range of the Red-billed Blue Magpie (adapted from del Hoyo et al. 1986–2008).

Subtropical/Intratropical Migrant

Members of this group appear to breed in the subtropical dry forest during the hot dry season and early rainy season (February–August), and migrate south to winter by the middle or end of the rainy season. Five species seem to fit this pattern based on transect data (Table 1) including the Black Drongo (Figure 4). Data from a single site are not sufficient to confirm most movement types, but especially this type of movement. In the case of the drongo, there are additional data from the literature, e.g., Smythies (1953, 174), who comments on apparent migration by Black Drongos from breeding areas in the central dry zone to wintering areas in southern Myanmar, as does King et al. (1987). As in the case of the Black Drongo, some individuals of other species in this group can be found throughout the year in Chatthin's dipterocarp forest, although at much lower numbers during the non-breeding period (September–January) than during the breeding period (February–July). This pattern is typical of many subtropical and intratropical migrants, much of whose breeding and non-breeding distributions overlap to a greater or lesser extent.

Local/Regional Migrant

The species that fit this category demonstrated very clear patterns of presence and absence at Chatthin, appearing mainly during the hot dry season (February–May)

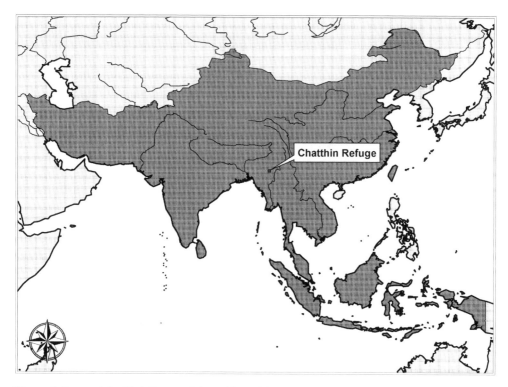

Figure 4. Range of the Black Drongo (adapted from del Hoyo et al. 1986–2008).

and, for some species, portions of the early rainy season (June–August), and then be-ing largely or entirely absent for the remainder of the year (September–January). This pattern is very similar to that of the Subtropical/Intratropical Migrant. The difference is in the reported evidence of long-distance migration from elsewhere in the range (i.e., disappearance from a breeding area in the north with concomitant appearance in non-breeding areas to the south). The species in our Local/Regional Migrant category have not been recorded in the literature as demonstrating long-distance migratory move-ments. Therefore, it seems likely that the movements that seem apparent at Chatthin are local or regional in nature. The Red-vented Bulbul is exemplary. Although reported as resident throughout its range (Figure 5), local seasonal movements have been re-ported in the literature (Fishpool and Tobias 2005). This species demonstrates distinct seasonal occurrence patterns at Chatthin as do five other resident species (Table 1).

CONCLUSIONS

There are obvious reasons for caution in interpreting our data as revealing the five movement patterns that we propose. For instance, despite a relatively uniform level of sampling effort extending across several years, there were marked variations in

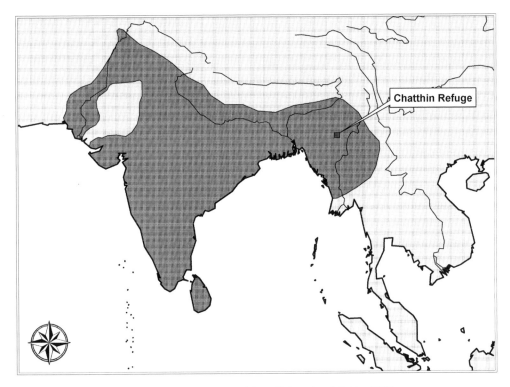

Figure 5. Range of the Red-vented Bulbul (adapted from Fishpool and Tobias 2005).

species detectability based on weather conditions (heavy rain or extreme heat are common occurrences during different times of the year), visibility (birds are easier to see when trees are leafless as in March and April), and behavior (singing birds during the breeding season are much easier to detect; Rappole et al. 1998). Also, many species in dipterocarp forest forage in mixed-species flocks during most of the year, the exception being during the nesting period (King and Rappole 2001a). When these birds are in flocks, they are easier to find than when they are solitary or in pairs. Nevertheless, we believe our data provide a sound basis for further investigation of avian seasonal movements in seasonally dry forest habitats of the South Asian tropics and subtropics.

The data presented in this chapter provide a very clear picture of the extraordinary diversity and complexity of avian seasonal movements in subtropical Asian dry forest. As shown in Table 1, it is not uncommon for the populations of a large number of the avian species found in this habitat to vary by one or more orders of magnitude from one season to the next. We have suggested five major movement types based on the seasonal abundance patterns observed, but it is probable that other types exist because different movement types could have similar seasonal abundance patterns. For instance, species that disappear from dry forest during the summer rainy period may have migrated elsewhere regionally, or simply have moved into a different habitat, e.g., rice fields. Obviously, development of an understanding of movement patterns requires long-term information from more than a single site.

We propose three possible causes for the fact that Asian subtropical dry forest is among the most diverse in the world in terms of avian movement patterns: (1) Continental climate patterns bringing migrants from the Palearctic, (2) Regional monsoon climate patterns involving semiannual alternation between wet and dry seasons, and 3) Strategic geographic location between the temperate and tropical regions. In addition, at Chatthin there is a fourth possible factor at work, namely, sharp seasonal changes in the relative availability of certain food types in neighboring habitats (e.g., dry forest and rice fields).

ACKNOWLEDGMENTS

We would like to acknowledge the help and support of the following individuals: C. Wemmer, C. Pickett, Uga, Lei Lei Hnin, Thida Swe, Thein Win, and Khyne U Mar. We would also like to thank in particular the Forestry Ministry and the Nature and Wildlife Conservation Division of the Forestry Department of Myanmar. This research was made possible by support from the Friends of the National Zoo, the National Geographic Society, and British Air.

REFERENCES

del Hoyo, J., J. A. Elliott, J. Sargatal, and D. Christie, eds. 1986–2008. *Handbook of the Birds of the World*, vols. 1–13. Barcelona: Lynx Edicions.

Fishpool, L. D. C., and J. A. Tobias. 2005. Red-vented Bulbul. In *Handbook of Birds of the World*, ed. J. del Hoyo, A. Elliott, J. Sargatal, and D. Christie, 10:184–85. Barcelona: Lynx Edicions.

Gill, F. B., and M. Wright. 2006. *Birds of the World*. Princeton, NJ: Princeton University Press.

Jones, P. 1998. Community Dynamics of Arboreal Insectivorous Birds in African Savannas in Relation to Seasonal Rainfall Patterns and Habitat Change. In *37th Symposium of the British Ecological Society*, ed. D. M. Newberry, H. H. T. Prins, and N. D. Brown, 421–47. London: Blackwell Science.

King, B., M. Woodcock, and E. C. Dickinson. 1987. *A Field Guide to the Birds of South-East Asia*, 2nd ed. London: Houghton.

King, D. I., and J. H. Rappole. 2001a. Mixed-Species Bird Flocks in Dipterocarp Forest of North-Central Burma (Myanmar). *Ibis* 143:380–90.

———. 2001b. Kleptoparasitism of Laughingthrushes *Garrulax* by Greater Racket-tailed Drongos *Dicrurus paradiseus* in Myanmar. *Forktail* 117:121.

———. 2002. Commensal Foraging Relationships of the White-browed Fantail (*Rhipidura aureola*) in Myanmar (Burma). *Journal of the Bombay Natural History Society* 99 (2): 308–12.

Lekagul, B., and P. D. Round. 1991. *A Guide to the Birds of Thailand*. Bangkok: Saha Karn Bhaet.

McClure, H. E., III. 1974. *Migration and Survival of the Birds of Asia*. Bangkok: U.S. Army Medical Component, SEATO Medical Project.

Morrison, R. I. G., and J. P. Myers. 1989. Shorebird Flyways in the New World. In *Flyways and Reserve Networks for Waterbirds*, ed. H. Boyd and J.-Y. Pirot, 85–96. Slimbridge, UK: International Waterfowl and Wetlands Research Bureau.

Nay Myo Shwe, Thida Swe, Lei Lei Hnin, Htein Win, and J. H. Rappole. 1999. *A Guide to the Birds of Chatthin Wildlife Sanctuary, Union of Myanmar*. Front Royal, VA: Smithsonian Conservation and Research Center.

Oberholser, H. C. 1974. *Bird Life of Texas*. Austin, TX: University of Texas Press.

Peacock, E. H. 1933. *A Game-Book for Burma & Adjoining Territories*. London: H.F. & G. Witherby.

Rappole, J. H. 1995. *The Ecology of Migrant Birds*. Washington, DC: Smithsonian Institution Press.

Rappole, J. H., and G. W. Blacklock. 1985. *Birds of the Texas Coastal Bend*. College Station: Texas A&M University Press.

Rappole, J. H., B. Helm, and M. Ramos. 2003. An Integrative Framework for Understanding the Origin and Evolution of Avian Migration. *Journal of Avian Biology* 34:124–28.

Rappole, J. H., and P. Jones. 2003. Evolution of Old and New World Migration Systems. *Ardea* 90 (3; special issue): 525–37.

Rappole, J. H., and C. Wemmer. 1998. Birds, Trees, and Biodiversity in Myanmar. *Myanmar Forestry Journal*, no. 2, 1998.

Rappole, J. H., K. Winker, and G. V. N. Powell. 1998. Migratory Bird Habitat Use in Southern Mexico: Mist Nets versus Point Counts. *Journal of Field Ornithology* 69:635–43.

Rasmussen, P., and J. Anderton. 2005. *Birds of South Asia*. Washington, DC: Smithsonian Institution Press.

Rasmussen, P., J. H. Rappole, and J. Anderton. 1995. New Records for Birds from the Union of Myanmar. *Oriental Bird Club Bulletin* (May 1995).

Robson, C. 2000. *A Guide to the Birds of Southeast Asia*. Princeton, NJ: Princeton University Press.

Salter, R. E., and J. A. Sayer. 1983. *Kyatthin Wildlife Sanctuary: Draft Management Plan*. Nature Conservation and National Parks Project FO:BUR/80/006. Field Document 6/83. Rangoon: Food and Agriculture Organization of the United Nations.

Smythies, B. E. 1953. *The Birds of Burma*. Edinburgh: Oliver and Boyd.

Taylor, P. B. 2006. Red-breasted Flycatcher. In *Handbook of Birds of the World*, ed. J. del Hoyo, A. Elliott, J. Sargatal, and D. Christie, 11:135–36. Barcelona: Lynx Edicions.

Vega, J. H., and J. H. Rappole. 1994. Composition and Phenology of an Avian Community in the Rio Grande Plain of Texas. *Wilson Bulletin* 106:366–80.

A Trip along River Cauvery

A. J. T. Johnsingh and R. Raghunath

Dry deciduous forests flank river Cauvery as it tumbles down the Deccan Plateau along scenic Hogenakal Falls to the plains of Tamil Nadu. The surrounding terrain is hilly, with numerous deep valleys and *nallahs*, and clouds swirl about the mountain tops fifteen hundred meters high. This landscape constitutes almost twelve hundred square kilometers of potential habitat for large mammals. It is part of one of the largest contiguous elephant habitats in Asia, and it stretches from Silent Valley National Park in the south and Aralam Wildlife Sanctuary in the west to the Bannerghatta forests in the north. On the Karnataka side there are the Cauvery forests administered by the Cauvery Wildlife Division and the Kanakapura and Kollegal Forest. The forests on the Tamil Nader side come under the Erode, Dharmapuri, and Kollegal Forest Divisions.

The landscape, which remains bone dry and hot in summer, is dotted with numerous human settlements and shrines. Thousands of pilgrims visit the popular Shiva temple in the Malai Mahadeshwara (MM) Hills, the Hanuman temple at Muthathi, and the Mariamman temple at Gopinatham. Unfortunately, the pilgrims leave behind an unsightly and polluting trail of garbage, mostly nonbiodegradable, such as plastic bottles and bags. The major tribal communities are the Sholigas, largely in Karnataka, and the Irulas, mainly in Tamil Nadu. The Sholigas are especially adept in the jungle and know the terrain around their homes.

In mid-July 2009, we were in the MM Hills on the right bank of river Cauvery to carry out a rapid survey of the landscape for large mammals. Around us, the dry deciduous forest was dominated by trees such as *Albizzia amara*, *Chloroxylon swietinia*, and *Hardwickia binata*; and shrubs such as *Dodonea viscosa*, *Carissa carandus*, *Capparis sepiaria, and Fluggea leucopyrus*. We tried to evade the hard, curved thorns of *Pterolobium indicum*—the most common thorny straggler here and one that can stop even elephants in their tracks. We heard grey partridges calling merrily from different directions. As the western sky turned orange, we descended to the Palar River, an ephemeral tributary of the Cauvery that forms the boundary between Karnataka and Tamil Nadu. We searched for animal signs along a three-kilometer stretch of the Palar's white, sandy river bed and found elephant tracks and dung leading to a sliver of water in the middle of the riverbed, hardly one hundred meters from the bridge across the river. There were many chital tracks and a few of sambar and wild pigs. There were also numerous tracks of cattle from the nearby Sholiga colony, hardly a kilometer from the interstate boundary. We were not looking for a tiger pugmark (footprint), which is exceedingly rare in this landscape, but we hoped to see signs of both leopard and sloth bear. We saw neither. As darkness descended, we left the river and climbed to the Salem-Mysore road. A common langur crossed the road, hurrying to the safety of its roost, as we did to our vehicle parked near the bridge. The staff at the forest checkpost there told us that some time after we had crossed the river bed, a big elephant had come to the river for a drink. Since it was big, alone, and tuskless, we assumed it was a *makhna* (tuskless male). On the following thirty-kilometer night-drive to our next destination, the forest rest-house in MM Hills, we didn't see a single large mammal.

Not long before, these forests had been the haunt of the dreaded brigand K. M. Veerappan. The many caves on the now quiet mountain tops had been a perfect shelter for this criminal who held sway over the Cauvery forests for over two decades. His "business" was the smuggling of precious sandalwood and the peddling of elephant tusks.

As we walked our survey routes, the forest staff accompanying us pointed out in a macabre sight-seeing tour the many places where this bandit had waylaid staff and murdered others. The forest guard in the MM Hills seemed to think that Veerappan's presence had deterred other poachers, and that chital had been abundant during his reign. In the opinion of the guard, Veerappan's death in 2004 had emboldened new poachers to systematically decimate the wildlife.

The next morning, koels called plaintively to the monsoon, as though pleading for a full rain. Although we were well into the rainy season, the landscape till then had received only a few light showers. After we had made a morning visit to Sorekaimadu and its grisly landmine pits, the forest guard took us to Aler halla where a year before ten gaur had been seen. We saw neither gaur nor any signs, but there were numerous local cattle with beautiful patterns of white on their red and brown skins. Enjoying the occasional drizzle, we walked about three kilometers along the mostly dry *nallah*, with some water in the pools, and the only wildlife sign we saw was the spoor of an otter that had walked along the *nallah*. We did see a lot of evidence of fishing by the local people, and most of the pools stank because fish entrails had been left rotting in the water.

At the Palar Forest bungalow we met C. Putta, a fifty-year-old forest watcher who narrated his encounter with a mother sloth bear and her two cubs nearly twenty years ago. He told of how he had made the mistake of trying to frighten away the bears by throwing a stone. The mother bear had charged and bitten him on the right knee. Putta retaliated by slashing at her with his knife, cutting her snout and nearly severing it. At this, she had screamed and run away. The bear, unable to eat or drink, was found dead a week later.

In the evening we drove to Hogenakal Falls and returned just as it was getting dark. Although the habitat all along the road to Hogenakal is potentially good for chital, sambar, and gaur, with dry deciduous forest on one side and the Cauvery on the other side, we didn't see a single mammal on this fifty-kilometer drive. Just before reaching the MM Hills forest rest-house, though, we chanced upon a pair of jackals that in the headlights of our car ran across the road.

Trees swayed in the howling wind as if seized by a demon as we descended from *Tala betta* to Gopinatham village. The sixteen-kilometer trail dropped from seven hundred to two hundred meters. The forest staff had told us that the valley along this trail would be the best place to see large mammals. A year ago Divisional Forest Officer Kumar Pushkar had seen a tiger pugmark on this trail. Spotted babblers sang cheerfully as we walked on the tracks of sloth bear, leopard, and elephant. On the last lap of our walk, a strong wind and heavy drizzle lashed us, forcing us to race along the edge of a reservoir to our destination for the night—a nearly century-old forest bungalow on its shore.

Having seen portions of the forests on the Karnataka side, we decided to drive back to Mysore through the forests on the left bank of river Cauvery. Near Hogenakal and Muthathi we saw numerous tourists camping and cooking by the side of the river using firewood. A few were even cooking huge quantities of fish obviously caught illegally in the river. Our overnight stay in a fishing camp, run by the Karnataka government, was exceedingly useful as we could discuss conservation issues with Sundar, the manager of the camp. Sundar knows the riverine tract like the back of his hand. We were delighted to see several endangered grizzled giant squirrels (*Ratufa macroura*) around the camp. This is a species endemic to southern India and Sri Lanka. On our way to the camp we had driven through nearly three hundred kilometers of forest, and walked another twenty-five kilometers, but had seen only six chital for all the effort. We were especially pleased, therefore, to sight nearly one hundred chital around the camp, evidently enjoying the protection offered by Sundar and his team who kept away poachers and eliminated village dogs that preyed on the chital. This demonstrated the potential of these forests to support a much higher density of chital, sambar, and gaur, though the limited information available shows gaur to be exceedingly rare in the landscape. The dry tracts on either side of river Cauvery can also be an excellent place for the *chowsingha*, or four-horned antelope, a low-density species that can't withstand enormous habitat disturbance, particularly grazing. It looks like the *chowsingha* is extremely rare in this landscape.

Sitting in the cool shade of the riverine forest, we discussed what could be done to improve the conservation value of this landscape. The forests suffer from a natural aridity in summer that limits the carrying capacity of the land, as well as from human activities such as poaching and the grazing of cattle throughout the year. We believe that the best way forward is to improve protection of the landscape's forests, ungulates, and elephant tuskers as well as the river that harbors *mahseer*—the famous sport fish of the Cauvery. It is also important to meet the growing firewood needs of the landscape's villagers, tourists, and pilgrims through fuel-wood plantations. The rivers Shimsha and Arkavathy, crucial for spawning of the endangered *mahseer*, should be freed from commercial fishing. As in the fishing camps we visited, the fishermen of Arkavathy and Shimsha could be employed as fish guards instead. Further, specific cooking places for tourists and pilgrims should be identified in places like Muthathi and Hogenakal to prevent the people from cutting wood and cooking anywhere on the bank of the river as they do now. Special efforts should also go into strengthening the existing connections between Hosur and Bannerghatta forests, and between Hanur Range and the Biligiri Rangan Temple Wildlife Sanctuary.

Each and every problem needs to be addressed if we are to revive and ensure the future of wildlife in the forests on either side of river Cauvery. In some ways, these forests and people have come far since the days of marauding brigands, but the biggest challenge is ahead of them: to bring their activities into harmony with the magnificent plants and animals we saw in our travels along river Cauvery. The recent decision of Tamil Nadu Government to declare nearly 300 sq.km on the left bank of Cauvery as a wildlife sanctuary is a much needed step in strengthening conservation in this landscape.

8

Asian Elephants and Seasonally Dry Forests

Prithiviraj Fernando and Peter Leimgruber

Elephants, the largest living terrestrial animals, have long captured popular imagination and sparked our fascination. Elephants top most lists of charismatic species, presumably because of their great size, but also for their complex social behavior and the dangers they pose to people (Leader-Williams and Dublin 2000).

Two characteristics of elephant feeding ecology set the species apart from most other herbivores. Elephants consume large quantities of vegetation on a daily basis (for an overview, see Sukumar 2003), and they are mixed feeders, easily switching between grazing and browsing (Sukumar 2003; Dierenfeld 2006). Combined, these factors make the elephant a formidable herbivore, potentially a *keystone species* that might be expected to have an effect disproportionately greater than its biomass (Paine 1995). Most of what scientists and the public know about elephants stems from observations of African savanna elephants (*Loxodonta africana*) in open woodlands, savannas, and grasslands of eastern and southern Africa. In these ecosystems, elephants commonly destroy trees by debarking, uprooting, and breaking branches (Laws 1970; Guy 1989; Holdo 2006; Mapaure and Moe 2009).

Scientists have hypothesized that African elephants are indeed a keystone species, maintaining the structure of open woodlands by destroying a significant proportion of the trees and allowing growth of grasses essential to sustain themselves and diverse communities of herbivore grazers (Dublin et al. 1990), and thus playing an important role in maintaining this ecosystem's biodiversity. This hypothesis has been amplified in recent years through research indicating complex interactions among rainfall, fire frequency, elephants, and other herbivore grazers (Dublin et al. 1990; Ben-Shahar 1996; Mapaure and Moe 2009). Elephant density might be one of the most crucial factors determining the species' ecological role (Holdo 2007; Holdo et al. 2009). At high density, elephants are considered detrimental to biodiversity (Baxter and Getz 2005; Whyte and Fayrer-Hosken 2008). Similar to the effects that overabundant deer populations have along forest ecotones (Alverson et al. 1988; McShea et al. 1997; Augustine and Jordan 1998; McShea 2005), increased densities of browsing elephants can become detrimental to woodland ecosystems by suppressing and reducing the diversity of plant species and other biodiversity components. However, because of the logistical, ethical, and political difficulties of conducting experiments with an endangered mega-vertebrate, the keystone ecological function of African elephants is often inferred without much proof of the concept, i.e., through exclusion experiments or other experimental designs to link canopy and woodland structure directly to elephant presence or absence and the extent of browsing at the landscape scale (but see Pringle et al. 2007; Asner et al. 2009).

The Asian elephant (*Elephas maximus*) is also frequently called a keystone species (Kurt 1974), but comparative studies and systematic evaluations of its effects on ecosystems are largely nonexistent (but see Mueller-Dombois 1971; Kurt 1974; Ishwaran 1983; Sukumar 1990; Pradhan et al. 2007). Because of the paucity of scientific studies on the Asian species, observations on African elephants are often generalized as holding true in Asia. Such generalization across species is not warranted, and the ecological role of elephants may be even more complex in Asia than in Africa. In Asia, people have played a large role in shaping the ecosystems occupied by elephants, and elephants may have adapted to anthropogenic change more than would be true in many African landscapes. These arguable differences in human-elephant interactions may be somewhat obscured, and further complicated, by the range of socioeconomic, cultural, and even religious differences characterizing human society across the Asian elephant's geographic range.

Indeed, Asian elephant populations may be impacted by their human-affected habitats rather than being themselves the main drivers of habitat successional processes. Asian elephant populations reach their highest density levels along forest–grassland or forest–agriculture ecotones where food plants become more abundant and accessible. These landscapes features are found within seasonally dry forests, where significant human agricultural development has occurred for millennia in conjunction with elephant populations. Perhaps considering humans', not elephants', keystone potential would be most appropriate within most Asian forests.

Generally, population densities of Asian elephants are thought to be higher in savanna-woodland habitats (i.e., seasonally dry forest) than in rain forests that are characterized by closed canopies and tall trees (Leimgruber et al. 2003; Sukumar 2003). This is ascribed to the greater carrying capacity of dry forest–savanna habitats due to a greater abundance of grasses and higher proportion of accessible and edible woody plant species (Sukumar 2003). According to Sukumar (2003) tropical rain forests provide little palatable food to elephants despite their high biomass productivity, whereas savanna-woodlands provide ample and diverse foods ranging from grasses to woody plants and their component parts. Sukumar (2003) states that savanna-woodlands can support elephant densities of 3–5 elephants/km², compared to only about 0.2 elephants/km² in rain forest habitats.

Significant elephant populations in remaining dry forests are largely restricted to Sri Lanka, India, and Nepal. In most other areas, seasonally dry forests are now among the most threatened forest ecosystems (Songer et al., this volume; Leimgruber et al. 2005). Remaining areas are often fragmented and have lost their original large- and medium-sized mammal components, including Asian elephants (Myint Aung et al. 2004; Loucks et al. 2009). As awareness increases about the need for conservation of Asian dry forests, we face squarely the question of how important Asian elephants may be as components of dry forest ecosystems.

To assess the relationship between Asian elephants and seasonally dry forests, we have reviewed existing literature and data, in combination with our own research and observations. Specifically, we have explored the following areas:

1. Past and present geographic distribution of Asian elephants relative to the current distribution of dry forest.
2. Asian elephant abundance patterns in dry forest and other habitats.
3. The feeding ecology of Asian elephants and how this relates to their use of dry forest.
4. The role of people in mediating Asian elephant effects in dry forest.

PAST AND PRESENT GEOGRAPHIC DISTRIBUTION OF ASIAN ELEPHANTS

The geographic range of the Asian elephant is thought to have once been contiguous from the Euphrates-Tigris in present-day Iraq through South and Southeast Asia to the Yangtze Kiang in China (Deraniyagala 1955; Olivier 1978; Santiapillai and Jackson 1990; Sukumar 2003). Such an extensive range would only have existed prehistorically, perhaps at the end of the last ice age, 15,000–20,000 years ago. It was then fragmented and reduced by the earliest human civilizations in Mesopotamia, the Indus and Ganges Valleys, and along the Yangtze River (Madella and Fuller 2006; Li et al. 2009). These civilizations established permanent agriculture, specifically rice cultivation, which started in south China along the Yangtze River and spread throughout Asia by 3,000 BCE (Li et al. 2009). Agricultural expansion likely had a significant impact on Asian elephant populations and habitats, and must have resulted in increased human-elephant conflict situations.

Asian elephant populations persisted through many millennia, but marked declines have occurred since the 1800s, attributable to excessive hunting during colonial periods (Lahiri-Choudhury 1999; Jayewardene 1994) and dramatic increases in agricultural activities (Flint 1994; Sodhi et al. 2004). Widespread agricultural conversions in the past two hundred years have been driven partly by locally increasing human populations requiring more rice and other subsistence crops, and partly by rising global demand for cash crops such as tea, coffee, teak, rubber, oil palm, and coconut (Corlett 1992; Flint 1994; Bryant 1997; Sodhi et al. 2004; Rasul 2007). After colonization, global market demand for tropical timber, especially for valuable dipterocarp species found in dry deciduous and mixed deciduous forests, resulted in large-scale logging and forest conversion to human habitations and cultivations that continues today (see Sodhi et al. 2004). In the early 2000s, deforestation rates in Asia reached a historic high of 1.4 percent annually (Sodhi et al. 2004).

Ultimately, human population growth has been the principal driver of elephant habitat loss by increasing local demand for food acreage and global demand for tropical timber and cash crops. This pattern continues today. Asian elephant range countries hold almost half the world's human population (3.13 billion, or 46 percent; U.S. Census Bureau 2009). Even after exclusion of China as a minor range state, Asian elephant range countries contain more than a quarter of the world's population (1.79 billion, or 26 percent; U.S. Census Bureau 2009).

To assess the effects of growing human populations and expanding agriculture on Asian elephants and their habitats, we reconstructed the geographic range of Asian elephants in the early 1900s and compared it with a recent range map from the IUCN Asian Elephant Specialist Group (2009). For the 1900s range estimate, we expanded the current range of elephants by adding historical range areas as shown in maps from Stracey (1963). To provide a conservative (i.e., inclusive) range estimate, we adjusted the range polygons to expand into areas that in the 1900s had natural vegetation and low human populations (less than 50 persons/km^2; History Database of the Global Environment, HYDE; Goldewijk 2001).

In the early 1900s, Asian elephant range probably covered about 2.87 million km^2 and was composed of several large yet discrete populations throughout Asia (Figure 1). In South Asia and mainland Southeast Asia, this range corresponds well with the climatic envelope for dry forests (compare Figure 1 of this chapter with Figure 2 in Leimgruber et al., this volume). By the early 2000s, the species' geographic range had declined to about 620,000 km^2 (Figure 1), a 78 percent decline in total area in only one hundred years. Asian elephant populations are now restricted to fragmented habitat

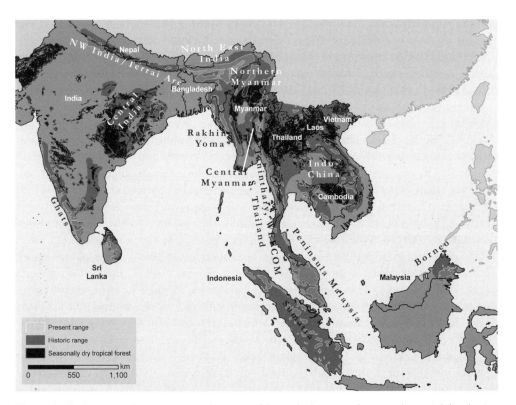

Figure 1. The historic and current geographic range of Asian elephants in relation to the spatial distribution of seasonally dry tropical forest.

islands dispersed across thirteen South and Southeast Asian states, namely, Sri Lanka, India, Nepal, Bhutan, Bangladesh, Myanmar, China, Thailand, Cambodia, Laos, Vietnam, Malaysia (peninsular and Sabah), and Indonesia (Sumatra and Kalimantan). Current Asian elephant range is highly fragmented and continues to decline in most countries (Leimgruber et al. 2003).

In South Asia and mainland Southeast Asia, elephants were once abundant in the dry forest ecosystems that spanned most of this region. However, human populations have grown more rapidly in these climatic zones than elsewhere, with mean human population densities as much as three times higher than in other forest ecosystems in the region (Leimgruber et al., this volume). Consequently, in these areas elephants and dry forest ecosystems have experienced the same fate, disappearing together. Notably, the species is still abundant in some dry forests such as the Eastern and Western Ghats of India, and Sri Lanka's dry zone.

ASIAN ELEPHANT ABUNDANCE PATTERNS

Most of the diverse definitions of Asian dry forests (see Bunyavejchewin et al., this volume) incorporate the element of seasonal rainfall mediated by monsoon weather patterns. Here we define *dry forest* as "natural vegetation" (including patches of savannah-grassland) occurring in areas with annual rainfall $\geq$ 1,600 mm and monthly rainfall $\leq$ 100 mm during at least six months of the year (Leimgruber et al., this volume). Using a map that combines climate with current forest cover extent (see Figure 2B in Leimgruber et al., this volume), we calculated the amount and percentage cover of dry forest in remaining Asian elephant range (Table 1). Although dry forests make up only 9–12 percent of Asia's land cover (Leimgruber et al., this volume), these forests often serve as important elephant habitat and make up approximately 21 percent of the remaining elephant range (Figure 1; Table 1). Some of the largest elephant populations (more than two thousand elephants) are found in areas possessing large percentages of dry forest, including Sri Lanka, the Ghats in Southern India, central India, the Rakhine Yoma mountain range in Myanmar, and the border region of Thailand and Myanmar (the Western Forest Complex, Tanintharyi, and southern Thailand). As much as 30–40 percent of all living elephants may be found in these areas (Santiapillai and Jackson 1990; Leimgruber et al. 2003; Sukumar 2003).

The estimates of dry forest in elephant ranges are very crude. In all of eastern Sri Lanka, for example, dry forest is the dominant forest type (P. Fernando, personal observation), yet our map shows dry forest only in the northeast and southeast of Sri Lanka. As a consequence, our estimate that 22 percent of the total Asian elephant range is covered by dry forest probably underestimates the extent of overlap between elephant and dry forest ranges.

Asian elephants are generally thought to occur at higher densities in dry forests than in rain forests (Sukumar 2003). These patterns in elephant density are paralleled by a dichotomy between the relative distribution of remaining elephant habitats and elephant populations (Leimgruber et al. 2003). Although systematic and range-wide population estimates don't exist (Blake and Hedges 2004), experts agree that southern India and Sri

Table I. Seasonally dry forest extent (km²) in the Asian elephant geographic range.

Region	Range Area	Dry Forest Area
Sri Lanka	33,241	9,818 (30%)
Ghats, India	43,833	9,744 (22%)
Central India	32,043	23,873 (75%)
Terrai Arc / Northwest India	27,628	8,671 (31%)
Northeast India	89,507	0 (0%)
Bangladesh	3,592	0 (0%)
Northwest Myanmar	28,583	10,227 (36%)
Rakhine Yoma	32,075	8,914 (28%)
Bago Yoma / Central Myanmar	21,461	11,459 (53%)
Northeast Myanmar	15,553	9,499 (61%)
Western Forest Complex / Tanintharyi / Southern Thailand	49,982	17,737 (35%)
Eastern Thailand	12,583	5,359 (43%)
Indochina	69,689	17,474 (25%)
Peninsula Malaysia	72,380	0 (0%)
Sumatra	62,284	0 (0%)
Borneo	21,975	0 (0%)
China	4,209	31 (1%)
Total	620,618	132,806 (21%)

Lanka support the largest elephant populations (up to 20 percent of all living wild Asian elephants). Paradoxically these areas are smaller, more fragmented, and more densely populated by people than large elephant habitats in Myanmar and Thailand (Leimgruber et al. 2003). These differences reflect the prevalence of open dry forest in southern India and Sri Lanka. However, these patterns are not simply a function of the differential carrying capacity of habitat types, but are also mediated by national and local differences in hunting and capture of elephants, agricultural practices, cultural attitudes and religious beliefs, and protection of wildlife along with wildlife law enforcement.

THE FEEDING ECOLOGY OF ASIAN ELEPHANTS

A key to understanding interactions between the environment and elephants is elephant feeding ecology. Elephants are hindgut fermenters with rapid passage times for food and low digestibility and energy intake (Dumonceaux 2006). This, combined with their body size, explains why elephants require such a large daily intake of food. In the wild, elephants have been observed to spend 75–85 percent of the day feeding (Vancuylenberg 1977). Digestive physiology also suggests that elephant feeding strategies consist of consuming enormous quantities of low-quality food, passing this food as rapidly as possible through the gastrointestinal tract, and gaining nutrition from the quantity, rather than quality, of foods (Dumonceaux 2006). Thus, elephants are generalized feeders, utilizing a vast number of plants and plant parts, ranging from grasses to leaves, branches, roots, seedlings, and fruit (McKay 1973; Sukumar 1989).

Where grass is available, elephants preferentially feed on it, switching to browse only when grasses are unavailable or cannot be consumed because they are mature and unpalatable (Sukumar 1990). Where grasses are unavailable, elephants feed exclusively on browse (Sukumar 1990, 2003). Elephants in mainland Southeast Asia may differ from those in India and rely less on grazing and more on browsing (Chen et al. 2006; Himmelsbach et al. 2006; Campos-Arceiz et al. 2008a, 2008b).

Elephants can be fairly selective when browsing. Although they have a wide breadth of diet, sometimes consuming over one hundred plant species at a site (McKay 1973; Sukumar 1990; Chen et al. 2006; Himmelsbach et al. 2006), usually only a few species are consumed in large quantities (Sukumar 1990).

Asian elephants strip bark (usually the cambium) off trees and may preferentially consume fruits or pods (Sukumar 1989; Campos-Arceiz et al. 2008a, 2008b). Similar to African elephants, Asian elephants can cause tree mortality (Ishwaran 1983; Sukumar 1989; Pradhan et al. 2007). They kill some trees by debarking, but based on our personal observations in Sri Lanka, such behavior appears to be commoner at high elephant densities brought about by restricting elephants to particular areas. Also, it is our observation that some tree species have evolved strategies to cope with intensive browsing by elephants. For example, indigenous trees in Sri Lanka such as *Bauhinia racemosa* are resistant to elephant browsing. Damaging the main trunk or debarking does not cause mortality. Instead, the tree responds by sprouting multiple trunks. Other species such as figs (*Ficus* sp.) have convoluted trunks and cannot be ring-barked by elephants. In species such as wood apple (*Limonia acidissima*), elephants strip small branches and twigs of their bark but not larger branches or trunks.

Many elements of Asian elephant feeding ecology resemble patterns observed in northern hemisphere herbivores such as white-tailed deer (Alverson et al. 1988; McShea et al. 1997; McShea 2005), such as switching from herbaceous to woody plant diets, and heavy browsing on trees and shrubs when resources become limited (McCullough 1985; Russell et al. 2001; Fulbright and Ortega-S. 2006). Similar to these deer species, elephants thrive in open forest habitats, early-successional forests, and along forest edges. These feeding strategies could also explain elephants' observed geographic distribution, population density, and habitat choice of dry forests. Grasses and shrubs in the understory are more-common and more-easily accessible in open canopy forests such as dry forest than in moist evergreen forests. Because of their mixed feeding strategy, effectively combining and opportunistically switching between grazing and browsing strategies, elephants are extremely well adapted to open canopy forests, disturbed and early-successional forests, and forest edges.

THE ROLE OF PEOPLE IN MEDIATING ASIAN ELEPHANT EFFECTS IN DRY FOREST

Increasing human densities cause the loss of elephant range, but the relationship is complex and depends on the interactions of people and elephants (Fernando et al. 2005; Fernando 2006). It is likely, though hard to prove, that declines in Asian elephant populations have occurred in tandem with dry forest declines. Negative effects of dry forest decline on elephant populations may include increased hunting (Hedges et al. 2005; Corlett 2007), conflict with humans, and habitat loss (Fernando et al. 2005; Hedges et al. 2005; Fernando 2006). Conversion of natural habitat to human-dominated habitat results in fragmentation and eventually loss of elephant range (Fernando et al. 2005; Hedges et al. 2005; Fernando 2006). Ironically, elephant populations may initially increase with habitat fragmentation due to increased forest disturbance and edge habitat (Fernando 2006). However, habitat fragmentation also leads to more-frequent encounters and interactions between people and elephants, and intensifying human-elephant conflict, which together with habitat loss ultimately leads to the extirpation of elephants from such areas.

Perhaps the best broad-scale natural experiment testing the hypothesis of Asian elephants as early-successional or edge-adapted species has been taking place in Sri Lanka where, during the past 2,400 years, humans have created thousands of reservoirs and utilized seasonal swidden agriculture (De Silva 1981; Fernando 2006). Today, many of Sri Lanka's rural dry zone landscapes consist of a heterogeneous mosaic of forest and successional lands and abundant edge habitat. As a consequence of these land-use strategies, Sri Lanka (65,610 km^2; comparable in size to West Virginia [62,755 km^2] or Lithuania [65,300 km^2]) has a wild elephant population that may exceed four thousand individuals (Hoffman 1978; Jayewardene 1994) while also supporting over 20 million people (for comparison, West Virginia's population is 1.2 million, and Lithuania's population is 3.3 million; U.S. Census Bureau 2009). In this context, humans are modifying the landscape and artificially increasing its carrying capacity for Asian elephants (Ishwaran 1993; Fernando et al. 2005; Fernando 2006). It is not a coincidence that Sri Lanka, together with India, has the highest yearly numbers of human fatalities from elephants (50–70 people/year; Perera 2009).

Generally, forest habitat modifications by humans, and specifically, land-use activities that create abundant early-successional habitat, produce optimal elephant habitat. Such habitats may be as important as, or more important than, dry forest availability per se. Elephant populations increase in these successional habitats because elephants are attracted to them and because reproductive rates may be higher. With increasing elephant population size, the risk for intense human-elephant conflict and associated elephant mortality also rises. Ultimately, humans take control of the area and elephants are removed, either by being driven out of the habitat or falling victim to human-elephant conflict. This connection is difficult to prove with experimental studies or large-scale habitat manipulations because these areas have been dominated by human activities for a long time. As a consequence, it may be impossible to distinguish whether elephant densities are determined more by ecosystem characteristics of dry forest or by human agricultural practices.

Habitat changes generally viewed as detrimental to conservation—such as swidden agriculture, logging, and forest fires—frequently benefit elephants, as these activities create new habitat similar to forest edge (Fernando 2006). The critical factor is that forest habitat is changed to early-successional habitat by such disturbance and not converted to a land use that excludes elephants. Where habitat loss is caused by conversion of forests to permanent human settlements and cultivations or commercial plantations such as for oil palm, tea and rubber, elephants will be eliminated (see Fernando et al. 2005; Hedges et al. 2005; Uryu et al. 2008). This is clearly evidenced by the dramatic twinned losses of forest and elephant populations in oil palm plantations in Sumatra (Hedges et al. 2005; Uryu et al. 2008). Conversion of forests to permanent crops, in addition to causing habitat loss, leads to escalation of human-elephant conflict due to crop raiding and increased interaction of elephants with humans, and hence to further loss of elephants (Blake and Hedges 2004; Fernando et al. 2005; Hedges et al. 2005). Increased habitat heterogeneity due to small-scale disturbance often results in better conditions for elephants, while large-scale and homogeneous changes are likely to be detrimental. Therefore, with regard to dry forest and elephants, we should distinguish between habitat disturbance and habitat conversion or loss.

KEYSTONE OR EDGE SPECIES? CONSERVATION IMPLICATIONS

Asian elephant geographic distribution coincides with the distribution of seasonally dry forests. Despite the scarcity of dry forests throughout Asia, these areas continue to harbor the largest wild elephant populations. As dry forests continue to decline, Asian elephants will probably also continue to decline; yet conservation of dry forests might not always benefit elephants.

The importance of habitat succession to the dynamics between forests and elephants, and the observation that habitat suitability for elephants increases with an intermediate disturbance regime, creates a dilemma for those intent on conserving dry forests and elephants. Managers need to assess whether elephants should be maintained as a low-density species in mature dry forests or whether elephant densities should be increased through habitat management, such as an intermediate disturbance regime.

Elephant conservation and management and human-elephant conflict mitigation across Asia have largely been based on the restriction of elephants to protected areas (Fernando 2006; Fernando et al. 2005). The main instrument of such restriction is the "elephant drive," which attempts to clear elephants out of large areas by driving them into protected areas to which they are subsequently restricted by electric fencing. Yet given the dominant effect of habitat (i.e., carrying capacity) on elephants, the number of elephants in a specific area cannot be increased significantly without a corresponding increase in carrying capacity (Fernando 2006). Habitat management specifically targeting elephants involves activities that may be detrimental to forest conservation, such as cutting and burning of forest (Ishwaran 1993). Thus, increasing the carrying capacity for elephants involves the destruction of forests, which may be counter to conservation objectives. Restriction of elephants to protected areas without a corresponding increase in the areas' carrying capacities causes habitat degradation and eventually the decline and extinction of those elephant populations (Fernando 2006). Habitat management specifically for elephants in protected areas is not practical in many cases due to logistical constraints and expenses imposed by the necessary scale of management (Fernando 2006), in addition to biodiversity concerns. In Sri Lanka and probably across the Asian elephant range, higher densities of elephants and a larger segment of the elephant population are found outside, rather than inside, protected areas (Fernando 2006). Therefore, those involved in or interested in conservation of Asian elephants must consider ranges outside the boundaries of protected areas and undisturbed forests.

Artificially high elephant densities, created by driving elephants into small protected areas and fencing them in, may lead to elephants having destructive effects on woodland communities. However, there is little evidence that destructive population levels are reached under natural conditions. Ongoing research in Sri Lanka suggests that artificially high densities often exceed carrying capacity for a park, and are ultimately detrimental to elephants. Elephants so confined suffer high mortality and decreased reproduction and recruitment, leading to population decline (P. Fernando, unpublished data). There is little evidence that Asian elephants serve as a keystone species in Asia's seasonally dry tropical forests. More likely they are a species whose abundance closely tracks the productivity of the habitat; in the context of Asian dry forest, these habitat attributes are mostly anthropogenic.

REFERENCES

Alverson, W. S., D. M. Waller, and S. L. Sohlheim. 1988. Forest Too Deer: Edge Effects in Northern Wisconsin. *Conservation Biology* 2:348–58.

Asner, G. P., S. R. Levick, T. Kennedy-Bowdoin, D. E. Knapp, R. Emerson, J. Jacobson, M. S. Colgan, and R. E. Martin. 2009. Large-Scale Impacts of Herbivores on the Structural Diversity of African Savanna. *Proceedings of the National Academy of Sciences, USA* 106:4947–52.

Augustine, D. J., and P. A. Jordan. 1998. Predictors of White-Tailed Deer Grazing Intensity in Fragmented Deciduous Forests. *Journal of Wildlife Management* 62:1076–85.

Baxter, P. W. J., and W. M. Getz. 2005. A Model-Framed Evaluation of Elephant Effects on Tree and Fire Dynamics in African Savannas. *Ecological Applications* 15:1331–41.

Ben-Shahar, R. 1996. Do Elephants Over-Utilize Mopane Woodlands in Northern Botswana? *Journal of Tropical Ecology* 12:505–15.

Blake, S., and S. Hedges. 2004. Sinking the Flagship: The Case of Forest Elephants in Asia and Africa. *Conservation Biology* 18:1191–1202.

Bryant, R. L. 1997. *The Political Ecology of Forestry in Burma, 1824–1994*. London: C. Hurst & Co.

Campos-Arceiz, A., Thin Zar Lin, Wan Htun, S. Takatsuki, and P. Leimgruber. 2008a. Working with Mahouts to Explore the Diet of Work Elephants in Myanmar (Burma). *Ecological Research* 28:1057–64.

Campos-Arceiz, A., A. R. Larrinaga, U. R. Weerasinghe, S. Takatsuki, J. Pastorini, P. Leimgruber, P. Fernando, and L. Santamaria. 2008b. Behavior rather than Diet Mediates Seasonal Differences in Seed Dispersal by Asian Elephants. *Ecology* 89:2684–91.

Chen, J., X. B. Deng, L. Zhang, and Z. L. Bai. 2006. Diet Composition and Foraging Ecology of Asian Elephants in Shangyong, Xishuangbanna, China. *Acta Ecologica Sinica* 26:309–316.

Corlett, R. T. 1992. The Ecological Transformation of Singapore, 1819–1990. *Journal of Biogeography* 19:411–20.

———. 2007. The Impact of Hunting on the Mammalian Fauna of Tropical Asian Forests. *Biotropica* 29:292–303.

Deraniyagala, P. E. P. 1955. *Some Extinct Elephants, Their Relatives, and the Two Living Species*. Colombo: National Museum of Ceylon.

De Silva, K. M. 1981. *A History of Sri Lanka*. London: C. Hurst & Co.

Dierenfeld, E. S. 2006. Nutrition. In *Biology, Medicine, and Surgery of Elephants*, ed. M. E. Fowler and S. K. Mikota, 57–65. Ames, Iowa: Blackwell Publishing.

Dublin, H. T., A. R. E. Sinclair, and J. McGlade. 1990. Elephants and Fire as Causes of Multiple States in the Serengeti-Mara Woodlands. *Journal of Animal Ecology* 59:1147–64.

Dumonceaux, G. A. 2006. Digestive System. In *Biology, Medicine, and Surgery of Elephants*, ed. M. E. Fowler and S. K. Mikota, 299–307. Ames, Iowa: Blackwell Publishing.

Fernando, P. 2006. Elephant Conservation in Sri Lanka: Integrating Scientific Information to Guide Policy. In *Principles of Conservation Biology*, ed. M. J. Groom, G. K. Meffe, and C. R. Carroll, 649–52. Sunderland, MA: Sinauer Associates.

Fernando, P., E. Wickramanayake, D. Weerakoon, L. K. A. Jayasinghe, M. Gunawardene, and H. K. Janaka. 2005. Perceptions and Patterns in Human-Elephant Conflict in Old and New Settlements in Sri Lanka: Insights for Mitigation and Management. *Biodiversity and Conservation* 14:2465–81.

Flint, E. P. 1994. Changes in Land Use in South and Southeast Asia from 1880 to 1980—A Data Base Prepared as Part of a Coordinates Research Program on Carbon Fluxes in the Tropics. *Chemosphere* 29:1015–62.

Fulbright, T. E., and J. A. Ortega-S. 2006. *White-Tailed Deer Habitat: Ecology and Management on Rangelands (Perspectives on South Texas)*. College Station: Texas A&M University Press.

Goldewijk, K. K. 2001. Estimating Global Land Use Change over the Past 300 Years: The HYDE Database Source. *Global Biogeochemical Cycles* 15:417–33.

Guy, P. R. 1989. The Influence of Elephants and Fire on a *Brchysteigia-Jullbernardia* Woodland in Zimbabwe. *Journal of Tropical Ecology* 5:215–26.

Hedges, S., M. J. Tyson, A. F. Sitompul, M. F. Kinnaird, D. Gunaryadi, and Aslan. 2005. Distribution, Status, and Conservation of Asian Elephants (*Elephas maximus*) in Lampung Province, Sumatra, Indonesia. *Biological Conservation* 124:35–48.

Himmelsbach, W., M. A. Gonzalez-Tagle, K. Fuldner, H. H. Hoefle, and W. Htun. 2006. Food Plants of Captive Elephants in the Okkan Reserved Forest, Myanmar (Burma), Southeast Asia. *Ecotropica* 12:15–26.

Hoffman, T. W. 1978. Distribution of Elephants in Sri Lanka. *Loris* 14:366.

Holdo, R. M. 2006. Elephant Herbivory, Frost Damage, and Topkill in Kalahari Sand Woodland Savanna Trees. *Journal of Vegetation Science* 17:509–18.

———. 2007. Elephants, Fire, and Frost Can Determine Community Structure and Composition in Kalahari Woodlands. *Ecological Applications* 17:558–68.

Holdo, R. M., R. D. Holt, and J. M. Fryxell. 2009. Grazers, Browsers, and Fire Influence the Extent and Spatial Pattern of Tree Cover in the Serengeti. *Ecological Applications* 19:95–109.

Ishwaran, N. 1983. Elephant and Woody-Plant Relationships in Gal Oya, Sri Lanka. *Biological Conservation* 26:255–70.

———. 1993. Ecology of the Asian Elephant in Lowland Dry Zone Habitats of the Mahaweli River Basin, Sri Lanka. *Journal of Tropical* Ecology 9:169–82.

IUCN Asian Elephant Specialist Group. 2009. IUCN Red List of Threatened Species, www.iucnredlist.org/initiatives/mammals/description/download-gis-data (accessed December 2009).

Jayewardene, J. 1994. *The Elephant in Sri Lanka*. Colombo: Wildlife Heritage Trust of Sri Lanka.

Kurt, F. 1974. Asiatic Elephants: Shapers of the Jungle. *Image Roche* 59:2–13.

Lahiri-Choudhury, D. K., ed. 1999. *The Great Indian Elephant Book: An Anthology of Writings on Elephants in the Raj*. New Dehli: Oxford University Press.

Laws, R. M. 1970. Elephants as Agents of Habitat and Landscape Changes in East Africa. *Oikos* 21:1–15.

Leader-Williams, N., and H. Dublin. 2000. Charismatic Megafauna as "Flagship Species." In *Priorities for the Conservation of Mammalian Diversity*, ed. A. Entwistle and N. Dunstone, 53–81. Cambridge: Cambridge University Press.

Leimgruber, P., J. B. Gagnon, C. Wemmer, D. S. Kelly, M. A. Songer, and E. R. Selig. 2003. Fragmentation of Asia's Remaining Wildlands: Implications for Asian Elephant Conservation. *Animal Conservation* 6:347–59.

Leimgruber, P., D. S. Kelly, M. Steininger, J. Brunner, T. Mueller, and M. A. Songer. 2005. Forest Cover Change Patterns in Myanmar (Burma) 1990–2000. *Environmental Conservation* 32:356–64.

Li, X., J. Dodson, J. Zhou, and X. Zhou. 2009. Increases of Population and Expansion of Rice Agriculture in Asia, and Anthropogenic Methane Emissions since 5,000 BP. *Quaternary International* 202:41–50.

Loucks, C., M. B. Mascia, A. Maxwell, K. Huy, K. Dong, N. Chea, B. Long, N. Cox, and T. Seng. 2009. Wildlife Decline in Cambodia, 1953–2005: Exploring the Legacy of Armed Conflict. *Conservation Letters* 2:82–92.

Madella, M., and D. Q. Fuller. 2006. Palaecology and the Harappan Civilization of South Asia: A Reconsideration. *Quaternary Science Review* 25:1283–1301.

Mapaure, I., and S. R. Moe. 2009. Changes in the Structure and Composition of Miombo Woodlands Mediated by Elephants (*Loxodonta africana*) and Fire over a 26-year Period in North-Western Zimbabwe. *African Journal of Ecology* 47:175–83.

McCullough, D. R. 1985. Variables Influencing Food Habits of White-Tailed Deer in the George Reserve. *Journal of Mammalogy* 66:682–92.

McKay, G. M. 1973. Behavior and Ecology of the Asiatic Elephant in Southeastern Ceylon. *Smithsonian Contributions to Zoology* 125:1–113.

McShea, W. J. 2005. Forest Ecosystems without Carnivores: When Ungulates Rule the World. In *Large Carnivores and the Conservation of Biodiversity*, ed. J. C. Ray, K. Redford, R. S. Steneck, and J. Berger, 138–53. Washington, DC: Island Press.

McShea, W. J., H. B. Underwood, and J. H. Rappole, eds. 1997. *The Science of Overabundance: Deer Ecology and Population Management*. Washington, DC: Smithsonian Institution Press.

Mueller-Dombois, D. 1971. Crown Distortion and Elephant Distribution in the Woody Vegetations of Ruhuna National Park, Ceylon. *Ecology* 53:208–26.

Myint Aung, Khaing Khaing Swe, Thida Oo, Kyaw Kyaw Moe, P. Leimgruber, T. Allendorf, C. Duncan, and C. Wemmer. 2004. The Environmental History of Chatthin Wildlife Sanctuary, a Protected Area in Myanmar (Burma). *Journal of Environmental Management* 72 (4): 205–16.

Olivier, R. 1978. Distribution and Status of the Asian Elephant. *Oryx* 14:379–424.

Paine, R. T. 1995. A Conversation on Refining the Concept of Keystone Species. *Conservation Biology* 9:962–64.

Perera, B. M. A. O. 2009. The Human-Elephant Conflict: A Review of Current Status and Mitigation Methods. *Gajah* 30:41–52.

Pradhan, N. M. B., P. Wegge, and S. R. Moe. 2007. How Does a Re-colonizing Population of Asian Elephants Affect the Forest Habitat? *Journal of Zoology* 273:183–91.

Pringle, R. M., T. P. Young, D. I. Rubinstein, and D. J. McCauley. 2007. Herbivore-Initiated Interaction Cascades and Their Modulation by Productivity in an African Savanna. *Proceedings of the National Academy of Sciences, USA* 104:193–97.

Rasul, G. 2007. Political Ecology of the Degradation of Forest Commons in the Chittagong Hill Tracts of Bangladesh. *Environmental Conservation* 34:153–63.

Russell, P. L., D. B. Zippin, and N. L. Fowler. 2001. Effects of White-Tailed Deer (*Odocoileus virginianus*) on Plants, Plant Populations and Communities: A Review. *American Midland Naturalist* 146:1–26.

Santiapillai, C., and P. Jackson. 1990. *The Asian Elephant: An Action Plan for Its Conservation*. Gland, Switzerland: IUCN/SSC Asian Elephant Specialist Group.

Sodhi, N. S., L. P. Koh, B. W. Brook, and P. K. L. Ng. 2004. Southeast Asian Biodiversity: An Impending Disaster. *Trends in Ecology and Evolution* 19:654–60.

Stracey, P. D. 1963. *Elephant Gold*. London: Weidenfeld and Nicolson.

Sukumar, R., 1989. *The Asian Elephant*: *Ecology and Management*. Cambridge: Cambridge University Press.

———. 1990. Ecology of the Asian Elephant in Southern India. II. Feeding Habits and Raiding Patterns. *Journal of Tropical Ecology* 6:33–53.

———. 2003. *The Living Elephants*. Oxford: Oxford University Press.

Uryu, Y., C. Mott, N. Foead, K. Yulianto, A. Budiman, Setiabudi, F. Takakai, Nursamsu, Sunarto, E. Purastuti, N. Fadhli, C. M. B. Hutajulu, J. Jaenicke, R. Hatano, F. Siegert, and M. Stüwe. 2008. *Deforestation, Forest Degradation, Biodiversity Loss, and CO_2 Emissions in Riau, Sumatra, Indonesia*. WWF Indonesia Technical Report, http://assets.panda.org/downloads/riau_co2_report__wwf_id_27feb08_en_lr_.pdf (accessed October 2009).

U.S. Census Bureau. 2009. International Data Base, Population Division, www.census.gov/ipc/www/idb/informationGateway.php (accessed August 2009).

Vancuylenberg, B. W. B. 1977. Feeding Behavior of the Asiatic Elephant in South-East Sri Lanka in Relation to Conservation, *Biological Conservation* 12:23–54.

Whyte, I., and R. Fayrer-Hosken. 2008. Playing Elephant God: Ethics of Managing Wild African Elephant Populations. In *Elephants and Ethics*, ed. C. Wemmer and C. Christen, 399–17. Baltimore: Johns Hopkins University Press.

9

Tropical Deer in the Seasonally Dry Forests of Asia
Ecology, Concerns, and Potential for Conservation

William J. McShea and Megan C. Baker

Large ungulates are a defining element of savannas and open woodlands throughout the world (Danell et al. 2006). For all open woodlands, a dominant ground cover of grasses and forbs is exploited by herbivores, whose partitioning of the resource is across a broad range of body sizes (Fritz and Loison 2006). Whereas these herbivores can include elephants, perissodactyls (rhinos, zebras, and horses), and artiodactyls (bovids, cervids, pigs, and camels), cervids and bovids are the most abundant herbivores in Asia (Fritz and Loison 2006). These herbivores have evolved to exploit the grasses and forbs that result from an open canopy and a frequent fire regime. Their abundance and diversity are the basis for supporting large carnivore species that can capture and kill large, and often swift, animals. Tiger (*Panthera tigris*), leopard (*Panthera pardus*), dhole (*Cuon alpinus*), and lion (*Panthera leo*) populations in Asia are dependent on large ungulates (Sunquist et al. 1999; Karanth et al. 2004; Smith et al. this volume), which are dependent on seasonally dry forests with abundant grass cover. Any disturbance or alteration that impacts grass cover or productivity will cascade through the food web, impacting large ungulates and their predators.

We will not discuss all large herbivores in this chapter: Elephant (*Elephas maximus*), banteng (*Bos javanicus*), and gaur (*Bos gaurus*), are discussed in other chapters of this volume (Fernando and Leimgruber, this volume; Bhumpakphan and McShea, this volume); rhinos (*Rhinoceros unicornis*) have been extirpated from most Asian systems; the wild boar (*Sus scrofa*) is a generalist species and not unique to seasonally dry forests; and two additional species, bluebuck (*Boselaphus tragocamelus*) and chousingha (*Tetracerus quadricornis*), can co-occur with cervids in dry forests but are not discussed in this chapter. There are six species of cervids that are found in Asian dry forests. The red muntjac (*Muntiacus vaginalis*) and sambar (*Rusa unicolor*) are generalist species found in the wetter forests; hog deer (*Axis porcinus*) and barasingha (*Rucervus duvaucelii*) are riverine species along the streams and rivers that crisscross dry forests; and chital (*Axis axis*) and Eld's deer (*Rucervus eldii*) are dry forest specialists. A third riverine species, Schomburgk's deer (*Rucervus schomburgki*), went extinct within the past eighty years (Duckworth et al. 2008c). No single location has all six species, as there appears to be a major geographic division within southern Asia; two species are confined to the Indian subcontinent (chital and barasingha), and one species is confined to Southeast Asia (Eld's deer, and previously Schomburgk's deer).

South Asian deer are a radiation of Old World deer that retain many primitive cervid characteristics (Geist 1998). All the deer have a relatively simple branching pattern for their antlers, and forage more on grasses than woody plants. This group includes the muntjacs with protoantlers and small stocky body form—a body form also seen in the hog deer. Chital, Eld's deer, and barasingha are larger, found in more-open habitat, and able to outrun their predators over short distances. Sambar is a large-bodied deer that does not readily enter open areas and relies more on rapid movement through dense understory. The cervids in Asia have a complex phylogeny that is still in debate (Groves 2006), with Eld's deer and barasingha recently revised from *Cervus* to *Rucervus* (Wilson and Reeder 2005). Based on analysis of mitochondrial DNA, the genera within Asia are non-monophyletic, with species forming clades that do not reflect their current classification (Pitra et al. 2004). For example, hog deer appear closely related to sambar, Eld's deer to Pere David's deer (*Elaphurus davidianus*), and chital to barasingha (Pitra et al. 2004). These results indicate that strong ecological pressures and limited dispersal between ecoregions have converged distantly related species into similar functional roles within each region. Eld's deer and barasingha may seem functionally similar, but they represent different lines of cervids that have each adapted to seasonally dry forests.

None of these cervid species are currently at high population densities across their range. All but the red muntjac and chital are considered endangered or vulnerable according to the 2008 IUCN Redlist (www.iucnredlist.org; Table 1). For red muntjac, this is possibly due to a territorial social organization that does not allow concentrations of individuals (Odden and Wegge 2007), but the other cervids are social animals. Hog deer, chital, and barasingha can be found in social groups throughout the year (Odden and Wegge 2007; Mishra 1982; Duckworth et al. 2008b). Eld's deer form social groups only during their breeding season (February–May; Myint Aung et al. 2001). Sambar generally form small family groups, though larger groups have been reported in India and Thailand (Timmins et al. 2008a). The social species have been reported at high densities; chital have been observed in large social groups of over 100 individuals (Mishra 1982; Karanth and Sunquist 1992) and local densities in India have been reported at greater than 50 deer/km^2 (Jathanna et al. 2003). Barasingha populations are not currently at high densities anywhere in its range (Duckworth et al. 2008b). The highest density reported for Eld's deer is only 9.2 deer/km^2 in Chatthin Wildlife Sanctuary, Myanmar (Myint Aung et al. 2004). With the exception of chital in some Indian reserves, all current populations appear to be well below carrying capacity, although this can not be precisely calculated.

There is an obvious need to develop techniques for estimating carrying capacity for prey populations in the dry forests of Asia so that current densities of cervids can be evaluated relative to the habitat potential. Stocking rates of ungulates have been calculated for African savanna systems, with rainfall being the driving force in

Table 1. Natural history attributes of cervid species found in seasonally dry forests of tropical Asia.

Attribute	South Asian Cervid Species						
	Red Muntjac[a] Muntiacus Vaginalis	Hog Deer[b,h,i] Axis Porcinus	Chital[b] Axis Axis	Eld's Deer[c] Rucervus Eldii	Schomburgk's Deer[d] Rucervus Schomburgki	Barasingha[e,i] Rucervus Duvaucelii	Sambar[a,d] Rusa Unicolor
Body weight (kg)[d]	20–28	70–95	70–90	70–130	100–120	170–180	180–260
Home range (km²)[f]	1.5–2.5	0.1–1.3	0.15–0.2	12–17	—	10	0.1–1
Mean group size	1–2	1–4	5–17	2–6	—	2–25	3–4
Birth season (months)	Nov.–May (8)	Feb.–May (4)	Oct.–Apr. (7)	Oct.–Nov. (2)	—	Aug.–Oct. (3)	May–Aug. (4)
Diet[d,g]	Browse, forbs	Grass, fruits	Grass, browse, fruits	Grass, browse, fruits	Grass, aquatic plants	Grass, aquatic plants	Browse, ferns, grass, forbs
Range extent in South Asia (× 10,000 km²); % protected	415.7; 0.09	37.7; 0.05	127.5; 0.05	78.5; 0.15	—	36.1; 0.11	438.8; 0.09
IUCN Redlist status[j]	LC	EN	LC	EN	EX	VU	VU

aMishra 1982
bDhungel and O'Gara 1991
cMyint Aung et al. 2001
dNowak 1999, and Francis 2008
eSchaller 1967
fbased on convex polygon method
gSmith and Xie 2008
hOdden et al. 2005
iWegge et al. 2006
jbased on 2008 listing, www.iucnredlist.org.

determining range capacity (Ward 2006). Karanth et al. (2004) found tiger densities to closely track ungulate densities in select reserves across the region, but could only find a weak relationship between prey density and rainfall, probably because of local differences in protection status.

The role of ungulates in African systems has been extensively modeled (Weisberg et al. 2006), but much less is known of Asian systems. Bagchi et al. (2003) estimated herbivore biomass in Ranthambhore Tiger Reserve, Rajasthan, India, and compared these estimates to other areas that differed in moisture and plant species. The authors concluded that grass productivity over the course of the year determined prey biomass. Cervids within this system did differentiate their foraging niche both spatially and with regard to plant composition (Bagchi et al. 2003), although the recent loss of many species may reduce evidence for competition within the ungulate assemblage (Bagchi 2006).

Ungulates move nutrients and plants across the landscape through nonrandom foraging and defecation/urination (Pastor et al. 2006). Prasad et al. (2006) found that seeds of an Asian tree species, *Phyllanthus emblica*, remain viable following regurgitation by chital in Rajaji National Park, India. Seeds of dry forest tree species were often observed in the feces of Eld's deer in Myanmar, and in preliminary tests these seeds could be germinated (W. J. McShea, personal observation). Chital were estimated to deposit thirteen tons of dung per month in a 10 km^2 study area in Nepal, with most of the nutrients moved from feeding sites within the grasslands to bedding sites within the dry forest (Moe and Wegge 2008). These observations indicate that the loss of cervids from dry forest ecosystems would have important ramifications for ecosystem functioning.

Seasonally dry forests are formed by the pattern of monsoon rains and the occurrence of fire (Bullock et al. 1995). All forests occupied by cervids in Asia have some temporal pattern of fire, with the drier forests experiencing annual fires and the wetter forests experiencing fire once each century (Kobsak Wanthongchai and Goldammer, this volume). Ungulates can influence the fire regime by altering the amount of flammable material on the forest floor and shifting plant composition (Hobbs 2006). The prevention of fire within dry forests does not maintain the integrity of the forest type, either with regard to ground cover or tree species composition (Murdiyarso and Lebel 2007).

The lack of fire does not impact all cervids equally. Sambar inhabit the moister dry forests in Southeast Asia, where bamboo is present in the understory, and individuals are large enough to move through younger bamboo stands. The small-bodied muntjacs are also able to move through the bamboo matrix in ways not possible for the medium-sized cervids (i.e., Eld's deer, hog deer, chital, and barasingha). These medium-sized cervids, chital and Eld's deer in particular, avoid forests with dense understory structure (Schaller 1967; McShea et al. 2001), seemingly as a means of predator avoidance. The prevention of fire in dry forests leads to an increase in understory density (Kobsak Wanthongchai and Goldammer, this volume), particularly of bamboo species, making the habitat less suitable for most cervids.

Mapping the extent of seasonally dry forest is extraordinarily difficult, and recent estimates diverge widely for the region (Leimgruber et al. 2003; Leimgruber et al., this volume). Regardless of the amount of forest, there is agreement on what causes loss of dry forest. In Asia, seasonally dry forests at lower elevations are excellent candidates for conversion to rice production (Murphy and Lugo 1986; Rundel and Boonpragob

1995) and not for inclusion in protected areas (McShea et al. 2005). Lowland forests are the more-valuable forests, as their soils have the best capacity for water to support grasses (Bullock et al. 1995). Dry forests of Asia are experiencing some of the highest percentage losses in the past twenty years, with the decline largest in the buffer zones around protected forests (DeFries et al. 2005). This isolation of Asian forests creates islands of wildlife that can be the focus of poaching activities (Myint Aung et al. 2004).

All Asian cervids are thought to rely on surface water during the dry season. The riverine species (i.e., hog deer and barasingha) are never found far from open water in streams and rivers (Timmins et al. 2008a; Duckworth et al. 2008b). Chital are observed to move daily to water sources (Schaller 1967). Eld's deer in both Laos and Cambodia make frequent use of manmade water holes during the dry season (W. J. McShea, personal observation). For Eld's deer in Myanmar, fruits are a significant component of diets during the dry season and may supplement waterhole use (Myint Aung et al. 2001). If access to water is restricted by villagers focused on livestock production, or diverted for water projects, it would have severe impacts on the suitability of dry forests for cervids.

All analyses of ungulate dynamics within Asian dry forests have to consider the role played by livestock, as these forests exist in a matrix of humans and their domestic animals. Wild bovids are matched in size by domestic cows (500–800 kg) and, along with cervids, form a guild of ruminant grazers focused on the same resource (i.e., grass). Madhusudan (2004) reported dramatic increases in ungulates following a 50 percent decline in livestock density in Bandipur National Park, India. Madhusudan attributed the increase in wild ungulates to an increased biomass of preferred grass and forb species. Khan et al. (1996) found a thirteenfold increase in chital in the nineteen years following removal of villagers and their livestock from around Gir Forest, India. For all studies of this nature, it is hard to isolate the impact of livestock removal from the impact of removing its owners. One example of combined impacts is that of domestic dogs, which often accompany humans and livestock in South Asia. Dhole, although now missing from many Asian forests, are a natural predator of all cervids, and it is our observation that cervids respond immediately to the presence of dogs, domestic or wild. Only sambar have been observed to defend themselves against wild and domestic dogs (Timmins et al. 2008b); all the other cervids rely on a flight-and-hide strategy (Geist 1998). Both chital and Eld's deer have been known to run into villages to avoid dholes (W. J. McShea, personal observation; Geist 1998), and domestic dogs are used to capture Eld's deer and hog deer in Myanmar (W. J. McShea, personal observation). With the exception of sambar, the inability of cervids to successfully defend themselves against dogs limits their abundance around human settlements. The exclusion of livestock, and its attendant humans and dogs, from reserves can be the most effective tool for cervid conservation beyond the prevention of poaching.

The primary focus in this chapter has been on habitat and landscape qualities needed to maintain cervids in Asian seasonally dry forests. The loss of habitat due to agriculture and hydroelectric activity is a real concern for conservationists, but relatively easy to measure compared to poaching activity. Ungulates are a common focus of poachers throughout Asia and Africa (Fa et al. 2005; Steinmetz et al. 2006; Corlett 2007), such that many Asian forests are absent large mammals (Datta et al. 2008).

Cervids in dry forests are often the last remaining large herbivores and therefore the primary focus of local poachers. For several cervids, such as barasingha, hog deer, and Eld's deer, it is hard to differentiate declines due to habitat loss and those due to poaching, since habitat losses have been remarkable during the twentieth century. For sambar, we agree with Timmins et al. (2008b) that poaching is the only plausible explanation for observed declines, since forest cover seems to be abundant for this generalist species. A similar argument could be made for all the muntjacs. If saving forests is the essential first step of cervid conservation, then protection from illegal harvest is the subsequent requirement. In the age of modern weaponry, protection is difficult for species that spend a portion of the year in open habitat, as is the case for all cervids in dry forests. Means to enhance protection of wildlife inside and outside protected areas should be a primary concern for these forests (Maxwell and Cox, this volume).

For Eld's deer, and to some extent barasingha and hog deer, remaining populations occupy so little of their former range that it is difficult to determine the range of forests suitable for the species. We are most familiar with Eld's deer and its adaptations to dry forests. There are three subspecies of Eld's deer, *Rucervus eldii siamensis, R. eldii eldi*, and *R. eldii thamin* (Balakrishnan et al. 2003), but only the last subspecies has been well studied in natural populations in central Myanmar (Myint Aung et al. 2001; McShea et al. 2001). The entire species is listed in CITES Appendix 1 (www.cites.org/eng/app/appendices.shtml) and as endangered on the IUCN Red List (Timmins and Duckworth 2008). The largest populations of Eld's deer still reside in Myanmar, but there has been a steady decline since the 1940s (Salter and Sayer 1986; McShea et al. 1999). The largest population remains in Chatthin Wildlife Sanctuary, but this population declined 40 percent between 1983 and 1995, and continued to decline through the most recent survey (Myint Aung et al. 2004). *Rucervus eldii thamin* are most often found in dry, deciduous dipterocarp forests with an open understory in central Myanmar (McShea et al. 1999). In this region the southwest monsoon results in variable annual rainfall from 500 cm on the fringes of Eld's deer habitat (i.e., the hilly borders of the Irrawaddy Plain) to 75 cm in Eld's deer primary habitats (Davis 1960; Salter and Sayer 1986). Pristine habitat is nonexistent for *R. eldii thamin*, and they now inhabit areas that range from dry scrub and thorn forest to open deciduous forest in various stages of secondary succession (McShea et al. 1999).

Although the primary forest type of most Eld's deer populations is dry dipterocarp (McShea et al. 2005), the fringe populations of this species occupy wetter ecosystems. The last remaining population of *R. eldii eldi* inhabits floating mats of dense vegetation within a small (less than 15 km^2) region in Manipur, India (Timmins and Duckworth 2008). The Chinese subpopulation of *R. eldii siamensis* is found on the tropical moist island of Hainan in shrub forest (Zeng et al. 2005). Isolated populations of *R. eldii siamensis*, both in southern Laos (Round 1998) and around Ang Trapeang Thmor Reservoir in northwestern Cambodia (W. J. McShea, personal observation), inhabit marshy areas similar to the conditions described by Lekagul and McNeely (1997) for extirpated deer in Thailand. These populations indicate that moister dipterocarp forests or grasslands are potentially productive Eld's deer habitat that is not realized in the deer's present distribution. Lekagul and McNeely (1977) proposed that the Eld's deer inhabited swampy areas and were more recently driven into the drier forests due to the pressures imposed by hunting and the expansion of agricultural areas.

A more-detailed study of suitable forest for Eld's deer (McShea et al. 2005) identi-fied 44,530 km^2 of potential habitat across Southeast Asia (the Indian subcontinent and China were not included in the survey). The proportion of suitable forest was ≤ 1 percent of the remaining forest within each country, with the exception of Cambodia (13 percent). The suitable forest was not in large patches, with the mean patch size for each country less than 100 km^2 (a minimum sustainable size), with the exception of Cambodia (263 km^2) and Myanmar (147 km^2). The protected status of this potentially suitable forest was also low, at less than 1 percent, with the exception of Cambodia (11.2 percent).

Comparable analyses are not available for the other cervid species, but we specu-late any analysis would yield similar results. Mapping the extent of a species is an important management tool, as range contraction is a measure of conservation status (Channell and Lomolino 2000; Laliberte and Ripple 2004). We attempted to estimate the current range and protected status of dry forest cervids using current maps and forest cover (Figure 1). We created a forest layer from the Global Land Cover Char-acteristics Database, vers. 2 (Loveland et al. 1999; http://edcsns17.cr.usgs.gov/glcc) and used the IUCN current distributions published in 2008 for each cervid (www.iucnredlist.org). The IUCN maps are relatively crude for the more-widespread species so we reduced the distribution to within 5 km of forest cover. We used the United Nations Environment Programme 2007 coverage for parks and protected areas in the region (http://protectedplanet.net). The results are listed in Table 1 and indicate all but sambar and red muntjacs have restricted ranges within the region, especially relative to the amount of remaining forest cover. For protected status, the two most endangered species, Eld's deer and barasingha, are the most protected (15 percent and 11 percent, respectively), but no species, in our opinion, is effectively protected.

The Southeast Asian region had the highest vulnerability to extinctions among continental regions (Mace and Balmford 2000) and the largest gaps in protection (MacKinnon 2000). The marked economic inequalities between nations within South-east Asia (Balmford and Long 1995), however, do not support the supposition that all forests are equal. Whereas Thailand and India might be in the best economic and bureaucratic position to protect deer, most dry forests reside within Cambodia and Myanmar, which are at the opposite end of the economic spectrum in terms of infra-structure and monetary support of conservation (Balmford and Long 1995). A major impediment to conservation is the lack of political will and funds in developing coun-tries like Myanmar (Brunner et al. 1998; Rao et al. 2002).

Whereas dry dipterocarp forests are both declining and seldom protected, the situation is worse for riverine systems (Dudgeon 2000). For both hog deer and baras-ingha, there is strong dependence on these seasonally flooded grasslands. In Nepal's Royal Bardia National Park, hog deer almost exclusively utilized the riverine grass, *Saccharum spontaneum*, for forage and cover (Odden et al. 2005). They appeared to be dependent on these grasslands and did not shift habitat with season or in response to disturbance due to humans or annual fires (Odden et al. 2005). The loss of habitat due to either agricultural conversion or hydroelectric construction and changes in flood would severely impact this species (Odden et al. 2005). Barasingha and hog deer forage on similar plant species, but the barasingha exploit a broader range of grasses, including those on upland soils and later successional species (Wegge et al. 2006).

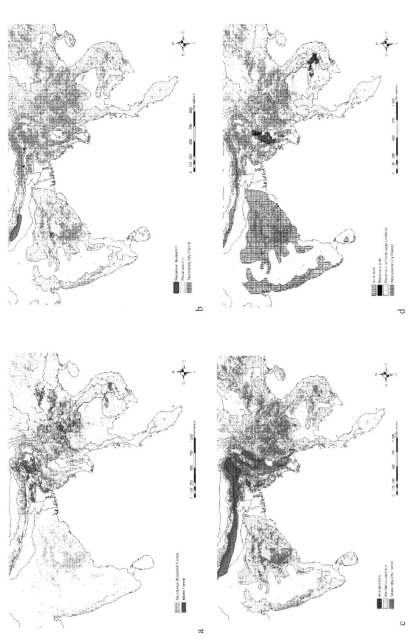

Figure 1. Distribution of cervids in South Asia. (A) Forest cover for range of seasonally dry forests in region based on the Global Land Cover Characteristics Database, vers. 2 (Loveland et al. 1999; http://edcsns17.cr.usgs.gov/glcc); the two forest types are combined for subsequent maps. (B) Range of *Rucervus duvaucelii* and *Rusa unicolor* in South Asia. For *Rusa unicolor*, the total range extends beyond region. (C) Range of *Axis porcinus* and *Muntiacus vaginalis* in South Asia. For *Muntiacus vaginalis* the total range extends beyond study area. (D) Range of *Axis axis*, *Rucervus eldii*, and *Rucervus schomburgki*. *Rucervus schomburgki* is extinct. Current ranges of the cervids were derived from the 2008 IUCN red list (www.iucnredlist.org). For widespread species (*Muntiacus vaginalis*, *Rusa unicolor* and *Axis axis*), we restricted the published IUCN range using forest cover.

Riverine grasslands for both species are being lost throughout South Asia due to water impoundment and diversion (Dudgeon 2000). Sarma et al. (2008) documented the conversion of alluvial grasslands into savanna grasslands within Manas National Park, India, due to changes in hydrology upstream, which resulted in a 47 percent decline in hog deer habitat within the park.

A major conservation effort in South Asia is the conservation of large carnivores, such as tigers (Seidensticker et al. 1999; Smith et al. this volume). Tiger densities can be predicted from prey densities, which can vary tenfold over Asian forests (Karanth et al. 2004), and large cervids were the main tiger prey in a comparison of reserves across the region (Sunquist et al. 1999; Smith et al. this volume). As mentioned earlier, it is both plant productivity and protection efficiency that drive ungulate densities (Karanth et al. 2004). Karanth et al. (2004) speculate that tiger declines are the indirect result of prey declines due to human impacts (i.e., habitat loss and poaching). From our perspective, there is a mutualistic relationship between large carnivores and ungulate populations in Asia. While attention has focused on how ungulates sustain large carnivores, the opposite may also be true. Large, diverse ungulate populations may only persist in the presence of significant carnivore populations. As reviewed here, large populations of chital only persist in parks maintained for tigers and lions; barasingha and hog deer are only abundant in wetlands of Nepal maintained for tigers. For Chatthin Wildlife Sanctuary in Myanmar, the loss of predators rendered the park safe for livestock, with a subsequent loss of banteng and decline of Eld's deer (Myint Aung et al. 2004). These carnivores serve as a functional patrolling system against human and livestock encroachment, the real danger to cervid populations. Human patrols do not provide the same level of protection, as their effectiveness is subject to logistic and monetary concerns. By maintaining large carnivores, parks protect forests and wetlands from conversion to agriculture, prevent the stocking of forests with livestock during the dry season, and prevent the monopolization of water sources for livestock and fish production. The sequence of species loss from protected areas rarely starts with the ungulates or the forest cover. It is the loss of carnivores that starts the cascade toward the domestification of the forest, and a concomitant loss of biodiversity (Ray et al. 2005).

Cervids form an important guild of ungulates within seasonally dry forests of Asia both for their role as ecosystem agents and as prey for large carnivores. Preservation or restoration of this guild relies on effective management of fire and water resources, as well as limitations on livestock and poaching. The loss of cervids from Asian forests will doom the sustainable management of large carnivores and probably the maintenance of this important ecosystem. Likewise, the loss of large carnivores makes protection of cervids and their critical resources more difficult.

REFERENCES

Bagchi, S. 2006. Assemblage Rules in Large Herbivores: A Null Model Analysis of Local and Regional Diversity Patterns of Ungulates in Dry Tropical Forests of Western India. *Acta Zoologica Sinica* 52:634–40.

Bagchi, S., S. P. Goyal, and K. Sankar. 2003. Niche Relationships of an Ungulate Assemblage in Dry Tropical Forest. *Journal of Mammalogy* 84:981–88.

Balakrishnan, C. N., S. L. Monfort, A. Gaur, L. Singh, and M. D. Sorenson. 2003. Phylogeography and Conservation Genetics of Eld's Deer (*Cervus eldi*). *Molecular Ecology* 12:1–10.

Balmford, A., and A. Long. 1995. Across-Country Analyses of Biodiversity Congruence and Current Conservation Efforts in the Tropics. *Conservation Biology* 9:1539–48.

Brunner, J., K. Talbot, and C. Elkin. 1998. *Logging Burma's Frontier Forests: Resources and the Regime*. Washington, DC: World Resources Institute Press.

Bullock, S. H., H. A. Mooney, and E. Medina, eds. 1995. *Seasonally Dry Tropical Forests*. Cambridge: Cambridge University Press.

Channel, R., and M. V. Lomolino. 2000. Trajectories to Extinction: Spatial Dynamics of the Contraction of Geographic Ranges. *Journal of Biogeography* 27:169–79.

Corlett, R. T. 2007. The Impact of Hunting on the Mammalian Fauna of Tropical Asian Forests. *Biotropica* 39:292–303.

Danell, K., R. Bergstrom, P. Duncan, and J. Pastor, eds. 2006. *Large Herbivore Ecology, Ecosystem Dynamics and Conservation*. Cambridge: Cambridge University Press.

Datta, A., M. O. Anand, and R. Naniwadekar. 2008. Empty Forests: Large Carnivore and Prey Abundance in Namdapha National Park, North-East India. *Biological Conservation* 141:1429–35.

Davis, J. H. 1960. *The Forests of Burma*. Gainesville: University of Florida Press.

DeFries, R., A. Hansen, A. C. Newton, and M. C. Hansen. 2005. Increasing Isolation of Protected Areas in Tropical Forests over the Past Twenty Years. *Ecological Applications* 15:19–26.

DeFries, R. S., M. C. Hansen, J. R. C. Townshend, A. C. Janetos, and T. R. Loveland. 2000. A New Global 1-km Dataset of Percentage Tree Cover Derived from Remote Sensing. *Global Change Biology* 6:247–54.

Dhungel, S. K., and B. W. O'Gara. 1991. Ecology of Hog Deer in Royal Chitwan National Park, Nepal. *Wildlife Monographs* 119:1–40.

Duckworth, J. W., N. S. Kumar, Md. Anwarul Islam, Hem Sagar Baral, and R. J. Timmins. 2008a. *Axis axis*. IUCN Red List of Threatened Species (2008), www.iucnredlist.org.

Duckworth, J. W., N. S. Kumar, Chiranjibi Prasad Pokheral, H. Sagar Baral, and R. J. Timmins. 2008b. *Rucervus duvaucelii*. IUCN Red List of Threatened Species (2008), www.iucnredlist.org.

Duckworth, J. W., W. G. Robichaud, and R. J. Timmins. 2008c. *Rucervus schomburgki*. IUCN Red List of Threatened Species (2008), www.iucnredlist.org.

Dudgeon, D. 2000. Large-Scale Hydrological Changes in Tropical Asia: Prospects for Riverine Biodiversity. *Bioscience* 50:793–806.

Fa, J. E., S. F. Ryan, and D. J. Bell. 2005. Hunting Vulnerability, Ecological Characteristics and Harvest Rates of Bushmeat Species in Afrotropical Forests. *Biological Conservation* 121:167–76.

Francis, C. M. 2008. *A Guide to the Mammals of Southeast Asia*. Princeton, NJ: Princeton University Press.

Fritz, H., and A. Loison. 2006. Large Herbivores across Biomes. In *Large Herbivore Ecology, Ecosystem Dynamics and Conservation*, ed. K. Danell, R. Bergstrom, P. Duncan, and J. Pastor, 19–49. Cambridge: Cambridge University Press.

Geist, V. 1998. *Deer of the World: Their Evolution, Behavior and Ecology*. Mechanicsburg, PA: Stackpole Books.

Groves, C. 2006. The Genus *Cervus* in Eastern Eurasia. *European Journal of Wildlife Research* 52:14–22.

Hobbs, N. T. 2006. Large Herbivores as Sources of Disturbance in Ecosystems. In *Large Herbivore Ecology, Ecosystem Dynamics and Conservation*, ed. K. Danell, R. Bergstrom, P. Duncan, and J. Pastor, 261–88. Cambridge: Cambridge University Press.

Jathanna, D., K. U. Karanth, and A. J. T. Johnsingh. 2003. Estimation of Large Herbivore Densities in the Tropical Forests of Southern India Using Distance Sampling. *Journal of Zoology* 261:285–90.

Karanth, K. U., J. D. Nichols, N. S. Kumar, W. A. Link, and J. E. Hines. 2004. Tigers and Their Prey: Predicting Carnivore Densities from Prey Abundance. *Proceedings of the National Academy of Sciences, USA* 101:4854–58.

Karanth, K. U., and B. M. Stith. 1999. Prey Depletion as a Critical Determinant of Tiger Densities. In *Riding the Tiger: Tiger Conservation in Human-Dominated Landscapes*, ed. J. Seidensticker, S. Christie, and P. Jackson, 100–103. Cambridge: Cambridge University Press.

Karanth, K. U., and M. E. Sunquist. 1992. Population Structure, Density and Biomass of Large Herbivores in the Tropical Forests of Nagarahole, India. *Journal of Tropical Ecology* 8:21–35.

Khan, J. A., R. Chellam, W. A. Rodgers, and A. J. T. Johnsingh. 1996. Ungulate Densities and Biomass in the Tropical Dry Deciduous Forests of Gir, Gujarat, India. *Journal of Tropical Ecology* 12:149–62.

Laliberte, A. S., and W. J. Ripple. 2004. Range Contraction of North American Carnivores and Ungulates. *Bioscience* 54:123–38.

Leimgruber, P., J. B. Gagnon, C. Wemmer, D. S. Kelly, M. A. Songer, and E. R. Selig. 2003. Fragmentation of Asia's Remaining Wildlands: Implications for Asian Elephant Conservation. *Animal Conservation* 6:347–59.

Lekagul, B., and J. A. McNeely. 1977. *Mammals of Thailand*. Bangkok: Kurusapha Ladprao Press.

Loveland, T. R., J. E. Estes, and J. Scepan. 1999. Global Land Cover Mapping and Validation. *Photographic Engineering and Remote Sensing* 65:1011–12.

Mace, G. M., and A. Balmford. 2000. Patterns and Processes in Contemporary Mammalian Extinction. In *Priorities for the Conservation of Mammalian Diversity: Has the Panda Had Its Day?* ed. A. Entwistle and N. Dunstone, 27–52. Cambridge: Cambridge University Press.

MacKinnon, K. 2000. Never Say Die: Fighting Extinction. In *Priorities for the Conservation of Mammalian Diversity: Has the Panda Had Its Day?* ed. A. Entwistle and N. Dunstone, 335–54. Cambridge: Cambridge University Press.

Madhusudan, M. D. 2004. Recovery of Wild Large Herbivores following Livestock Decline in a Tropical Indian Wildlife Reserve *Journal of Applied Ecology* 41:858–69.

McShea, W. J., Myint Aung, D. Poszig, C. Wemmer, and S. L. Monfort. 2001. Forage, Habitat Use, and Sexual Segregation by a Tropical Deer (*Cervus eldi thamin*) in a Dipterocarp Forest. *Journal of Mammology* 82:848–57.

McShea, W. J., K. Koy, T. Clements, A. Johnson, C. Vongkhamheng, and Myint Aung. 2005. Finding a Needle in the Haystack: Regional Analysis of Suitable Eld's Deer (*Cervus eldi*) Forest in Southeast Asia. *Biological Conservation* 125:101–11.

McShea, W. J., L. Leimgruber, Myint Aung, S. L. Monfort, and C. Wemmer. 1999. Range Collapse of Tropical Cervid (*Cervus eldi*) and the Extent of Remaining Habitat in Central Myanmar. *Animal Conservation* 2:173–83.

Mishra, H. R. 1982. The Ecology and Behavior of Chital (*Axis axis*) in the Royal Chitwan National Park, Nepal, with Comparative Studies of the Hog Deer (*Axis porcinus*), Sambar (*Cervus unicolor*) and the Barking Deer (*Muntiacus muntjak*). PhD diss., University of Edinburgh.

Moe, S. R., and P. Wegge. 2008. Effects of Deposition of Deer Dung on Nutrient Redistribution and on Soil and Plant Nutrients on Intensively Grazed Grasslands in Lowland Nepal. *Ecological Research* 23:227–34.

Murdiyarso, D., and L. Lebel. 2007. Local to Global Perspectives on Forest and Land Fires in Southeast Asia. *Mitigation and Adaptations Strategies for Global Change* 12:3–11.

Murphy, P., and A. E. Lugo. 1986. Ecology of Tropical Dry Forests. *Annual Review of Ecology and Systematics* 17:67–88.

Myint Aung, W. J. McShea, S. Htung, T. M. Soe, S. Monfort, and C. Wemmer. 2001. Ecology and Social Organization of a Tropical Deer (*Cervus eldi thamin*). *Journal of Mammalogy* 82:836–47.

Myint Aung, Khaing Khaing Swe, Thida Oo, Kyaw Kyaw Moe, P. Leimgruber, T. Allendorf, C. Duncan, and C. Wemmer. 2004. The Environmental History of Chatthin Wildlife Sanctuary, a Protected Area in Myanmar (Burma). *Journal of Environmental Management* 72 (4): 205–16.

Nowak, R. M. 1999. *Walker's Mammals of the World*, 6th ed., 2:1091–1132. Baltimore: Johns Hopkins University Press.

Odden, M., and P. Wegge. 2007. Predicting Spacing Behavior and Mating Systems of Solitary Cervids: A Study of Hog Deer and Indian Muntjac. *Zoology* 110:261–70.

Odden, M., P. Wegge, and T. Storaas. 2005. Hog Deer *Axis porcinus* Need Threatened Tallgrass Floodplains: A Study of Habitat Selection in Lowland Nepal. *Animal* Conservation 8:99–104.

Pastor, J., Y. Cohen, and N. T. Hobbs. 2006. The Roles of Large Herbivores in Ecosystem Nutrient Cycles. In *Large Herbivore Ecology, Ecosystem Dynamics and Conservation*, ed. K. Danell, R. Bergstrom, P. Duncan, and J. Pastor, 289–325. Cambridge: Cambridge University Press.

Pitra, C., J. Fickel, E. Meijaard, and P. C. Groves. 2004. Evolution and Phylogeny of Old World Deer. *Molecular Phylogenetics and Evolution* 33:880–95.

Prasad, S., J. Krishnaswamy, R. Chellam, and S. P. Goyal. 2006. Ruminant-Mediated Seed Dispersal of an Economically Valuable Tree in Indian Dry Forests. *Biotropica* 38:679–82.

Rao, M., A. Rabinowitz, and S. T. Khaing. 2002. Status Review of the Protected-Area System in Myanmar, with Recommendations for Conservation Planning. *Conservation Biology* 16:360–68.

Ray, J. C., K. H. Redford, R. S. Steneck, and J. Berger, eds. 2005. Large Carnivores and the Conservation of Biodiversity. Washington, DC: Island Press.

Round, P. D. 1998. Wildlife, Habitats, and Priorities for Conservation in Dong Khanthung Proposed National Biodiversity Conservation Area, Champasak Province, Lao PDR. Vientiane, Lao People's Democratic Republic: Champasak Province Agency Wildlife Management / Wildlife Conservation Society Cooperative Program.

Rundel, P. W., and K. Boonpragob. 1995. Dry Forest Ecosystems of Thailand. In *Seasonally Dry Tropical Forests*, ed. S. H. Bullock, H. A. Mooney, and E. Medina, 93–124. Cambridge: Cambridge University Press.

Salter, R. E., and J. A. Sayer. 1986. The Brow-Antlered Deer in Burma: Its Distribution and Status. *Oryx* 20:241–45.

Sarma, P. K., B. P. Lacar, S. Ghosh, A. Rabha, P. Jyoti, J. P. Das, N. K. Nath, S. Santanu Dey, and N. Brahma. 2008. Land-Use and Land-Cover Change and Future Implication Analysis in Manas National Park, India Using Multi-Temporal Satellite Data. *Current Science* 95:223–27.

Schaller, G. B. 1967. *The Deer and the Tiger*. Chicago: University of Chicago Press.

Seidensticker, J., S. Christie, and P. Jackson, eds. 1999. *Riding the Tiger: Tiger Conservation in Human-Dominated Landscapes*. Cambridge: Cambridge University Press.

Smith, A. T., and Y. Xie, eds. 2008. *A Guide to the Mammals of China*. Princeton, NJ: Princeton University Press.

Steinmetz, R., W. Chutipong, and N. Seuaturien. 2006. Collaborating to Conserve Large Mammals in Southeast Asia. *Conservation Biology* 20:1391–1401.

Sunquist, M., K. U. Karanth, and F. Sunquist. 1999. Ecology, Behavior, and Resilience of the Tiger and Its Conservation Needs. In *Riding the Tiger: Tiger Conservation in Human-Dominated Landscapes*, ed. J. Seidensticker, S. Christie, and P. Jackson, 5–18. Cambridge: Cambridge University Press.

Timmins, R. J., and J. W. Duckworth. 2008. *Rucervus eldii*. IUCN Red List of Threatened Species (2008), www.iucnredlist.org.

Timmins, R. J., J. W. Duckworth, N. Samba Kumar, Md. Anwarul Islam, H. Sagar Baral, B. Long, and A. Maxwell. 2008a. *Axis porcinus*. IUCN Red List of Threatened Species (2008), www.iucnredlist.org.

Timmins, R. J., R. Steinmetz, H. Sagar Baral, N. S. Kumar, J. W. Duckworth, Md. Anwarul Islam, B. Giman, S. Hedges, A. J. Lynam, J. Fellowes, B. P. L. Chan, and T. Evans. 2008b. *Rusa unicolor*. IUCN Red List of Threatened Species (2008), www.iucnredlist.org.

Ward, D. 2006. Long-Term Effects of Herbivory on Plant Diversity and Functional Types in Arid Ecosystems. In *Large Herbivore Ecology, Ecosystem Dynamics and Conservation*, ed. K. Danell, R. Bergstrom, P. Duncan, and J. Pastor, 142–69. Cambridge: Cambridge University Press.

Wegge, P., A. K. Shrestha, and S. R. Moe. 2006. Dry Season Diets of Sympatric Ungulates in Lowland Nepal: Competition and Facilitation in Alluvial Tall Grasslands. *Ecological Research* 21:698–706.

Weisberg, P. J., M. B. Coughenour, and H. Bugmann. 2006. Modeling on Large Herbivore-Vegetation Interactions in a Landscape Context. In *Large Herbivore Ecology, Ecosystem Dynamics and Conservation*, ed. K. Danell, R. Bergstrom, P. Duncan, and J. Pastor, 348–82. Cambridge: Cambridge University Press.

Wilson, D. E., and D. A. Reeder. 2005. *Mammal Species of the World: A Taxonomic, and Geographic Reference*, 3rd ed. Baltimore: Johns Hopkins University Press.

Zeng, Z. G., Y. L. Song, J. S. Li, L. W. Teng, Q. Zhang, and F. Guo. 2005. Distribution, Status and Conservation of Hainan Eld's Deer (*Cervus eldi hainanus*) in China. *Folia Zoologica* 54:249–57.

10

Ecology of Gaur and Banteng in the Seasonally Dry Forests of Thailand

Naris Bhumpakphan and William J. McShea

There are 140 species of bovids worldwide, with most in Africa and Eurasia (Wilson and Reeder 2005). Within Asia, the goats and the bovids are the dominant subfamilies, with species found from tropical islands to alpine meadows. Within the Indian subcontinent and Southeast Asia, bovids are the largest terrestrial mammals, with the exception of elephants (*Elephas maximus*) and rhinos (*Rhinoceros unicornis* and *R. sondaicus*). Of the four Asian species of genus *Bos*, the yak (*B. grunniens*) is confined to high-elevation grasslands, and the remaining species are found in the dry forests and lowland hills of the Indian subcontinent and Southeast Asia. This chapter will focus on two *Bos* species that are found within seasonally dry forests of this region: gaur (*Bos gaurus* C. H. Smith) and banteng (*Bos javanicus* D'Alton). These dry forest ungulates are not unique, but lend themselves to closer examination. On the Indian subcontinent, the nilgai (*Boselaphus tragocamelus*) and chousingha (*Tetracerus quadricornis*) are bovids that can also be found in drier forests, but their habitat and management needs are less known. Additional bovids in the region include multiple species of water buffalo (*Bubalus* spp.) and oxen (*Bos* spp.) that are similar in body configuration and weight to gaur and banteng, but are either extirpated or confined to islands, riparian marshes, or human-modified habitats (Heinen and Srikosamatara 1996). The kouprey (*B. sauveli*) may or may not have been a separate species (Galbreath et al. 2006; Hedges et al. 2007; Hassanin and Robiquet 2007), but regardless has not been sighted within its native range of Cambodia since the 1970s. It is important to note that domesticated cattle are derived from ancestors within this group, and thus there is a high degree of overlap in diet, habitat needs, and disease transmission between domestic and wild bovids (Chaiyarat and Srikosamatara 2009). Gaur and banteng are the two wild bovids whose ecology is most integrated into the dry forests of Asia.

The gaur and banteng share similar ecological traits in that they are both large (400–1,000 kg) bovids that, as grazers, have a physiology, gut morphology, and tooth structure specialized for foraging on grasses (McNaughton and Georgiadis 1986; Hofmann 1989). Both species are adapted for xeric conditions, but appear to need surface water for drinking. Wild banteng are entirely confined to deciduous dry forests in

Southeast Asia (including Borneo), while gaur are more widespread in both distribution (India through the Malay Pennisula) and habitat (both deciduous and evergreen forests; Karanth et al. 2009). Banteng have been domesticated (Bali cattle) and spread to islands throughout the Pacific, including Australia and New Zealand (Bradshaw et al. 2006).

Due to continuous threats from habitat loss and uncontrolled hunting, banteng and gaur have decreased precipitously in numbers during the past decades and occupy fragmented subpopulations throughout their historical ranges. Gaur is currently listed as vulnerable by the IUCN (Duckworth et al. 2008) and banteng as endangered (Timmins et al. 2008). These trends are alarming because of the rapid fall in these two species that were well distributed as recently as fifty years ago. Their period of decline matches that of the dry forests where they are found (Leimgruber et al., this volume).

Turning to the distribution of banteng and gaur in Thailand, we find that both are distributed mainly in the dry deciduous and evergreen forest habitats of northern and western Thailand, while only gaur can be found in the dry and moist evergreen forests of the southern peninsula (Lekagul and McNeely 1988). Both species were once relatively widespread across Thailand (Srikosamatara and Suteethorn 1995), but currently there are seven large subpopulations of gaur (*B. g. laosiensis*) and four subpopulations of banteng (*B. javanicus birmanicus*; Figure 1).

In this chapter, we concentrate on two study sites with both gaur and banteng: Haui Kha Khaeng Wildlife Sanctuary (HKWS) and Ta Phrya National Park (TPNP). Our focus is on both the ecology and conservation of gaur and banteng, including the importance of intact dry forest for their persistence in Thailand. TPNP is located in the southeast of Thailand in what was previously a large plain of dry forest and open grasslands, but today is primarily dedicated to rice and crop production. HKWS is within the Western Forest Complex (WEFCOM), which is a large mountainous region along the Thailand-Myanmar border that contains a mosaic of dry, wet, and montane forests. Srikosamatara (1993) estimated that banteng and gaur (along with sambar, *Rusa unicolor*) represented 70 percent of the large-mammal biomass within HKWS. Gaur and banteng share habitat in Thailand with other large mammals, such as the Asian elephant, water buffalo, and sambar. In recent history, gaur and banteng shared habitat with three deer species (*Rucervus eldii*, *Rucervus schomburgki*, and *Axis porcinus*) of which wild populations are now extirpated from Thailand. Natural predators of gaur and banteng in Thailand include tigers (*Panthera tigris*), leopards (*Panthera pardus*), and dhole (*Cuon alpinus*), with all predators currently at low densities.

GAUR

Current Distribution

Until recently, there were three recognized subspecies and two extinct subspecies of gaur: (1) *Bos gaurus gaurus*, which inhabits seasonally dry forest habitat in India, Nepal, and Bhutan; (2) *Bos gaurus laosiensis* (formerly *Bos gaurus readei*), which is found in the dry and evergreen forest habitats of Southeast Asia north of the Isthmus of Kra (lat. 11°N); (3) *Bos gaurus hubbacki*, which is found in the evergreen forest

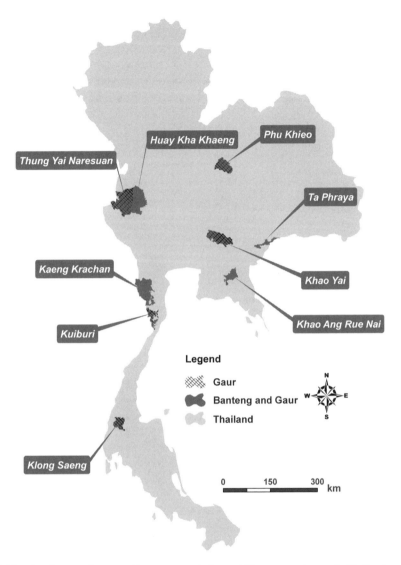

Figure 1. The distribution of gaur and banteng populations within protected areas in Thailand as of 2009.

habitat of the southern peninsula of Thailand and western Malaysia (Lekagul and McNeely 1988; National Research Council 1983); and two extinct Pleistocene and historical forms, namely, (1) *Bos gaurus grangeri*, ca. 200,000 years ago, which inhabited southwestern China (Corbet and Hill 1992); and (2) *Bos gaurus sinhaleyus*, which survived in Sri Lanka into historical times. In a recent revision (Grubb 2005), the subspecies *B. g. hubbacki* was subsumed within *B. g. laosiensis*, and we will retain that nomenclature, with all populations within Thailand considered *B. g. laosiensis*. There are currently five subpopulations of gaur located above the Isthmus of Kra and two populations south of the isthmus (see Figure 1).

Habitat Use and Forage Selection

Gaur use both deciduous and evergreen dry forest across a broad altitudinal range from mean sea level to the peak of Inthanon Mountain (2,560 m elevation; Lekagul and McNeely 1988). The gaur of TPNP (Figure 1) in southeast Thailand use dry evergreen forest and mixed deciduous forest, as well as manmade secondary forest and grasslands (Faculty of Forestry 2004). This habitat distribution agrees with what Choudhury (2002) observed for the Indian subcontinent.

Dry forest provides both forage species and mineral licks for gaur. Gaur in HKWS use deciduous forest less during the dry season, except immediately following fires when there is a flush of new grass. This ability to forage broadly across forest types is reflected in their broad diet composition in HKWS (169 forage species; Bhumpakphan 1997). Gaur in TPNP have no access to mixed deciduous forests, and their diet from dry evergreen forest and manmade grasslands includes only 52 forage species (Bhumpakphan 2004; Table 1).

Gaur forage for various plant species, but *Pollinia* spp. and the tender shoots of *Saccharum* sp. and *Imperata* sp. are preferred foods (Bhumpakphan 1997, 2004). Bamboo shoots are eaten readily when they sprout in July and August. Bamboo leaves, the leaves and shoots of herbaceous plants, and the bark of some trees are also eaten in Myanmar (Peacock 1933). Schaller (1967) and Krishnan (1972) list the forage species used by gaur in dry forests of India, and these lists include all parts of woody species such as *Bauhinia racemosa*, *B. vahlii*, *Shorea robusta*, *Bridelia squamosa*, *Cassia fistula*, *Gmelina arborea*, and *Terminalia tomentosa*. Grass and grasslike species consumed by gaur in India include *Imperata cylindrica*, *Themeda triandra*, *Coix lachrymajobi*, and *Dendrocalamus strictus*. In Malaysia, gaur feeding-site examinations showed that its diet included 87 plant species, with *Paspalum conjugatum*, *Mikania chordata*, and *Centrosema pubescens* considered important food items (Conry 1981). In WEFCOM, gaur feed on leaves and twigs of *Bauhinia*, *Diospyros*, *D. ehretioides*, *Hymenodictyon exelsum*, *Anogeissus acuminata*, *Butea superba*, and *Saccharum spontaneum*; the leaves and shoots of *Bambusa nutans*, *Dendrocalamus strictus*, *Gigantochloa albociliata*, and *Thyrsostachys siamensis*; and the fruits of *Dillenia parviflora* and *D. indica* (Nakhasathien et al. 1987). In summary, gaur are primarily grazers on forest grasses, with increased consumption of forbs, fruits, tender bamboo shoots, and woody leaves and stems during the rainy season.

Mineral licks are frequently used by gaur, possibly as often as every night (Lekagul and McNeely 1988). Herds and individual gaur visit several licks in the mixed deciduous forest of HKWS, possibly to obtain different essential elements (Bhumpakphan 1997). We do not know if the minerals bovids require are limited within dry forests; the need to regularly drink water is usually associated in ungulates with relatively inefficient kidney filtration, which results in minerals lost through urination (Cain et al. 2006).

Home Range

There have been no home-range studies of gaur within Thailand, and no comparisons of gaur across forest types, making broad generalizations difficult. Conry (1989) reports that the annual home range of gaur (*B. g. hubbacki*) inhabiting an evergreen forest of the

Table 1. Comparison of major seasonal forage species eaten by gaur[a] and banteng.[b]

Botanical Name	Family	Part Eaten[c]	Habitat Type[d]	Gaur Season Dry	Gaur Season Wet	Banteng Season Dry	Banteng Season Wet
Herbaceous							
Elephantopus scaber	Compositae	L	GL		x		
Euphatorium sp.	Compositae	L	GL		x		
Cyperus spp.	Cyperaceae	L	RP	x	x		
Cyperus compectus	Cyperaceae	L	RP	x			
Corchorus sp.	Tiliaecae	L	GL		x		
Elatostema cf. acuminofa	Urticaceae	L	DE		x		
Grasses							
Apluda mutica	Graminae	L	DD	x	x	x	
Arundinaria spp.	Graminae	L	MD				x
Arundo donex	Graminae	L St	RP	x	x		
Chrysopogon orientalis	Graminae	L	DD			x	
Cynodon dactylon	Graminae	L	GL		x		
Desmodium hecterocarpon	Graminae	L	DD			x	
Heteropogon triticeus	Graminae	L	DD	x		x	x
Harrisonia perforata	Graminae	L	MD				x
Imperata cylindrica	Graminae	L	MD, GL	x		x	
Microstegium ciliatum	Graminae	L	MD		x	x	x
Muhlenbergia ramosa	Graminae	L	MD	x	x		
Neyraudia sp.	Graminae	L	RP		x		
Neyraudia renandiana	Graminae	L St	RP		x		
Panicum cambogiense	Graminae	L	MD		x		
Panicum montanum	Graminae	L	MD			x	x
Paspalum conjugatum	Graminae	L	RP		x		
Pennisetum sp.	Graminae	L	GL	x			
Saccharum spontaneum	Graminae	L St	RP	x	x	x	
Sporobolus indicus	Graminae	L	MD	x			
Bamboos							
Bambusa arundinacea	Graminae	L S Sl	MD	x	x	x	x
Dendrocalamus strictus	Graminae	L S	MD		x	x	x
Vietnamosasa pulsilla	Graminae	L	DD		x		
Thyrostachy siamensis	Graminae	L S	MD		x	x	x
Climbers							
Artabotrys siamensis	Annonaceae	L	DE	x			
Gmelina paniculata	Labiatae	L	MD	x			
Pueraria barbata	Papilionaceae	L	MD		x	x	x
Sphenodesma pentandra	Symphoremataceae	L	MD				x
Dioscorea burmanica	Dioscoreae	L	MD	x	x		
Dioscorea hispida	Dioscoreae	L	MD		x	x	
Cajanus scarabaeoides	Leguminosae	L	GL		x		
Hedyotis capitellata	Rubiaceae	L	DE	x			

(Continued)

Table 1. Comparison of major seasonal forage species eaten by gaur[a] and banteng.[b] *(Continued)*

Botanical Name	Family	Part Eaten[c]	Habitat Type[d]	Gaur Season Dry	Gaur Season Wet	Banteng Season Dry	Banteng Season Wet
Under shrub							
Desmodium spp.	Papilionaceae	L	DD		x	x	
Globba spp.	Zingiberaceae	L	DE		x	x	
Shrubs							
Melodorum fruticosum	Annonaceae	L	DE	x			
Leea indica	Leeaceae	L	GL		x		
Helicteres sp.	Sterculiaceae	L	MD, GL		x		
Triumfetta bartramia	Tiliaecae	L	MD	x			
Clerodendrum viscosum	Verbenaceae	L	DE	x	x		
Suregada multiflorum	Euphorbiaceae	L	DE	x			
Gmelina elliptica	Labiatae	L	MD	x			
Leucaena leucocephala	Leguminosae	L	GL		x		
Melientha suavis	Opiliaceae	L	DD	x			
Understory Trees							
Bauhinia glauca	Caesalpinaceae	L	MD		x		
Bauhinia malabarica	Caesalpinaceae	L	MD	x	x		
Bauhinia saccocalyx	Caesalpinaceae	L	MD	x	x		
Bauhinia sp.	Caesalpinaceae	L	MD	x		x	
Colona auriculata	Tiliaecae	L	MD		x		
Canopy Trees							
Wrigthia sp.	Apocynaceae	L	MD		x		
Shorea obtusa	Dipterocarpaceae	Sl	DD	x		x	
Diospyros sp.	Ebenaceae	L	MD	x			
Afzelia xylocarpa	Fabaceae	B	DE, MD	x			
Sindora siamensis	Fabaceae	L	MD, DD	x			
Cratoxylon formosum	Hyperaceae	L	MD, DD		x		
Cratoxylon sp.	Hyperaceae	L	MD, DD		x		
Xylia xylocarpa	Mimosaceae	L	MD, DD	x			
Toona ciliata	Meliaceae	L	DE			x	
Streblus asper	Moraceae	L	MD	x			
Dalbergia sp.	Papilionaceae	L	MD	x			
Hadina cordifolia	Rubiaceae	L B	MD			x	
Hymenodictyon exelsum	Rubiaceae	L	MD	x			
Mitragyna brunonis	Rubiaceae	L	DD			x	x
Mitragyna javanica	Rubiaceae	L	MD		x		
Microcos tomentosa	Tiliaecae	L	MD, DE	x			
Vitex peduncularis	Verbenaceae	L	MD	x			

[a]Major forage species from a total of 185 species eaten by gaur in Huai Kha Khaeng and Taphraya; available data from Bhumpakphan (1997) and observations by authors.
[b]Major forage species from a total of 150 species eaten by banteng in Huai Kha Khaeng; available data from Prayurasithi (1997) and observations by authors.
[c]L = Leaf; B = Bark; S = Shoot; Sl = Seedling; St = Stem.
[d]DD = Dry deciduous dipterocarp forest; MD = Mixed deciduous forest; DE = Dry evergreen forest; GL = Grassland; RP = Riparian.

Malaysian peninsula varied from 16.9 to 137.3 km^2, while Prayurasithi (1997) reports that the home range of Indochinese gaur (*B. g. laosiensis*) was similar for dry seasons (27.8 km^2) and wet seasons (39.1 km^2). In the same study, there were large seasonal shifts in habitat so that the mean annual home range was 65.5 km^2 ± 27.8.

Gaur are mostly nocturnal, resting and sleeping throughout most of the daylight hours (Lekagul and McNeely 1988). This daily activity cycle is similar to that of elephants, and gaur often follow elephants through their feeding routes (Tun Yin 1967; Lekagul and McNeely 1988). We speculate that elephants may also benefit gaur by opening new mineral licks and pathways through the dense understory of evergreen forests.

Population and Herd Size

Gaur usually form herds led by an adult cow (the matriarch), while adult bulls are usually solitary (Lekagul and McNeely 1988). The gaur herd structure at HKWS was estimated from direct sightings, and the average herd composition was 1 bull, 4 cows, 1.6 subadults, 1.2 juveniles, and 1 calf. This estimate was similar for adults and calves to an estimate that was based on the number of tracks and bed-size examination, which showed 6 adults, 3.8 subadults, 4 juveniles, and 1 calf per group (Bhumpakphan 1997). The reason for the larger number of subadults and juveniles in the indirect measures is not known. The average herd size observed at HKWS was 8.15 individuals ± 6.2 SD (range: 2–42 animals, n = 40 herds; Bhumpakphan 1997). No seasonal differences in herd size or composition were observed at HKWS (Bhumpakphan 1997).

BANTENG

Current Distribution

Banteng is a wild bovid of Southeast Asia (Figure 2) with three subspecies: (1) *Bos javanicus javanicus*, found only on the island of Java; (2) *Bos javanicus lowi*, found only on the island of Borneo (Sabah and Kalimantan); and (3) *Bos javanicus birmanicus*, found throughout the mainland of Southeast Asia including Myanmar, Thailand, Laos, Vietnam, Cambodia, and Malaysia (Lekagul and McNeely 1988).

Thailand has only four populations of the banteng *Bos javanicus birmanicus*, which are located in HKWS and Khaeng Krachan National Park in the west, and in TPNP and Khao Runai Wildlife Sanctuary in southeastern Thailand. HKWS is the only protected area where this species is more abundant than the other bovids (Prayurasithi 1997).

Habitat and Forage Use

Banteng typically inhabit forests that are more open than those occupied by gaur. Their main habitat in Thailand is lowland dry forest, especially dry deciduous dipterocarp and mixed deciduous forest types (Prayurasithi 1987; Nakhasathien et al. 1987), and this habitat selection seems consistent across their range (Tun Yin 1967; Steinmetz 2004; Pedrono et al. 2009). They are frequently observed feeding within open grasslands or dry forest, but they retreat to denser forest when disturbed (Wharton 1968; Lekagul and McNeely 1988).

Prayurasithi (1997) reported that banteng had been directly observed to consume 143 plant species in the wet season and 82 species in the dry season at HKK. The top 25 plants eaten during both seasons included *Barleria siamensis*, *Commelina* spp., *Bauhinia* spp., *Bambusa arundinaria*, *Thyrsostachys siamensis*, *Apluda mutica*, and *Sorghum nitidum*—all forage species common to dry deciduous forests. The predominance of grasses and forbs in the diet of banteng is in contrast to gaur and probably limit banteng range beyond the open canopy forests.

Home Range

As with gaur, there are few studies of banteng that allow for broad generalizations concerning their movements. Banteng's home-range size at HKWS, as estimated through observation, was 20.8–43.9 km^2 in the dry season and 30.0–44.8 km^2 in the wet season (Prayurasithi 1997).

Daily Activity

Banteng feed in open grasslands, and dry forests with dense grass cover. Local villagers near HKK and TPNP reported frequent use of rice paddy fields prior to the decline in numbers of banteng (Naris Bhumpakphan, personal observation). Banteng are observed at saltlicks and water holes regularly (Tun Yin 1967; Lekagul and McNeely 1988: Prayurasithi 1997). In HKWS, after ground fires in February and March, banteng were observed to forage in newly burned areas with flushes of grasses and to bed down within dry evergreen forest during the daytime (Bhumpakphan 1997). During the wet season, banteng remained within deciduous dipterocarp forest throughout the day (Bhumpakphan 1997).

Population and Herd Size

Banteng have been observed to form herds of 2–25 that are led by a mature cow, while mature bulls are usually solitary (Lekagul and McNeely 1988). For banteng at HKWS, the herd size observed was 2–20 individuals, with occasional sightings of up to 30 animals (Nakhasathien et al. 1987). In the open dipterocarp forest of Myanmar, several herds have been reported to stay in the same locality for long periods (Tun Yin 1967). Banteng in the wild have no specific rutting season, and calves were seen throughout the year (Peacock 1933). In HKWS, the mean ratio for herd structure (bull:cow:juvenile/calf) was 2:5:2 in the wet season and 3:6:3 in the dry season (Prayurasithi 1997).

INTERACTIONS WITH OTHER HERBIVORE SPECIES AND PREDATORS

The seasonal and elevational movements of the three sympatric bovids in HKWS differ between the dry and wet seasons (Ucharoensak 1992; Chaiyarat 2004; Bhumpakphan

1997; Prayurasithi 1997). Wild water buffalo used a narrow strip of riverine habitat along the lower Huai Kha Khaeng River (100–300 m elevation), while banteng used the lowland valley (160–600 m elevation). Gaur habitat overlaps with banteng habitat (200–600 m elevation), but their range extends to the highest elevations in the park (over 1,660 m). Both mixed deciduous and dry deciduous dipterocarp forest in HKWS can support seasonal foraging by gaur and banteng populations in the wet season. Thus, both species were observed visiting saltlicks and water holes together and resting within close proximity (Prayurasithi 1997; Bhumpakphan 1997). Mixed herds of bovids were also reported in Cambodia, where banteng and kouprey were reported in mixed herds within dry deciduous dipterocarp forests (Lekagul and McNeely 1988). Gaur have been found with wild water buffalo during the dry season at HKWS, but this is the only season when they overlap in habitat use (Chaiyarat 2004).

The diets of gaur and banteng overlap with those of many large ungulates within Thailand, but the habitat overlap among these species is most extensive with gaur. In HKWS, banteng are almost exclusive to lowland deciduous dry forest, which in the past was probably also utilized by Eld's deer (McShea and Baker, this volume). Gaur use of dry and moist evergreen forests and mixed deciduous forest would bring them in contact with elephants, Malayan tapirs (*Tapirus indicus*), and sambar during both dry and wet seasons. Gaur and banteng overlap in forage composition for 17 percent of plant species, while gaur and wild elephant overlap for 20 percent of their consumption of plant species (Bhumpakphan 1997; Prayurasithi 1987; Sukmasuang 1993). Steinmetz (2004) reported similar habitat segregation for gaur and banteng within the Xe Pian Biodiversity Conservation Area in Lao PDR.

Large herds of domestic cattle and water buffalo can be found in many reserves in Thailand (Chaiyarat and Srikosamatara 2009). These incursions are generally limited to the boundaries for most reserves, but quantifying this impact is not often done. Chetri (2003) reported that gaur and domestic cattle foraged in the same habitats, and domestic cattle shared 24 species (63.2 percent) of 38 plant species eaten by gaur in Parsa Wildlife Reserve in Nepal. Diseases of domestic cattle and buffalo such as rinderpest, anthrax, and foot-and-mouth disease can be transmitted to wild cattle (Chaiyarat and Srikosamatara 2009). Rinderpest has been a widespread cause of death in gaur in India (Schaller 1967; Lekagul and McNeely 1988). Gaur are considered especially susceptible to disease transmission from domestic cattle (as reported in Duckworth et al. 2008).

Heinen and Srikosamatara (1996) consider interbreeding of domestic and wild bovids a serious concern for the region. Male gaur and banteng can mate with domestic cows (Lekagul and McNeely 1988), based on reports of wild cattle interbreeding with domestic cattle in Cambodia (Wharton 1957), of banteng mating with domestic cattle in Myanmar (Tun Yin 1967), and of gaur mating with domestic cattle in Malaysia (Ngha et al. 1985). In Thailand, offspring produced from gaur mating with domestic cattle have been photographed at two reserves: Khao Yai National Park and Eastern Thung Yai Naresuan Wildlife Sanctuary (Prachak Baukaew, personal communication). Whether these matings are an artifact of gross difference in population density (i.e., high domestic cattle densities and low densities of wild bovids) is unknown.

Predators of gaur, banteng, and wild water buffalo in WEFCOM are tiger, leopard, and reportedly dhole. Gaur is one of 18 species consumed by tigers and composed

9 percent of tiger fecal remains, while banteng composed 29 percent of tiger-food items identified in feces from HKWS (Petdee 2000). Elsewhere within WEFCOM (Tung Yai Wildlife Sanctuary), gaur were one of 15 prey species and composed 19 percent of tiger food items (Phrommakul 2003). Gaur are considered the primary food item for tigers throughout their range overlap (Smith et al., this volume), but the same has not been reported for banteng, possibly due to their limited distribution. Leopards in HKWS prey upon at least 25 mammal species, and banteng (6.8 percent) and young gaur (0.45 percent) are both found within their feces (Simcharoen 2008). We have no records of dhole consuming gaur and banteng in HKWS, and the few dhole in TPNP have not been observed to prey upon these bovids.

CONSERVATION OF WILD BOVIDS IN THAILAND

Gaur is listed in CITES Appendix I (www.cites.org/eng/app/appendices.shtml), and although banteng is not listed in CITES, they are listed as endangered on the IUCN Redlist (Timmins et al. 2008). Within Thailand, gaur, banteng, and wild water buffalo are protected by the Wild Animal Reservation and Protection Act B.E. 2535 (A.D. 1992). Gaur and banteng are classified as a "protected" species and wild water buffalo as a "reserved" species within this system, which means they cannot be killed for any purpose and can not be kept as domestic animals. The main threats to gaur and banteng populations in Thailand are poaching by shooting and snaring, loss of dry forest habitat, and livestock and human activities within suitable forests. Three steps are essential, in our opinion, to ensure the persistence of banteng and gaur in Thailand: enhanced reserve protection, habitat management, and reduction of human-wildlife conflicts.

Reserve Protection

Long-term protection of wildlife sanctuaries and national parks serves the national goal of biodiversity conservation in Thailand. To mitigate threats due to humans, the wildlife sanctuaries and parks were established to protect wildlife habitat, and the main mechanism of this protection is fencing and patrolling. Given the current human population densities in Thailand, it is doubtful that additional reserves will be established in habitat suitable for gaur and especially banteng. Increased numbers of these species, therefore, require increased protection of current populations. The value of effective patrolling in increasing wildlife populations has been documented throughout the region (Martin and Vigne 1996; Myint Aung et al. 2004; Steinmetz 2004). Effective patrolling systems have been established in some wildlife sanctuaries in Thailand such as HKWS, and Phu Khieo and Khao Ang Ru Nai wildlife sanctuaries. For example, the patrolling teams at HKWS, supported by the Wildlife Conservation Society–Thailand, monthly collects geocoded data on threats from poachers, human activity, and the distribution of wildlife (Chayawat 2007; S. Chayawattana and A. Pattanavibool, personal communication). This information provides a visual and spatial record that is persuasive to local and national policy makers and quantifies patrolling activities for management. These efforts are funded almost entirely through concern for tigers but serve to benefit major prey species such as gaur and banteng. Effective patrolling is the primary conservation tool for the bovids in Thailand.

Habitat Protection and Manipulation

Dry forest is a fire-dependent ecosystem, and it should be properly managed as a priority conservation goal for bovids as well as many of the other animals and plants outlined in this volume. Wild fires in dry forests have been strongly discouraged within Thailand, as this approach has not been considered a means to preserve biodiversity by the forestry community (Kobsak Wanthongchai and Goldammer, this volume), but it is our belief that this policy is degrading suitable habitat for banteng and possibly gaur. A comparison of burned and unburned deciduous forest at HKWS found burned areas are utilized by banteng, sambar deer, and barking deer more than unburned areas, but no response was noted by gaur (Sirimagorn 2009). Gaur restricted their activity mostly to higher elevation evergreen forests that were not burned in the study. Fire as a management tool has been long advocated by wildlife experts (Wharton 1968) and traditionally practiced by local villagers (MacInnes, this volume), yet has been slow to be adopted by the foresters who maintain control over the Department of Natural Resources in Thailand. Deciduous dipterocarp forest and open grassland habitat should be manipulated for gaur, banteng, and other ungulates through prescribed burning in the early dry season. Political or logistical constraints can limit prescribed burning as a management tool, and some reserves include shrub mitigation by hand cutting as a means to maintain open habitat. It is difficult with hand cutting to clear a sufficient area to significantly impact wildlife populations, and this may serve primarily to increase wildlife viewing opportunities.

Creating water sources and saltlicks has been tried in both focal reserves, HKWS and TPNP. As with many efforts within the region, the success of these procedures has not been summarized in the scientific literature, but both reserves consider the projects successful because of increased animal tracks and sign around the project sites (Naris Bhumpakphan, personal observation). Increases in population numbers would be a validation of reserve staff observations.

Mitigation of Human-Bovid Conflicts

Whereas patrolling activities are focused on the prevention of poaching, there is a second area of human-wildlife conflict that needs attention. With increasing gaur and banteng populations within some protected areas in Thailand, there are conflicts developing between farmers and wild bovids. These conflicts work both ways as domestic and wild bovids move across the park boundaries. There are reports from TPNP of subadult gaur dispersing outside the reserve to raid crops (corn and cassava), and of three cases of local villagers injured by gaur in 2008 as they defended their crops. Park rangers have been tasked with driving gaur back into TPNP and have conducted two training workshops for villagers affected (July 2009 and August 2009). The workshops provided an opportunity for discussion of a new vision in resolving gaur-human conflicts, and local people and their leaders agreed to become involved in their own protection by informing park staff when gaur were sighted outside the park. Education programs for local people, including school students, have been implemented in HKWS and TPNP to obtain better understanding and good support from the local communities.

To promote local involvement in the conservation of wild cattle and other wildlife species, some Department of National Park reserves (i.e., Khao Yai National Park, Kui Buri National Park, and TPNP) are promoting gaur as part of their ecotourism efforts. Villagers living around these reserves have received education and training in promoting gaur conservation. Increased ecotourism, which includes gaur sightseeing, has the potential to supplement the income of local villages. Large mammal–based ecotourism has worked in other regions (Goodwin and Leader-Williams 2000), but there has been no concerted government effort to deal with the logistic and economic mechanisms needed to make this goal a reality. Chaiyarat and Srikosamatara (2009) also advocate ecotourism for protected areas within WEFCOM that contain either gaur or banteng. It is encouraging that guard stations at both Khao Yai and Kui Buri national parks have been tasked with involving local villagers in ecotourism activities.

We believe the majority of human-bovid conflicts occurring in Thailand are due to the straying of domestic cattle into protected areas. Domestic livestock in protected forests can be a serious detriment to populations of wild bovids due to direct competition for food and water, and disease transmission (Chaiyarat and Srikosamatara 2009). As mentioned at the outset of this chapter, the close evolutionary history between domestic and wild bovid species makes disease transmission a significant issue throughout the world (Daszak et al. 2000; Palmer 2007), and also as noted, diseases such as rinderpest, anthrax, and foot-and-mouth disease, transmitted by domestic cattle, are a potential threat (see Timmins et al. 2008; Duckworth et al. 2008). Disease is common in domestic livestock in Thailand. For example, foot-and-mouth disease was present in up to 50 percent of the herds examined by Chaiyarat and Srikosamatara (2009) in WEFCOM. Most protected areas in Thailand have regulations against livestock entry, but the level of enforcement is low. Chaiyarat and Srikosamatara (2009) estimated 50 percent of protected forests in WEFCOM were impacted by grazing of domestic livestock. Effective patrolling as advocated earlier should result in reduced entry of livestock into these parks. Further, vaccinating cattle adjoining protected areas that contain wild bovids has been tried in other regions (Palmer 2007) and possibly could be applied to Thailand as there are so few wild populations of each species here. However, the potential threat of wildlife disease is less urgent than the need of direct evidence of disease transmission, which is lacking at this time. Fencing to prevent livestock moving into reserves has been successful in Vietnam (Pedrono et al. 2009) and is currently being used along the boundaries of HKWS, TPNP, Khao Yai National Park, and Khao Ang Ru Nai Wildlife Sanctuary. There have been no systematic efforts to determine if these fences are effective.

Reintroduction

The breeding and transport of bovids in captivity is probably one of the best understood processes within agriculture. These skills can be used to maintain captive, genetically diverse populations of both banteng and gaur for the foreseeable future. Banteng were introduced successfully into Australia (Bradshaw et al. 2006), and these expatriated populations may serve as a valuable conservation hedge against increasing human-wildlife conflicts in Southeast Asia (Bradshaw et al. 2006). It is possible that animals from current populations in Thailand or Cambodia could serve as source animals for re-introduction efforts within Thailand and other range countries, such as Laos or Vietnam.

The government of Thailand has not formulated a national reintroduction plan for either banteng or gaur. One recent project for banteng involved Zoological Park Organization staff releasing thirteen captive-born banteng into Khao Khiew-Khao Chompu Wildlife Sanctuary, and these increased to thirty-four individuals over a 4-year period (Sangpong et al. 2009). According to radiotracking and observation data from the first 2.5 years, the released banteng formed a single herd and utilized a 100–600 m elevation range of the reserve. Their home range was estimated at greater than 7.96 km², and they foraged on at least 23 plant species (Sangpong et al. 2009), which is similar to the population at HKWS (Bhumpakphan 1997; Prayurasithi 1987). This pilot study reinforces our belief that there is strong potential to reestablish banteng and gaur populations across their former range in Thailand, if habitats can be restored and effective patrolling established.

CONCLUSIONS

Gaur populations from each area currently under protection have seemingly increased from 10–25 animals to more than 50 animals in Khao Yai and Kui Buri national parks and TPNP. For banteng, HKWS staff report their population has apparently increased during the past twenty years; and Khao Ang Runai Wildlife Sanctuary staff report increasing numbers within the past ten years (Sawai Wanghongsa, personal communication). These apparent increases, if true, may be due to any of the recommendations outlined in this chapter, as all the reserves have tried a combination of remedies to halt the population declines. The good news is they may have been successful, but the bad news is we do not know the critical factors responsible for these increases, nor do we have confirmed population estimates. Indexes for animal populations should be possible using the geocoded sign-and-track data being collected at HKWS and other reserves, but this information has yet to be analyzed, and will provide only relative abundances.

There are no populations of banteng and gaur existing outside protected areas in Thailand, indicating the importance of active management to preserve these species within the country. Rangewide declines in both species (Duckworth et al. 2008; Timmins et al. 2008) seem to be buffered by successful protection within suitable habitat. Fire policies must be changed, however, to allow suitable habitat to persist for banteng, and protected areas must enforce livestock prohibitions in response to mounting pressure from local villagers for grazing land and to guard against the serious potential for disease transmission.

REFERENCES

Bhumpakphan, N. 1997. Ecological Characteristics and Habitat Utilization of the Gaur (*Bos gaurus* H. Smith) in Different Habitat Sites. PhD diss., Kasetsart University, Bangkok.

Bradshaw, C. J. A., Y. Isagi, S. Kaneko, D. M. J. S. Bowman, and B. W. Brook. 2006. Conservation Value of Non-native Banteng in Northern Australia. *Conservation Biology* 20:1306–11.

Cain, J. W., III, P. R. Krausman, S. S. Rosenstock, and J. L. Turner. 2006. Mechanisms of Thermoregulation and Water Balance in Desert Ungulates. *Wildlife Society Bulletin* 34:570–81.

Chaiyarat, R. 2004. Ecology of the Wild Water Buffalo (*Bubalus bubalis* Linn.) in Huai Kha Khaeng Wildlife Sanctuary. PhD diss., Kasetsart University, Bangkok.

Chaiyarat, R., and S. Srikosamatara. 2009. Populations of Domesticated Cattle and Buffalo in the Western Forest Complex of Thailand and Their Possible Impacts on the Wildlife Community. *Journal of Environmental Management* 90:1448–53.

Chetri, M. 2003. Food Habits of Gaur (*Bos gaurus gaurus* Smith, 1827) and Livestock (Cows and Buffaloes) in Parsa Wildlife Reserve, Central Nepal. *Himalayan Journal of Sciences* 1 (1): 31–36.

Choudhury, A. 2002. Distribution and Conservation of the Gaur *Bos gaurus* in the Indian Subcontinent. *Mammalian Review* 12:199–226.

Conry, P. J. 1989. Gaur, *Bos gaurus*, and Development in Malaysia. *Biological Conservation* 49:47–65.

Corbet, G. B., and J. E. Hill. 1992. *The Mammals of the Indohimalayan Region: A Systematic Review*, Natural History Museum Publications. New York: Oxford University Press.

Daszak, P., A. A. Cunningham, and A. D. Hyatt. 2000. Emerging Infectious Diseases of Wildlife: Threats to Biodiversity and Human Health. *Science* 21:443–49.

Duckworth, J. W., R. Steinmetz, R. J. Timmins, A. Pattanavibool, Than Zaw, Do Tuoc, and S. Hedges. 2008. *Bos gaurus*. IUCN Red List of Threatened Species (2009), www.iucnredlist.org.

Galbreath, G. J., J. C. Mordacq, and F. H. Weiler. 2006. Genetically Solving a Zoological Mystery: Was the Kouprey (*Bos sauveli*) a Feral Hybrid? *Journal of Zoology* 270:561–64.

Goodwin, H. L., and N. Leader-Williams. 2000. Tourism and Protected Areas—Distorting Conservation Priorities toward Charismatic Megafauna? In *Priorities for the Conservation of Mammalian Diversity: Has the Panda Had Its Day?* ed. A. Entwistle and N. Dunstone, 257–75. Cambridge: Cambridge University Press.

Grubb, P. 2005. Order Artiodactyla. In *Mammal Species of the World: A Taxonomic and Geographic Reference*, 3rd ed., ed. D. E. Wilson and D. M. Reeder, 637–722. Baltimore: Johns Hopkins University Press.

Hassanin, A., and A. Ropiquet. 2007. Resolving a Zoological Mystery: The Kouprey Is a Real Species. *Proceedings of the Royal Society—Biological Sciences* 274 (1627): 2849–55.

Hedges, S., C. P. Groves, J. W. Duckworth, E. Meijaard, R. J. Timmins, and J. A. Burton. 2007. Was the Kouprey a Feral Hybrid? A Response to Galbreath et al. (2006). *Journal of Zoology* 271:242–45.

Heinen, J. T., and S. Srikosamatara. 1996. Status and Protection of Asian Wild Cattle and Buffalo. *Conservation Biology* 10:931–34.

Hofmann, R. R. 1989. Evolutionary Steps of Ecophysiological Adaptation and Diversification of Ruminants: A Comparative View of Their Digestive Systems. *Oecologia* 78:443–57.

Karanth, K. K., J. D. Nichols, J. E. Hines, K. U. Karanth, and N. L. Christensen. 2009. Patterns and Determinants of Mammal Species Occurrence in India. *Journal of Applied Ecology*, doi10.1111/j.1365-2664.2009.01710.x.

Krishnan, M. 1972. An Ecological Survey of the Large Mammals of Peninsular India. *Journal of the Bombay Natural History Society* 69 (2): 322–49.

Lekagul, B., and J. A. McNeely. 1988. *Mammals of Thailand*, 2nd ed. Bangkok: Darnshutha Press.

Martin, E. B., and L. Vigne. 1996. Nepal's Rhino: One of the Greatest Conservation Success Stories. *Pachyderm* 21:10–26.

McNaughton, S. J., and N. J. Georgiadis. 1986. Ecology of African Grazing and Browsing Ungulates. *Annual Review of Ecology and Systematics* 17:39–65.

Myint Aung, Khaing Khaing Swe, Thida Oo, Kyaw Kyaw Moe, P. Leimgruber, T. Allendorf, C. Duncan, and C. Wemmer. 2004. The Environmental History of Chatthin Wildlife Sanctuary, a Protected Area in Myanmar (Burma). *Journal of Environmental Management* 72 (4): 205–16.

Nakhasathien, S., N. Bhumpakphan, and S. Simcharoen. 1987. *Ecology of Forest and Wildlife in Thung Yai Narae-suan and Huai Kha Khaeng Wildlife Sanctuaries.* Bangkok: Royal Forest Department.

National Research Council. 1983. *Little-Known Asian Animals with a Promising Economic Future.* Washington, DC: National Academy Press.

Ngha, N. A., R. Awaluddin, and A. Mahmood. 1985. Selembu-Malaysia's New Animal. *Kajian Vegerinar* 17:157–58.

Palmer, M. V. 2007. Tuberculosis: A Reemerging Disease at the Interface of Domestic Animals and Wildlife. *Current Topics in Microbiology and Immunology* 315:195–215.

Peacock, E. H. 1933. *A Game-Book from Burma & Adjoining Territories.* London: H.F. & G. Witherby.

Pedrono, M., Ha Mihn Tuan, P. Chouteau, and F. Vallejo. 2009. Status and Distribution of the Endangered Banteng *Bos javanicus birmanicus* in Vietnam: A Conservation Tragedy. *Oryx* 43:618–25.

Petdee, A. 2000. Feeding Habits of the Tiger (*Panthera tigris* [Linnaeus]) in Huai Kha Khaeng Wildlife Sanctuary by Fecal Analysis. Master's thesis, Kasetsart University, Bangkok.

Phrommakul, P. 2003. Habitat Utilizations and Prey of the Tiger (*Panthera tigris* [Linnaeus]) in the Eastern Thung Yai Naresuan Wildlife Sanctuary. *Journal of Wildlife in Thailand* 11 (1): 1–12.

Prayurasithi, T. 1987. Ecological Banteng (*Bos javanicus*) in Huai Kha Khaeng Wildlife Sanctuary. Master's thesis, Kasetsart University, Bangkok.

———. 1997. The Ecological Separation of Gaur (*Bos gaurus*) and Banteng (*Bos javanicus*) in Huai Kha Khaeng Wildlife Sanctuary, Thailand. PhD diss., University of Minnesota.

Sangpong, S., W. Tunwattana, R. Chaiyarat, T. Siripiyanak, J. Taewnern, P. Dunriddach, S. Pannim, and A. Onpunya. 2009. Ecology of Banteng (*Bos javanicus*) after Reintroduction in Khao Kheow Open Zoo, Chonburi Province.

Schaller, G. B. 1967. *The Deer and the Tiger: A Study of Wildlife in India.* Chicago: University of Chicago Press.

Simcharoen, S. 2008. Ecology of the Leopard in Huai Kha Khaeng Wildlife Sanctuary. PhD diss., Kasetsart University, Bangkok.

Sirimagorn, R. 2009. The Effects of Prescribed Burning on Habitat Use of Large Herbivores in Huai Kha Khaeng Wildlife Sanctuary, Uthai Thani Province. Master's thesis, Kasetsart University, Bangkok.

Srikosamatara, S. 1993. Density and Biomass of Large Herbivores and Other Mammals in a Dry Tropical Forest, Western Thailand. *Journal of Tropical Ecology* 9:33–43.

Srikosamatara, S., and V. Suteethorn. 1995. Populations of Gaur and Banteng and Their Management in Thailand. *Natural History Bulletin of the Siam Society* 43:55–83.

Steinmetz, R. 2004. Gaur (*Bos gaurus*) and Banteng (*Bos javanicus*) in the Lowland Forest Mosaic of Xe Pian Protected Area, Lao PDR: Abundance, Habitat Use and Conservation. *Mammalia* 68:141–57.

Sukmasuang, R. 1993. Ecology of the Wild Elephant (*Elephas maximus* Linn.) in Huai Kha Khaeng Wildlife Sanctuary. Master's thesis, Kasetsart University, Bangkok.

Timmins, R. J., J. W. Duckworth, S. Hedges, R. Steinmetz, and A. Pattanavibool. 2008. *Bos javanicus. IUCN Red List of Threatened Species* (2009), www.iucnredlist.org.

Tun Yin. 1967. *Wild Animals of Burma.* Rangoon, Burma: Rangoon Gazette.

Ucharoensak, W. 1992. *Ecology and Distribution of Wild Water Buffalo (Bubalus bubalis* Linn.) *in Huai Kha Khaeng Wildlife Sanctuary*. Special problem, Kasetsart University, Bangkok.

Wharton, C. 1957. An Ecological Study of Kouprey, *Novibos sauveli* (Urbain). Institute of Science and Technology, Manila, Philippines.

Wharton, C. H. 1968. Man, Fire, and Wild Cattle in South-East Asia. *Proceedings of the Annual Tall Timbers Fire Ecology Conference* 8:107–67.

Wilson, D. E., and D. A. M. Reeder. 2005. *Mammal Species of the World*. Washington, DC: Smithsonian Institution Press.

Rainfall Patterns and Unpredictable Fruit Production in Seasonally Dry Evergreen Forest and Their Effects on Gibbons

Warren Y. Brockelman

The seasonal evergreen forests of continental Southeast Asia are rich in species of animals and plants, and threatened by exploitation and incursion of fires (Ashton 1990). These forests are delicately balanced between moist and dry conditions, and the loss of some canopy cover or increases in temperature can turn them into deciduous forests, scrub vegetation, or even grassland. Less dramatically, subtle changes in phenology may trigger extensive changes throughout the community (Tutin and White 1998; Corlett and LaFrankie 2004).

The term *dry forest* generally refers to deciduous forest, with a period of several months in which all trees are leafless. However, the term may also encompass evergreen forests with a severe dry season, or "dry evergreen" forests (Rundel and Boonpragob 1995), which may receive approximately 1,200–1,500 mm of rainfall annually. Such forests border on deciduous forests at lower elevations and wetter evergreen forests at higher elevations (Ogawa et al. 1961). The forest on the Huai Kha Khaeng 50 ha forest dynamics plot (FDP), referred to as "seasonal dry evergreen forest" (Bunyavejchewin et al. 2001), apparently is near the evergreen–deciduous transition as fire occasionally invades this forest (Baker et al. 2005). This plot is approximately 600 m above mean sea level (msl) and receives an average of about 1,475 mm of rainfall annually. Dry or seasonally dry evergreen forests have high percentages of deciduous species (Rundel and Boonpragob 1995), however, and are not sharply distinct from deciduous forests, despite the seeming rigidity of forest classifications (Ashton 1990). A category of dry evergreen forest called "mixed evergreen + deciduous forest" (Maxwell and Elliott 2001) recognizes that *evergreen* and *deciduous* refer more precisely to species than to forests as a whole. Van de Bult and Greijmans (2006) have carried out an altitudinal transect in Mae Wong National Park, western Thailand, and found that the percentage of deciduous species declines gradually from 85 percent to 15 percent, from 600 m to 1,100 m msl.

The confused terminology of seasonally dry forests suggests that we need to base our classifications on direct measures of structure, cover and climate, and seasonal variation. The evergreen forest of Khao Yai National Park featured in this chapter is a relatively humid example of seasonal evergreen forest, and has been referred to as an outlier of "tropical wet forest" by Rundel and Boonpragob (1995). Holdridge et al. (1971) classified this forest as "subtropical wet" and estimated (without any rainfall data) that it should have only one to two effectively dry months, probably reflecting their belief that forest on the mountain should have a more perhumid zonal climate than forest in the lowland. However, there is in fact a "dry atmospheric association" in the Holdridge scheme. Rainfall at all elevations in Khao Yai is markedly seasonal. This forest is at the northwest end of what Ashton (1990) refers to as the "south-eastern seasonal evergreen forests" that extend into southwestern Cambodia. In this chapter I show that rainfall seasonality in Khao Yai Park is associated with highly seasonal availability of fruits for wildlife (in this case gibbons), and that considerable year-to-year variation exists, which also strongly affects food and diet. Describing the rainfall regime by using monthly means underestimates seasonal variation and ignores inter-annual variation. Exactly how year-to-year variation in weather affects interannual fruiting variation in each species cannot be answered here, and will require longer-term phenology data.

STUDY AREA AND METHODS

The Mo Singto area of Khao Yai National Park is located around lat. 14° N and at 725–815 m msl. Annual rainfall averages about 2,200 mm at park headquarters (approximately 750 m msl, 1994–2007, National Park Division), and occurs mostly during the southwest monsoon between May and October. Unfortunately, reliable temperature data are available for only a few years, and only daily maximum and minimum temperatures are recorded, but the average annual temperature is about 22°C–23°C.

The Mo Singto area has been the focus of research on the ecology and behavior of gibbons (*Hylobates lar*) since about 1980 (J. Raemaekers and P. Raemaekers 1985; J. Raemaekers et al. 1984; Brockelman et al. 1998; Bartlett 2009). In 1996 I began to survey the forest in order to establish an FDP to facilitate study of frugivore diets, foraging, seed dispersal, and tree (and liana) recruitment. The first complete census of all stems ≥ 1 cm dbh (diameter at breast height, taken as 1.3 m) on the 30 ha Mo Singto plot was completed in 2005, and all lianas ≥ 3 cm dbh were also inventoried (Brockelman et al. 2010). The plot contains 262 species of trees and shrubs ≥ 1 cm dbh and 204 species ≥ 10 cm dbh. The tree density on the plot averages 4,295 stems ≥ 1 cm dbh/ha and 514 stems ≥ 10 cm dbh/ha. Approximately 120 species of large lianas have been found on the plot, many of which provide food for gibbons and other wildlife, especially in the dry season. The density of liana genets ≥ 3 cm dbh is approximately 270 genets/ha.

Gibbons (*Hylobates lar*) occur at a density of about 4 groups/km^2 in the Mo Singto area. They live in monogamous territorial groups with an average of 4 individuals per group (Brockelman et al. 1998; Brockelman 2004). Gibbon group A, with 5 individuals, occupies most of the plot. We tracked the foraging path of this group using the plot census database and ArcView GIS software for 6 days each month for three years (2003–2005), and at least 45 days continuously in April and May, the fruiting season of the tree *Nephelium melliferum* Gagnep (Sapindaceae), during 2006, 2007, and 2008. While tracking the group, one observer followed one of the adults as a focal animal (usually the female) and recorded all behaviors, while other persons read tree numbers from the tree tags and collected feces when the animals defecated. *N. melliferum*, which starts fruiting in mid-April, is favored over all other foods when available but fruits synchronously only in one year out of two or three (W. Brockelman, unpublished data). A particular question of interest was what the gibbons eat in years when *N. melliferum* fruit is not available.

Phenology of a sample of sixty common species of trees has been followed since 2003, including all major species consumed by gibbons. Ten reproductive-sized individuals of each species were checked twice a month and scored for presence of flowers, ripe fruit, or deciduous. These major food species have medium-to-large fleshy fruits that are preferred over other species when available.

Food species use is measured here by the number of visits made to each fruit source during the period in question. A gibbon group typically travels directly to and enters a fruit tree or liana and feeds for a period of 15–45 minutes, usually with pauses or rest periods, until all individuals appear to be satiated or until the ripe fruits have all been harvested. Visits to sources are independent events distributed in time, and although they do not measure consumption quantitatively, they serve as a convenient measure of the importance of the species in the diet over the year. Estimates of actual mass or energy intake cannot be made during daily follows, mainly because individual gibbons are out of view much of the time, and because other behavioral data are being recorded at the same time. The number of visits, however, is a less-suitable measure of insect and leaf consumption, which take place more intermittently and opportunistically.

RESULTS

Seasonality

The distribution of average monthly rainfall shows a climate with approximately five months of dry weather (Figure 1), indicating a highly seasonal forest typical of lowland continental Thailand (Holdridge et al. 1971). Rainfall less than 100 mm in a month is taken as a crude indication of a "dry" month because it is close to the potential evapotranspiration (PET; Walsh 1996; van Schaik and Pfannes 2005). Holdridge et al. (1971) estimated PET to range from approximately 80 to 120 mm per month in most of Thailand (although their treatment of temperatures above 30 °C in calculating PET is problematic). When the PET/rainfall ratio exceeds 1.0 in a month, a water deficit exists and soil moisture declines. Of course, the rate of drying is affected by wind, soil water-holding capacity, and the presence of a water table or lateral ground flow (Walsh 1996).

Rainfall aggregated by month will underestimate the length of the actual dry season because the beginning and the end of the dry season will likely fall in months having more than the 100 mm average (Walsh 1996). A forest with an average of five

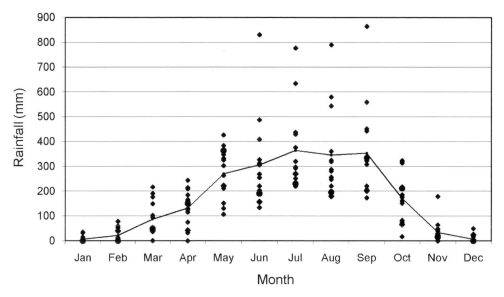

Figure 1. Monthly rainfall and its variability from 1994 to 2007 at Khao Yai National Park headquarters (about 1 km from Mo Singto Plot). The line segments connect mean values (data from Department of National Parks, Wildlife and Plant Conservation, Thailand).

months of less than 100 mm of rainfall could have a dry season ranging from 3.5 to 6.5 months, depending on when in the month the dry season begins and ends. Rainfall also varies greatly from year to year. At Mo Singto in Khao Yai National Park, December, January, and February have been consistently dry, and November usually so, but March, April, and October have been dry in approximately half of the years (Figure 1). In an extreme year the dry season could last from September to May. The rainfall from June through September is highly skewed upward, indicating that in very wet years, most water becomes surface runoff. From the fourteen-year monthly rainfall averages, the site has a perhumidity index, a measure of rainfall seasonality used by Walsh (1996), of 3.5, placing it in the semi-evergreen seasonal forest category but near the transition to evergreen seasonal forest.

In most years annual rainfall is in the range of 2,000–3,000 mm. Rainfall over the fourteen-year period (Figure 2) shows a slight declining trend, but the most notable feature in the graph is the sudden drop by over half in the El Niño Southern Oscillation (ENSO) year of 1997, and in the following year when it reached a low of 1,264 mm. The year 2006, another ENSO year, was also followed by low rainfall.

In order to examine the actual onset, length, and end of the dry season, a running average of rainfall was plotted by day, summing the rainfall over the previous 15 days (Figure 3). I considered that a 15-day sum of 50 mm or less indicated a "dry" day. Figure 3 presents results for five years beginning with mid-1994. Each year is started with June (always a wet month) so as not to interrupt the dry season. The first three years illustrate the length and variability of the dry season, which was about six months in the first two years, but was cut short in 1997 by a rainy spell in April. The ENSO event appeared to affect rainfall first in August 1997, which saw a marked decline.

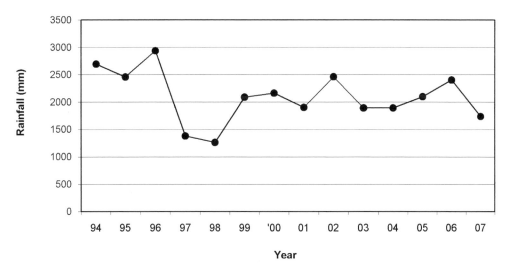

Figure 2. Mean annual rainfall at Khao Yai National Park headquarters from 1994 to 2007.

September had only 221 mm. October had only 80 mm but this caused the 15-day average to exceed 50 mm until after the middle of October. The dry season was seven months and lasted until the end of May 1998. The 1998–99 dry season started in early October (66 mm) and "normal" rainfall did not resume until April 1999.

These limited data illustrate that the dry season in the seasonal evergreen forest is both longer and more variable than traditional climate statistics suggest. It is inevitable that this variability will influence the phenology of woody plants, by affecting moisture availability and solar radiation (van Schaik et al. 1993; Wright and van Schaik 1994). In the sections below I will illustrate some of the variability in fruiting behavior in the forest, but it will not be possible without longer-term data to determine the proximal stimuli of flowering or fruiting success.

Seasonality of Fruit Availability

The number of individual fruit sources (including fig species), and the number of species used, during the monthly 5-day periods with complete foraging data in 2005, indicate that the gibbons' major energy sources are unevenly available over the year (Figure 4a and 4b). Numerous species and sources are available during April through June, but other months are less predictable. Gibbons typically made 10 to 12 visits to fruit sources per day and visited about 50 sources during the 5-day periods in months of high availability, and utilized 18–20 species. Most sources were visited only once in one day but were visited repeatedly on different days. July was a low-fruiting month in some years, but in all three years, fruit availability declined from September to December. November and December are months of food stress, as the most dependable late-season fruiting tree (= *Choerospondias*) *axillaris* (Anacardiaceae), finishes fruiting during November, and no other reliable sources are available. The female of group A utilized only 11 sources including 5 species in November. Fruiting of figs

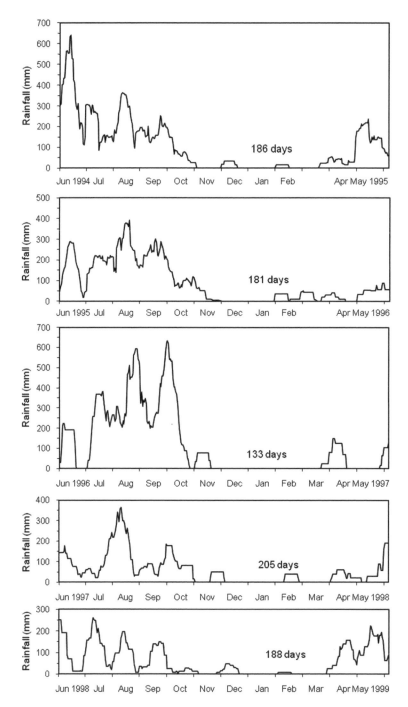

Figure 3. Trends in rainfall from June 1994 to May 1999, plotted as a daily moving average of the previous 15 days' accumulation. Each year begins with June so as not to interrupt the dry season. The length of the dry season (a "dry" day signifies less than 50 mm accumulation over the preceding 15-day period) is given for each year. An El Niño Southern Oscillation event began in 1997.

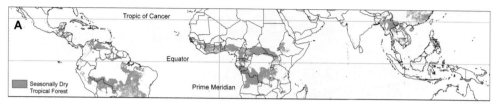

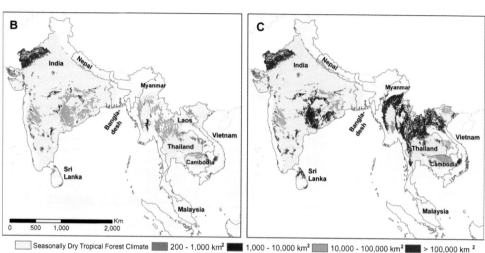

Seasonally Dry Tropical Forest Climate ☐ 200 - 1,000 km² ▓ 1,000 - 10,000 km² ░ 10,000 - 100,000 km² ▓ > 100,000 km²

Figure 1. Distribution of seasonally dry tropical forest. A: Estimated global distribution of seasonally dry forests . This estimate was developed by integrating climate and forest cover from global data sources for the geographic region between 30° North and 20° South. We used Wordclim data (www.worldclim.org) to delineate a seasonal dry forest climate envelope (envelope thresholds: 900–2,200 mm annual precipitation and four to six months with less than 100 mm precipitation). Forest cover was derived from the International Geosphere-Biosphere Programme (IGBP) DISCover (Belward et al. 1999; Loveland et al. 2000). Please refer to Chapter 3 for complete citations. B and C: Estimated distribution of seasonally dry forest areas from South Asia to Indochina. These estimates were developed by combining Worldclim data with forest cover information from four existing land cover data sets: International Geosphere-Biosphere Programme (IGBP) DISCover (Belward et al. 1999; Loveland et al. 2000), MODIS/Terra Land Cover Classification (MLCC; Strahler et al. 1999; Friedl et al. 2002), Global Land Cover 2000 (GLC2000; Bartholome and Belward 2005; Stibig et al. 2007), and Global Distribution of Tropical Dry Forest (GDTDF; Miles et al. 2006). To adjust for regional characteristics, the climate envelope was adjusted to 1,600–2,000 mm annual precipitation and four to six months with less than 100 mm precipitation. The more conservative B estimate included only forests that were identified as seasonally dry forests by at least one of the forest cover data sets. The C estimate included all forested areas delineated by at least two of the forest cover data sets. Forest patches that were less than 200 km² were excluded from both B and C. For more detailed description, complete citations, and analysis of these estimates, refer to Chapter 3.

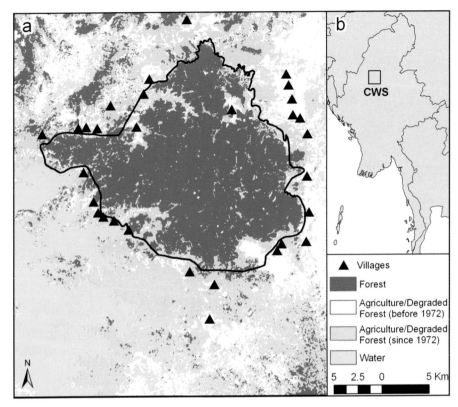

Figure 2. Distribution of forests, agriculture and human settlements around Chatthin Wildlife Sanctuary (CWS), Kanbalu District, Myanmar : (a) CWS and surrounding villages based on land cover change analysis comparing satellite imagery from 1973 and 2005; (b) location of CWS in Myanmar.

Legend:
▲ Villages
Forest
Agriculture/Degraded Forest (before 1972)
Agriculture/Degraded Forest (since 1972)
Water

5 2.5 0   5 Km

Figure 3. Deciduous dipterocarp forest. (Photograph by Stuart J. Davies.)

Figure 4. Pine forest. (Photograph by Stuart J. Davies.)

Figure 5. Mixed deciduous forest in Huai Kha Khang Wildlife Reserve, Thailand. (Photograph by Stuart J. Davies.)

Figure 6. Lower montane forest. (Photograph by Stuart J. Davies.)

Figure 7. Wet seasonal evergreen forest. (Photograph by Stuart J. Davies.)

Figure 8. Dry seasonal evergreen forest. (Photograph by Stuart J. Davies.)

Figure 9. Deciduous dipterocarp forest during a dry season ground fire, Nepal. (Photograph by Teri D. Allendorf.)

Figure 10. Deciduous dipterocarp forest following a dry season ground fire. (Photograph by Stuart J. Davies.)

Figure 11. Market day in Chatthin village outside Chatthin Wildlife Sanctuary, Kanbalu District, Myanmar. (Photograph by William J. McShea.)

Figure 12. Representative gingers with examples from both dry and wet forest habitats. Top row (left to right): *Hitchenia glauca* Wall. (dry forest); *Globba wengeri* (C.E.C. Fischer) K. J. Williams (dry forest); *Curcuma attenuata* Wall. (dry forest). Center (left to right): *Kaempferia candida* Wall. (dry forest); *Zingiber kerrii* Craib. (wet forest). Bottom row (left to right): *Hemiorchis rhodorrahchis* K. Schum. (dry forest); *Smithatris myanmarensis* W. J. Kress (wet forest); *Alpinia officinarum* Hance (wet forest). (All photos by W. J. Kress).

Figure 13. Leopard (*Panthera pardus*) in Mudumalai Wildlife Reserve, India. (Photograph by Stuart J. Davies.)

Figure 14. Rambutan (*Nephelium melliferum* Sapindaceae) fruits on the ground, dropped by pigtail monkeys in Khao Yai National Park, Thailand. The fruits are eaten by gibbons, as well as by many terrestrial mammals such as sambar and muntjac. (Photograph by Warren Brockelman.)

Figure 15. *Ficus* fruits, in this instance partially consumed by Asiatic black bears (*Ursus thibetanus*), in the Western Forest Complex, Thailand. (Photograph by Robert Steinmetz.)

Figure 16. Tiger (*Panthera tigris*) in Panna National Park, Central India. (Photograph by John Seidensticker.)

Figure 17. Male Eld's deer, or brow-antlered deer (*Rucervus eldii*). The male is in velvet antler stage and will shed the velvet and enter mating season, which extends from January through April. (Photograph by Lisa Ware.)

Figure 18. Guar (*Bos gaurus*) feeding on new grass following a dry season fire in the forest of Nagarhole National Park, India. (Photograph by Naris Bhumpakphan.)

Figure 19. Male banteng (*Bos javanicus*) in Huai Kha Khaeng Wildlife Reserve, Thailand. (Photograph by Naris Bhumpakphan.)

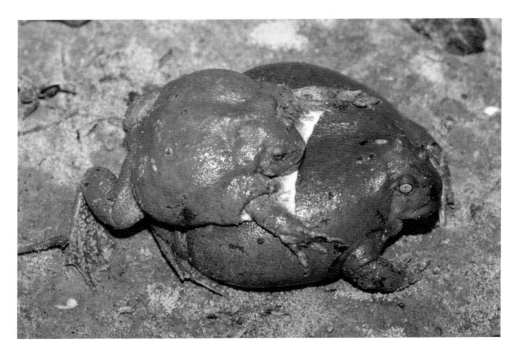

Figure 20. Box-chinned burrowing frog, *Glyphoglossus molossus* (Microhylidae). This subterranean frog is strongly fossorial, emerging above ground in the early wet season to reproduce. The female is twice as large as the male. Because he cannot grasp her tightly with his forearms in a typical amplexus squeeze, he glues himself to her back with a skin secretion. Residents of the shrubby grassland and dry forest from central Burma through Thailand and Laos to southern Vietnam. (Photograph by George R. Zug.)

Figure 21. Elephants (*Elephas maximus*) feeding on early succession forest in Sri Lanka. (Photograph by Prithiviraj Fernando.)

Figure 22. Women collecting medicinal plants during March inside Chatthin Wildlife Sanctuary, Kanbalu District, Myanmar. (Photograph by Melissa A. Songer.)

Figure 23. Women collecting thatch during August inside the buffer zone of Chatthin Wildlife Sanctuary, Kanbalu District, Myanmar. (Photograph Melissa A. Songer.)

Figure 24. Male white-handed gibbon (*Hylobates lar*) feeding on *Ficus* fruits in Khao Yai National Park, Thailand. (Photograph by Warren Y. Brockelman.)

Figure 25. People carting bamboo stems in March out of Chatthin Wildlife Sanctuary, Kanbalu District, Myanmar. (Photograph by Melissa A. Songer.)

Figure 26. Sambar (*Cervus unicolor*) use a watering hole in Pukeo Wildlife Sanctuary, Thailand. (Photograph by Naris Bhumpakphan.)

Figure 27. Green pea fowl (*Pavo muticus*) in mixed deciduous forest in Huai Kha Khaeng Wildlife Sanctuary, Thailand. (Photograph by Ronglarp Sukmasuang.)

Figure 28. Asiatic wild dog, or dhole (*Cuon alpines*), in Huai Kha Khaeng Wildlife Sanctuary, Thailand. (Photograph by Ronglarp Sukmasuang.)

Figure 29. Sloth bear (*Ursus ursinus*) digging for ants in Panna National Park, Central India. (Photograph by K. Yoganand.)

Figure 30. Clouded leopard (*Neofelis nebulosa*) captured by trip-camera in Huai Kha Khaeng Wildlife Reserve, Thailand. (Photograph by Ronglarp Sukmasuang.)

(*Ficus* spp.) is also not dependable in the winter, and during 2005 only one fig source was utilized in October and November and two in December. The average number of fig sources utilized during the year was 4.5 per 5-day period. Food availability was relatively high in January of 2006 but none of the major tree fruit species (Table 1) fruits in that month. Several species of lianas, however, commonly bear fruit in January and February.

The ranging behavior of the gibbons is strongly affected by fruit availability. While following group A about its range, we recorded the number of all trees entered so that the daily range could be plotted. This is not a perfect measure of the daily range because in months of high activity, many of the trees the gibbons move through are not recorded because of the speed of movement of the group. Nevertheless, the data show that in months of reduced food availability the gibbons reduced their activity and traveled less (Figure 4c). The number of trees entered by gibbons and recorded is correlated with the number of fruit sources used (product moment correlation coefficient $r = 0.814$, $p < 0.01$). Ranging distance in times of fruit abundance is approximately four times that in the time of fruit shortage.

Interannual Variation in Fruiting

The phenology list (Table 1) includes 15 major fruit species of gibbons and 9 minor species. The list is not comprehensive as many other species of trees provide minor fruit sources, and about 12 species of lianas are also major fruit sources. The list gives the number of trees ≥ 10 cm dbh on the plot, and the number of adult-sized trees (the average beginning size of fruiting varies among species) within the range of the gibbon group. The number of trees within the range, however, is not an accurate indication of the amount of fruit available, as most trees of a given species may not fruit even in a good year. Some factors that affect fruit productivity include the position of the tree crown with respect to neighboring trees, presence of lianas in the crown, bole breakage by storms, and presence of dioecy. Gibbons must also compete for fruit with other species of animals. The sample of phenology study trees is also not a random sample of all trees; phenology trees were randomly selected from a sample of trees that were in good condition with visible crowns, and few or no lianas. The problem of inferring fruit availability from tree numbers will be analyzed further below for one of the preferred species for which fruiting data are available.

I considered a fruiting species to "fail" in a given year if only 20 percent or less of the phenology sample produced fruit. Fruiting of individual trees tended to be all or none, and variation between trees was great in most species. Only 6 out of 15 (40 percent) of the major food species, but 8 (80 percent) of the minor species, fruited in all six years. Three species (*Cinnamomum subavenium*, *Prunus javanica*, and *Nephelium melliferum*) failed to fruit in four of the six years.

The standard deviation (SD), and not the coefficient of variation, is used here as a measure of fruiting variability because the absolute variability, and not the relative variation, more closely relates to the variation in actual food supply. There is a correlation between the SD of fruiting percentage and the median time of fruiting of species (Figure 5). The product moment r is -0.620, which is very close to the value for $p = 0.01$;

Table 1. Phenology scores (percentage of trees with ripe fruit) for fruiting species consumed by gibbons for past six years on Mo Singto forest dynamics plot.

No.	Gibbon Fruit Species	Trees[a]	Year						SD[b]	Months[c]
			2003	2004	2005	2006	2007	2008		
	Major fruits									
1	Cinnamomum subavenium	441/179	0	0	100	0	0	40	40.8	2/1–3/2
2	Bridelia insulana*	55/55	100	0	90	30	80	80	39.3	3/1–4/2
3	Prunus javanica*	100/43	20	100	0	0	20	10	37.8	3/2–4/1
4	Eugenia syzygioides	150/58	100	100	100	0	80	70	38.9	3/2–4/2
5	Baccaurea ramiflora*	27/24	29	0	40	40	60	30	19.7	4/1–5/2
6	Polyalthia simiarum*	272/142	60	100	100	30	100	80	28.6	4/1–5/2
7	Nephelium melliferum*	728/308	11	90	10	10	100	0	45.3	4/2–5/1
8	Podocarpus neriifolius*	38/21	90	40	90	80	80	70	18.7	5/1–7/1
9	Eugenia cerasoides	366/67	0	100	90	60	100	50	38.8	6/1–7/1
10	Balakata baccata	55/46	56	60	60	30	20	10	22.2	6/1–7/2
11	Sandoricum koetjape	18/13	40	80	90	90	90	60	20.7	6/1–8/1
12	Alphonsea boniana	54/36	78	100	100	50	100	100	20.6	6/2–8/1
13	Choerospondias axillaris	169/139	100	100	100	90	100	100	4.1	6/2–11/1
14	Aphananthe cuspidata	23/5	40	40	60	50	50	60	8.9	7/1–10/2
15	Garcinia benthamii	109/66	43	100	50	20	100	0	41.0	8/1–9/2
	Mean		51.1	67.3	72.0	38.7	72.0	50.7	13.8	
	Minor fruits									
1	Gironniera nervosa*	725/344	40	40	40	40	60	60	10.3	1/1–12/2
2	Gomphandra tetrandra*	84/79	60	40	40	70	70	80	16.7	2/1–5/1
3	Knema elegans*	367/429	30	30	30	20	22	30	4.7	3/1–5/1
4	Aglaia elaeagnoidea*	213/844	82	36	36	18	18	64	25.7	4/1–5/2
5	Beilschmiedia glauca*	340/141	88	0	0	70	0	0	41.2	5/1–8/2
6	Diospyros glandulosa	28/27	60	50	70	40	66	50	11.3	7/1–9/1
7	Aidia densiflora	391/373	100	90	90	100	100	100	5.2	7/1–10/1
8	Ilex chevalieri	998/335	20	30	30	70	50	50	18.3	7/2–9/2
9	Elaeocarpus sphaericus	57/37	30	80	60	60	80	70	18.6	7/2–11/1
10	Anthocephalus chinensis	11/11	78	100	100	89	100	89	9.0	8/2–11/2
	Mean		58.8	49.6	49.6	57.7	56.6	59.3	4.5	

*Species fruiting during the April–May period included in Table 2 and Figure 6.
[a]Trees greater than 10 cm diameter at breast height (dbh) on plot/adult-size trees within gibbon range.
[b]Standard deviation.
[c]Month/first or second half for beginning and end of main fruiting season, in typical heavy-fruiting years.

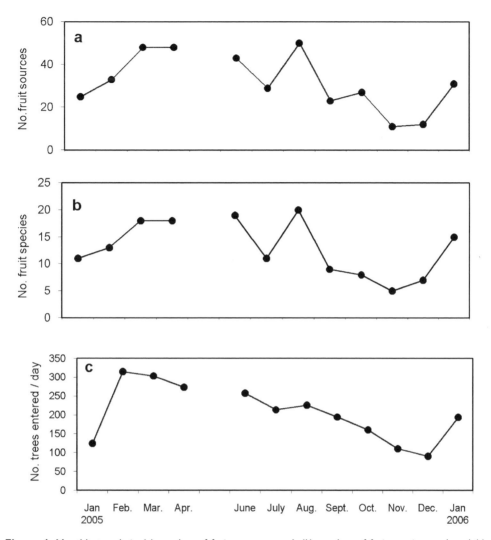

Figure 4. Monthly trends in (*a*) number of fruit sources used, (*b*) number of fruit species used, and (*c*) number of trees entered per day by the adult female of gibbon group A. Each month's data are based on five complete days spent following the group. Data were not collected during May.

however, the data are clearly not normally distributed. The Spearman rank correlation coefficient is not significant (p < 0.20, 2-tailed). More species are needed to test this relation. A cluster of five species with high SD that fruit before May are primarily responsible for this correlation. We may conclude that although species fruiting in the dry season all have rather high interannual SD values, most, but not all, species fruiting during the rainy season have lower SD values. It is also noteworthy that no major fruit species fruits in December. The dry season variability of fruiting could be related to the uncertainty of weather in October and November, the months when many dry season fruiting species should be in flower.

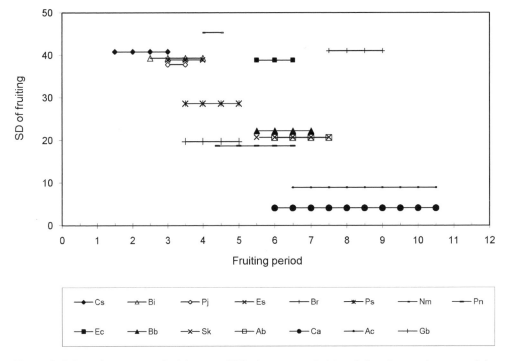

Figure 5. Relation between standard deviation (SD) of percentage fruiting of phenology study trees and the season of fruiting, given in two-week periods. Abbreviations (genus and species) refer to consecutive major fruit species in Table 1.

Fruiting in major food species was not strongly synchronized between species, as they varied in the frequency of fruiting and pattern across years. The SD values for years (columns) were relatively high, indicating lack of synchrony between species. There were, nevertheless, relatively "good" years and "poor" years of fruiting; 2006 had significantly lower fruiting than the other years.

The minor food species showed a similar lack of synchrony between species but in general, less variation among years than was evident in the major food species. The average fruiting of species showed virtually no variation among years (Table 1, SD = 3.0). Many of the minor species fruited at the same level in all years, and could be considered dependable alternate food species for gibbons. It is unclear why the major and minor food species differed in their regularity.

Interannual Variation in Gibbon Diet

In the April–May gibbon-following data, *Nephelium melliferum* fruited in only one year (2007) out of three (Table 2). In the other years, alternate species were consumed, but these differed between 2006 and 2008. In all years, lianas (WC) constituted 30–40 percent of the top ten fruit species in the diet. In *Nephelium* off-years, the gibbons also utilized tree species in the understory (UT) below 10 m, and consumed greater numbers of figs. One species eaten in 2008, *Rourea minor* (Connaraceae), has dehiscent

Table 2. Top ten fruit species, and all fig species, in the diet of adult female of gibbon group A at Mo Singto study area from 10 April to near the end of May in three different years.

2006

No.	Fruit Species	Habit[a]	Days	Visits
1	Podocarpus neriifolius	T	27	62
2	Erycibe elliptilimba	WC	17	72
3	*Gnetum macrostachyum	WC	17	34
4	Gironniera nervosa	T	15	19
5	Baccaurea ramiflora	T	11	14
6	Knema elegans	UT	10	11
7	Polyalthia simiarum	T	6	6
8	*Ardisia sanguinolenta	UT	4	4
9	*Aglaia edulis	T	3	3
10	*Desmos dumosa	WC	3	3

No.	Fig Species	Habit[a]	Days	Visits
1	Ficus altissima	Str	19	34
2	Ficus kurzii	Str	12	19
3	Ficus sagittata	HEp	12	16
4	Ficus nervosa	T	5	5

2007

No.	Fruit Species	Habit[a]	Days	Visits
1	*Nephelium melliferum	T	42	232
2	Polyalthia simiarum	T	39	80
3	*Ziziphus attopensis	WC	18	38
4	Gironniera nervosa	T	16	23
5	Erycibe elliptilimba	WC	13	28
6	*Eugenia cerasoides	T	9	21
7	*Diploclisia glaucescens	WC	9	14
8	Podocarpus neriifolius	T	9	15
9	Baccaurea ramiflora	T	5	5
10	*Gnetum montanum	WC	4	5

No.	Fig Species	Habit[a]	Days	Visits
1	Ficus kurzii	Str	10	15
2	Ficus sagittata	HEp	9	12
3	Ficus punctata	WC	7	7
4	Ficus stricta	Str	7	7
5	Ficus nervosa	T	6	6
6	Ficus villosa	WC	3	3

2008

No.	Fruit Species	Habit[a]	Days	Visits
1	Polyalthia simiarum	T	31	88
2	*Bridelia insulana	T	22	37
3	Erycibe elliptilimba	WC	20	63
4	Podocarpus neriifolius	T	16	25
5	*Salacia chinensis	WC	15	57
6	*Aglaia elaeagnoides	UT	10	11
7	Gironniera nervosa	T	8	8
8	*Rourea minor	WC	7	9
9	Knema elegans	UT	5	7
10	*Walsura robusta	T	4	4

No.	Fig Species	Habit[a]	Days	Visits
1	Ficus sagittata	HEp	23	38
2	Ficus nervosa	T	16	19
3	Ficus annulata	T	13	22
4	Ficus villosa	WC	5	6
5	Ficus altissima	Str	4	4
6	Ficus punctata	WC	4	5

Note: Nephelium melliferum (Sapindaceae) was the preferred species during study period but fruited in only one year out of three.
*Species appearing on only one top ten list.
[a]T = tree; HEp = hemi-epiphyte; UT = understory tree; Str = strangler fig; WC = woody climber.

capsules and seeds with a thin bright red fatty aril, a fruit-type typically consumed by birds. This was the first time gibbons had ever been seen to consume this species in about twenty years of observation. Half of the major non-fig species consumed each year were not on the list for other years and overall, 14 out of the 20 species on the top ten lists appeared on only a single list.

There were, however, dependable species that were consumed in all three years: *Podocarpus neriifolius*, *Erycibe elliptilimba*, *Gironniera nervosa*, and *Polyalthia simiarum*. *P. simiarum* had a weak fruiting year in 2006 (Table 1) and was in seventh place in that year. The rank order of utilization therefore may depend on availability as well as on preference.

The species of figs utilized were also unpredictable, and gibbons cannot depend on any species being available in a particular month. About 20 species of *Ficus* occur on the plot, nearly all of which are available and consumed in unpredictable months. The rank order of use of *Ficus* species also changed greatly from year to year. The largest "strangler" trees on the plot, *Ficus altissima* and *F. kurzii*, each provided fruit in only two of the three years. Although species of *Ficus* are a regular part of the diet of the gibbons, none of them matches *Nephelium melliferum* in popularity.

The number of visits to *Nephelium melliferum* in 2007 was the same as the number of visits to all other non-fig sources combined, and similar to the total number of visits to non-fig fruit sources in 2006 and 2008. *N. melliferum* has a relatively large-sized fruit with seeds 2–3 cm long, and the group defecated more of its seeds (3,206) than those of all other species combined (2,945 seeds). The fruits, although bulky, are high in water and sugars and are easily digested. In 2008, the group defecated more seeds of the somewhat smaller *Ericibe elliptilimba* fruits (4,971) than all other species combined, indicating that they ingested more fruits of this species per visit than all other species. Seed output is not such a good indication of food consumption across species because fruits vary in size and the seeds of some species, such as *Podocarpus neriifolius* and *Polyalthia simiarum*, are not usually swallowed.

Fruit is not normally limiting during April and May, and in these three years major sources were available. It is possible, however, in some years of unusual weather, that all major sources might by chance fail to flower and fruit.

Does Availability Determine Use?

A question that needs answering is whether the relative use of different fruit species correlates with their availability in the forest as reflected crudely by the phenology samples. If it does not, this may indicate that our measure of availability is too crude, the species did not vary significantly in availability, or that the food in question is neither preferred nor limiting in the diet (Hemingway and Bynum 2005). We may begin to test for correlations by comparing the phenology of fruits in the three last years of phenology data (Table 1) with relative use during the same three years (Table 2). Eleven species in the phenology samples (six major and five minor species) fruited during the April–May period of the foraging study.

Comparisons of use (number of visits to sources) with putative availability yields mixed results, and suggests three different food categories based on preference

(Figure 6). First, three of the fruit species do show positive correlations, led by *Nephelium melliferum*, a species both preferred and common. This was also observed to be the case in previous years; the last year of fruiting prior to 2006 was in 2004, when gibbons also relied heavily on it (W. Brockelman, unpublished observation). From 10 April to 28 May 2008, the home range of group A included 30.1 ha based on the number of 20 × 20 m quadrats entered. This includes 130 quadrats (5.2 ha) of included quadrats not actually entered, mostly in gaps and regenerating forest, plus an area of approximately 2.5 ha lying off the plot that was used. A total of 308 *Nephelium* trees of adult size (18 cm dbh) grew within the range of the group on the plot (27.56 ha), at a density of 11.2 trees/ha. It is known that 75 (25 percent) of these trees fruited in 2007, and 45 of these (plus 5 trees off the plot) were visited by the gibbon group.

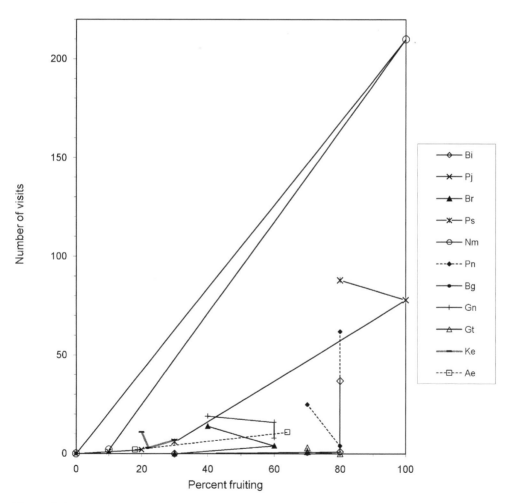

Figure 6. Relation between number of gibbon visits in April and May and percentage fruiting of phenology study trees for 2006, 2007, and 2008. Abbreviations (genus and species) refer to consecutive major and minor fruit species in Table 1.

Trees were used very unevenly; the number of visits per fruiting tree ranged from 0 to 30, depending on tree size and fruit characteristics. Most of the fruit on the trees is not harvested by gibbons; they must compete with three species of squirrels and pig-tailed monkeys (W. Brockelman, unpublished observation) for *Nephelium* fruit. These calculations show that gibbons can take advantage of a relatively abundant species of fruit. They can also obtain significant quantities of fruit from large trees of rare species, some of which have fewer frugivore competitors.

The second-most preferred species, *Polyalthia simiarum*, also shows a positive relationship between availability and use. It fruited heavily in two years and was utilized heavily in both. A third species, the understory tree *Aglaia elaeagnoides*, also shows a positive relationship although usage was much lower. This relatively small tree is quite abundant, but its fruiting characteristics have not been studied in detail.

Another group of two species, *Bridelia insulana* and *Podocarpus neriifolius*, was utilized in some years of heavy fruiting but not in others. These may be considered to be nonpreferred, but alternate food sources. *P. neriifolius* appears to be an important source used most heavily when *Nephelium melliferum* and *Polyalthia simiarum* are not so available (as in 2006). A third group of species made up of most of the remaining ones was used only occasionally and opportunistically. The use of these species did not correlate with availability. Some of these, such as *Gironniera nervosa*, *Knema elegans*, and *Baccaurea ramiflora*, were available in all or nearly all years.

These detailed observations cover only a sustained six-week period during April and May; however, they are corroborated by other (unpublished) observations made during 2003–2005 of the use of preferred species that have highly variable fruiting in other seasons. For example, we have noted that gibbons spend much feeding time on *Cinnamomum subavenium*, *Prunus javanica*, *Eugenia syzygioides*, *Balakata baccata*, and *Garcinia benthamii* in the years in which they are available, but must find alternate foods in other years.

DISCUSSION

The term *dry forest* potentially covers a wide variety of structural and climatic forest types. Dry evergreen forests experience seasonal droughts that trigger conditions that limit the number of animal and plant species. Some animal species, particularly terrestrial mammals, thrive under these conditions and benefit from the increased availability of ground forage as a result of increased insolation (McShea and Baker, this volume). Primates, however, are limited to forests with relatively high forest cover and high plant diversity. They are limited primarily by food availability in the dry season. Although several species of macaques may occur in moist (mixed) deciduous forest, few occur in dry deciduous dipterocarp forest (Eudey 1980). None survive throughout the year in this forest type in Thailand. Predominantly arboreal primates that depend on a closed canopy for locomotion, such as many leaf monkeys and all species of gibbons, can survive in dry evergreen, moist evergreen, or semi-evergreen forest. There is growing evidence (discussed below), however, that the dry season tests the flexibility and survival ability of primates that occur in seasonal forests. Some species that live in large home ranges can retreat into gallery forests or uphill into moister forests. Others, such as gibbons that live in relatively small territories or primates that live in restricted patchy habitats such as karst outcrops, must try to survive where they are.

Every primate species that has ever been studied in the wild has been found to change its diet seasonally. During the season of low food availability, most frugivorous species switch to alternative or lower-quality fruits (see Oates 1977; Milton 1980, 1982; Terborgh 1983; Galdikas 1988; Goldizen et al. 1988; Tutin et al. 1997; Conklin-Brittain et al. 1998; Nunes 1998; McConkey et al. 2003; Marshall and Leighton 2006; Hohmann et al. 2006; Dela 2007; Marshall and Wrangham 2007), or are forced to consume more leaves, seeds, or other plant parts from which they may not be able to obtain a sustainable supply of energy (Struhsaker 1975; Kool 1993; Peres 1994, 1996; Yamagiwa and Mwanza 1994; Doran 1997; Knott 1998; Tutin and White 1998; Wrangham et al. 1998; Davies et al. 1999; Hemingway and Bynum 2005; Stevenson 2005; Wright et al. 2005; Suarez 2006; Dela 2007). Climatic variations, especially ENSO events, have already been found to affect fruit availability at some sites (Foster 1982; van Schaik 1986; Tutin et al. 1997; Wright et al. 1999; Wich and van Schaik 2000; Chapman et al. 2005b; Milton et al. 2005).

During times of stress primates are seen to rely mostly on "fallback foods," species that are available in most or all months but are ignored when preferred foods are available (Peres 1994; Conklin-Brittain et al. 1998; Tutin and White 1998; Wrangham et al. 1998; Hemingway and Bynum 2005; Marshall and Wrangham 2007). One fruit that is available in all months in Khao Yai National Park and may serve as a fallback food is *Gironniera nervosa*; however, this fruit (and its leaves) is consumed to some extent in nearly all months, and never becomes the major source of energy. Many species of fruits cannot be shown to be nonpreferred, or fallback, foods because they have not been seen to be available at the same time as the preferred foods. Therefore, I prefer to call them "alternative foods" because it is difficult to determine how preferred they are relative to other foods. Most such alternative foods found in this study have high interannual variability themselves. Because different fruits have different nutrient content, they are not necessarily 100 percent substitutable for other "preferred" foods. Gibbons appear to require a diverse diet of fruits, leaves, and insects, so no food can be shown to be completely substitutable or expendable.

Evidence that the season of low food availability affects the behavior of primates and causes stress in times of food shortage is difficult to obtain (van Schaik et al. 1993), but has been found in some species (Goldizen et al. 1988; review in Hemingway and Bynum 2005). White-handed gibbons (*Hylobates lar*) and siamang (*Symphalangus syndactylus*) have been found to reduce travel in times of low food abundance in the relatively nonseasonal rain forest of peninsular Malaysia (J. Raemaekers 1980) as well as in more seasonal forests (Bartlett 2003, 2009; this chapter). Fan and Jiang (2008) have found the same in *Nomascus concolor* in highly seasonal montane forest. Reduction in travel distance with lower fruit availability may reflect an awareness that increased travel will not be rewarded and a decision to utilize lower-quality foods. Evidence that ranging is reduced in months of preferred food shortage comes from other territorial (mantled howler monkeys: Milton 1980; Verreaux's sifaka: Norscia et al. 2006; chimpanzees: Doran 1997) and non-territorial (capped langurs: Stanford

1991; white-bellied spider monkeys: Nunes 1996; lowland gorillas: Yamagiwa and Mwanza 1994) primates as well. In other species, it is unclear whether food quality or changes in location of food sources is responsible for seasonal changes in ranging pattern (Terborgh 1983; Hemingway and Bynum 2005). Many species cope with reduced availability of food by shifting their ranging patterns to other parts of the total range (Kinzey 1977; Leighton and Leighton 1983; Boinski 1987; Overdorff 1993; Soini 1993; Peres 1994; Zhang 1995). Hemingway and Bynum (2005) reported fifteen instances of such seasonal "habitat shifting" occurring in their survey of 157 studies reporting seasonal data for one hundred primate species. There is anecdotal evidence that seasonal shifts in range use occur in *Hylobates lar* in forests at the dry seasonal evergreen–deciduous dipterocarp forest ecotone in Huai Kha Khaeng Wildlife Sanctuary, western Thailand, but there is still no published account.

Nomascus concolor gibbons at the northern extreme of gibbon distribution in subtropical China also appear to be exceptional in their ranging behavior (Fan et al. 2009). Their ranging pattern shifts slightly month by month to different parts of their relatively large home ranges, which may be greater than 100 ha. Tree diversity in the montane forests of central Yunnan is lower than in warm tropical forests, and food trees appear to be more patchy in distribution. The diet of these gibbons also includes more leaves and flowers in dry months with low fruit availability (Fan et al. 2009).

It is likely that changes in ranging behavior are associated with changes in reproductive fitness. Bartlett reported that social activities such as play, grooming, and singing behavior are reduced in months of food stress at Mo Singto (Bartlett 2003). Gursky (2000) reported an increase in territorial conflicts in spectral tarsiers in times of food scarcity. A few studies have reported weight loss in species subject to seasonal stress (Goldizen et al. 1988; Knott 1998; Wright et al. 2005), and changes in mortality (Milton 1982). There is a large literature on reproductive seasonality in primates (see reviews by Brockman and van Schaik 2005; Janson and Verdolin 2005), which points to the importance of seasonal variations in energy and protein availability. Gibbons at Mo Singto also show some seasonality in births (Savini et al. 2008), although as is the case for most species, the important selective factors acting on female reproductive timing are still somewhat speculative.

As more long-term studies accumulate, high interannual variation in fruiting (Newstrom et al. 1994) is increasingly seen as the normal pattern in both the Old and New World tropics (McClure 1966; Medway 1972; Frankie et al. 1974; Leighton and Leighton 1983; Howe 1983; van Schaik 1986; van Schaik and van Noordwijk 1985; Galdikas 1988; Tutin and Fernandez 1993; Newbery et al. 1998; Sakai 2001; Wright et al. 1999; Wright et al. 2005; Chapman et al. 2005a; Milton et al. 2005; Marshall and Leighton 2006; Norden et al. 2007). The term *masting* has been used to describe variation within species in which very distinct fruiting peaks occur at intervals of at least two years (Norden et al. 2007), which would apply to several species in the present study. However, the distinction between masting and high interannual variability is a relative one, and cannot be made without observations on large numbers of individual trees in populations. High interannual variability is not confined to the seasonal forests. Sakai (2001) reported that more than 60 percent of 187 tree species studied for four years at Lambir, Sarawak, showed supra-annual flowering patterns. Many of these species participated in general flowering, or synchronized flowering among species, usually referred to as mast flowering and fruiting (Appanah 1985; Janzen 1974; Ashton et al. 1988). General flowering is not prominent at Khao Yai, although, as in all forests, there are "good" and "poor" years of fruiting. Multi-annual datasets are too few to test whether interannual variation in fruiting varies with seasonality in evergreen forests.

Although the rainfall data shown in detail above depict the worst ENSO event in recent history, it is clear from the fourteen-year record that the beginning and the end of the dry season in Khao Yai are unpredictable. Unpredictability in the length of the dry season is a feature characteristic of many seasonal forests near the evergreen/semi-evergreen boundary (Walsh 1996). Because flowering and fruiting in evergreen tropical forests appears to be controlled primarily by seasonal cues in solar radiation, and to a lesser extent by rainfall and temperature (van Schaik et al. 1993; Wright and van Schaik 1994), we may expect to find interannual variability to be most common in the seasonal forests. Where such variability is high, we may expect that the uncommon "extreme years" will be most important in determining the limits of survival of species at the edge of their tolerances (Murphy and Lugo 1986).

The major threat to frugivores such as gibbons, however, is that interannual variation is likely to cause more stress in the dry season when fewer species fruit, and when more species vary interannually. As there is little interannual correlation among species in phenology, eventually there will come a year when no species, preferred or alternate, fruit during the season of stress. If we introduce into this scenario increased global warming and more variable ENSO events, we can predict increased threats to the survival of frugivores in seasonal environments at the margins of their ranges.

ACKNOWLEDGMENTS

I am grateful to our field workers Amnart Boonkongchart, Saiwaroon Chongko, Jantima Santorn, Jantima Saentorn, Ratasart Somnuek, Umaporn Martmoon, and Wisanu Chongko, for their hard work in following the gibbons and other tasks, and students Chanpen Wongsriphuek, Wirong Chanthorn, and Petchprakai Wongsorn for their field help. Anuttara Nathalang has provided constant help in management of the plot data, and Onuma Petrmitr in herbarium support. J. F. Maxwell is thanked for his invaluable botanical help. We thank Mr. Prawat Wohandee and Mr. Narong Mahunnop, past directors of Khao Yai National Park, for their constant support. Kim McConkey and several anonymous reviewers provided valuable comments on the manuscript. This research was funded by Biodiversity and Training Program grants BRT 239001, BRT 242001, BRT R_346005 and BRT R349009.

REFERENCES

Appanah, S. 1985. General Flowering in the Complex Rain Forests of Southeast Asia. *Journal of Tropical Ecology* 1:225–40.

Ashton, P. S. 1990. Thailand: Biodiversity Center for the Tropics of Indo-Burma. *Journal of the Science Society of Thailand* 16:107–16.

Ashton, P. S., T. J. Givnish, and S. Appanah. 1988. Staggered Flowering in the Dipterocarpaceae: New Insights into Floral Induction and the Evolution of Mast Fruiting in the Aseasonal Tropics. *American Naturalist* 132:44–66.

Baker, P. J., S. Bunyavejchewin, C. D. Oliver, and P. S. Ashton. 2005. Disturbance History and Stand Dynamics of a Seasonal Forest in Western Thailand. *Ecological Monographs* 75:317–43.

Bartlett, T. Q. 2003. Intragroup and Intergroup Social Interactions in White-Handed Gibbons. *International Journal of Primatology* 24:239–59.

———. 2009. *The Gibbons of Khao Yai: Seasonal Variation in Behavior and Ecology*. Upper Saddle River, NJ: Pearson Prentice Hall.

Boinski, S. 1987. Habitat Use by Squirrel Monkeys in a Tropical Lowland Forest. *Folia Primatologica* 49:151–67.

Brockelman, W. Y. 2004. Inheritance and Selective Effects of Color Phase in White-Handed Gibbons (*Hylobates lar*) in Central Thailand. *Mammalian Biology* 69:73–80.

Brockelman, W. Y., A. Nathalang, and G. A. Gale. 2010. The Mo Singto Forest Dynamics Plot, Khao Yai National Park, Thailand. *Natural History Bulletin of the Siam Society* (in press).

Brockelman, W. Y., U. Reichard, U. Treesucon, and J. J. Raemaekers. 1998. Dispersal, Pair Formation and Social Structure in Gibbons (*Hylobates lar*). *Behavioral Ecology and Sociobiology* 42:329–39.

Brockman, D. K., and C. P. van Schaik. 1995. Seasonality and Reproductive Function. In *Seasonality in Primates: Studies of Living and Extinct Human and Non-human Primates*, ed. D. K. Brockman and C. P. van Schaik, 269–305. New York: Cambridge University Press.

Bunyavejchewin, S., P. J. Baker, J. V. LaFrankie, and P. S. Ashton. 2001. Stand Structure of a Seasonal Dry Evergreen Forest at Huai Kha Khaeng Wildlife Sanctuary, Western Thailand. *Natural History Bulletin of the Siam Society* 49:89–106.

Chapman, C. A., L. J. Chapman, T. T. Struhsaker, A. E. Zanne, C. J. Clark, and J. R. Poulsen. 2005a. A Long-Term Evaluation of Fruiting Phenology: Importance of Climate Change. *Journal of Tropical Ecology* 21:1–14.

Chapman, C. A., L. J. Chapman, A. E. Zanne, J. R. Poulsen, and C. J. Clark. 2005b. A 12-year Phonological Record of Fruiting: Implications for Frugivore Populations and Indicators of Climate Change. In *Tropical Fruits and Frugivores: The Search for Strong Interactors*, ed. J. L. Dew and J. P. Boubli, 75–92. Dordrecht: Springer.

Conklin-Brittain, N. L., R. W. Wrangham, and K. D. Hunt. 1998. Dietary Response of Chimpanzees and Cercopithecines to Seasonal Variation in Fruit Abundance. II. Macronutrients. *International Journal of Primatology* 19:971–98.

Corlett, R. T., and J. V. LaFrankie. 2004. Potential Impacts of Climate Change on Tropical Asian Forests through an Influence on Phenology. *Climatic Change* 39:439–53.

Davies, A. G., J. F. Oates, and G. L. DaSilva. 1999. Patterns of Frugivory in Three West African Colobine Monkeys. *International Journal of Primatology* 20:327–57.

Dela, J. D. S. 2007. Seasonal Food Use Strategies of *Semnopithecus vetulus nestor*, at Panadura and Piliyandala, Sri Lanka. *International Journal of Primatology* 28:607–26.Doran, D. 1997. Influence of Seasonality on Activity Patterns, Feeding Behavior, Ranging, and Grouping Patterns in Tai Chimpanzees. *International Journal of Primatology* 18:183–206.

Eudey, A. A. 1980. Pleistocene Glacial Phenomena and the Evolution of Asian Macaques. In *The Macaques: Studies in Ecology, Behavior and Evolution*, ed. D. G. Lindberg, 52–83. New York: Van Nostrand Reinhold.

Fan, P.-F., and X.-L. Jiang. 2008. Effects of Food and Topography on Ranging Behavior of Black Crested Gibbon (*Nomascus concolor jingdongensis*) in Wuliang Mountain, Yunnan, China. *American Journal of Primatology* 70:871–78.

Fan, P.-F, Q. Y. Ni, G. Z. Sun, B. Huang, and X. L. Jiang. 2009. Gibbons under Seasonal Stress: The Diet of the Black Crested Gibbons (*Nomascus concolor*) on Mt. Wuliang, Central Yunnan. *Primates* 50:3744.

Foster, R. B. 1982. Famine on Barro Colorado Island. In *The Ecology of a Tropical Forest: Seasonal Rhythms and Long-Term Changes*, ed. E. G. Leigh, A. S. Rand, and D. M. Windsor, 151–72. Washington, DC: Smithsonian Institution Press.

Frankie, G. W., H. G. Baker, and P. A. Opler. 1974. Comparative Phenological Studies of Trees in Tropical Wet and Dry Forests in the Lowlands of Costa Rica. *Journal of Ecology* 62:881–919.

Galdikas, B. M. F. 1988. Orangutan Diet, Range, and Activity at Tanjung Putting, Central Borneo. *International Journal of Primatology* 9:1–35.

Goldizen, A. W., J. Terborgh, F. Cornejo, D. T. Porras, and R. Evans. 1988. Seasonal Food Shortage, Weight Loss, and the Timing of Births in Saddle-Backed Tamarins (*Saguinus fuscicollis*). *Journal of Animal Ecology* 57:893–901.

Gursky, S. 2000. Effect of Seasonality on the Behavior of an Insectivorous Primate, *Tarsius spectrum*. *International Journal of Primatology* 21:477–95.

Hemingway, C. A., and N. Bynum. 2005. The Influence of Seasonality on Primate Diet and Ranging. In *Seasonality in Primates: Studies of Living and Extinct Human and Non-human Primates*, ed. D. K. Brockman and C. P. van Schaik, 57–104. Cambridge: Cambridge University Press.

Hohmann, G., A. Fowler, V. Sommer, and S. Ortmann. 2006. Frugivory and Gregariousness of Solonga Bonobos and Gashaka Chimpanzees: The Influence of Abundance and Nutritional Quality of Fruit. In *Feeding Ecology in Apes and Other Primates: Ecological, Physical, and Behavioral Aspects*, ed. G. Hohmann, M. M. Robbins, and C. Boesch, 123–59. Cambridge: Cambridge University Press.

Holdridge, L. R., W. C. Grenke, W. H. Hathaway, T. Liang, and J. A. Tosi, Jr. 1971. *Forest Environments in Tropical Life Zones: A Pilot Study*. Oxford: Pergamon Press.

Howe, H. F. 1983. Annual Variation in a Neotropical Seed-Dispersal System. In *Tropical Rain Forest: Ecology and Management*, ed. S. L. Sutton, T. C. Whitmore, and A. C. Chadwick, 211–27. Oxford: Blackwell Scientific.

Janson, C., and J. Verdolin. 2005. Seasonality of Births in Relation to Climate. In *Seasonality in Primates: Studies of Living and Extinct Human and Non-human Primates*, ed. D. K. Brockman and C. P. van Schaik, 307–50. New York: Cambridge University Press.

Janzen, D. H. 1974. Tropical Blackwater Rivers, Animals, and Mast Fruiting by the Dipterocarpaceae. *Biotropica* 6:69–103.

Kinzey, W. G. 1977. Diet and Feeding Behavior of *Callicebus torquatus*. In *Primate Ecology: Studies of Feeding and Ranging Behaviour in Lemurs, Monkeys and Apes*, ed. T. H. Clutton-Brock, 127–51. London: Academic Press.

Knott, C. D. 1998. Changes in Orangutan Caloric Intake, Energy Balance, and Ketones in Response to Fluctuating Fruit Availability. *International Journal of Primatology* 19:1061–79.

Kool, K. M. 1993. The Diet and Feeding Behavior of the Silver Leaf Monkey (*Trachypithecus auratus sondaicus*) in Indonesia. *International Journal of Primatology* 14:667–700.

Leighton, M., and D. Leighton. 1983. Vertebrate Responses to Fruiting Seasonality within a Bornean Rain Forest. In *Tropical Rain Forest: Ecology and Management*, ed. S. L. Sutton, T. C. Whitmore, and A. C. Chadwick, 181–96. Oxford: Blackwell Scientific.

Marshall, A. J., and M. Leighton. 2006. How Does Food Availability Limit the Population Density of White-Bearded Gibbons? In *Feeding Ecology in Apes and Other Primates: Ecological, Physical, and Behavioral Aspects*, ed. G. Hohmann, M. M. Robbins, and C. Boesch, 313–35. Cambridge: Cambridge University Press.

Marshall, A. J., and R. W. Wrangham. 2007. Evolutionary Consequences of Fallback Foods. *International Journal of Primatology* 28:1219–35.

Maxwell, J. F., and S. Elliott. 2001. Vegetation and Vascular Flora of Doi Sutep National Park, Northern Thailand. *Thai Studies in Biodiversity*, no. 5.

McClure, H. E. 1966. Flowering, Fruiting and Animals in the Canopy of a Tropical Forest. *Malayan Forester* 29 (3): 182–203.

McConkey, K. R., A. Ario, F. Aldy, and D. J. Chivers. 2003. Influence of Forest Seasonality on Gibbon Food Choice in the Rain Forests on Barito Ulu, Central Kalimantan. *International Journal of Primatology* 24:19–33.

Medway, F. L. S. 1972. Phenology of a Tropical Rain Forest in Malaya. *Biological Journal of the Linnean Society* 4:117–46.

Milton, K. 1980. *The Foraging Strategy of Howler Monkeys*. New York: Columbia University Press.

———. 1982. Dietary Quality and Demographic Regulation in a Howler Monkey Population. In *The Ecology of a Tropical Forest: Seasonal Rhythms and Long-Term Changes*, ed. E. G. Leigh, A. S. Rand, and D. M. Windsor, 273–89. Washington, DC: Smithsonian Institution Press.

Milton, K., J. Giacalone, S. J. Wright, and G. Stockmayer. 2005. Do Frugivore Population Fluctuations Reflect Fruit Production? In *Tropical Fruits and Frugivores: The Search for Strong Interactors*, ed. J. L. Dew and J. P. Boubli, 5–35. Dordrecht: Springer.

Murphy, P. C., and A. E. Lugo. 1986. Ecology of Tropical Dry Forest. *Annual Review of Ecology and Systematics* 17:67–88.

Newbery, D. M., N. C. Songwe, and G. B. Chuyong. 1998. Phenology and Dynamics of an African Rainforest at Korup, Cameroon. In *Dynamics of Tropical Communities*, ed. D. M. Newbery, H. H. T. Prins, and N. D. Brown, 267–308. London: Blackwell Science.

Newstrom, L. E., G. W. Frankie, H. G. Baker, and R. K. Colwell. 1994. Diversity of Long-Term Flowering Patterns. In *La Selva: Ecology and Natural History of a Neotropical Rain Forest*, ed. L. A. McDade, K. S. Bawa, H. A. Hespenheide, and G. S. Hartshorn, 142–60. Chicago: University of Chicago Press.

Norden, N., J. Chave, O. Belbenoit, A. Caubère, P. Châtelet, P.-M. Forget, and C. Thébaud. 2007. Mast Fruiting Is a Frequent Strategy in Woody Species of Eastern South America. *PLoS One* 2 (10): e1079, doi:10.1371/journal.pone.0001079.

Norscia, I., V. Carrai, and S. M. Borgognini-Tarli. 2006. Influence of Dry Season and Food Quality and Quantity on Behavior and Feeding Strategy of *Propithecus verrearxi* in Kirindy, Madagascar. *International Journal of Primatology* 27:1001–22.

Nunes, A. 1996. Foraging and Ranging Patterns in White-Bellied Spider Monkeys. *Folia Primatologica* 65:85–99.

———. 1998. Diet and Feeding Ecology of *Ateles belzebuth belzebuth* at Maraca Ecological Station, Roraima, Brazil. *Folia Primatologica* 69:61–76.

Oates, J. F. 1977. The Guereza and Its Food. In *Primate Ecology*, ed. T. H. Clutton-Brock, 276–321. London: Academic Press.

Ogawa, H., K. Yoda, and T. Kira. 1961. A Preliminary Survey on the Vegetation of Thailand. *Nature and Life in Southeast Asia* 1:21–157.

Overdorff, D. J. 1993. Similarities, Differences and Seasonal Patterns in the Diet of *Eulemur rubiventer* and *Eulemur fulvus rufus* in the Ranomafana National Park, Madagascar. *International Journal of Primatology* 14:721–53.

Peres, C. A. 1994. Primate Responses to Phenological Changes in an Amazonian Terra Firma Forest. *Biotropica* 26:98–112.

———. 1996. Food Patch Structure and Plant Resource Partitioning in Interspecific Associations of Amazonian Tamarins. *International Journal of Primatology* 17:695–723.

Raemaekers, J. 1980. Causes of Variation between Months in the Distance Traveled Daily by Gibbons. *Folia Primatologica* 34:46–60.

Raemaekers, J. J., and P. M. Raemaekers. 1985. Field Playback of Loud Calls to Gibbons (*Hylobates lar*): Territorial, Sex-Specific, and Species-Specific Responses. *Animal Behavior* 33:481–93.

Raemaekers, J. J., P. M. Raemaekers, and E. H. Haimoff. 1984. Loud Calls of the Gibbon (*Hylobates lar*): Repertoire, Organization and Context. *Behaviour* 91:146–89.

Rundel, P. W., and K. Boonpragob. 1995. Dry Forest Ecosystems of Thailand. In *Seasonally Dry Forests*, ed. S. H. Bullock, H. A. Mooney, and E. Medina, 93–123. Cambridge: Cambridge University Press.

Sakai, S. 2001. Phenological Diversity in Tropical Forests. *Population Ecology* 43:77–86.

Savini, T., C. Boesch, and U. H. Reichard. 2008. Home-Range Characteristics and the Influence of Seasonality on Female Reproduction in White-Handed Gibbons (*Hylobates lar*) at Khao Yai National Park, Thailand. *American Journal of Physical Anthropology* 135:1–12.

Singhrattna, N., B. Rajagopalan, K. K. Kumar, and M. Clark. 2005. Interannual and Interdecadal Variability of Thailand Summer Monsoon Season. *Journal of Climate* 18:1697–1708.

Soini, P. 1993. The Ecology of the Pygmy Marmoset, *Cebuella pygmaea*: Some Comparisons with Two Sympatric Tamarins. In *Marmosets and Tamarins*: *Systematics, Behaviour, and Ecology*, ed. A. B. Rylands, 257–61. Oxford: Oxford University Press.

Stanford, C. 1991. The Capped Langur in Bangladesh: Behavioral Ecology and Reproductive Tactics. *Contributions to Primatology* 26:1–179.

Stevenson, P. 2005. Potential Keystone Species for the Frugivore Community at Tinigua Park, Colombia. In *Tropical Fruits and Frugivores*: *The Search for Strong Interactors*, ed. J. L. Dew and J. P. Boubli, 37–57. Dordrecht: Springer.

Struhsaker, T. T. 1975. *The Red Colubus Monkey*. Chicago: University of Chicago Press.

Suarez, S. A. 2006. Diet and Travel Costs for Spider Monkeys in a Nonseasonal, Hyperdiverse Environment. *International Journal of Primatology* 27:411–36.

Terborgh, J. 1983. *Five New World Primates*: *A Study in Comparative Ecology*. Princeton, NJ: Princeton University Press.

Tsuji, Y., and T. Takatsuki. 2009. Effects of Yearly Change in Nut Fruiting on Autumn Home-Range Use by *Macaca fuscata* on Kinkazan Island, Northern Japan. *International Journal of Primatology* 30:169–81.

Tutin, C. E. G., and M. Fernandez. 1993. Relationship between Minimum Temperature and Fruit Production in Some Tropical Forest Trees in Gabon. *Journal of Tropical Ecology* 9:241–48.

Tutin, C. E. G., R. M. Ham, L. J. T. Lee, L. J. T. White, and M. J. S. Harrison. 1997. The Primate Community of the Lopé Reserve, Gabon: Diets, Responses to Fruit Scarcity, and Effects on Biomass. *American Journal of Primatology* 42:1–24.

Tutin, C. E. G., and L. J. T. White. 1998. Primates, Phenology and Frugivory: Present, Past and Future Patterns in the Lopé Reserve, Gabon. In *Dynamics of Tropical Communities*, ed. D. M. Newbery, H. H. T. Prins, and N. D. Brown, 309–37. London: Blackwell Science.

van de Bult, M., and M. Greijmans. 2006. Vegetation Types and the Deciduous-Evergreen Forest Continuum along an Elevation Gradient in Mae Wong National Park, Western Thailand. *Natural History Bulletin of the Siam Society* 54:27–74.

van Schaik, C. P. 1986. Phenological Changes in a Sumatran Rain Forest. *Journal of Tropical Ecology* 2:327–47.

van Schaik, C. P., and K. R. Pfannes. 2005. Tropical Climates and Phenology: A Primate Perspective. In *Seasonality in Primates*: *Studies of Living and Extinct Human and Non-human Primates*, ed. D. K. Brockman and C. P. van Schaik, 23–54. Cambridge: Cambridge University Press.

van Schaik, C. P., J. W. Terborgh, and S. J. Wright. 1993. The Phenology of Tropical Forests: Adaptive Significance and Consequences for Primary Consumers. *Annual Review of Ecology and Systematics* 24:353–77.

van Schaik, C. P., and M. A. van Noordwijk. 1985. Interannual Variability in Fruit Abundance and the Reproductive Seasonality in Sumatran Long-Tailed Macaques (*Macaca fascicularis*). *Journal of Zoology, London*. 206:533–49.

Walsh, R. P. D. 1996. Climate. In *The Tropical Rain Forest: An Ecological Study*, 2nd ed., by P. W. Richards with contributions by R. P. D. Walsh, I. C. Ballie, and P. Greig-Smith, 159–205. Cambridge: Cambridge University Press.

Wich, S. A., and C. P. van Schaik. 2000. The Impact of El Niño on Mast Fruiting in Sumatra and Elsewhere in Malesia. *Journal of Tropical Ecology* 16:563–77.

Wrangham, R. W., N. L. Conklin-Brittain, and K. D. Hunt. 1998. Dietary Response of Chimpanzees and Cercopithecines to Seasonal Variation in Fruit Abundance. I. Antifeedants. *International Journal of Primatology* 19:949–70.

Wright, P. C., V. R. Razafindratsita, S. T. Pochron, and J. Jernvall. 2005. The Key to Madagascar Frugivores. In *Tropical Fruits and Frugivores: The Search for Strong Interactors*, ed. J. L. Dew, and J. P. Boubli, 121–38. Dordrecht: Springer.

Wright, S. J., C. Carrasco, O. Calderón, and S. Paton. 1999. The El Niño Southern Oscillation, Variable Fruit Production and Famine in a Tropical Forest. *Ecology* 80:1632–47.

Wright, S. J., H. C. Muller-Landau, O. Calderón, and A. Hernandéz. 2005. Annual and Spatial Variation in Seedfall and Seedling Recruitment in a Neotropical Forest. *Ecology* 86:848–60.

Wright, S. J., and C. P. van Schaik. 1994. Light and the Phenology of Tropical Trees. *American Naturalist* 143:192–99.

Yamagiwa, J., and N. Mwanza. 1994. Day-Journey Length and Daily Diet of Solitary Male Gorillas in Lowland and Highland Habitats. *International Journal of Primatology* 15:207–24.

Zhang, S.-Y. 1995. Activity and Ranging Patterns in Relation to Fruit Utilization by Brown Capuchins (*Cebus apella*) in French Guiana. *International Journal of Primatology* 16:489–507.

Sloth Bears Living in Seasonally Dry Tropical and Moist Broadleaf Forests and Their Conservation

John Seidensticker, K. Yoganand, and A. J. T. Johnsingh

The tropical dry and moist broadleaf forests that extend along a great arc from Sri Lanka through the Indian subcontinent, Myanmar, and into Southeast Asia have been reduced in area, are highly fragmented, and are battered by extensive and excessive anthropogenic burden. These forests, however, are essential habitats for sloth bears (*Ursus ursinus*)[1] living in Sri Lanka, India, and Nepal, and for Asiatic black bears (*Ursus thibetanus*) and sun bears (*Ursus malayanus*) in South and Southeast Asia.

Following a brief summary in which we position the sloth bear in a phylogenetic and morphological context within the Ursidae, we explore the extent of occurrence of sloth bears, their area of occupancy, sloth bear foraging in a seasonally dry forest site, and the survival risks sloth bear populations face, with particular reference to forest fragmentation and human–sloth bear conflict.

ASIAN BEAR BASICS

The study of phylogenetic relationships among the eight species in the Ursidae has a contentious history. For example, Bininda-Emonds et al. (1999) used a procedure that merged phylogenetic trees derived from various data sources to conclude that the sun bear and sloth bear were sister species as were brown bears and polar bears. Brown bears, polar bears, and Asiatic black bears formed the sister clade to the sloth bear and sun bear clade. However, in the interpretation of the molecular phylogeny (Li Yu et al. 2007) and the fully resolved phylogeny of the Ursidae prepared by Krause et al. (2008), within the Ursinae the sloth bear is the sister taxa to the other five ursine bears: the sun bear/Asiatic black bear–American black bear (*Ursus americanus*) clade, and the brown bear– (*Ursus arctos*) polar bear (*Ursus maritimus*) clade. This ursine bear radiation occurred near the Miocene–Pliocene boundary 5.3 million years ago, with the exception of the brown bear and polar bear separation, which occurred mid-Pleistocene (Krause et al. 2008). The Miocene–Pliocene transition was a period of major climate change with a significant drop in temperature, increased seasonality, reduced forest cover, the spread of arid habitats in North America and Eurasia, and the emergence of extensive grasslands.

Van Valkenburgh (1999) found that the evolution of the Carnivora generally proceeds by the loss of generalized features, such as small body mass and versatile teeth, in favor of more specialization, such as large size and teeth adapted to a particular diet. Once specialized, species rarely revert back to generalists. The sloth bear stands alone as the ursid most specifically adapted as an insectivore, based on patterns of morphological variation in the skulls of extant bears (Sacco and Van Valkenburgh 2004). It is a medium-sized bear (adult males 75–140 kg; females 55–95 kg) with a suite of adaptations for feeding on ants and termites (myrmecophagy) that includes (1) reduced post-canine dentition with relatively small cheek teeth that may function to fracture the exoskeletons of its prey; (2) reduced or absent maxillary incisors; (3) a vaulted palate; (4) protrusable mobile lips; (5) a mobile snout; (6) nostrils that can be closed voluntarily; (7) a long, shaggy, typically black coat with a long dorsal ruff of hairs; and (8) reduced hair on the muzzle that probably is an adaptation to cope with the sticky defensive secretions of some termites. Several features of sloth bear physiology, morphology, and behavior are convergent with those of other mammalian anteaters (Laurie and Seidensticker 1977) including a low basal metabolic rate (McNab 1992), low reproductive rate, solitary habits, extended parental care, and extensive carrying of young by the mother. However, Joshi et al. (1999) and Garshelis (2004) argue these are general features of the family Ursidae except for the carrying of young, which is unique and likely related to defending cubs from predation in an environment with few trees.

Sun bears, Asiatic black bears, and American black bears cluster together based on their craniodental features as omnivorous bears, compared to the herbivorous giant panda (*Ailuropoda melanoleuca*) and the large carnivorous brown bears and polar bears. Carnivorous bears have reduced grinding areas on the molars, relatively small carnassial blades, and low mandibular rigidity. The omnivorous bears have blade-shaped canines and moderately enlarged grinding areas on the molars (Sacco and Van Valkenburgh 2004).

Among the driving forces in evolution are predation, competitive pressure among species, and the spatial-temporal patterning of resources. Morphology and behaviors that contribute to food finding must be adapted to cope with hard times when limiting conditions are approached and competition is greatest. The sloth bear's ancestors evolved into a niche where the efficient harvesting of ants and termites was the defining resource dimension, and this shaped their dentition, morphology, physiology, and behavior as they moved away from a more omnivorous or generalized diet. However, the maximum body mass that can be sustained on an invertebrate diet in mammals is about 21.5 kg, a manifestation of the simple energetic constraint of relying on small-particle food (Carbone et al. 1999). Sloth bears are three to seven times this mass. This apparently exceptional characteristic of sloth bears could be partly explained by the fact that they prey mostly on social insects. Social insects have been described as super-organisms (Holldobler and Wilson 2009), with the biomass of a single colony more than one thousand to one million times larger (Holldobler and Wilson 1991) than the single-prey insect body mass that Carbone et al. (1999) considered in their analysis. Large body size is a deterrent to smaller predators and potential competitors either through direct (predation, interference competition) or indirect (exploitative competition, trophic cascades) interactions (Palomares and Caro 1999).

We see a combination of traits in sloth bears, including a relatively large body size and a suite of morphological, behavioral, and physiological adaptations that are specializations toward efficient myrmecophagy. We posit that adaptations to efficient myrmecophagy evolved to cope with critical periods when other food resources are limited. However, sustaining a large body requires the availability of sugar-rich fruits. By efficiently harvesting social insects such as termites and ants from excavated colonies throughout the year, sloth bears survive in habitats where insects are abundant, and in a range of habitats where fruits are available for only a portion of the year or not at all during years of fruiting-failure (Yoganand 2005).

EXTENT OF OCCURRENCE

The sloth bear is the bear of the Indian subcontinent and island of Sri Lanka, usually occurring in habitats below 1,000 m in India and Nepal (Garshelis et al. 1999b), and below 300 m in Sri Lanka (Ratnayeke et al. 2007). Johnsingh (personal observation) has observed sloth bear scats at 1,370 meters in the Kalakad Hills (Western Ghats, India) rain forest. Bears visit these forests to feed on wild mango (*Mangifera indica*) and jackfruit (*Artocarpus heterophyllus*) in season. Yoganand (personal observation) observed sloth bear scats in rain forest patches, tea garden edges, and forest and grassland habitats up to 2,000 m in the Nilgiris and Anamalai hills (Western Ghats, India). Sloth bears today, and historically, occur almost exclusively in the dry zone of Sri Lanka but not in the central hills and wet zone (Ratnayeke et al. 2007). The southern, eastern, and western limits of the sloth bear's range on the mainland Indian subcontinent are established by geographical/ecological barriers: oceans and deserts. The northwest extent of sloth bear occurrence abuts India's deserts. Sloth bears occurred once in Bangladesh but are now thought to have been extirpated (Sarker 2006).

The sun bear and the Asiatic black bear overlap extensively in the lowland and montane tropical moist and dry broadleaf forests of Southeast Asia (Saw Htun 2006; Vinitpornsawan et al. 2006). Sloth bears apparently were never part of the Southeast Asian large-mammal assemblage, which includes many of the same large mammals that live in India, Nepal, and Bhutan today. For example, fossils recovered from the Lang Trang limestone cave system in Vietnam dating to 150,000–500,000 years ago include tiger (*Panthera tigris*), leopard (*Panthera pardus*), and dhole (*Cuon alpinus*), species that also occur in South Asia, and sun bear, Asiatic black bear, and giant panda, but no sloth bear fossils have been recovered (Ciochon and Olson 1991).

Based on information from local correspondents, Garshelis et al. (1999b) reported that sloth bears, sun bears, and Asiatic black bears co-occur in the forests of northeast India. The definition of co-occurrence depends on the scale at which you frame the

co-occurrence question. Unfortunately, a recently conducted study of the status of tigers, copredators, and prey in India (Jhala et al. 2008) does not help resolve the extent
of occurrence of the three bear species in this region because observers employed to do
the on-the-ground searches for tracks and sign could not reliably distinguish the species
of bear from the tracks and other sign they were observing (Maraj and Seidensticker
2006). In a camera-trapping study in India's Namdapha Tiger Reserve, a 1,985 km²
tract of evergreen broadleaf rain forest that extends to the border with Myanmar, a
single sun bear photo was obtained, but none of sloth bears or Asiatic black bears
(Karanth and Nichols 2000). In more-recent camera-trapping studies in northeastern
India, photographs of sun bears and Asiatic black bears were obtained in Namdapha
(A. Dutta, personal communication) and sun bears in eastern Assam (K. Kakati, personal communication), but no sloth bears. Recent field surveys in Mizoram and Tripura
states in the same region have yielded evidence of the presence of Asiatic black and sun
bears in many localities, but no sloth bears (K. Yoganand, unpublished data). Sun bears
are frequently found in Chin Hills-Arakan Yoma montane rain forest and Mizoram-
Manipur-Kachin rain forest ecoregions at the western edge of their range (Wikramanayake et al. 2002). These observations suggest that the sloth bear would not have existed
in the hilly habitats of northeast India eastwards beyond the lowland deciduous forests
of Assam. Therefore, sloth bears and sun bears apparently do not overlap in any meaningful (measurable) way in their local distributions in India's northeast.

In the northern extent of the sloth bear's range, along the Himalayan foothills, it
and the Asiatic black bear overlap at the edges of their ranges. Spoor of both species
were found in Chilla Range of Rajaji National Park (K. Yoganand and A. J. T. Johnsingh, personal observation), and both species were recently photographed at the same
camera traps during camera-trapping surveys of tigers (B. Pandav, personal communication). Black bears come down to the Himalayan foothills (Uttarakhand) in winter to
raid crops and feed on fruits of *Zizyphus mauritiana* (Johnsingh 2003). Based on their
surveys of bear signs in the Nepal Terai, Garshelis et al. (1999b) found that sloth bears
and Asiatic black bears do not overlap there. What food resources and habitat features
facilitate the coexistence of these species in these habitats, or limit the extension of one
or another's distribution, is a question for further study. We do not know the historical
distribution of sloth bears in the middle Himalayas in India, Nepal, and Bhutan, or if
they occasionally were found in the riparian forest zone along the rivers in this region
(the eastern and western Himalayan broadleaf forest ecoregions). Generally, Asiatic
black bears live in Himalayan mountain forests (Sathyakumar 2006), and sloth bears
in the foothills and lowlands, at the base of the mountains (Yoganand et al. 2006); the
Himalayan alpine meadows are brown bear habitat.

Sloth bears coexist with the largest Asian cats—tigers and leopards—and dholes.
This is facilitated, at least in part, by the sloth bear's body size and the fact that they
usually don't eat carrion (large, dead ungulates), which is usually the remains of a large
cat's prey (Laurie and Seidensticker 1977). Tigers occasionally kill sloth bears, and
leopards kill young sloth bears. Tigers and sloth bears do fight frequently (Yoganand
2005). In Chitwan National Park, Nepal, sloth bears were observed to repel tigers on a
number of occasions (Joshi et al. 1999). Sloth bears are not sympatric with India's last
remaining population of Asiatic lions (*Panthera leo*) in the dry, teak-dominated Gir

Forest. Forests in the same ecoregion (Kathiawar-Gir dry deciduous forests) without lions do have sloth bears living in them. Lions have been extirpated from the other remaining tropical dry forests of northern India (summarized in Seidensticker 2008). How lions may have restricted the local distribution of sloth bears will remain a subject of conjecture. Indian wolves (*Canis pallipes*) broadly overlap with sloth bears in western and central India, but wolves live in more open, drier habitats, and usually outside protected areas, while sloth bears prefer areas with denser cover.

The general forest types, particularly seasonally dry forests, that support sloth bears in South Asia extend far to the east beyond the edge of the sloth bear's range in India. What limits sloth bear distribution to the east? Broadly, geographic and habitat barriers, interspecific competition, predation, infectious diseases, foraging success, genetics, and human activities are among the factors that could limit the sloth bear's eastern extent of occurrence. There is no evidence that disease or a genetic factor limits the extent of sloth bear occurrence. There is also no evidence that a single large predator or even predator guild limits the extent of occurrence of sloth bears. However, historically, lions probably influenced, perhaps greatly restricted, sloth bear range occupancy in much of the dry forests of India. While sloth bears occur in evergreen broadleaf rain forests in southern India, they do not in Sri Lanka. That they live at such a low density in this forest type suggests that environmental conditions are limiting year-round food availability and reducing sloth bear foraging success. In India's northeastern moist evergreen broadleaf hill forests, Asiatic black bears and sun bears are present to compete for food resources that are already limiting sloth bears in this habitat. This suggests that ecological factors driving the availability of food, bear foraging efficiencies, and exclusion by competing bear species determine the northern and northeastern extent of sloth bear occurrence and restrict the occurrence of Asian black bears and sun bears in sloth bear range.

AREAS OF OCCUPANCY AND HABITAT PREFERENCE

Within the extent of sloth bear range today, human activities are the predominant factors that determine areas of occupancy. The recent assessment of the status of tigers, copredators, and prey in India (Jhala et al. 2008) is the first on-the-ground, forest-beat by forest-beat record of carnivores across India. Jhala et al. (2008) reported that sloth bears occupied 215,466 km² (32 percent) of India's 678,333 km² total forested area. By landscape, sloth bears occurred in 4,515 km² of the Shivalik Hills-Gangetic Plains, which is composed of tropical moist deciduous forest, alluvial tall-grass savanna, and tropical dry forest; in 169,016 km² of the central India and Eastern Ghats landscape, composed of dry tropical forest and tropical moist deciduous forest; in 40,877 km² in the Western Ghats landscape composed of mostly tropical moist deciduous forest and tropical evergreen forest, with some tropical dry deciduous forest; and in 1,058 km² in the northeast hills and Brahmaputra landscape, composed of tropical moist evergreen forests, tropical moist deciduous forest, and alluvial tall-grass savanna. Sloth bears occupy about 11,810 km² or 17 percent of Sri Lanka's total land area (Ratnayeke et al. 2007). Comparable occupancy estimation data are not available for Nepal and Bhutan.

Garshelis et al. (1999b), using expert-based density estimates from twenty-three Indian national parks and wildlife sanctuaries, found a mean of 22 and a median of 12 sloth bears/100 km². Estimations of sloth bear density are available from five areas composed of four habitat types. In Chitwan National Park, Nepal, Garshelis et al. (1999a) estimated 68 bears/100 km² in alluvial tall-grass savanna in the dry season and 27 bears/100 km² in the upland sal (*Shorea robusta*) forests during the wet season (Terai-Duar savanna and grasslands ecoregion). Panna National Park, India, situated in the Narmada Valley dry deciduous forest ecoregion, supported 6–8 bears/100 km² (Yoganand 2005). In the disturbed and unprotected forests interspersed with agricultural fields and villages of North Bilaspur Forest Division, Chhattisgarh, located in the Chhota-Nagpur dry deciduous forest ecoregion, Akhtar et al. (2008) reported 23 bears/100 km². In Sri Lanka, Ratnayeke et al. (2006) reported that sloth bear densities may exceed 100 bears/100 km² in some protected areas in the Sri Lankan dry-zone dry evergreen forest ecoregion. However, except for Garshelis et al. (1999a), these density estimates are not based on scientifically rigorous methods, thus only indicate relative abundances of sloth bears in the different forest types. There is a great need to rigorously monitor sloth bear densities in these different habitats to inform management interventions in habitats critical for sloth bear survival (Yoganand et al. 2006).

FORAGING DYNAMICS

Descriptions of sloth bear food habits are now available for several sites in moist and dry broadleaf forests in India, Nepal, and Sri Lanka (Table 1). Sloth bears obtain a large portion of their annual diet from insects (12–83 percent) or fruits (14–88 percent) comprising a variety of species. Sloth bear diet composition varies with forest type, forest conditions, and season, as well as from year to year, indicating considerable dietary plasticity.

In this section, we report on how sloth bears select their foods in the challenging, highly seasonal environment of a tropical dry broadleaf forest. We use as our case study observations made in Panna National Park, northern Madhya Pradesh, central India. Details of this study are reported in Yoganand (2005).

Panna National Park (543 km²) is in the Vindhyan Plateau, a landscape of plateau-terraces separated by escarpments rising south of the Gangetic Plains and bisected by Ken River, a tributary of the Yamuna River. Panna receives approximately 1,100 mm rainfall annually during the southwest monsoon from July to September. During the hot and dry season (March–June), mean temperatures are 30°C–40°C; during the cold season (November–February) they fall to a low of 5°C–15°C. In the dry season, water is available only in the river, seepages at the bases of the escarpments, and a few artificial water holes. Thirteen villages existed within Panna when this work was done, with human and cattle populations of about 6,000 and 9,500 respectively (Chundawat et al. 1999). Panna was declared a national park in 1981, but final notification is pending because many villages remain within its boundary.

The vegetation in Panna can be broadly classified into (1) closed-canopy, high tree-density forests that occur along the escarpments, stream courses, and less-disturbed

Table 1. Sloth bear diets from sites in India, Nepal, and Sri Lanka in various habitats expressed as relative composition (by frequency), relative volume, relative dry weight of scat, or relative ingested biomass.

Ecoregion/Site	Method; Period (number of scats examined)	Fruit %	Insect %	Other %	Study
Eastern Highland Moist Deciduous Forest					
Kanha	% composition; annual (92)	61	39	0	Schaller (1967)
Southwestern Ghats Moist Deciduous Forest					
Bandipur	% composition; annual (95)	37	53	10	Johnsingh (1981)
Mudumalai	% dry weight; annual (567)	88	12	trace	Baskaran et al. (1997)
Chhota-Nagpur Dry Deciduous Forest					
Bilaspur	% dry weight; annual (568)	85	10	5	Bargali et al. (2004)
Terai-Duar Savanna and Grassland					
Chitwan	% composition; annual (139)	42	52	7	Laurie and Seidensticker (1977)
Chitwan	% composition; annual (627)	14	83	3	Joshi et al. (1997)
Narmada Valley Dry Deciduous Forest					
Panna	% ingested biomass; annual (410)	56	44	trace	Yoganand (2005)
Sri Lanka Dry Zone Dry Evergreen Forest					
Wasgomuwa	% composition; annual (N/A)	30	65	5	Ratnayeke et al. (2006)

Note: Ecoregion classification based on Wikramanayake et al. (2002).

localities; (2) open forests with grass and shrub understory; (3) short-grass/open savanna habitat on the shallow-soil, drained plateaus; (4) tall grasslands that grow on old village sites; (5) dense shrub habitat dominated by the exotic invasive *Lantana camara*; and (6) open scrub with sparse patches of *L. camara*, open areas, and degraded vegetation. The latter two types occur mostly on the peripheries of the park and around villages. The forest is thought to have been dominated by *Terminalia tomentosa* and *Anogeissus latifolia* but was transformed into a forest dominated by teak (*Tectona grandis*) through forestry activities in past centuries. Some parts of the area were protected as hunting preserves before Indian independence in 1947; forestry operations were carried out for decades before full wildlife protection was established in 1981. Panna supports a diverse mammalian assemblage that includes tigers (later extirpated and now reintroduced), leopards, striped hyenas (*Hyaena hyaena*), sloth bear, dhole, and wolf. Ungulates include sambar (*Rusa unicolor*), chital (*Cervus axis*), wild pig (*Sus scrofa*), nilgai (*Boselaphus tragocamelus*), chinkara (*Gazella bennetti*), and chousingha (*Tetracerus quadricornis*).Twelve sloth bears were captured and fitted with radio collars to determine bear activity and movements. Untagged bears were followed opportunistically. A habitat map of the study site was developed from satellite images, and the quality of habitats available for sloth bears in different areas of the study site was quantified. Fruit biomass available as food was estimated using food plant densities, proportion of plants that fruited in a sample period, proportion of plants in each of the fruit abundance classes, fruit crop size, and fruit weights to calculate the total biomass of fruits that were produced in the study area in 1999–2000. Social insect colony abundance was estimated in 100 m² plots; colony sizes of insect prey were determined by excavating and weighing sample colonies. Detailed methods are described in Yoganand (2005).

Yoganand (2005) reported that an adult male sloth bear tracked for more than a year had an 129 km² total home range; an adult female had a 23 km² home range. These are three to six times larger than the mean home range sizes of 9 km² and 14 km² for females and males, respectively, in the tall-grass savanna habitat/sal forest in Chitwan National Park, Nepal (Garshelis et al. 1999b). Bears preferred to use open and dense forest habitats in the core of the park; at the edge of the park, bears used dense and open shrub habitat; short-grassland/savanna on plateaus and degraded scrubland were avoided. Sloth bears foraged at night and in the early morning and evening.

Usually bears ate ripe fruits that had fallen to the ground from trees. Bears did feed on ripe fruit directly from shrubs. Fruits contributed 56 percent, ants 29 percent, termites 10 percent, and other foods (honey and bees, Carabid beetles, dung beetle larvae, buried monitor lizard eggs) 4 percent of annual biomass consumed (Figure 1). Fruits contributed 75 percent of the diet in the hot and dry season (March–June), 37 percent in the wet season (July–October), and 52 percent in the cold season (November–February). During the dry season, ants and termites contributed about 11 percent of the diet each. During the wet season, ants contributed 47 percent of the diet, termites 7 percent, and other food, mainly honey, contributed 9 percent. During the cold season, ants and termites made up 35 percent and 11 percent of the diet, respectively.

Sloth bears ate fruits as they became available during the annual cycle. For example, bears consumed the fruits of *Diospyros melanoxylon* from April to June, *Ziziphus mauritiana* during November and December, *Cassia fistula* from June through August, and *Lantana camara* in August and from October to January. Fruits of some species were selected over others that were available. *D. melanoxylon* made up 22 percent of the diet, yet this was only 10 percent of the fruit available; *Z. mauritiana* contributed 11 percent to the sloth bear's diet but composed only 2 percent of the fruit available. *Aegle marmelos*, *Buchanania lanzan*, and *L. camara* were consumed in proportion to their availability. *Z. oenoplia* and *Madhuca longifolia* fruits were consumed in amounts that were less than available fruit biomass. The shrubby climber *Z. oenoplia*

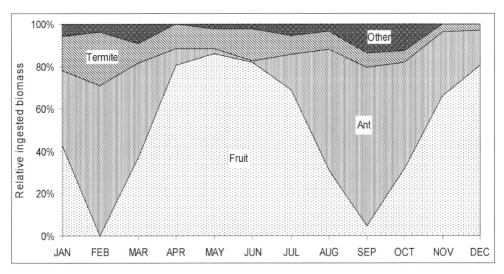

Figure 1. Relative contribution of food types to sloth bear diet in different months in Panna National Park, India (Yoganand 2005).

has small, thorn-protected fruits and low rates of fruit fall after ripening. *M. longifolia* produces crops of large fleshy flowers (more than 50,000/tree) but they were available for only a month.

During peak periods of fruit production, sloth bears could have met all of their energy needs from fruit but they fed on insects irrespective of the availability of fruits, presumably to meet other dietary requirements such as for calcium, protein, specific amino acids, and other nutrients (Rode and Robins 2000), as do other bear species. The ant *Camponotus compressus* contributed about 13 percent of the annual sloth bear diet; termites (mainly *Odontotermes* spp.) contributed about 10 percent of the food biomass; and another ant, *C. irritans*, contributed 8 percent. Other ant species contributed the remainder. *C. compressus* was consumed in lower amounts relative to the biomass available; the other insects were consumed relative to their available biomass.

HUMAN CONFLICT WITH SLOTH BEARS

People Killing or Removing Sloth Bears from the Wild

People kill sloth bears that cause human injuries and raid crops, and this may be the greatest source of mortality. Sloth bears are legally protected throughout their range and are listed in CITES Appendix I (www.cites.org/eng/app/appendices.shtml), but are poached to supply the traditional Asian medicinal markets. And sloth bears are captured as cubs to supply "dancing bear" street shows. The latter has been a matter of recent government and nongovernmental organization (NGO) attention in India, and "dancing bears" are being gathered and placed in rehabilitation centers. We have no estimate of the extent to which these factors affect the persistence of sloth bear populations.

Human Casualties from Sloth Bear Encounters

Sloth bears have a reputation for being aggressive (Garshelis et al. 1999b; Joshi et al. 1999; Yoganand 2005), and there are numerous accounts scattered throughout the Indian hunting literature of sloth bears injuring humans. In response to disturbance, sloth bears usually run away. Neither adults nor cubs usually climb trees, although they are excellent climbers. In sudden, short-range encounters, the sloth bear's response is a spectacular charge and/or bipedal display. In the early 1970s, working in Chitwan National Park, Nepal, Laurie and Seidensticker (1977) reported on their close encounters with sloth bears while they were on foot and while riding trained elephants. The response of the bear depended on the abruptness of the encounter. At distances of more than 20 m, the bear retreated, sometimes vocalizing, sometime not. In an abrupt encounter at short range (3 m) while they were on foot, the bear stood up, roared, then turned and ran. But in an encounter between a villager and an adult female bear in a maize field at night, the bear stood, roared, and then mauled the man. Laurie and Seidensticker heard of four other injuries caused by encounters with sloth bears near the study area. All of the bears' actions in these sudden encounters with people appeared to be defensive not offensive. In encounters observed between tigers and sloth bears, sloth bears appeared to be on the offensive (Joshi et al. 1999).

Since Chitwan's establishment as a national park in 1973, the buffer zone on its northern edge has changed from a badly degraded commons to a lush, community-managed forest and grazing area (Dinerstein 2003). At the interface, the park and the villages and agriculture areas were originally separated by barbwire fences and trenches, which a sloth bear could easily cross; now they are separated by an electric fence. There are no longer reports of human–sloth bear encounters like those that Laurie and Seidensticker (1977) found in the early-1970s (J. Seidensticker, personal observation).

The 277 victims of sloth bear attacks in Sri Lanka interviewed by Ratnayeke et al. (2006) reported most attacks occurred between 9 a.m. and 4 p.m. when, far from villages, the individuals entered the forest to hunt, gather forest products, or tend their plots of shifting slash-and-burn agriculture. There were no reports of sloth bears raiding agricultural crops.

The extent of human injuries and deaths caused by sloth bears in India has been documented for some Indian states (Wildlife Institute of India database). During a five-year period (April 1989 to March 1995), 735 (67 percent) of casualties (deaths and injuries) were due to sloth bears out of the 1,094 reported casualties caused by large mammals in Madhya Pradesh. During a similar period, 47 casualties (14 percent) of 344 recorded for Bihar and 67 (29 percent) of 227 casualties in Orissa were by sloth bears, with a ratio of about 1 death to 14 injuries. The ratio of casualties in protected areas to casualties outside protected areas in these same Indian states was about 1 to 7 (Rajpurohit and Krausman 2000).

Eighty casualties caused by sloth bears were reported in and around Panna National Park between 1981 and 2000 (Yoganand 2005); one of these was fatal. Monthly frequencies of casualties were significantly positively correlated with the intensity of human use of the forest, daytime activity of bears, and use of *Lantana* habitat for day-resting by bears. Most casualties (80 percent) occurred equally often in the wet and cool seasons; 20 percent occurred in the dry season; and 70 percent were in evenings and mornings. All were in dense vegetative cover, scrubby patches of *Lantana*, or in escarpment habitat. Forty percent of the people were attacked while herding cattle, usually in *Lantana* scrub habitat; 30 percent were collecting non-timber forest products (NTFPs); and the rest when traveling along forest foot paths. Most people (80 percent) were alone when the encounter occurred. Solitary bears were reported in half the casualties; the other half involved females with cubs, or more than one bear. About half the encounters occurred when the bear was resting, usually in the escarpment habitat where the bears use caves as dens for day beds, and in day beds located in *Lantana* scrub; the other half involved foraging bears.

One human–sloth bear conflict hot spot is in the Tumkur district, 100 km northwest of Bangalore, Karnataka, where the dry forest has been degraded and the only remaining shelter for sloth bears is rocky hillocks. Sloth bears feed on crops (mango, jackfruit, and groundnuts) and this leads to substantial conflict. Amita Mitra (personal communication) gathered evidence of 99 casualties (4 deaths) from 2004 to 2009.

A second human–sloth bear conflict hot spot in Chhattisgarh, involving 137 casualties (11 deaths) from 1998 to 2000, has been investigated in detail by Bargali et al. (2005). (This area was part of Madhya Pradesh when Rajpurohit and Krausman [2000] conducted the study on sloth bear casualties described above.) This area of about 1,400 km² is a matrix of degraded patchy forests and forest fragments totaling 337 km². The forest is highly degraded dry tropical broadleaf, with the forests on hillocks with rocky outcrops. People farm only during the monsoon rains (July–October); they collect non-timber forest products and fuelwood and graze cattle in the forest fragments as their main occupation at other times. Bargali et al. (2005) reported that there is intense competition between people and bears, especially for fruit. Here, bears raid crops. More than half of the casualties occurred during the wet season when people were working their fields; more occurred between 4 a.m. and 8 a.m. than later in the day; and more occurred when people were defecating in the bushes than engaged in other activities, e.g., walking within villages, in fields, herding cattle, and collecting non-timber forest products. About 28 percent of the casualties occurred in close encounters when the bear stood on its hind legs then knocked the person down; 36 percent of the attacks occurred when the bear ran after the person. In two incidents the bear climbed a tree after the person. Attacks occurred in kitchen gardens, in fields, and in adjoining forests.

A LAYERED CONSERVATION RESPONSE

It is our view that the consequences of sloth bear attacks on humans are a critical obstacle to sloth bear conservation and place the sloth bear's long-term persistence at risk. Humans and predators have coexisted for millennia but the frequency of conflicts between them has grown because of the increase in human populations, increasingly constricted habitats, competition with people for food resources, and expansion of human activities. Conflicts between humans and large predators in general are the product of contentious socioeconomic and natural systems because the resources concerned have economic value and the predators involved are high profile and legally protected, but are not usually highly valued locally (Damania et al. 2008).

In developed countries where there is a strong commitment to carnivore conservation, schemes have been established to compensate farmers and livestock owners for losses to depredation and to compensate the families of people who are injured or killed. This is coupled with selective removal of individual problem predators. In many developing countries, lack of attention by governments, or low and delayed compensation payments, discourage sufferers from participating in similar schemes. The only perceived solution is to kill the offending predator, and many non-offending animals are killed in the process. People who fear for their safety or perceive economic risk to their lives and livelihoods will not support conservation efforts (Kellert 1996).

Defining the sloth bear as a problem requiring a solution depends both on where the bear is and how it is behaving. It would be self-defeating to assume that any sloth bear outside a protected area is a problem animal. A long-standing strategy for large carnivore conservation has been to separate large carnivores and people through the creation of reserves, but

> human-induced mortality contributes more to the extinction of populations of large carnivores isolated in small reserves than do stochastic environmental process. Conservation measures which aim only to combat stochastic process are therefore unlikely to avert extinctions. Instead, priority should be given to measures which seek to maximize reserve size or to mitigate persecution on reserve borders and in buffer zones (Woodroffe and Ginsberg 1998, 2128).

With the small average size of discrete sloth bear populations, removal of animals found inside or outside protected areas is risky for population persistence. Conservation strategies will be more effective working at the landscape rather than just the reserve scale.

We believe that a maxim in the conservation management of any forest-dwelling bear is that the bear is less likely to be at the edge if the interior is good habitat. This is the lesson we derive from the case studies reported above from Sri Lanka, India, and Nepal. But as long as we're not going to build huge walls around all wildlife areas and separate them entirely from human activities, then there is always going to be "the edge." Also, there are sloth bear populations surviving in some much-degraded forests where the bears have adapted their foraging to include a large proportion of agriculture crops; this results in hot spots of human–sloth bear conflict. Conflict issues are always going to arise in multiple-use forests, at the "edges," and even within protected areas when people enter to collect forest products and graze their livestock, as our case studies have also shown. We need a policy framework to address them.

Enforced zoning schemes can be used to prohibit certain human activities, but much more attention must be directed to improving their effectiveness in limiting the intersection of human and sloth bear activities, as we show in our Panna example above, and this can be achieved, as we have shown in our Nepal example. Protected areas are a form of zoning, as is voluntary resettlement of human populations. Resettlement has been employed in India and Nepal for three decades to move people out of Asian lion and tiger habitat in India and tiger and one-horned rhinoceros (*Rhinoceros unicornis*) habitat in Nepal, a practice that benefits the sloth bears also living in those areas, resulting in substantial reductions in conflicts. However, successful relocation schemes must be truly voluntary and incentive-driven, with participants gaining net benefits including improved access to jobs and essential services. Resettlement schemes may face political opposition from groups wishing to protect local land claims or culture, and these concerns have to be addressed (Karanth and Gopal 2005).

Ultimately, a local population or governing body responsible for sloth bear management labels a sloth bear as a problem because of a real or perceived threat to human life and livelihood. A comprehensive response has two components: (1) removal of the real or perceived threat through a set of actions that entail minimal financial and logistic costs, with (2) provision of a maximum probability of retaining the problem animal as a reproductive member of the wild population. The appropriate response to a problem sloth bear should always achieve the first component, but the second becomes more and more unlikely as the severity of the problem increases.

Steps in a potential framework for implementing a comprehensive response to problem bears include (1) creating professional, well-trained, and adequately equipped response teams with the experience and confidence to handle the range of situations that are encountered and who fully understand the consequences of their actions; and (2) crafting national policies and protocols for response to human–sloth bear conflicts.

Many states have no written policies for dealing with problem bears; situations are handled on a case-by-case basis with no guidelines and without the appropriate staff training, equipment, and resources. National policies are needed that provide protocols that clearly delineate appropriate actions for the various situations likely to be encountered. While every situation is unique, a general protocol that empowers local wildlife managers to make decisions, makes them accountable, and provides response teams with guidelines to follow for deciding the appropriate course of action is a critical first step in dealing quickly and efficiently with problem animals.

Because local support for, or at least tolerance of, sloth bears is one of the key factors determining the fate of all sloth bear populations, elimination of real or perceived threats is pivotal. It is important to make a distinction here between problem animals and areas where bear problems occur. The ability to retain a problem animal within the wild population will depend on the abilities of the response personnel and the severity of the problem. Steps in the response include keeping the local people informed and involved to minimize misinformation and misunderstanding of sloth bear biology; and inducing the problem animal to leave the area through some disturbance. Both of these are very difficult to achieve on the basis of individual bears deemed problem animals. Rather, we must focus on problem areas versus nonproblem areas. Ultimately, successful conservation interventions come from recognizing and minimizing the factors in an area that lead to conflict in the first place, and not blaming the problem on individual animals.

Crop depredation and personal injury are problems where sloth bear ranges overlap with agricultural areas and where people go into the forest to graze cattle and collect non-timber forest products. There is a graded response to these. Quick and efficient programs to compensate villagers financially when casualties occur and livelihoods are threatened by sloth bears can be critical to reducing local animosity toward sloth bears. But compensation schemes often fail because of the large sums involved, the low value of the crops lost relative to the expense of getting claims verified, corruption in the official machinery and among claimants, and a general lack of rural financial mechanisms to enable quick transactions (Karanth and Gopal 2005). Compensation schemes must also be sustainable. For example, an insurance plan for wildlife-related human injuries and crop damage is being proposed in Bhutan (S. Wang, personal communication) but has not been implemented. This would be capitalized from outside sources in the beginning, with start-up funding based on known long-term average casualties and livelihood losses. Capital gains above those required for the maintenance of the program would be invested in low- or no-interest loans to help farmers improve husbandry techniques and reduce depredation rates.

Ideally, what is needed in any "edge" area such as a buffer zone or mixed-land-use region is that the sloth bears living there do not bother the villagers. Consistently removing sloth bears that cause problems is unsatisfactory because another sloth bear will likely move into the vacated area, with an equal probability of trouble, until such time as sloth bear density is too low to provide a replacement, or there are none left at all. Modifying habitats so sloth bears do not cause problems is required, rather than removal of offending sloth bears. A potential solution is the creation and maintenance of buffer habitat between forest cores and villages where the sloth bears are maintained at a low density; our Nepal example shows this is feasible. The Panna case study shows that bears prefer the core habitat, and it is when people enter this core that conflict occurs. Electric fencing may be an effective option in such problem areas. We recognize that modifying sloth bears' behavior so they do not cause problems will be difficult. Research on how to modify predator behavior, such as that of bears and wolves, has been ongoing in North American wildlife management circles. Methods of non-lethal deterrence of predators through disruptive stimuli (e.g., *fladry*—a line of flags hung on the outside of a pasture, electronic guards, and radio-activated guards) and aversive stimuli (e.g., electronic training collars and shooting with less-than-lethal ammunition) are in various stages of development and deployment (Shivik 2006). These can be tested and adapted as needed for sloth bears, but they are costly and we believe they will be difficult to implement.

If a sloth bear persists in destroying crops or attacking humans, there are essentially three options in the short term: translocate it, capture and keep it in captivity, or kill it. Translocation should be attempted only if it can be shown there is sufficient habitat and food available at the release site, it is sufficiently far from human habitation that further conflict is unlikely, and bear density at the site is low. Capture and removal of problem animals reduces the viability of wild populations. Availability of resources to appropriately house a captured sloth bear and include it in a managed breeding program aimed at generating the maximum possible conservation support for wild sloth bears, and the suitability for captivity of the sloth bear in question, must be major factors in a decision to capture, rather than to translocate or kill it. Killing the offending animal has been the historical and usual reaction to a sloth bear perceived as a problem. While this response is still widely used and will remain appropriate in some instances, the sloth bear's low numbers and protected status make it a response to be avoided whenever possible. It is critical for local officials to demonstrate to their rural constituency that protecting them against wildlife attacks on their lives and livelihood is a priority, and this position requires acknowledging that it is appropriate for some sloth bears to be removed from the wild. For animals that are unsuitable for captivity, death may be the only sensible option. Official protocols must clearly delineate specific guidelines for granting permission to kill. We must recognize that it is not the people's behavior that has led to the problem but rather the bear's behavior, which is in response to its diminishing habitat. An appropriate ultimate management intervention is to establish how human–sloth bear interactions can be reduced through habitat manipulations.

SLOTH BEAR RESPONSE TO FOREST FRAGMENTATION

While the forest areas still occupied by sloth bears in India are extensive (215,466 km², Jhala et al. 2008; more than 295,000 km², Garshelis et al. 1999b; or 260,000 km², Yoganand et al. 2006), this forest cover is highly fragmented, no single patch is greater than a few thousand square kilometers in size, and most patches are small (Yoganand et al. 2006). It is remarkable that the extent of occupied sloth bear habitat is large, despite the bear's coexistence today with 1.2 billion people in India. However, even this already fragmented habitat is at risk of further fragmentation or being lost altogether because of extensive infrastructure (mines, roads, and dams) development that is being undertaken at a massive scale to support a rapidly expanding Indian economy (K. Varma, World Bank Group, personal communication). Added to this loss are the incursions of human habitation into forest habitats, illegal land conversions, forest encroachment, and land claims by local people.

Sloth bears have disappeared from parts of their former geographic range in India, Nepal, and Sri Lanka, even while those areas retained forest cover. A survey across the lowland of Nepal found sloth bears were either absent or occurred at low densities in areas with high human use, despite having high termite densities (Garshelis et al. 1999a). Yoganand et al. (2006) examined the pattern and extent of sloth bear extirpation reported in forest areas in India. In recent times, sloth bears were reported locally extinct in at least thirty-five protected areas, forest patches, or forest administrative blocks in India, mostly in areas in the northern Western Ghats, northwestern semi-arid areas, and the western Shivalik/Terai. Most of these areas of extirpation are on the periphery of sloth bear range. Of forest patches where sloth bears had been extirpated, 71 percent were less than 300 km² in size; no bears were reported extirpated from forest patches larger than 800 km². Degree of isolation of forest patches is related to the risk of sloth bear extirpation, with 69 percent of reported extirpations occurring in forest blocks isolated from others by 5 km or more (Yoganand et al. 2006). In Sri Lanka, sloth bear habitat retains a high degree of connectivity, but occurrence is negatively correlated with increasing human and road density (Ratnayeke et al. 2007).

These findings are consistent with conservation theory (Soulè and Terborgh 1999): we would expect higher risk of extirpation in smaller forest patches as a result of pathogens, poachers, demographic stochasticity, genetic deterioration, and catastrophe (fires, floods, and droughts) combined with the barriers that restrict the immigration of bears from adjacent areas. This highlights the need to maintain and reestablish forest connectivity for the long-term persistence of sloth bear populations.

SEASONALLY DRY TROPICAL FOREST AS ESSENTIAL SLOTH BEAR HABITAT

When we (Yoganand et al. 2006) asked Indian forest managers, biologists, and naturalists from approximately three hundred forest areas across India, through a questionnaire survey, about their perception of habitat quality for sloth bears, they reported that sloth bears occurred more frequently and in higher abundance in moist deciduous forests followed by seasonally dry tropical forests (SDTFs). Seasonally dry tropical forests compose approximately 45 percent and moist forests approximately 35 percent of Indian forests, of which approximately 40 percent are degraded. In the same survey we found that observers reported more moist forest areas with extirpated sloth bear populations (45 percent) than dry deciduous forest areas (32 percent), which we posit is a factor of the extent of forest fragmentation: dry forests are fragmented but less so than moist forests. We believe that, taken together, dry and moist broadleaf forests are essential habitat for sloth bears, of which less than 25 percent is in some form of protected status.

A key factor that has ensured protection of sloth bear habitat over the past four decades is the attention given to tiger conservation. India began protecting core tiger breeding areas with eight tiger reserves in 1972 and had expanded this to thirty-eight tiger reserves by 2009. Sloth bears lived in all eight original tiger reserves, and we believe they live in at least thirty-five of the tiger reserves today. (Sloth bears do not occur in Sariska, from which tigers were extirpated, the Sundarbans, the mangrove forest at the mouths of the Ganges, Brahmaputra, and Meghna rivers, or in Namdapha, as we report above.)

Ranganathan et al. (2008) used a landscape-scale, density-based population model as a tool to assess whether reasonably effective management of current reserves is adequate to ensure the survival of tiger populations in the Indian subcontinent. They found that just 21 of 150 reserves or complexes of reserves could hold 58–95 percent of the approximately 3,500–6,500 tigers that potentially could be supported in the subcontinent. Significantly, these all have sloth bears living in them. Thirteen of these protected area complexes include tropical dry forest, seven of which are more than 1,000 km². Sloth bears live at approximately four times the density of tigers living in Chitwan's tall-grass savanna (18 tigers/100 km², Barlow et al. 2009; 68 sloth bears/100 km², Garshelis et al. 1999a). In the dry forests of Panna (before tigers were extirpated), tigers and sloth bears occurred at about the same densities (approximately 7 tigers and approximately 7 sloth bears/100 km²; Karanth et al. 2004; Yoganand 2005). Tigers can occur in moist forests at densities similar to those in tall-grass savanna but we do not have sloth bear density data for this forest type to make comparisons. The 14,065 km² of tropical dry forests in the reserve complexes identified by Ranganathan et al. (2008) could be expected to support about 1,000 sloth bears. Sanderson et al. (2006) estimate there are 110,572 km² of tropical dry broadleaf forests in the Indian subcontinent (not including Sri Lanka), of which about 60,000 km² is potential habitat for tigers. Potentially this would support 4,200 sloth bears as well as tigers, if it were to be managed as effectively as was Panna during our study there. The future of tropical dry forests will have a major bearing on the persistence of sloth bear populations, but depending on protected areas alone in the dry forests as the foundation of sloth bear conservation will result in a further diminished sloth bear population.

A POLICY FRAMEWORK FOR HUMAN–SLOTH BEAR COEXISTENCE

In Table 2 we propose an ecologically based policy framework for human–sloth bear coexistence. As we have noted, core sloth bear breeding areas will depend on protected areas, but Indian subcontinent protected areas are generally small and isolated. Ideally, the conservation of sloth bears will include plans to manage sloth bears on lands outside protected areas too. The successful coexistence of sloth bears and people will depend on how effectively conflicts between people and sloth bears can be minimized. It is unlikely that conflicts can be eliminated as long as people and sloth bears use the

Table 2. Proposed policy framework for human–sloth bear coexistence proposed in sloth bear range states. This framework prioritizes the needs to keep sloth bears spatially separate from incompatible human land use at the scale of protected reserves, while at the same time mitigating conflicts by prioritizing human need at larger landscape scales.

	Conflict Type		
Land Use	Crop Depredation	Accidental Killing of Humans	Persistent Killing of Humans
Protected area for sloth bear conservation	Toleration; relocation of people	Compensation; relocation of people	Compensation; relocation of people; lethal control
Sloth bear habitat in multiple-use forest	Toleration; compensation	Compensation	Compensation; lethal control
Sloth bear habitat in privately owned land	Compensation	Compensation; capture-captivity	Compensation; lethal control
Unsuitable sloth bear habitat in public or private land	Compensation; translocation of sloth bear	Compensation; translocation of sloth bear; lethal control	Compensation; lethal control
Human habitation; crop fields	Compensation; translocation of sloth bear; barriers	Compensation; translocation of sloth bear; capture-captivity; lethal control	Compensation; lethal control

Note: The management context (each box in the table) is defined by the nature of the conflict (three right-hand columns) and the priority land use at the site of the conflict (rows). The suggested conflict-mitigation tactics (not mutually exclusive) are listed in order of priority within each relevant box. These tactics include passive tolerance of conflict, financial compensation, sloth bear capture followed by translocation into the source population or captivity, lethal control of sloth bears, and incentive-driven relocation of human settlements. This framework is adapted from that proposed for tigers by Karanth and Gopal (2005).

same land, but conflict can be reduced to a level that does not threaten sloth bear population viability and is acceptable to the people of the forest communities who incur the costs of living with sloth bears. Successful solutions to these conflicts will require partnerships that combine policy changes by states and national administrations, and local grassroots efforts that demonstrate ways to effectively implement conflict resolution. Although examples of true partnerships between governments and NGOs may be rare, they are a pivotal component of far-reaching sloth bear conservation planning and successful conservation interventions.

NOTE

1. *Ursus* is the genus name for all the species in the subfamily Ursinae now that we have fully resolved the phylogeny of the Ursidae (Krause et al. 2008).

REFERENCES

Akhtar, N., H. S. Bargali, and N. P. S. Chauhan. 2008. Distribution and Population Abundance of Sloth Bear (*Melursus ursinus*) in Disturbed and Unprotected Habitat of North Bilaspur Forest Division, Chhattisgarh. *Tigerpaper* 35 (3): 15–21.

Bargali, H. S., N. Akhtar, and N. P. S. Chauhan. 2004. Feeding Ecology of Sloth Bears in a Disturbed Area in Central India. *Ursus* 15:212–17.

———. 2005. Characteristics of Sloth Bear Attacks and Human Casualties in North Balispur Forest Division, Chhattisgarh, India. *Ursus* 15:263–76.

Barlow, A. C. D., C. McDougal, J. L. D. Smith, B. Gurung, S. R. Bhatta, S. Kumal, B. Majato, and D. B. Tamang. 2009. Temporal Variation in Tiger (*Panthera tigris*) Populations and Its Implications for Monitoring. *Journal of Mammalogy* 90:472–78.

Baskaran, N., V. Sivaganesan, and J. Krishnamooty. 1997. Food Habits of Sloth Bear in Mudumalai Wildlife Sanctuary, Tamil Nadu, Southern India. *Journal of the Bombay Natural History Society* 94:1–9.

Bininda-Emonds, O. R. P., J. L. Gittleman, and A. Purvis. 1999. Building Large Trees by Combining Phylogenetic Information: A Complete Phylogeny of the Extant Carnivores (Mammalia). *Biological Reviews* 74:143–75.

Carbone, C., G. M. Mace, S. C. Roberts, and D. W. Macdonald. 1999. Energetic Constraints of Terrestrial Carnivores. *Nature* 402:286–88.

Chundawat, R. S., N. Gogate, and A. J. T. Johnsingh. 1999. Tigers in Panna: Preliminary Results from an Indian Tropical Forest. In *Riding the Tiger: Tiger Conservation in Human-Dominated Landscapes*, ed. J. Seidensticker, S. Christie, and P. Jackson, 123–29. Cambridge: University of Cambridge Press.

Ciochon, R. L., and J. W. Olsen. 1991. Paleoanthropological and Archaeological Discoveries from Lang Trang Caves: A New Middle Pleistocene Hominid Site from Northern Viet Nam. *Indo-Pacific Prehistory Association Bulletin* 10:59–73.

Damania, R., J. Seidensticker, and T. Whitten et al. 2008. *A Future for Wild Tigers*. Washington, DC: World Bank.

Dinerstein, E. 2003. *The Return of the Unicorns*. New York: Columbia University Press.

Garshelis, D. L. 2004. Variation in Ursid Life Histories—Is There an Outlier? In *Giant Pandas: Biology and Conservation*, ed. D. Lindburg and K. Baragona, 53–73. Berkeley: University of California Press.

Garshelis, D. L., A. R. Joshi, and J. L. D. Smith. 1999a. Estimating Density and Relative Abundance of Sloth Bears. *Ursus* 11:87–98.

Garshelis, D. L., A. R. Joshi, J. L. D. Smith, and C. G. Rice. 1999b. Sloth Bear Conservation Action Plan. In *Bears*: *Status Survey and Conservation Action Plan*, ed. C. Servheen, S. Herrero, and B. Peyton, 55–50. Gland, Switzerland: IUCN/SSC Bear Specialist Group.

Hölldobler, B., and E. O. Wilson. 1991. *The Ants*. Cambridge, MA: Belknap Press.

———. 2009. *The Superorganism*: *The Beauty, Elegance, and Strangeness of Insect Societies*. New York: W.W. Norton & Co.

Jhala, Y. V., R. Gopal, and Q. Qureshi, eds. 2008. *Status of Tigers, Co-predators and Prey in India*. Dehra Dun: National Tiger Conservation Authority / Wildlife Institute of India.

Johnsingh, A. J. T. 1981. Ecology and Behavior of Dhole (*Cuon alpinus)*, with Special Reference to Predator-Prey Relations in Bandipur. PhD diss., Madurai University, Madurai, India.

———. 2003. Bear Conservation in India. *Journal of the Bombay Natural History Society* 100:192–201.

Joshi, A. P., D. L. Garshelis, and J. L. D. Smith. 1997. Seasonal and Habitat-Related Diets of Sloth Bears in Nepal. *Journal of Mammalogy* 78:584–97.

Joshi, A. P., J. L. D. Smith, and D. L. Garshelis. 1999. Sociobiology of the Myrmecophagous Sloth Bear in Nepal. *Canadian Journal of Zoology* 77:1690–1704.

Karanth, K. U., R. S. Chundawat, J. D. Nichols, and N. S. Kumar. 2004. Estimation of Tiger Densities in the Tropical Dry Forests of Panna, Central India, Using Photographic Capture-Recapture Sampling. *Animal Conservation* 7:285–90.

Karanth, K. U., and R. Gopal. 2005. An Ecology-Based Policy Framework for Human-Tiger Coexistence in India. In *People and Wildlife*: *Conflict or Coexistence?* ed. R. Woodroffe, S. Thirgood, and A. Rabinowitz, 373–87. Cambridge: Cambridge University Press.

Karanth, K. U., and J. D. Nichols. 2000. *Ecological Status and Conservation of Tigers in India*. New York: Wildlife Conservation Society.

Kellert, S. R. 1996. *The Value of Life*: *Biological Diversity and Human Society*. Washington, DC: Island Press.

Krause, J., T. Unger, and A. Nocon et al. 2008. Mitochondrial Genomes Reveal an Explosive Radiation of Extinct and Extant Bears near the Miocene-Pliocene Boundary. *BMC Evolutionary Biology* 8:200, doi:10.1186/1471-1248-8-220.

Laurie, A., and J. Seidensticker. 1977. Behavioural Ecology of the Sloth Bear (*Melursus ursinus)*. *Journal of Zoology, London* 182:187–204.

Li Yu, Yi-Wei Li, O. A. Ryder, and Ya-Pin Zhang. 2007. Analysis of Complete Mitochondrial Genome Sequences Increases Phylogenetic Resolution of Bears (Ursidae), a Mammalian Family. *BMC Evolutionary* Biology 7:198, doi:10.1186/1471-2148-7-198.

Maraj, R., and J. Seidensticker. 2006. *Assessment of a Framework for Monitoring Tiger Population Trends in India*. Bangkok: IUCN, World Conservation Union.

McNab, B. K. 1992. Rate of Metabolism in the Termite-Eating Sloth Bear (*Ursus ursinus)*. *Journal of Mammalogy* 73:168–72.

Palomares, F., and T. M. Caro. 1999. Interspecific Killing among Mammalian Carnivores. *American Naturalist* 153:492–508.

Rajpurohit, K. S., and P. R. Krausman. 2000. Human–Sloth Bear Conflicts in Madhya Pradesh, India. *Wildlife Society Bulletin* 28:393–99.

Ranganathan, J., K. M. A. Chan, K. U. Karanth, and J. L. D. Smith. 2008. Where Can Tigers Persist in the Future? A Landscape-Scale, Density-Based Population Model for the Indian Subcontinent. *Biological Conservation* 141:67–77.

Ratnayeke, S., F. T. van Manen, R. Pieris, and V. S. J. Pargash. 2007. Landscape Characteristics of Sloth Bear Range in Sri Lanka. *Ursus* 18:189–202.

Ratnayeke, S., S. Wijeyamohan, and C. Santiapillai. 2006. The Status of Sloth Bears in Sri Lanka. In *Understanding Asian Bears to Secure Their Future*, 35–40. Ibaraki, Japan: Japan Bear Network.

Rode, K. D., and C. T. Robbins. 2000. Why Bears Consume Mixed Diets during Fruit Abundance. *Canadian Journal of Zoology* 78:1640–45.

Sacco, T., and B. Van Valkenburgh. 2004. Ecomorphological Indicators of Feeding Behaviors in the Bears (Carnivora: Ursidae). *Journal of Zoology, London* 263:41–54.

Sanderson, E., J. Forrest, and C. Loucks et al. 2006. *Setting Priorities for the Conservation and Recovery of Wild Tigers: 2005–2015. The Technical Assessment.* Washington, DC: Wildlife Conservation Society / WWF / Smithsonian National Zoological Park Conservation & Research Center.

Sarker, M. S. U. 2006. The Status and Conservation of Bears in Bangladesh. In *Understanding Asian Bears to Secure Their Future*, 41–44. Ibaraki, Japan: Japan Bear Network.

Sathyakumar, S. 2006. The Status of Asiatic Black Bears in India. In *Understanding Asian Bears to Secure Their Future*, 12–19. Ibaraki, Japan: Japan Bear Network.

Saw Htun. 2006. The Status and Conservation of Bears in Myanmar. In *Understanding Asian Bears to Secure Their Future*, 45–49. Ibaraki, Japan: Japan Bear Network.

Schaller, G. B. 1967. *The Deer and the Tiger: A Study of Wildlife in India*. Chicago: University of Chicago Press.

Seidensticker, J. 2008. Ecological and Intellectual Baselines and Saving Lions, Tigers, and Rhinos in Asia. In *Foundations of Environmental Sustainability: The Coevolution of Science and Policy*, ed. L. Rockwood, R. Stewart, and T. Dietz, 98–117. New York: Oxford University Press.

Shivik, J. A. 2006. Tools for the Edge: What's New for Conserving Carnivores? *Bioscience* 56:253–59.

Soulè, M. E., and J. Terborgh, eds. 1999. *Continental Conservation: Scientific Foundations of Regional Reserve Networks*. Washington, DC: Island Press.

Van Valkenburgh, B. 1999. Major Patterns in the History of Carnivorous Mammals. *Annual Review of Earth and Planetary Science* 27:463–93.

Vinitpornsawan, S., R. Steinmetz, and B. Kanchanasakha. 2006. The Status of Bears in Thailand. In *Understanding Asian Bears to Secure Their Future*, 45–49. Ibaraki, Japan: Japan Bear Network.

Wikramanayake, E., E. Dinerstein, C. J. Loucks, D. M. Olson, J. Morrison, J. Lamoreux, M. McKnight, and P. Hedao. 2002. *Terrestrial Ecoregions of the Indo-Pacific: A Conservation Assessment*. Washington, DC: Island Press.

Woodroffe, R., and J. R. Ginsberg. 1998. Edge Effects and the Extinctions of Populations inside Protected Areas. *Science* 280:2126–28.

Yoganand, K. 2005. Behavioural Ecology of Sloth Bear (*Melursus ursinus*) in Panna National Park, Central India. PhD diss., Saurashtra University, Rajkot, India.

Yoganand, K., C. G. Rice, A. J. T. Johnsingh, and J. Seidensticker. 2006. Is the Sloth Bear in India Secure? A Preliminary Report on Distribution, Threats and Conservation Requirements. *Journal of the Bombay Natural History Society* 103:57–66.

13

Seasonally Dry Tropical Forest Is Essential Tiger Habitat

James L. David Smith, Saksit Simchareon, Achara Simchareon, Peter Cutter, Bhim Gurung, Raghunandan Chundawat, Charles McDougal, and John Seidensticker

Tigers (*Panthera tigris*) are habitat generalists as evidenced by their once-wide distribution in multiple vegetation types through Central, East, South, and Southeast Asia. Tigers are the largest obligate meat-eaters living in Asian wildlands. They kill prey ranging in mass from 20 kg to more than 1,000 kg as encountered, but also selectively seek out and kill large-bodied ungulate prey—large deer (*Cervus* spp., *Axis* spp., *Rucervus* spp., *Rusa* sp.), wild cattle (*Bos* spp., *Bubalus* sp.), and wild pigs (*Sus* sp.)—thereby gaining access to a major percentage of potential prey biomass that is contributed by relatively few individuals (Seidensticker et al. 2010a). In South and Southeast Asia, tigers historically occupied vast expanses of tropical and subtropical moist forest, seasonally dry tropical and subtropical forests (SDTFs), and tropical and subtropical tall-grass savanna biomes (Wikramanayake et al. 2002). They occupy less than 7 percent of their historical range today. The tiger's geographical range is collapsing and its numbers are declining rapidly due to poaching of tigers and their prey, retaliatory killing of tigers, and habitat fragmentation, degradation, and loss (Kenney et al. 1995; Dinerstein et al. 2007; Gurung et al. 2008; Seidensticker 2010).

In the absence of human-induced mortality, prey density and distribution, rather than habitat vegetation parameters, explain tiger density (Miquelle et al. 1999; Karanth et al. 2004). Broadly, however, tiger prey densities vary across the different forest types because the production and availability of food for ungulates varies in different forest types (Eisenberg and Seidensticker 1976). The territory size of a reproducing female is a base measure of the carrying capacity of tiger habitat (Smith et al. 1987). In this chapter, we review and compare territory sizes of reproducing females in two SDTF sites with territory sizes in two other highly productive vegetative cover types in South Asia. We relate territory size of reproducing females to measures of prey abundance. We conclude with a discussion of the importance of SDTF in the stabilization and recovery of tiger numbers.

TIGER TERRITORY SIZE

A reproducing tiger's home range is defended and, thus, is its territory (Sunquist 1981). Breeding males defend their territories from other breeding males, and each male territory overlaps those of several breeding females. Male territory size is dependent on the assertiveness of the resident and adjacent territorial males. Reproducing female tigers exhibit strong intrasexual territorial behavior, excluding other breeding females from their territories. Female territory size tracks available prey biomass (Sunquist 1981; Smith et al. 1987).

A reproducing female tiger's challenge is to optimize her territory size relative to available prey (Smith et al. 1987). A territory size that does not support an adequate number of prey or a territory in which prey is depleted due to human hunting manifests as reduced or failed reproductive output (Karanth and Stith 1999; Chapron et al. 2008). If a territory is larger than needed to supply a female's food requirement, the female may incur an increased energetic cost due to increased travel time and an increased risk of injury in territorial defense; territorial fighting can lead to injuries and loss of ability to hunt and defend a territory. However, there is no a priori reason a female has a fixed territory size in areas where other breeding females in a population have been depleted due to poaching or other sources of mortality.

A female's territory must be large enough to support enough prey to feed her and her offspring as they grow and until they disperse at 19–28 months of age. The energetics of this has not been calculated for tigers but it has been calculated in another large cat, pumas (*Puma concolor*). Over the course of gestation, lactation, and dependence of 2.6 cubs, a reproducing female puma's energy requirements are about twice what she needs when she is not breeding and rearing cubs (Ackerman et al. 1986; Laundré 2005). A breeding female tiger's territory must also be large enough to provide some of the prey the resident breeding male requires. And, further, it must be large enough to supply prey for the nonreproducing, transient (dispersing and nonreproducing adults) component in each tiger population, which can form approximately 18 percent of the total population (Karanth et al. 2006).

Establishing tiger territory size has been a demanding enterprise for scientists. After more than thirty years of effort by several teams of investigators, we have now measured tiger territory size at four sites in South and Southeast Asia: Chitwan National Park, Nepal (Sunquist 1981; Smith et al. 1999; this chapter); Panna National Park, India (Chundawat et al. 1999; Chundawat and van Gruisen 2004); Nagarahole National Park, India (Karanth and Sunquist 2000); and Huai Kha Khaeng Wildlife Sanctuary (HKK), Thailand (this chapter). Studies of territory size are currently underway in the Russian Far East (D. Miquelle and J. Goodrich, personal communication) and in Kanha National Park, India (Y. Jhala, personal communication).

Tigers at these sites were captured by field-anesthesia, i.e., darted from a tree during a drive (Smith et al. 1983), darted from trained elephant's back (Seidensticker et al. 1974), or darted when the tiger was captured in a foot snare (Goodrich et al. 2001). Once anesthetized, tigers were fitted with VHF radio-collars. Positions and locations of tigers with VHF transmitters were obtained by triangulation or by approaching within a distance of 100–200 m and partially circling the animal to obtain a geographic location, or "fix" (Smith 1993). Animals in this study at HKK and one in Chitwan were also fitted with satellite GPS instruments. Individual animals were tracked for periods of more than one year (six years for two females in Nepal); tracking in all cases indicated that the full extent of the territory being used had been measured. Territory sizes were calculated using several methods in these studies, but only convex minimum polygon (Mohr 1947) estimates were common to all studies; these we report here.

Female tiger territory size varied from 14 km^2 in Chitwan to 78 km^2 in HKK (Table 1). There was an approximately 3–7 percent overlap between adjacent female territories in Chitwan (Smith et al. 1987). Panna and HKK tigresses had a similarly low overlap between adjacent territories. Male territories overlap female territories and are three to nearly five times larger than the territories of female tigers (Table 2). Therefore, in addition to supplying a female and young's energetic needs, each female territory supplies approximately 20–33 percent of the energy needed to support a resident male. We estimate that a female's territory on average feeds approximately

Table 1. Territory sizes of seventeen breeding female tigers in four protected areas in South and Southeast Asia. Territory size is the 100% minimum convex polygon (MCP). Number-letter codes identify individual tigers.

Site/Tiger Number	MCP (km^2)
Huai Kha Khaeng, Thailand	
TF4	70
TF5	78
	Mean = 74
Panna, India	
F 118	35
F 111	40
F 113	54
F 120	40
	Mean = 42
Nagarahole, India	
FT2	18
Chitwan, Nepal	
101	34.7
103	16.5*
106	17.7
109	19*
115	19.6*
JP	14*
122	20.8*
118	17.9*
PP	28.4
BP	20
	Mean = 21.8

Sources: Chitwan National Park, Nepal (Sunquist 1981; Smith et al. 1999; this study); Panna National Park, India (Chundawat et al. 1999; Chundawat and van Gruisen 2004); Nagarahole National Park, India (Karanth and Sunquist 1995); and Huai Kha Khaeng Wildlife Sanctuary, Thailand (this study).
*Territories used to calculate mean territory size in riverine habitat.

Table 2. Ratio of breeding male-tiger territory size to breeding female-tiger territory size at four sites, and their weighted mean ratio.

Location	Ratio of Male to Female Territory Size	Females	Males
Huai Kha Khaeng, Thailand	4.0	3	1
Panna, India	4.7	4	1
Nagarahole, India	3.4	1	1
Chitwan, Nepal	3.0 (2–8)*	11	9
Weighted mean	3.5		

*Range of female territories within a male's territory.

3.1–3.3 adult-sized individuals, based on our experience and energetic calculations by Ackerman et al. (1986) and Laundré (2005).

FEMALE TERRITORY SIZE AND PREY ABUNDANCE

At the HKK, Nagarahole, and Panna sites, prey abundance was estimated where female tiger territory sizes were recorded; in Nepal, female territory size was measured in Chitwan while prey data are from the similar alluvial grasslands of Bardia National Park. Following Karanth et al. (2004), we report prey numbers rather than biomass. We converted prey numbers of all species to *sambar units* (SU). A sambar (*Rusa unicolor*) is a large deer present at all study sites and a preferred tiger-prey species. The SU is based on the ratio of average mass of each prey species to an equivalent number of sambar with the same biomass. Thus, [(number of prey) x (mean prey-species mass)]/1 mean sambar mass = x SU.

Sambar units provide a standard measurement to compare different sites, and they are easier to relate to tiger feeding ecology than measures of biomass (abundance in kilograms). For Nagarahole, we used two estimates of SU, based on different estimates of the average mass of different species. In HKK, we estimated mean prey density from sixty randomly selected transects in the territories of two females (Table 3). In Chitwan, ten tigresses were radio-collared across a range of different habitats that varied in terms of prey availability. We used data from only six females with territory sizes of approximately 18.1 km² each that lived primarily in the tall-grass alluvial savanna

Table 3. Prey abundance at four sites. Number of animals of each species was converted to number of equivalent sambar, which we express as sambar units (SU) per km², facilitating comparisons across sites.

Location	SU/km²	Method[a]	Data Source
Nepal	57	FAR/BC	Wegge and Storaas (2009)
Panna	29	DS	Chundawat et al. (1999)
Nagarahole (1)	52	DS	Karanth and Sunquist (1995)
Nagarahole (2)	58	DS	Karanth and Sunquist (1995)
Huai Kha Khaeng	7.4	FAR	This study

[a]Data obtained from block counts (BC), distance sampling (DS), and fecal accumulation rate (FAR).

habitat where we had estimated abundance of prey to examine the relation between territory size and prey abundance (Figure 1).

A variety of approaches, including distance sampling, block counts, and fecal accumulation rate, were used to estimate prey abundance at the different sites. In HKK, we used fecal accumulation over a period of thirty days and then calculated absolute abundance based on known daily defecation rates. In Chitwan and Bardia, we used data based on block drives and distance sampling from trained elephants (Seidensticker 1976; Wegge and Storaas 2009). In Nagarahole and Panna, prey data were obtained using distance sampling on foot (Chundawat et al. 1999; Karanth et al. 2004).

Mean territory size of reproducing females had a strong negative relationship to mean prey abundance (Figure 1). Earlier research in Chitwan showed a similar relationship between territory size and relative prey abundance for seven resident, territorial females (Smith et al. 1987). Our analysis extends the finding of Smith et al. (1987) to other forest types, specifically SDTF.

We found considerable variability in prey availability in female territories when we multiplied SU/km² by mean female territory size in km² at each site: HKK: 548 SU; Chitwan: 1,288 SU; Panna: 1,218 SU; Nagarahole: 884–986 SU. On average, tigers are estimated to remove 10 percent of all available prey within their territories each year; the average kill rate is estimated to be approximately 50 ungulates per year; thus, 500 individuals can be considered the prey base needed to support a tiger (studies summarized in Karanth et al. 2004). Considering 500 individual prey animals as the base ungulate prey requirement for tiger maintenance energetics, we found that reproducing female tigers had a ratio of prey available to base need that varied from a low of 1.1 in HKK to a high of 2.6 in Chitwan.

The Nagarahole and Chitwan tiger populations have some of the highest densities known (Karanth et al. 2006; Barlow et al. 2009; Smith et al. 2010), while tiger density in HKK is four times lower (Simcharoen et al. 2007; Somphot, personal communication). The estimated prey consumed to prey base ratio in Panna (2.4) and in Chitwan (2.6) suggests that both areas could support a higher density of tigers. This seems clearly to be the case in Panna, as this population was expanding at the time of Chundawat's study and has subsequently gone extinct due to poaching. The Chitwan population, in contrast, may have reached a *social* saturation limit in which female

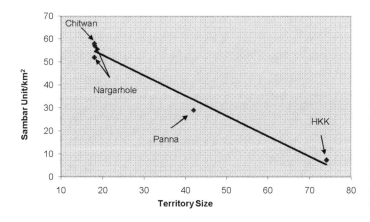

Figure 1. Relationship between mean breeding female tiger territory size (km²) and prey abundance measured in sambar units (SU) per km². Sambar units are the number of animals of each species converted to a number of equivalent sambar.

territory size did not shrink below some point even though prey availability appeared to be high enough for females to have reproduced on smaller territories. Occasionally a female tiger has successfully raised young on a territory measuring approximately 12 km^2 (C. McDougal, unpublished data), but based on data over a period of seven years, during which six females occupied an area of approximately 100 km^2, we concluded that female tigers had reached a lower size limit when territory was approximately 17.5 km^2—a 5 × 3.5 km area (Barlow et al. 2009).[1]

UNGULATE CARRYING CAPACITY AND FOREST TYPE

What determines prey carrying capacity of a particular forest type? Productivity of large ungulates is a manifestation of the vegetative response to soil quality and rainfall (Eisenberg and Seidensticker 1976). For example, seasonal flooding contributes to the formation of gallery forest and alluvial plains along rivers. The fertility of the soil in alluvial plains—as in Chitwan—is recharged on an annual basis and the productivity of the grasses is greatly increased, and thus supports the highest ungulate densities in South and Southeast Asia.

> As a generalization, when proceeding from a dry thorn forest to a moist deciduous forest, the mammalian biomass increases as [one] moves along an increase in rainfall gradient . . . If the forest becomes continuous and there is very little grassland area, then the forest will tend to support a very low density of terrestrial herbivores so that, at the extreme wet end of our imaginary vegetation gradient, the biomass of terrestrial herbivores will decline again . . . within limits, the absolute amount of rain is not as important as how evenly the rain is distributed throughout the annual cycle. (Eisenberg and Seidensticker, 298–99)

In their comparison of South and Southeast Asian ungulate assemblages, Eisenberg and Seidensticker found that mixed-browsers and grazers or grazers alone—the large deer and wild cattle that tigers select as prey—overall contribute the greatest percentage to terrestrial mammalian biomass in any given habitat and that grass and the interspersion of resources (grass, shrubs, and scrub, as well as low-stature trees) create the optimum habitat for utilization by a diverse terrestrial mammalian herbivore community.

Seasonal rainfall with a distinct wet and dry season annually is a key ecological driver of SDTF. An extended annual dry season results in a deciduous forest that has a more open canopy than found, for example, in an everwet evergreen forest. Increased light at the ground layer produces fast-growing shrubs, herbs, and grasses with high protein content. The general consequence of these ecological differences between dry deciduous and moist evergreen tropical forests is that SDTF produces a much higher biomass of the large mammals upon which tigers depend.

IMPORTANCE OF SDTF MANAGEMENT IN STABILIZING AND RECOVERING TIGER NUMBERS

Most of the tall-grass savannas—like those found in our Chitwan study site—in South and Southeast Asia were converted to agriculture prior to the twentieth century and are now the rarest and most endangered vegetation cover type in the region (Wikramanayake et al. 2002). The small amount that remains, however, supports some of the highest densities of tiger prey and tigers (more than 15 tigers over one year old per 100 km²; Carbone et al. 2001). The tropical and subtropical moist broadleaf forest biomes through Southeast Asia support comparatively low tiger densities of approximately 0.5–3.0 tigers/100 km², even where tigers are protected, for reasons we outline above (Carbone et al. 2001; Kawanishi and Sunquist 2004). South Asian moist deciduous forests, such as Nagarahole, support high densities of approximately 20 tigers over one year old per 100 km² where there is an interspersion of forest and meadows that supports high prey densities (Karanth et al. 2006).

SDTF covers approximately 600,000–1,000,000 km² in South and Southeast Asia (Leimgruber et al., this volume). Where forest cover and prey are protected, SDTF can support tiger densities of up to 7 tigers/100 km² (Karanth et al. 2004; Chundawat and van Gruisen 2004). The tiger density estimate for HKK is about 4 tigers/100 km² (Simcharoen et al. 2007).

Tigers occur at very low density or have been extirpated through most of the SDTF of Vietnam, Cambodia, Laos, and Myanmar, and the density of ungulate prey is very low in these forests because of intensive hunting (Sanderson et al. 2006). Further, the economic corridors being planned in the Greater Mekong Subregion will heavily impact these forests. Fortunately, issues of sustainability, biodiversity protection, and habitat connectivity are recognized and are being studied through the Asian Development Bank's core environmental program (Asian Development Bank 2008). If managed to their full potential to support tigers, these forests could support several thousand tigers; now these forest support a few hundred tigers at best, most of which occur in Thailand (Seidensticker et al. 2010b).

Sanderson et al. (2006) estimate there are 110,472 km² of SDTF in the Indian subcontinent, of which 14,065 km² is in thirteen tiger reserves or protected area complexes. However, these are managed well below their potential to support tigers. Managed to their full potential for the benefit of tigers, these forests could support several thousand tigers (Ranganathan et al. 2008).

Ranganathan et al. (2008) concluded that most tiger populations today are too isolated and small to allow the long-term persistence of tigers. Mean female territory size in a given locality is a base measure of reproducing tiger carrying capacity in areas where the tiger and prey populations are not suppressed by poaching or other mortality factors. In protected tiger populations, prey abundance is the primary determinant of breeding tiger territory size.

Tiger population persistence and resilience, however, is determined by the extent and connectivity of habitat units. Blocks of habitat, when separated from other blocks by inhospitable habitat that dispersing tigers are not likely to cross, define the extent and bounds of individual tiger populations. Blocks of discrete potential tiger habitat—termed *tiger conservation landscapes*—have been mapped and classified by Sanderson et al. (2006). Other ecological variables being equal, the number of breeding females within the habitat block is an ecological measure of a population's viability or resilience. The more breeding females there are, the more resilient the population and the higher the probability of long-term tiger population persistence. Larger populations exhibit less inbreeding depression and lower demographic stochasticity, and are less likely to be extirpated by a single catastrophic ecological event such as an epidemic, flood, drought, fire, or poaching incident.

To reverse the downward trend in tiger numbers across SDTF, and in all forest types, two primary actions are required: (1) stop the direct loss of tigers and prey from poaching, and (2) connect habitat blocks. The impact of poaching was clearly demonstrated in 2006 when poachers extirpated tigers in the Sariska Tiger Reserves in India, and, even more poignant, when all tigers were poached from Panna National Park after the work we describe here was conducted (Government of India 2005; Gopal et al. 2010; R. Chundawat, unpublished data); both extirpations occurred at SDTF sites.

Because functional habitat patches that do support tigers are usually small and increasingly fragmented in SDTF, augmenting connectivity between isolated reserves and increasing the quality of tiger habitat outside reserves through protection of tiger prey are necessary to stabilize and recover tiger populations; both require including tiger conservation in the rural development agenda (Damania et al. 2008). For example, across the lowlands of Nepal, local community forest user groups, acting to further local human well-being, are creating tiger habitat by restoring forests for the ecological services they provide to these people. The result of user-group management of their own resources is that tigers are currently breeding outside reserves (Gurung et al. 2006; Wikramanayake et al. 2010). However, local and government management in buffer zones needs to become more concentrated and participatory because human-tiger conflict increases in restored forests with increased tiger density (Gurung et al. 2008).

Two decades ago, management of forests by local people was unforeseen in the Nepal lowlands, or Terai, but today there are about 1,500 community forest user groups in the lowland forests. As local people manage their forests, habitat for tigers and other biodiversity is being restored. While tigers have continued to decrease during the past decade over most of their range, it is heartening to see examples where range countries with local stakeholders take leadership in conservation actions that can stabilize and recover tiger populations.

NOTE

1. Data from current research (2010) from HKK suggest that there is considerable variability in prey density across different parts of the sanctuary. Prey density is much higher in the central valley of HKK, and data from a recently collared female in this valley appear to show a territory size about half the size of the territories we report here for HKK.

REFERENCES

Ackerman, B. B., F. G. Lindzey, and T. P. Hemker. 1986. Predictive Energetics Model for Cougars. *Cats of the World: Biology, Conservation, and Management*, ed. S. D. Miller and D. D. Everett, 333–52. Washington, DC: National Wildlife Federation.

Asian Development Bank. 2008. *Biodiversity Conservation Corridor Initiative: Pilot Site Implementation Status Report 2007*. Bangkok: Asian Development Bank.

Barlow, A. C. D., C. McDougal, J. L. D. Smith, B. Gurung, S. R. Bhatta, S. Kumal, B. Mahato, and D. B. Tamang. 2009. Temporal Variation in Tiger *(Panthera tigris)* Populations and Its Implications for Monitoring. *Journal of Mammalogy* 90:472–78.

Carbone, C., S. Christie, K. Conforti, T. Coulson, N. Franklin, J. Ginsberg, M. Griffiths, J. Holden, K. Kawanishi, M. Kinnaird, R. Laidlaw, A. Lynam, D. Macdonald, D. Martyr, C. McDougal, L. Nath, T. O'Brien, J. Seidensticker, D. J. L. Smith, M. Sunquist, R. Tilson, and W. N. Wan Shahruddin. 2001. The Use of Photographic Rates to Estimate Densities of Tigers and Other Cryptic Mammals. *Animal Conservation* 4:75–79.

Chapron, G., D. G. Miquelle, A. Lambert, J. M. Goodrich, S. Legendre, and J. Colbert. 2008. The Impact on Tigers of Poaching versus Prey Depletion. *Journal of Applied Ecology* 45:1667–74.

Chundawat, R. S., N. Gogate, and A. J. T. Johnsingh. 1999. Tigers in Panna: Preliminary Results from an Indian Tropical Forest. In *Riding the Tiger: Tiger Conservation in Human-Dominated Landscapes*, ed. J. Seidensticker, S. Christie, and P. Jackson, 123–29. Cambridge: Cambridge University Press.

Chundawat, R. S., and J. van Gruisen. 2004. Tigers in the Dry Forest. In *Tiger: The Ultimate Guide*, ed. V. Thapar, 85–87. New York: CDS Books.

Damania, R., J. Seidensticker, A. Whitten, G. Sethi, K. Mackinnon, A. Kiss, and A. Kushlin. 2008. *A Future for Wild Tigers*. Washington, DC: World Bank.

Dinerstein, E., C. Loucks, E. Wikramanayake, J. Ginsberg, E. Sanderson, J. Seidensticker, J. Forrest, G. Bryja, A. Heydlauff, S. Klenzendorf, P. Leimgruber, J. Mills, T. O'Brien, M. Shrestha, R. Simons, and M. Songer. 2007. The Fate of Wild Tigers. *BioScience* 57:508–14.

Eisenberg, J. F., and J. Seidensticker. 1976. Ungulates in Southern Asia: A Consideration of Biomass Estimates for Selected Habitats. *Biological Conservation* 10:293–308.

Goodrich, J. M., L. L. Kerley, B. O. Schleyer, D. G. Miquelle, K. S. Quigley, E. N. Smirnov, E. N. Nikolaev, H. B. Quigley, and M. G. Hornocker. 2001. Capture and Chemical Anesthesia of Amur (Siberian) Tigers. *Wildlife Society Bulletin* 29:533–42.

Gopal, R., Q. Qureshi, M. Bhardwaj, R. K. J. Singh, and Y. V. Jhala. 2010. Evaluating the Status of the Endangered Tiger *Panthera tigris* and Its Prey in Panna Tiger Reserve, Madhya Pradesh, India. *Oryx* 44:383–89.

Government of India. 2005. *Joining the Dots*. New Delhi: Government of India.

Gurung, B., J. L. D. Smith, C. McDougal, J. Karki, and A. Barlow. 2008. Factors Associated with Man-Eating Tigers in Chitwan National Park. *Biological Conservation* 141:3069–78.

Gurung, B., J. L. D. Smith, and M. Shrestha. 2006. Using a "Bagh Heralu" Network to Map the Metapopulation Structure of Tigers in Nepal. In *Conservation Biology in Asia*, ed. J. A. McNeely, T. M. McCarthy, A. Smith, L. Olsvig-Whittaker, and E. D. Wikramanayake, 214–31. Kathmandu: Resources Himalaya Foundation.

Karanth, K. U., J. D. Nichols, N. S. Kumar, and J. E. Hines. 2004. Tigers and Their Prey: Predicting Carnivore Densities from Prey Abundance. *Proceedings of the National Academy of Sciences, USA* 101:4854–58.

———. 2006. Assessing Tiger Population Dynamics Using Photographic Capture-Recapture Sampling. *Ecology* 87:2925–37.

Karanth, K. U., and B. M. Stith. 1999. Prey Depletion as a Critical Determinant of Tiger Population Viability. In *Riding the Tiger: Tiger Conservation in Human-Dominated Landscapes*, ed. J. Seidensticker, S. Christie, and P. Jackson, 100–113. Cambridge: Cambridge University Press.

Karanth, K. U., and M. E. Sunquist. 1995. Prey Selection by Tiger, Leopard and Dhole in Tropical Forests. *Journal of Animal Ecology* 64:439–50.

———. 2000. Behavioural Correlates of Predation by Tigers (*Panthera tigris*), Leopard (*Panthera pardus*) and Dhole (*Cuon alpinus*) in Nagarahole, India. *Journal of Zoology, London* 250:255–65.

Kawanishi, K., and M. E. Sunquist. 2004. Conservation Status of Tigers in a Primary Rainforest of Peninsular Malaysia. *Biological Conservation* 120:329–44.

Kenney, J. S., J. L. D. Smith, A. M. Starfield, and C. W. McDougal. 1995. The Long-Term Effects of Tiger Poaching on Population Viability. *Conservation Biology* 9:1127–33.

Laundré, J. W. 2005. Puma Energetics: A Recalculation. *Journal of Wildlife Management* 69:723–32.

Miquelle, D. G., E. N. Smirnov, T. W. Merrill, A. E. Myslenkov, H. B. Quigley, M. G. Hornocker, and B. Schleyer. 1999. Hierarchal Spatial Analysis of Amur Tigers' Relationships to Habitat and Prey. In *Riding the Tiger: Tiger Conservation in Human-Dominated Landscapes*, ed. J. Seidensticker, S. Christie, and P. Jackson, 71–99. Cambridge: Cambridge University Press.

Mohr, C. O. 1947. Table of Equivalent Populations of North American Small Mammals. *American Midland Naturalist* 37:223–49.

Ranganathan, J., K. M. A. Chan, K. U. Karanth, and J. L. D. Smith. 2008. Where Can Tigers Persist in the Future? A Landscape-Scale, Density-Based Population Model for the Indian Subcontinent. *Biological Conservation* 141:67–77.

Sanderson, E., J. Forrest, C. Loucks, J. Ginsberg, E. Dinerstein, J. Seidensticker, and P. Leimgruber et al. 2006. *Setting Priorities for the Conservation and Recovery of Wild Tigers: 2005–2015. The Technical Assessment*. Washington, DC: Wildlife Conservation Society / Smithsonian National Zoological Park Conservation & Research Center.

Seidensticker, J. 1976. Ungulate Populations in Chitawan Valley, Nepal. *Biological Conservation* 10:183–210.

———. 2010. Saving Wild Tigers: A Case Study in Biodiversity Loss and Challenges to be Met for Recovery beyond 2010. *Integrative Zoology* 5:285–93.

Seidensticker, J., E. Dinerstein, S. P. Goyal, B. Gurung, A. Harihar, A. J. T. Johnsingh, A. Mandandhar, C. McDougal, B. Pandav, S. Shrestha, D. Smith, M. Sunquist, and E. Wikramanayake. 2010a. Tiger Range Collapse at the Base of the Himalayas: A Case Study. In *Biology and Conservation of Wild Felids*, ed. D. W. Macdonald and A. J. Loveridge, 305–23. New York: Oxford University Press.

Seidensticker, J., B. Gratwicke, and M. Shrestha. 2010b. How Many Wild Tigers Are There? An Estimate for 2008. In *Tigers of the World: The Science, Politics, and Conservation of* Panthera tigris, ed. R. Tilson and P. Nyhus, 2nd ed., 293–98. New York: Elsevier / Academic Press.

Seidensticker, J., K. M. Tamang, and C. W. Gray. 1974. The Use of CI-744 to Immobilize Free-Ranging Tigers and Leopards. *Journal of Zoo Animal Medicine* 5:22–25.

Simcharoen, S., A. Pattanavibool, K. U. Karanth, J. D. Nichols, and N. Samba Kumar. 2007. How Many Tigers *Panthera tigris* Are There in Huai Kha Khaeng Wildlife Sanctuary, Thailand? An Estimate Using Photographic Capture-Recapture Sampling. *Oryx* 41:447–53.

Smith, J. L. D. 1993. The Role of Dispersal in Structuring the Chitwan Tiger Population. *Behaviour* 124:165–95.

Smith, J. L. D., C. McDougal, S. C. Ahearn, A. R. Joshi, and K. Conforti. 1999. Metapopulation Structure of Tigers in Nepal. In *Riding the Tiger: Tiger Conservation in Human-Dominated Landscapes*, ed. J. Seidensticker, S. Christie, and P. Jackson, 178–89. Cambridge: Cambridge University Press.

Smith, J. L. D., C. McDougal, B. Gurung, N. Shrestha, M. Shrestha, T. Allendorf, A. Joshi, and N. Dhakal. 2010. Securing the Future for Nepal's Tigers: Lessons from the Past and Present. In *Tigers of the World: The Science, Politics, and Conservation of* Panthera tigris, ed. R. Tilson and P. Nyhus, 2nd ed., 329–42. New York: Elsevier / Academic Press.

Smith, J. L. D., C. W. McDougal, and M. E. Sunquist. 1987. Female Land-Tenure System in Tigers. In *Tigers of the World: The Biology, Biopolitics, Management, and Conservation of an Endangered Species*, ed. R. L. Tilson and U. S. Seal, 97–109. Park Ridge, NJ: Noyes Publications.

Smith, J. L. D., M. E. Sunquist, K. M. Tamang, and P. B. Rai. 1983. A Technique for Capturing and Immobilizing Tigers. *Journal of Wildlife Management* 47:255–59.

Sunquist, M. E. 1981. The Social Organization of Tigers (*Panthera tigris*) in Royal Chitwan National Park, Nepal. *Smithsonian Contributions to Zoology* 336:1–98.

Wegge, P., and T. Storaas. 2009. Sampling Tiger Ungulate Prey by the Distance Method: Lessons Learned in Bardia National Park, Nepal. *Animal Conservation* 12:78–84.

Wikramanayake, E., E. Dinerstein, C. J. Loucks, D. M. Olson, J. Morrison, J. Lamoreux, M. McKnight, and P. Hedao. 2002. *Terrestrial Ecoregions of the Indo-Pacific: A Conservation Assessment*. Washington, DC: Island Press.

Wikramanayake, E., A. Manandhar, S. Bajimaya, S. Nepal, G. Thapa, and K. Thapa. 2010. The Terai Arc Landscape: A Tiger Conservation Success Story in Human-Dominated Landscapes. In *Tigers of the World: The Science, Politics, and Conservation of* Panthera tigris, ed. R. Tilson and P. Nyhus, 2nd ed., 163–80. New York: Elsevier / Academic Press.

Ecology and Distribution of Sympatric Asiatic Black Bears and Sun Bears in the Seasonally Dry Forests of Southeast Asia

Robert Steinmetz

The sun bear (*Helarctos malayanus*) and Asiatic black bear (*Ursus thibetanus*) are sympatric throughout most of Myanmar, Thailand, Lao PDR, Cambodia, Vietnam, and portions of northeast India, Bangladesh, and southern China (Servheen 1999; Chauhan 2006; Sarker 2006). Sun bears, at 40–60 kg, are about half the size and mass of black bears (65–150 kg), but both species are opportunistic omnivores that seem ecologically and behaviorally similar (Lekagul and McNeely 1988). The widespread geographic overlap of black bears and sun bears coincides largely with the dry forest ecosystem of Southeast Asia (Figure 1). The dry forest ecosystem is defined by a pronounced seasonal climate that, overlain upon diverse local conditions of soil and topography, produces a landscape mosaic of closely juxtaposed seasonal evergreen and deciduous forest types (van Steenis 1950; Rundel and Boonpragob 1995). These distinct plant communities—known as semi-evergreen, mixed deciduous, and dry dipterocarp forest—are the shared habitat of black bears and sun bears. To the north of this region in China, where temperate forests replace tropical forests, only black bears occur. To the south in peninsular Thailand and Malaysia, where rain forests replace seasonal dry forests as a result of consistent monthly rainfall, only sun bears are found. Here I explore the ecology and distribution of sympatric sun bears and black bears in the dry forest mosaic of Southeast Asia.

The term *dry forest* in this region is often used to describe *dry dipterocarp forest*, a distinctive open plant community dominated by deciduous trees of the Dipterocarpaceae family. I used a more-inclusive definition of dry forest that includes other natural habitat types such as semi-evergreen and mixed deciduous forests. From an ecological perspective these forest types are integral to the landscape, especially for large wide-ranging mammals that are habitat generalists, such as bears.

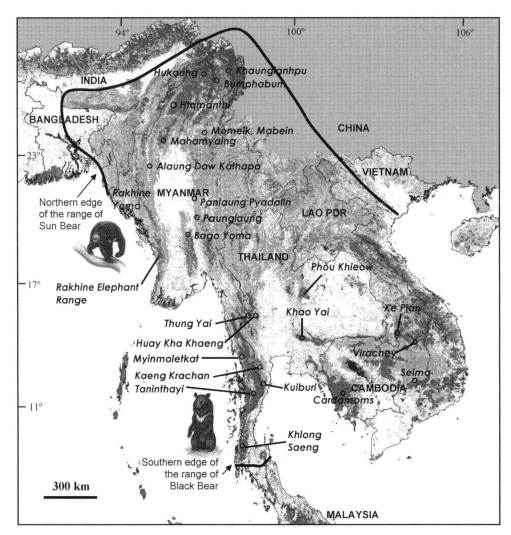

Figure 1. Forest cover map of mainland Southeast Asia, showing distribution limits and geographic overlap of sun bears and black bears. Study sites referred to in the text are shown. This map shows the landscape mosaic of contrasting evergreen and deciduous forest types that characterizes this seasonal tropical region. Dark gray indicates evergreen forest types, and light gray indicates deciduous forest (mixed deciduous and dry dipterocarp). Forest cover base map courtesy of H-J. Stibig.

Dry tropical forests have received little research attention compared with rain forests, despite their ecological uniqueness and the high degree of threat they face from commercial logging and agricultural conversion (Murphy and Lugo 1986; Rundel and Boonpragob 1995; Trisurat 2007). In a similar sense, little field research has been directed at sun bears and black bears in this region, and their ecologies remain poorly known. Both bear species are threatened with extinction due to forest loss and commercial poaching for the wildlife trade (IUCN 2008).

I used two sources of information to investigate bear ecology and distribution. First, I drew from a recent field study of habitat use and feeding ecology of sympatric sun bears and black bears in Thailand to describe and compare patterns of habitat use, then discuss feeding ecology with particular emphasis on foods in mixed deciduous forest. Second, I examined mostly unpublished camera-trap records of bears from sites across the dry forest region. I used this information to explore patterns in distribution and abundance of each species in relation to environmental variables.

METHODS

Field Study in Thailand

Study Site

The 3,622 km² Thung Yai Naresuan Wildlife Sanctuary (Thung Yai) is in western Thailand adjacent to Myanmar (Figure 1). The sanctuary is mountainous with elevations up to 1,811 m. There are three seasons: cool and dry (November to February), hot and dry (March to May), and rainy (June to October). Mean annual rainfall ranges from 1,700 to 2,400 mm, concentrated between July and September. Temperatures reach 34°C in April and descend to 12°C in January.

Thung Yai is predominantly a mosaic of mixed deciduous forest (MDF) and semi-evergreen forest (SEF), so it is a useful representation of Southeast Asia's dry forest mosaic. Semi-evergreen forest is tall, with a closed canopy at 25–40 m formed predominantly by evergreen tree species (Maxwell 1995). Main tree families in this species-rich forest type include Annonaceae, Meliaceae, and Myristicaceae (van de Bult 2003). Mixed deciduous forest (Figure 2c), in contrast, is characterized by deciduous tree species, especially from Leguminosae, Combretaceae, Dilleniaceae, and Lythraceae (Rundel and Boonpragob 1995; van de Bult 2003). Canopy height can reach 30 m, but tree density is lower than in SEF.

I conducted field work from 2001 to 2003 at three study sites (500–1,000 m elevation) with mosaics of SEF and MDF. Sites were 15 to 30 km apart, and encompassed 30–100 km². I established multiple sites to maximize the number of individual bears that would be reflected in these population-level data. Annual home ranges of adult sun bears and black bears encompass 6–21 km² (Wong et al. 2004) and 30–150 km² (Reid et al. 1991; Hazumi and Koyama, in Garshelis 2004), respectively, and individual home ranges overlap widely. Thus, I expected the size and distribution of study sites to reflect the activities of many individuals of each species. Sampling covered each of three seasons and data were pooled for analysis.

Observation of Bear Sign

Bears are rarely observed in the wild but leave abundant sign that is conspicuous, long lasting, and related mostly to feeding (Figure 2). Habitat use and diet of bears was determined from observations of bear sign. Similar sign-based methods have been used

Figure 2. Bears in mixed deciduous forest: (*a*) Sun bear in riparian semi-evergreen forest along stream in mixed deciduous forest (note background); (*b*) Asiatic black bear in mixed deciduous forest; (c) Mixed deciduous forest; (*d*) Fresh and old bear claw marks on a climbed tree; (e) Fruits of *Schleichera oleosa*, a dry forest species eaten by bears; (f, g) *Cassia fistula* pods and bear scat from feeding on them; (h) Raided stingless bee nest (*Trigona* sp.); (*i*) Broken branch from bear feeding on acorns of *Quercus kerrii*; (j) Fruits of *Syzygium cumini*. (Photos were taken in Thung Yai Naresuan Wildlife Sanctuary, Thailand, and are courtesy of the author; bear photos were taken by Wanlop Chutipong.)

to assess habitat use and relative abundance of bears elsewhere (Garshelis et al. 1999; Cuesta et al. 2003; Poscillico et al. 2004).

Bear sign was recorded in straight, 300 m strip transects dispersed throughout the study sites using a systematic random design. I conducted 38 transects in SEF and 27 in MDF, covering 29.4 ha in which approximately 14,000 trees were examined for claw marks and all ground area was searched for terrestrial bear signs. Transects were 10 m wide in SEF (0.3 ha) and 20 m wide in MDF (0.6 ha). Wider transects were used in MDF because tree density was lower in that habitat.

Main categories of bear sign were claw marks on climbed trees, terrestrial insect-feeding signs (holes dug for nests of wasps, ants, and termites; raided termite mounds; broken-apart rotten logs), and scats. Climbed trees were identified, and canopies examined with binoculars to determine presence of fruit as well as feeding sign such as broken branches or excavated nests of stingless bees (*Trigona* spp.) (see the appendix). I collected scats opportunistically, and calculated frequency of occurrence (FO) of different food items as FO_i (%) = $(n_i/N) \times 100$, where N was total number of scats and n_i the number of samples containing food item i.

Assessing Bear Frugivory

Claw marks on climbed trees were the predominant sign type. Sun bears and black bears may climb trees to feed, rest, or escape danger. I considered climbing to be associated with feeding if (1) multiple ages of claw marks were present (indicating seasonal revisiting), (2) broken branches occurred in the canopy (branches are broken to reach fruit), or (3) fresh claw marks coincided with fruiting in the climbed tree. Bears also feed on fallen fruits on the ground (Fredriksson et al. 2006); since this behavior leaves little trace, my results represent an incomplete subset of resource use by bears, though scats helped to fill this gap. The number of new families recorded approached an asymptote after 12 transects; thus, survey effort was probably sufficient to record the diversity of trees climbed by bears.

I conducted a comprehensive literature review of direct feeding evidence (scats, sightings of bears feeding) for sun and black bears, augmented with field observations compiled from biologists in the region. I used this information to help interpret patterns in my data, especially where direct evidence (e.g., scats) was lacking.

Distinguishing Bear Species and Aging Sign

Claw marks from Asiatic black bears tend to be larger than those of sun bears. The widths of sets of hind-foot claw marks (a group of 3–5 claw marks from the foot of a bear; Figure 2d) were measured and classified as either black bear, sun bear, or indeterminate, based on a discriminant function model developed from captive bears (Steinmetz and Garshelis 2008). Bear marks that were old and stretched with tree growth, incomplete (less than 3 claws imprinted on the bark), or from front feet (diagonally aligned on the trunk; hind feet marks are horizontal) were not identified to species.

This method resulted in an overall claw mark sample of bear-climbed trees, and a subsample of mostly within-year marks in which the bear species could be identified. Bear footprints at insect-feeding sites, where present, were considered to be from black bears if hind pad width was greater than 10 cm and total length was greater than 17 cm; from sun bears if measurements were below these thresholds; and indeterminate if length and width matched different species (R. Steinmetz, unpublished data from research with captive animals). Ages of claw marks on trees were assessed as fresh (less than 3 months old), recent (3–10 months), or old (more than 10 months old) based on the results of a mark aging experiment (2001–2002) that identified physical changes in claw marks over time (Steinmetz and Garshelis 2010). Thus, I could separately analyze sign that was very fresh versus older.

Data Analysis

Resource use by bears was expressed in terms of density of climbed trees and insect-feeding events per hectare. Many tree families used by bears were represented by just one or two climbed trees (these may have been rest or refuge trees) so I focused on frequently climbed families that comprised more than 2 percent of all records in a forest type. Overall differences in proportions of sun bear to black bear climbing activity among tree families and genera were tested using Fisher exact tests.

Habitat use was assessed using combined sign density, which I considered a measure of foraging activity. To describe bear habitat use at different temporal scales I analyzed fresh signs and recent signs together (encompassing foraging activity within the year), and just the subset of fresh signs (representing foraging activity within the previous 3 months). I compared sign density using Mann-Whitney U-tests and by examining 90 percent confidence intervals; I inferred statistical significance if intervals did not overlap. Results are presented as mean ± standard deviation.

Regional Patterns of Occurrence and Relative Abundance

Camera traps can indicate relative abundance of a species with the assumption that photo detection rates reflect animal abundance (Carbone et al. 2001; Moruzzi et al. 2002). Although the relationship between photograph rate and animal abundance is untested (Jennelle et al. 2002), I considered it worthwhile to explore patterns in available camera-trap data since so little is known about the distribution and abundance of bears in this region. I obtained data from twenty-four sites in Myanmar, Thailand, Cambodia, and Lao PDR (Table 1; Figure 1) where recent camera-trap surveys were conducted (1994–2009; Figure 2). Mean estimated area surveyed at a site was 337 km^2 (range: 88–2,000 km^2); mean months surveyed was 6.6 (range: 1.5–36); and mean total number of camera locations was 94 (range: 30–220). Most sites had mosaics of deciduous and evergreen forest or were set within larger landscapes of deciduous forest. Sites in southern Myanmar and Thailand were predominantly evergreen and mesic; I included them as geographical and ecological outliers for comparison. Baits were not used in any study except at a few camera locations in Xe Pian. Nonconsecutive pho-

Table 1. Characteristics of camera-trap surveys and survey sites in Southeast Asia (arranged by decreasing black bear:sun bear photograph ratio). Elevation refers to midpoint of highest and lowest camera stations.

Survey Site	Latitude	Mean Annual Rainfall (mm)	Elevation (m)	Season Surveyed	Main Habitat Surveyed	Trap Nights	Number of Photos		Black:Sun Photo Ratio	Source
							Black Bear	Sun Bear		
Alaung Daw Kathapa NP, Myanmar	22.35	1,507	640	wet	mosaic	1,621	7	2	3.5	Lynam 2003; WCS unpub. data
Phou Khieow WS, Thailand	16.33	1,400	950	both	eg	1,224	7	3	2.3	Grassman et al. 2006
Htamanthi WS, Myanmar	25.33	3,491	245	wet	eg	1,875	4	2	2.0	Lynam 2003; WCS unpub. data
Khao Yai NP, Thailand	14.30	2,270	800	both	eg	6,172	20	18	1.1	Jenks and Damrongchainarong 2006
Huay Kha Khaeng WS, Thailand	15.42	1,500	700	dry	mosaic	5,469	5	5	1.0	Conforti 1996; Sukmasuang 2001[a]
Thung Yai Naresuan WS, Thailand	15.42	2,200	923	both	mosaic	4,569	5	6	0.8	Chutipong, unpub. (2008)
Kaeng Krachan NP, Thailand	12.88	1,737	457	both	eg	4,493	8	10	0.8	Ngoprasert 2004
Panlaung catchment and Pyadalin Cave WS, Myanmar	20.00	1,905	1,130	wet	decid	1,292	4	7	0.6	Lynam 2003; WCS unpub. data
Klong Saeng WS, Thailand	9.27	2,718	400	both	eg	2,002	13	40	0.3	Kanchanasakha and Buanum 2002
Kuiburi NP, Thailand	12.20	1,737	400	dry	eg	1,907	1	6	0.2	Steinmetz et al. 2007
Cardamom Protected Forest, Cambodia	10.50	3,000	665	dry	eg	4,090	2	21	0.1	Namyi et al. 2006
Virachey NP, Cambodia	14.50	1,700	475	dry	eg	2,361	0	13	0.0	Seng Teak 2001
Rakhine Yoma Elephant Range, Myanmar	21.30	—	380	dry	eg	991	0	29	0.0	Lynam 2003; WCS unpub. data

(Continued)

Table 1. Characteristics of camera-trap surveys and survey sites in Southeast Asia (arranged by decreasing black bear:sun bear photograph ratio). Elevation refers to midpoint of highest and lowest camera stations. (Continued)

Survey Site	Latitude	Mean Annual Rainfall (mm)	Elevation (m)	Season Surveyed	Main Habitat Surveyed	Trap Nights	Number of Photos		Black:Sun Photo Ratio	Source
							Black Bear	Sun Bear		
Bago Yoma, Myanmar	19.13	3,235	390	dry	decid	987	0	1	0.0	Lynam 2003; WCS unpub. data
Bumphabum WS, Myanmar	26.50	2,339	775	dry	eg	980	0	8	0.0	Lynam 2003; WCS unpub. data
Myinmoletkat Taung Foothills, Myanmar	11.68	4,127	225	dry	eg/sec	959	0	2	0.0	Lynam 2003; WCS unpub. data
Rakhine Yoma, Myanmar	18.07	—	505	dry	eg	931	0	10	0.0	Lynam 2003; WCS unpub. data
Khaunglanhpu, Myanmar	26.82	—	1,355	dry	eg	896	0	1	0.0	Lynam 2003; WCS unpub. data
Hukaung Valley, Myanmar	26.60	2,339	260	dry	eg	881	0	5	0.0	Lynam 2003; WCS unpub. data
Taninthayi, Myanmar	13.50	4,127	90	dry	eg/sec	786	0	1	0.0	Lynam 2003; WCS unpub. data
Momeik and Mabein, Myanmar	23.83	1,338	460	wet	decid	618	0	1	0.0	Lynam 2003; WCS unpub. data
Mahamyaing RF, Myanmar	23.60	1,460	470	dry	mosaic	542	0	3	0.0	Lynam 2003; WCS unpub. data
Xe Pian Protected Area, Lao PDR	14.35	—	275	dry	mosaic	200	0	1	0.0	Steinmetz 1997
Paunglaung Catchment, Myanmar	21.07	1,905	715	wet	decid	838	0	0	n/a	Lynam 2003; WCS unpub. data

Note: eg = evergreen; decid = deciduous; mosaic = both evergreen and deciduous habitat; sec = secondary forest; NP = national park; WS = wildlife sanctuary; RF = reserve forest.
[a]Data pooled from these sources.

tos were considered independent, and bear photos were obtained so sparsely in most studies that almost all could be considered independent.

Although camera traps accumulate the large observational effort needed to detect low-density, difficult-to-observe species like bears (see Zaw et al. 2008), inferences about bear relative abundance among sites are hindered by environmental and survey-design differences (e.g., season, forest type, food availability, camera placement) between sites and survey periods that could affect detection rates. Therefore, I used the ratio of the number of black bear photographs to sun bear photographs at each site as an index of site-specific relative abundance between the species (instead of photos per trap-night of effort); such ratios should minimize bias from site or survey differences since both bear species would have faced similar conditions at a site. However, an assumption remains that detection probabilities of the species at each site are similar. This may be reasonable since sun bears and black bears have broadly similar diets, track the same sequence of fruit availability, forage extensively in trees, and tend not to use trails extensively (R. Steinmetz, personal observation, and see Results below); these shared behaviors should cause detection probabilities to converge. However, larger bears tend to move greater distances and thereby tend to have higher encounter rates with cameras (Mace 1994). Also, sun bears might be relatively more arboreal than black bears (Lekagul and McNeely 1988). For these reasons black bears might be expected to have somewhat higher detection probabilities than sun bears overall.

I used multiple linear regression to explore the relationship between relative abundance and three spatial or environmental variables: latitude, mean annual rainfall, and elevation (midpoint of elevation range at which cameras were set). I examined all possible combinations of these predictors (n = 7 models), and compared models using their coefficients of determination (r^2). Eight sites had three or fewer photographs of bears in total. Such small samples suffer from high probabilities of obtaining zero photographs of a species even if it is present (for example, if two photos were obtained, possible ratios would be 1:1, 0:2, or 2:0; if the actual ratio in the population was 1:1, the odds of getting either 0:2 or 2:0 would be equal to the odds of getting 1:1). Therefore, I limited analysis to sites with at least five photographs of bears (n = 16 sites); at this threshold, the probability of zero photographs of a species was reasonably low (i.e., 6 percent, if species are equally abundant). Independent variables were uncorrelated (Spearman's r < 0.32, p > 0.24). Data (photo ratios) were positively skewed, but residuals were reasonably normally distributed and had equal variance. Thus, regression assumptions were met.

RESULTS

Habitat Use by Bears in Thung Yai

I recorded 657 bear signs in Thung Yai: 555 climbed trees, 25 opened *Trigona* bee nests, 44 terrestrial insect-feeding signs (27 within transects, 17 observed opportunistically off transects), and 33 scats. Claw marks were the predominant sign in each habitat,

comprising 84 percent of the sample in MDF and 94 percent in SEF. Forty-seven percent of the claw marks (n = 270 trees, including trees climbed for *Trigona* bees) were
judged to have been created within the year (fresh and recent); and 20 percent were fresh
(created within 3 months). Forty-six percent of claw marks (n = 267) were sufficiently
distinct and complete to measure for species classification: 155 were identified as sun
bear, 104 as black bears, and 8 were indeterminate. Thirty (43 percent) insect-feeding
signs (including trees climbed for insects) could be identified. This identified subsample
was used to make inferences about habitat use and feeding ecology of each bear species.

Foraging activity of each bear species was significantly higher in SEF than in
MDF: fresh signs were about twice as abundant, and within-year signs were 3–4
times more abundant in SEF, reaching 9.2 ± 7.3 signs/ha for sun bears and 5.8 ±
5.7 for black bears (Figure 3). Old signs (not shown), which reflect multiple years of
foraging prior to the study period, were also significantly more abundant in SEF. Sun
bears produced 19–37 percent more signs per hectare than black bears on average in
each habitat (Figure 3); this probably reflects somewhat higher sun bear abundance,
given that foraging behavior (from which signs are produced) of the species was
similar. However, confidence intervals between species overlapped widely in most
cases (Figure 3).

Evidence of feeding (broken branches, multiple claw-mark ages on a tree, fresh
climbing on a fruiting tree) occurred on 70 percent of freshly climbed trees (n = 86/123),
indicating that bears climbed mostly to feed on fruits. Likewise, most scats contained
only fruit (n = 26/33). This evidence indicates that both species were mainly frugivorous in Thung Yai. Bears climbed trees from 29 families in MDF in Thung Yai; 14 of
these were frequently climbed (more than 2 percent of records), including at least 32
genera and 47 species (Table 2). Similar genera or species from most of these same
families had been documented as producing bear foods in other studies (Table 2).

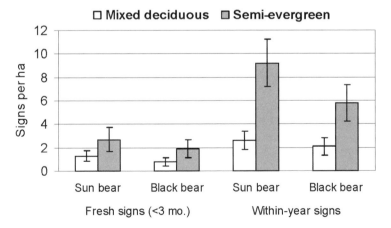

Figure 3. Mean density (± 90% confidence intervals) of sun bear and black bear signs in mixed deciduous
and semi-evergreen forest, Thung Yai Naresuan Wildlife Sanctuary, Thailand, 2001–2003. Fresh signs (less
than 3 months old) are a subset of within-year signs (less than 1 year old).

Table 2. Fruit characteristics and bear feeding evidence of plant families used by sun bears and black bears in mixed deciduous forest, Thung Yai Naresuan Wildlife Sanctuary, Thailand, 2001–2003. Bold font indicates frequently climbed tree families (i.e., accounting for more than two percent of climbed trees).

Tree Family	Fruit Type	Part Eaten	Feeding Evidence[a]	Sun Bear	Black Bear	Bear sp.[b]
Anacardiaceae	drupe	fruit	scat, obs.	3	7	16
Bignoniaceae	capsule	unknown	claw marks	16	16	
Burseraceae	drupe	fruit	scat, obs.	5, 9, 3		19
Celastraceae	aril	unknown	claw marks	16	16	
Combretaceae	drupe	fruit, leaf, flower	scat, obs., branch	16	16	4[c]
Dilleniaceae	aril	fruit	obs.	9	23	
Dipterocarpaceae	n/a	nesting bees	raided bee nests			16
Ebenaceae	berry	fruit	scat, obs.	3, 16	7, 13	16, 18
Elaeocarpaceae	drupe	fruit	scat			6[c]
Euphorbiaceae	drupe, berry	fruit	scat, obs.	3, 9	7	16, 18
Fagaceae	nut	nuts	scat, obs.	3, 9	1, 7, 2, 8, 12, 13, 16	16
Labiatae	drupe	fruit	scat, obs., branch		16, 22	16
Lauraceae	berry	fruit	scat, obs.	3, 9	7, 10, 12, 13	16, 18
Leguminosae	pod	fruit	scat, obs.	3, 16	16, 24, 8	16, 14[c]
Magnoliaceae	aril	fruit, flower, bark	obs.	3	15	
Meliaceae	aril	fruit	scat	3, 9	20	16
Moraceae	berry	fruit	scat, obs.	3, 5, 9	7	16
Myrtaceae	berry	fruit	scat, obs., branch	3, 9, 16		16, 18, 14[c]
Oleaceae	drupe	unknown	claw marks			16
Pandanaceae	compound	shoot	footprints, feeding sign	3, 16	16	
Rhamnaceae	drupe	fruit	scat		17	16
Rosaceae	drupe	fruit	scat, obs.	21	7, 2, 11	
Rubiaceae	compound	fruit	scat			16
Sapindaceae	aril	fruit	scat, obs., branch	3, 9	7, 16	16
Sonneratiaceae	capsule	unknown	claw marks			16
Theaceae	berry	unknown	claw marks		16	
Tiliaceae	berry	fruit	scat, obs.	3		18, 14[c]

Sources: (1) Sathyakumar and Viswanath 2003; (2) Manjrekar 1989; (3) Fredriksson et al. 2006; (4) Chhangani 2002; (5) McConkey and Galetti 1997; (6) Sreekumar and Balakrishnan 2001; (7) Hwang et al. 2002; (8) Saberwal 1989; (9) Wong et al. 2002; (10) Kitamura 2000; (11) Huygens and Hayashi 2001; (12) Reid et al. 1991; (13) Schaller et al. 1989; (14) Joshi et al. 1997; (15) Nozaki et al. 1983; (16) this study. (17) Corlett 1998; (18) Kitamura et al. 2002; (19) Kitamura et al. 2005; (20) Prawing Kinklay, personal communication; (21) Wangwon Sangkametawee, personal communication; (22) Loong Nong, personal communication; (23) Mon Tu-U, personal communication; (24) Naim Ahktar, personal communication.
[a]Obs. = direct observation of a bear feeding; branch = broken branches in tree canopy; claw mark evidence includes multiple claw mark ages on the same tree and/or fresh marks that coincided with fruiting and is only reported if direct feeding evidence (scat, broken branches) does not exist for that family.
[b]Feeding evidence not directly attributable to a species, and could pertain to sun bear, black bear, or both.
[c]Data for sloth bear (*Melursus ursinus*) included for comparison.

Diet Overlap between Sun Bears and Black Bears

Black bears and sun bears used similar tree families and genera in MDF. All frequently climbed families were climbed by both species (no signs could be identified in Euphorbiaceae; Figure 4), and I detected no differences in the ratio of black bear to sun bear climbing activity among tree families (Fisher's exact test: p = 0.27) or genera (p = 0.73). Additionally, tree families climbed most by one species tended to be selected at similar rates by the other species (Spearman's r = 0.6, p = 0.02). As in MDF, tree use by bears in SEF overlapped extensively (100 percent overlap among families).

What Fruits Do Bears Eat?

Most of the frequently climbed tree families produce fleshy fruits, typically berries, drupes, or arillate capsules (Table 2). The most commonly climbed families in MDF were Leguminosae (1.2 climbed trees/ha), Labiatae (1.1 climbed trees/ha), and Dilleniaceae (0.8 climbed trees/ha; Figure 4); these accounted for 24 percent of all climbed trees. The most commonly climbed genus in MDF was *Dillenia* (Dilleniaceae, 0.86 climbed trees/ha), which produces sweet, medium-sized (1–4 cm) fruits eaten by bears in the dry season. Other MDF fruits eaten included *Cassia fistula* (Leguminosae), *Syzygium cumini* (Myrtaceae; Figure 2) and two species of *Vitex* (Labiatae: *V. peduncularis*, *V. quinata*). Three oak species (Fagaceae) were climbed in MDF, most commonly *Quercus kerrii* and *Q. brandisiana* (0.49 climbed trees/ha, combined). Bears sometimes broke branches to access acorns of *Q. kerrii* in the rainy season (Figure 2).

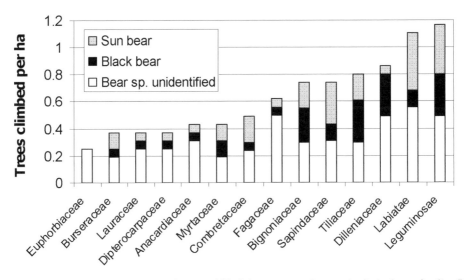

Figure 4. Tree climbing activity by sun bears and black bears among frequently climbed tree families (less than 2 percent of records) in mixed deciduous forest, Thung Yai Naresuan Wildlife Sanctuary, 2001–2003. Bear species was identified from fresh or recent (less than 1 year old) claw marks on climbed trees. The unidentified portion of the data (Bear spp.) was mostly old (more than 1 year old). Most families were climbed to feed on fruit, except Dipterocarpaceae, which was climbed to access stingless-bee nests.

One family, Rhamnaceae (*Ziziphus rugosa*), was not climbed but I found its fruits in scats; bears probably obtained these from the ground. Trees in this family appear too small to climb.

Based on numbers of fruit-tree species used for food, SEF was a somewhat richer habitat for bears than MDF, with 15 frequently climbed families, 49 genera, and 67 species. Mean fruit productivity was slightly higher in SEF (7 ± 6 fruiting trees/ha) than in MDF (5 ± 4 fruiting trees/ha; Steinmetz 2009). In SEF both bear species climbed Lauraceae most (1–2 climbed trees/ha).

Insect-Feeding

Raided stingless-bee nests (n = 25 trees) were more common in MDF (1.1 ± 1.8 raided nests/ha) than in SEF (0.7 ± 1.7 raided nests/ha; $U = 416$, p = 0.06), whereas ground-feeding signs (deadwood opened, nests of ants and termites excavated; n = 27 signs) were three times denser (though highly clumped) in SEF (1.6 ± 3.4) than in MDF (0.6 ± 1.2; $U = 503.5$, p = 0.7). Most identifiable insect-feeding signs (27/30 = 90 percent), whether arboreal (stingless-bees) or terrestrial (digging, log-opening), were from sun bears.

Insect remains occurred in 18 percent (n = 6) of scats. Dung beetles (Scarabaeidae, FO = 9 percent) and ants (Formicidae, FO = 6 percent) were eaten most. Ants included *Cerebera* sp., *Tetraponera ruflonigra*, and *Camponotus* sp.; the latter is a widespread genus in the dry forest ecosystem that inhabits leaf-litter on the forest floor. Bears also dug for larva of underground-nesting wasps (Vespidae, FO = 3 percent) in both MDF and SEF.

Regional Patterns of Occurrence and Relative Abundance

In 46,684 camera-trap nights of survey effort, accumulated over 165 survey months and 7,900 km², sun bears and black bears were photographed 195 and 76 times, respectively (Table 1). On average, 407 ± 307 trap nights were required to obtain a photo of a sun bear, and 874 ± 749 for a black bear. Neither species was photographed at Panlaung Pyadalin, despite bear signs there (A. Lynam, personal communication). Sun bears were recorded at all but 1 site (n = 23), whereas black bears were recorded at less than half (n = 11) of the 24 sites. Site and survey characteristics do not explain the black bear absences, since all sites were within the known distribution of the species, and surveys covered all seasons and a range of habitats that black bears use. The most likely explanation is that black bear density was much lower than sun bears at these sites, and survey effort was insufficient to detect them given this lower density. Indeed, additional camera-trap surveys in Myanmar (not analyzed here) have subsequently recorded black bears at a number of these sites (Saw Htun 2006). Numbers of photos increased significantly with survey effort (trap nights) across all sites for black bear (Spearman's r = 0.68, p < 0.001), but not sun bear (r = 0.31, p = 0.14); this pattern would be expected if black bears were significantly less abundant.

Bear relative abundance (photograph ratios) was not significantly related to latitude, elevation, or rainfall (range of r² for all models: 0.07–0.11, p > 0.32). However,

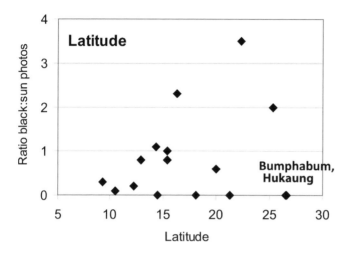

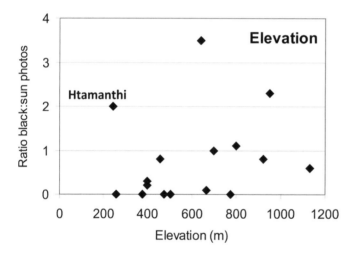

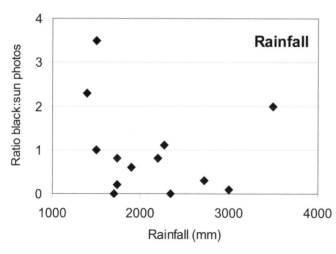

Figure 5. Scatterplots of the relationship between relative abundance of black bears and sun bears (expressed as a ratio of black bear photographs to sun bear photographs obtained by camera trapping) and latitude, elevation, and mean annual rainfall, at sixteen sites in Southeast Asia where camera trapping was conducted (1994–2008). Named points are outliers from apparent positive trends between photograph ratios and latitude and elevation.

scatterplots showed generally positive relationships between photograph ratios and latitude and elevation (Figure 5), but with prominent outliers. These results suggest possible trends, with black bears relatively more common at higher elevations and latitudes, and sun bears more abundant at lower elevations and latitudes. Ratios peaked at Alaungdaw Kathapa in northwest Myanmar (black bear photos to sun bear photos ratio: 3.5), and Phou Khieow (ratio: 2.3), a high-elevation site (950 m; Table 1). Ratios were low at the southernmost sites, Kuiburi (ratio: 0.2) and Khlong Saeng (ratio: 0.3). Major exceptions to these trends (named points in Figure 5) were Htamanthi—a low-elevation site (245 m) with abundant black bears (ratio: 2.0)— and Bumphabum and Hukaung, both of which are high-latitude sites with abundant sun bears (ratio: 0.0).

I detected no pattern of temporal partitioning between sun and black bears from surveys in which I examined photographs (Khao Yai, Thung Yai, Kuiburi, Xe Pian); both species were photographed during day and night.

DISCUSSION

Dry Forests from a Bear's Perspective

Sun bears and black bears co-occurred in MDF and SEF habitats of Thung Yai's dry forest mosaic. Both species were primarily frugivorous in each habitat and overlapped extensively in fruit selection. Despite these diet similarities, sun bears appeared to be more insectivorous than black bears, consistent with previous reports (Lekagul and McNeely 1988). Foraging for insects by bears was a relatively large proportion of the sign sample in MDF compared to SEF, perhaps compensating for lower fruit availability in MDF. However, density of diggings and broken-open logs was higher in SEF than in MDF. This may stem from (1) higher tree density in SEF, which produces more logs; and (2) the occurrence of annual burning in MDF, which reduces the availability of logs as a substrate for insects in MDF.

Contrary to sun bear diets elsewhere in Southeast Asia, the predominant food of sun bears in Thung Yai was fruit. In Borneo rain forests, the primary food of sun bears throughout the year is insects, except during mast-fruiting events, when bears are almost completely frugivorous (Wong et al. 2002; Fredriksson et al. 2006). These divergent foraging patterns result from profound underlying differences in climate and phenology between mainland and Sundaic portions of Southeast Asia. Sun bears in the aseasonal rain forests of Malaysia and Indonesia are mainly insectivorous because fruits are only available every three to seven years during synchronous mast-fruiting events (Primack and Corlett 2005). In contrast, most tree species in seasonal tropical forests such as Thung Yai have consistent, asynchronous patterns of annual fruiting that make fruit available through much of each year (Steinmetz 2009; Elliot et al. 1994). As a result, the relative contribution of insects and fruit to bear diets is essentially reversed between the dry forest and rain forest ecosystems of Southeast Asia.

Dry forest plant communities differ substantially in tree density, diversity, and structure. For example, in Thung Yai tree density and basal area in SEF (559 trees/ha,

43 m²/ha) are almost twice that in MDF (Webb 2007; Steinmetz 2009). Despite these differences, tree climbing was the main foraging activity in both MDF and SEF because each habitat provides numerous fruit-tree species eaten by bears.

Bear foraging sign was more abundant in SEF than in MDF, probably because SEF has a higher density and diversity of fruit-tree species (Dokrak et al. 1999; Bunyavejchewin et al. 2001, 2002; Webb 2007). However, within the shorter time frame reflected by fresh signs, differential habitat use was less pronounced, with each species creating about 1–3 signs/ha in MDF and SEF (Figure 3). This more equable short-term habitat use was reflected in habitat selection models from a concurrent study (Steinmetz et al. in press) in which neither habitat composition (percentage SEF within home-range sized circles around transects) nor forest type predicted habitat use of either bear species; density of fruiting trees was the sole significant predictor. These results may indicate the importance of habitat heterogeneity for these frugivorous bears, and the complementary nature of evergreen and deciduous forest types. Only 21 percent of fruit-tree genera used by bears, and 9 percent of species, were shared between MDF and SEF. Thus, each habitat provides a unique assemblage of fruit, and in combination they offer a more-diverse food supply than any one habitat. Habitat heterogeneity is important for frugivorous animals because it mitigates within- and between-year periods of fruit scarcity that typify tropical forest (Leighton and Leighton 1983; Whitmore 1998).

MDF fruit species eaten by bears in Thung Yai, such as the pods and fleshy fruits from Leguminosae, Dilleniaceae, and Labiatae, are generally widespread and abundant in MDF and dry dipterocarp forest throughout mainland Southeast Asia (Vidal 1979; Rundel 1999). For example, at another MDF site in Thailand, bear food-tree species (identified from Thung Yai) had a combined density of 41 trees/ha (24 percent of all stems) and comprised 34 percent of total basal area (Dokrak et al. 1999).

Stingless bee feeding in Thung Yai was higher in MDF than in SEF. This might reflect a higher availability of nests in MDF, though no data on bee abundance exists for dry forests. Stingless bee abundance is determined by the amount of pollen and nectar that bees feed on, and sufficient tree cavities for nesting (Eltz et al. 2002). Stingless bees often nest in the hollow trunks of *Lagerstroemia* spp., a common genus in MDF (R. Steinmetz, personal observation). Opened nests in Xe Pian (Lao PDR, Figure 1) were common in MDF and tended to be in *Lagerstroemia* trees; a sun bear I observed there appeared to be investigating a stingless bee nest at the base of a large *Lagerstroemia* (Steinmetz 1997).

To access a stingless bee nest, a bear must bite and tear deep into a tree trunk. This process requires many hours of strenuous effort, so the nutritional reward from honey and larva must be very high. All bear foraging signs on stingless bee nests I could identify were from sun bears; correspondingly, all direct observations of bears feeding on stingless bees I could trace have been of sun bears (D. Ngoprasert and W. Sangkhamethawee, personal communication). Sun bears have disproportionately large canines (Christiansen 2007), a very long tongue, and smaller body size compared to black bears. These attributes probably render sun bears more capable predators of stingless bees than black bears, by enabling them to bite through wood and access the nest while clinging for long periods to the side of a tree.

Regional Distribution and Relative Abundance of Bears

Black bears required over twice the survey effort to detect with camera traps (e.g., trap nights to obtain a photo) as sun bears. This finding is a strong indication that black bears generally occur at lower densities than sun bears, given that the detection rate for black bears was expected to be higher. This disparity in density may reflect differential effects of the long-standing commercial trade in bear parts in Asia. Asiatic black bear is the most valued bear species in traditional Chinese medicine (Servheen 2001), and there is a vigorous international trade in bear parts (paws, gall bladders, canines) in Southeast Asia that preferentially targets black bears (Shepherd and Nijman 2008). Nonetheless, results from this analysis should be viewed cautiously because photograph ratios are unadjusted for potentially different detection probabilities between the species.

Relationships between bear relative abundance (photo ratios) and independent variables were not statistically significant, due mainly to the wide scatter of points in each case (Figure 5), but there stood out visual trends and conspicuous outliers that warranted a closer look. Black bears tended to be more abundant at higher elevations whereas sun bears predominated at lower elevations. This result accords with other recent observations. For example, in the Cardamom Mountains of southwest Cambodia (Figure 1), the only black bears photographed were above 800 m elevation, but only one sun bear was photographed that high, despite being relatively common at lower elevations (Namyi et al. 2006). Second, in a seven-year camera-trapping effort (1999–2006) in Seima Conservation Area (eastern Cambodia, Figure 1), most of which lies below 600 m, sun bears were photographed fifty-five times, and black bears just twice (H. O'Kelly, personal communication). Both black bear photos were taken toward the upper elevation limits of Seima, in SEF at 480 and 540 m. Lastly, at my study site in Thung Yai, 91 percent of bear signs in montane evergreen forest above 1,000 m were from black bears (Steinmetz 2009), in contrast to the more equable sign abundance between species in lower SEF and MDF (Figure 3). Historical accounts by naturalists and hunters also describe black bears as preferring forested hills and mountains, but note that both bear species can occur at all elevations and forest types (Peacock 1933; Lekagul and McNeely 1988).

In Sundaic Southeast Asia, where black bears are absent, sun bears occur up to at least 2,000 m (Tumbelaka and Fredriksson 2006), and were frequently camera-trapped between 700 and 1,940 m in Sumatra, Indonesia (Linkie et al. 2007). Thus, the relative scarcity of sun bears at upper elevations in mainland Southeast Asia does not reflect any inherent constraint of the species. Sun bears might be limited by the predominance of a larger competitor (black bears), perhaps combined with reduced insect availability at higher elevations (Collins 1989; Steinmetz 2009).

The main outlier in the elevation trend, Htamanthi, is a northerly site whose proximity to the edge of the range of sun bears might naturally result in relatively low sun bear abundance regardless of elevation (if sun bears decline toward their range edge; see below). Conversely, the Cardamom Mountains are fairly high elevation yet had few black bears, but this site is quite far south (Table 1). These patterns suggest that the effects of elevation on bear relative abundance are mediated by latitude (i.e., there is an interaction between elevation and latitude).

Scatterplots suggested a latitudinal gradient in the relative abundance of black and sun bears, with black bears the more commonly photographed species at northern sites, and sun bears more common at southern sites. This pattern suggests declining abundance of black and sun bears toward their respective range edges, which is consistent with the idea that range peripheries support lower densities of some species compared to range centers due to less-favorable biotic or abiotic conditions (Guo et al. 2005). The main outliers in this trend were Hukaung and Bumphabum, where only sun bears were photographed despite the very northerly latitude of these sites and their proximity to the edge of sun bear range. It is plausible that these sites have been depleted of black bears because of their proximity to China, a major consumer of bear parts. Indeed, black bears (and not sun bears) were one of the preferred species of hunters interviewed by Rao et al. (2005) in this region.

Data from many more sites along range edges would be needed to adequately assess the effect of latitude on bear abundance. Nonetheless, it is relevant to explore the environmental transitions that coincide with the range edges of black bears and sun bears. The southern range edge for black bears occurs at about 9° N on the Thai peninsula (Vinitpornsawan et al. 2006; Figure 1). Camera-trap photo ratios were low at the southernmost sites in this analysis, Kuiburi and Khlong Saeng, perhaps reflecting declining black bear abundance near its range edge. This region, near the Isthmus of Kra, marks a phytogeographical ecotone between seasonal dry forests to the north and aseasonal evergreen rain forest to the south (Woodruff 2003). This division is probably important for frugivorous black bears because the mast-fruiting cycles of rain forests result in long periods of fruit scarcity (Primack and Corlett 2005). Black bears, with larger body size and presumably greater food requirements than sun bears, might be unable to survive under the less-predictable, resource-poor conditions that prevail in Sundaic rain forests.

Sun bears are very rare at the northern and western edges of their range in southern Yunnan (China), northeast India, and Bangladesh (Chauhan 2006; Gong and Harris 2006; Sarker 2006). The particularly high black bear to sun bear ratios at Alaungdaw Kathapa and Htamanthi in northwest Myanmar near the Indian border (Figure 1; Table 1) might reflect declining ecological suitability for sun bears near their biogeographic boundary. Interestingly, the rarity of sun bears in northwest Myanmar and nearby northeast India was apparent in historical times (Wroughton 1916; Higgins 1932), so is probably not related to human influence or recent habitat changes. The range limit for sun bears coincides closely with the edge of the Indochinese biogeographic subregion; to the north is the Himalayan subregion, and to the northwest lies the Indian subregion (Corbet and Hill 1992). The sun bear might be blocked from the Himalayan subregion by cold climate, and from the Indian subregion by competition with the endemic sloth bear (*Melursus ursinus*), which is highly insectivorous and frugivorous, and can occur at very high local densities (Garshelis et al. 1999; Joshi et al. 1997).

In sum, camera-trap data suggest that (1) black bears occur at lower densities than sun bears overall; (2) sun bears are more abundant at lower elevations, except perhaps near the edge of their range; (3) black bears predominate in montane forests; and (4) the effects of elevation might be mediated by proximity to species range edges. However, there was substantial variation in these trends, probably related to unmeasured effects from hunting and other variables, and perhaps specific field-design aspects such as camera placement.

CONCLUSIONS

Sun bears and black bears co-occur throughout the seasonally dry forest ecosystem of mainland Southeast Asia. This overlap occurs at a broad regional scale, but also locally within the major habitat types that define the landscape. Observations from a field study in Thung Yai suggest that this ecological overlap extends, even more profoundly, down to the level of individual foods within habitat types. Both bears are primarily frugivorous and select similar fruit species in both mixed deciduous and semi-evergreen forests. For bears, these dry forest types are complementary because they provide almost completely different species of fruits, and in combination provide a richer food supply than either one habitat could in isolation. Despite a remarkable degree of overlap, there is evidence for near-exclusive use of resources—sun bears eat more insects whereas black bears predominate in montane forest. This may represent niche partitioning that promotes the coexistence of sun and black bears in mainland Southeast Asia.

ACKNOWLEDGMENTS

My study in Thung Yai was made possible with the field assistance of Wanlop Chutipong, Naret Seuaturien, and Kietiphum Kaewplang; and with permission and support from the Department of National Parks, Wildlife and Plant Conservation, and the National Research Council of Thailand. I thank Paiboon Sawetmelanon and Erb Chirgsaard, previous superintendents of Thung Yai Naresuan Wildlife Sanctuary. I am grateful for the support of a doctoral dissertation fellowship from the University of Minnesota. This chapter would not have been possible without the generous collaboration of field biologists who shared their camera-trap data with me; they are Wanlop Chutipong, Kate Jenks, Budsabong Kanchanasakha, Tony Lynam, Hannah O'Kelly, Annette Olsson, and Seng Teak. Special thanks to Hans-Jurgen Stibig for providing the forest cover map of Southeast Asia, and to the International Bear Association for the bear illustrations. Finally, I thank Bill McShea for inviting me to write this chapter, and David Garshelis, Liu Fang, Karl Malcolm, and Bill McShea for thoughtful reviews and discussion of earlier drafts that much improved the final version.

Appendix. Trees climbed by sun bears and black bears in mixed deciduous forest, Thung Yai Naresuan Wildlife Sanctuary, Thailand, 2001–2003. Frequently climbed families (greater than 2% of records) are in bold.

Family	Genera	Species
Anacardiaceae	Lannea	coromandelica
	Semecarpus	cochinchinensis
	Spondias	pinnata
Bignoniaceae	Fernandoa	adenophylla
	Stereospermum	colais
	Stereospermum	neuranthum
Bombacaceae	Bombax	anceps
Burseraceae	Canarium	subulatum
	Garuga	pinnata
Celastraceae	Lophopetalum	wallichii
Combretaceae	Terminalia	alata
	Terminalia	bellirica
	Terminalia	chebula
	Terminalia	mucronata
Dilleniaceae	Dillenia	ovata
Dipterocarpaceae	Shorea	obtusa
	Shorea	siamensis
Ebenaceae	Diospyros	ehretioides
Elaeocarpaceae	Eleocarpus	sp.
Euphorbiaceae	Aporosa	dioica
	Bridelia	anceps
	Bridelia	glauca
	Macaranga	denticulata
Fagaceae	Castanopsis	tribuloides
	Quercus	brandisiana
	Quercus	kerrii
Labiatae	Callicarpa	arborea
	Gmelina	arborea
	Premna	latifolia
	Premna	villosa
	Vitex	peduncularis
	Vitex	quinata
Lauraceae	Cinnamomum	caudatum
	Cinnamomum	sp.
	Litsea	cf. salicifolia
	Phoebe	lanceolata
Leguminosae	Albizia	chinensis
	Cassia	fistula
	Dalbergia	cf. cana
	Dalbergia	sp.
	Millettia	brandisiana
	Pterocarpus	macrocarpus
Lythraceae	Lagerstroemia	tomentosa
Magnoliaceae	Michelia	baillonii
Meliaceae	Dysoxylum	sp.
Moraceae	Artocarpus	lakoocha
	Ficus	cf. lacor
	Ficus	sp.

Family	Genera	Species
Myrtaceae	Syzygium	cumini
	Syzygium	sp.
Oleaceae	Chionanthes	ramiflorus
Rhamnaceae	Ziziphus	rugosa
Rosaceae	Prunus	cerasoides
Rubiaceae	Wendlandia	tinctoria
Sapindaceae	Dimocarpus	longan
	Litchi	chinensis
	Schleichera	oleosa
Sonneratiaceae	Duabanga	grandiflora
Sterculiaceae	Sterculia	pexa
	Sterculia	urena
Theaceae	Ternstroemia	gymnanthera
Tiliaceae	Berrya	mollis
	Colona	cf. winitii
	Grewia	eriocarpa

REFERENCES

Bunyavejchewin, S. 1983. Analysis of Tropical Dry Deciduous Forest of Thailand. I. Characteristics of Dominance Types. *Natural History Bulletin of the Siam Society* 31:109–22.

Bunyavejchewin, S., P. J. Baker, J. V. LaFrankie, and P. Ashton. 2001. Stand Structure of a Seasonal Dry Evergreen Forest at Huai Kha Khaeng Wildlife Sanctuary, Western Thailand. *Natural History Bulletin of the Siam Society* 49:89–106.

———. 2002. Floristic Composition of a Seasonal Dry Evergreen Forest at Huai Kha Khaeng Wildlife Sanctuary, Western Thailand. *Natural History Bulletin of the Siam Society* 50:125–34.

Carbone, C., S. Christie, K. Conforti, T. Coulson, N. Franklon, J. R. Ginsberg, M. Griffiths, J. Holden, K. Kawanishi, M. Kinnaird, A. Lynam, D. W. Macdonald, D. Martyr, C. McDougal, L. Nath, T. O'Brien, J. Seidensticker, D. J. L. Smith, M. Sunquist, R. Tilson, and W. N. Wan Shahruddin. 2001. The Use of Photographic Rates to Estimate Densities of Tigers and Other Cryptic Mammals. *Animal Conservation* 4:75–79.

Chauhan, N. S. 2006. The Status of Malayan Sun Bears in India. In *Understanding Asian Bears to Secure Their Future*, 50–56. Ibaraki, Japan: Japan Bear Network.

Chhangani, A. K. 2002. Food and Feeding of Sloth Bear (*Melursus ursinus*) in Aravalli Hills of Rajasthan, India. *Tigerpaper* 29:1–6.

Christiansen, P. 2007. Evolutionary Implications of Bite Mechanics and Feeding Ecology in Bears. *Journal of Zoology* 272:423–43.

Collins, N. M. 1989. Termites. In *Tropical Rain Forest Ecosystems: Biogeographical and Ecological Studies*, ed. H. Leith and M. J. A. Werger, 455–71. Amsterdam: Elsevier Science Publishers.

Conforti, K. 1996. The Status and Distribution of Small Carnivores in Huai Kha Khaeng / Thung Yai Naresuan Wildlife Sanctuaries, West-Central Thailand. Master's thesis, University of Minnesota.

Corbet, G. B., and J. E. Hill. 1992. *The Mammals of the Indomalayan Region: A Systematic Review*. New York: Oxford University Press.

Corlett, R. T. 1998. Frugivory and Seed Dispersal by Vertebrates in the Oriental (Indomalayan) Region. *Biological Review* 73:413–48.

Cuesta, F., M. F. Peralvo, and F. van Manen. 2003. Andean Bear Habitat Use in the Oyacachi River Basin, Ecuador. *Ursus* 14:198–209.

Dokrak, M., U. Kutintara, C. Yarwudhi, H. Tanaka, and T. Nakashizuka. 1999. Structural Dynamics of a Natural Mixed Deciduous Forest in Western Thailand. *Journal of Vegetation Science* 10:777–86.

Elliott, S., S. Promkutkaew, and J. F. Maxwell. 1994. Flowering and Seed Production Phenology of Dry Tropical Forest Trees in Northern Thailand. In *Proceedings of an International Symposium on Genetic Conservation and Production of Tropical Forest Tree Seeds*, 52–62. Chiang Mai, Thailand: ASEAN-Canada Forest Tree Seed Project.

Eltz, T., C. A. Bruhl, S. van der Kaars, and K. E. Linsenmair. 2002. Determinants of Stingless Bee Nest Density in Lowland Dipterocarp Forests of Sabah, Malaysia. *Oecologia* 131:27–34.

Fredriksson, G. M., S. A. Wich, and Trisno. 2006. Frugivory in Sun Bears (*Helarctos malayanus*) Is Linked to El Niño–Related Fluctuations in Fruiting Phenology, East Kalimantan, Indonesia. *Biological Journal of the Linnean Society* 89:489–508.

Garshelis, D. L. 2004. Variation in Ursid Life Histories—Is There an Outlier? In *Biology and Conservation of the Giant Panda*, ed. D. Lindburg and K. Baragona, 53–73. Berkeley: University of California Press.

Garshelis, D. L., A. R. Joshi, and J. L. D. Smith. 1999. Estimating Density and Relative Abundance of Sloth Bears. *Ursus* 11:87–98.

Gong, J., and R. B. Harris. 2006. The Status of Bears in China. In *Understanding Asian Bears to Secure Their Future*, 50–56. Ibaraki, Japan: Japan Bear Network.

Grassman, L. I., J. E. Haines, J. E. Jenecka, and M. E. Tewes. 2006. Activity Periods of Photo-Captured Mammals in North-Central Thailand. *Mammalia* 70:306–9.

Guo, Q., M. Taper, M. Schoenberger, and J. Brandle. 2005. Spatial-temporal population dynamics across species ranges: From centre to margin. *Oikos* 108:47–57.

Higgins, J. C. 1932. The Malay Bear. *Journal of the Bombay Natural History Society* 35:673–74.

Huygens, O. C., and H. Hayashi. 2001. Use of Stone Pine Seeds and Acorns by Asiatic Black Bears in Central Japan. *Ursus* 12:47–50.

Hwang, M., D. L. Garshelis, and Y. Wang. 2002. Diets of Asiatic Black Bears in Taiwan, with Methodological and Geographical Comparisons. *Ursus* 13:111–25.

IUCN. 2008. IUCN Red List of Threatened Species (2008), www.iucnredlist.org.

Jenks, K. E., and K. Damrongchainarong. 2006. *Dong Phayayen—Khao Yai Carnivore Conservation Project, Summary report, 2003–2006*. Smithsonian National Zoological Park / WildAid Foundation / Khao Yai National Park / Department of National Parks, Wildlife and Plant Conservation, Bangkok, Thailand.

Jennelle, C. S., M. C. Runge, and D. I. MacKenzie. 2002. The Use of Photographic Rates to Estimate Densities of Tigers and Other Cryptic Mammals: A Comment on Misleading Conclusions. *Animal Conservation* 5:119–20.

Joshi, A. R., D. L. Garshelis, and J. L. D. Smith. 1997. Seasonal and Habitat-Related Diets of Sloth Bears in Nepal. *Journal of Mammalogy* 78:584–97.

Kanchanasakha, B., and S. Buanum. 2002. Monitoring Relative Abundance and Distribution of Large Mammals in Khlong Saeng Forest Reserves, Surat-Thani Province. *Journal of Wildlife in Thailand* 10:39–50 (in Thai).

Kitamura, S. 2000. Seed Dispersal by Hornbills in a Tropical Rain Forest in Khao Yai National Park, Thailand. Master's thesis, Kyoto University.

Kitamura, S., S. Suzuki, T. Yumoto, P. Chuailua, K. Plongmai, P. Poonswad, N. Noma, T. Maruhashi, and C. Suckasam. 2005. A Botanical Inventory of a Tropical Seasonal Forest in Khao Yai National Park, Thailand: Implications for Fruit-Frugivore Interactions. *Biodiversity and Conservation* 14:1241–62.

Kitamura, S., T. Yumoto, P. Poonswad, P. Chuailua, K. Plongmai, T. Maruhashi, and N. Noma. 2002. Interactions between Fleshy Fruits and Frugivores in a Tropical Seasonal Forest in Thailand. *Oecologia* 133:559–72.

Leighton, M. and D. R. Leighton. 1983. Vertebrate Responses to Fruiting Seasonality within a Bornean Rain Forest. In *Tropical Rainforest: Ecology and Management*, ed. S. L. Sutton, T. C. Whitmore, and A. C. Chadwich, 181–96. London: Blackwell Science Publishing.

Lekagul, B., and J. A. McNeely. 1988. *Mammals of Thailand*, 2nd ed. Bangkok: Darnshutha Press.

Linkie, M., Y. Dinata, A. Nugroho, and I. Achmad Haidir. 2007. Estimating Occupancy of a Data Deficient Mammalian Species Living in Tropical Rainforests: Sun Bears in the Kerinci Seblat Region, Sumatra. *Biological Conservation* 137:20–27.

Lynam, A. J. 2003. *A National Action Tiger Plan for the Union of Myanmar*. Myanmar Forest Department and Wildlife Conservation Society.

Mace, R. D., T. L. Manley, and K. E. Aune. 1994. Factors Affecting the Photographic Detection Rate of Grizzly Bears in the Swan Mountains, Montana. *International Conference on Bear Research and Management* 9:245–52.

Manjrekar, N. 1989. Feeding Ecology of the Himalayan Black Bear (*Selenarctos thibetanus* Cuvier) in Dachigam National Park. Master's thesis, Saurashtra University, Rajkot, India.

Maxwell, J. F. 1995. Vegetation and Vascular Flora of the Ban Seneh Pawng Area, Lai Wo Subdistrict, Sangklaburi District, Kanchanaburi Province, Thailand. *Natural History Bulletin of the Siam Society* 43:131–70.

McConkey, K., and M. Galetti. 1999. Seed Dispersal by the Sun Bear *Helarctos malayanus* in Central Borneo. *Journal of Tropical Ecology* 15:237–41.

Moruzzi, T. L., T. K. Fuller, R. M. DeGraaf, R. T. Brooks, and W. Li. 2002. Assessing Remotely Triggered Cameras for Surveying Carnivore Distribution. *Wildlife Society Bulletin* 30:380–86.

Murphy, P. G., and A. E. Lugo. 1986. Ecology of Tropical Dry Forest. *Annual Review of Ecology and Systematics* 17:67–88.

Namyi, H., H. Kimchlay, and A. Olsson. 2006. *The Status of Asiatic Black Bears and Malayan Sun Bears in the Central Cardamom Protected Forest, Southwest Cambodia*. Conservation International / Cambodia Forestry Administration, Phnom Penh, Cambodia.

Ngoprasert, D. 2004. Effects of Roads, Selected Environmental Variables and Human Disturbance on Asiatic Leopard (*Panthera pardus*) in Kareng Krachan National Park. Master's thesis, King Mongkut's University of Technology, Thonburi, Thailand.

Nozaki, E., S. Azuma, T. Aoi, H. Torii, T. Ito, and K. Maeda. 1983. Food Habits of Japanese Black Bear. *International Conference on Bear Research and Management* 5:106–9.

Peacock, E. H. 1933. *A Game-Book for Burma & Adjoining Territories*. London: H.F. & G. Witherby.

Poscillico, M., A. Meriggi, E. Pagnin, S. Lovari, and L. Russo. 2004. A Habitat Model for Brown Bear Conservation and Land Use Planning in the Central Apennines. *Biological Conservation* 118:141–50.

Primack, R., and R. Corlett. 2005. *Tropical Rain Forests, and Ecological and Biogeographical Comparison*. Malden, MA: Blackwell Publishing.

Rao, M., T. Myint, T. Zaw, and S. Htun. 2005. Hunting Patterns in Tropical Forests adjoining the Hkakaborazi National Park, North Myanmar. *Oryx* 39:292–300.

Reid, D., M. Jiang, Q. Teng, Z. Qin, and J. Hu. 1991. Ecology of the Asiatic Black Bear (*Ursus thibetanus*) in Sichuan, China. *Mammalia* 55:221–37.

Rundel, P. W. 1999. *Forest Habitats and Flora in Lao PDR, Cambodia, and Vietnam*. Hanoi, Vietnam: WWF–Indochina.

Rundel, P. W., and K. Boonpragob. 1995. Dry Forest Ecosystems of Thailand. In *Seasonal Dry Tropical Forests*, ed. S. H. Bullock, H. Mooney, and E. Medina, 93–123. New York: Cambridge University Press.

Saberwal, V. 1989. Distribution and Movement Patterns of the Himalayan Black Bear (*Selenarctos thibetanus* Cuvier) in Dachigam National Park, Kashmir. Master's thesis, Saurashtra University, Rajkot, India.

Sarker, M. S. U. 2006. The Status and Conservation of Bears in Bangladesh. In *Understanding Asian Bears to Secure Their Future*, 50–56. Ibaraki, Japan: Japan Bear Network.

Sathyakumar, S., and S. Viswanath. 2003. Observations on Food Habits of Asiatic Black Bear in Kedarnath Wildlife Sanctuary, India: Preliminary Evidence on Their Role in Seed Germination and Dispersal. *Ursus* 14:99–103.

Saw Htun. 2006. The Status and Conservation of Bears in Myanmar. In *Understanding Asian Bears to Secure Their Future*, 45–49. Ibaraki, Japan: Japan Bear Network.

Schaller, G. B., Q. T. Teng, K. G. Johnson, X. M. Wang, H. M. Shen, and J. C. Hu. 1989. The Feeding Ecology of Giant Pandas and Asiatic Black Bears in the Tangjiahe Reserve, China. In *Carnivore Behavior, Ecology and Evolution*, ed. J. L. Gittleman, 212–41. New York: Cornell University Press.

Seng Teak. 2001. *Surveys for Tigers and Other Large Mammals in Virachey National Park (2000–2001)*. Phnom Penh: WWF–Cambodia.

Servheen, C. 1999. Sun Bear Conservation Action Plan. In *Bears: Status Survey and Conservation Action Plan*, ed. C. Servheen, S. Herrero, and B. Peyton, 219–23. Gland, Switzerland, and Cambridge: IUCN/SSC Bear and Polar Bear Specialist Groups.

————. 2001. The Status of the Bears of the World, with Emphasis on Asia. In *Proceedings of the Third International Symposium on the Trade in Bear Parts*, ed. D. Williamson and M. J. Phipps, 4–9. Hong Kong: TRAFFIC East Asia.

Shepherd, C. R., and V. Nijman. 2008. The Trade in Bear Parts from Myanmar: An Illustration of the Ineffectiveness of Enforcement of International Wildlife Trade Regulations. *Biodiversity and Conservation* 17:35–42.

Sreekumar, P. G., and M. Balakrishnan. 2001. Seed Dispersal by the Sloth Bear (*Melursus ursinus*) in South India. *Biotropica* 34:474–77.

Steinmetz, R. 1997. *A Survey of Mammals and Habitats in Xe Piane National Biodiversity Conservation Area, Lao PDR*. Bangkok: WWF–Thailand.

————. 2009. Ecological Overlap of Sympatric Asiatic Black Bears and Sun Bears in a Tropical Forest Mosaic, Thailand. PhD diss., University of Minnesota.

Steinmetz, R., and D. Garshelis. 2008. Distinguishing Asiatic Black Bears and Sun Bears by Claw Marks on Climbed Trees. *Journal of Wildlife Management* 72:814–21.

————. 2010. Estimating ages of bear claw marks in Southeast Asian tropical forests as an aid to population monitoring. *Ursus* 21:143–53.

Steinmetz, R., D. Garshelis, W. Chutipong, and N. Seuaturien. In press. Public Library of Science One.

Steinmetz, R., N. Seuaturien, W. Chutipong, C. Chamnanjit, and R. Phoonjampa. 2007. The Ecology and Conservation of Tigers, Other Large Carnivores, and Their Prey in Kuiburi National Park, Thailand. Bangkok: WWF–Thailand.

Stibig, H.-J., and R. Beuchle. 2003. *Forest Cover Map of Continental Southeast Asia at 1:4,000,000 Derived from SPOT-VEGETATION Satellite Images*. Brussels: Institute for Environment and Sustainability.

Sukmasuang, R. 2001. Ecology of Barking Deer (*Muntiacus* spp.) in Huai Kha Khaeng Wildlife Sanctuary. PhD diss., Kasetsart University, Bangkok.

Trisurat, Y. 2007. Applying Gap Analysis and a Comparison Index to Evaluate Protected Areas in Thailand. *Environmental Management* 39:235–45.

Tumbelaka, L., and G. M. Fredriksson. 2006. The Status of Sun Bears in Indonesia. In *Understanding Asian Bears to Secure Their Future*, 73–78. Ibaraki, Japan: Japan Bear Network.

van de Bult, M. 2003. *The Vegetation and Flora of the Western Forest Complex*. Bangkok: National Park, Wildlife and Plant Conservation Department.

van Steenis, C. G. 1950. The Delimitation of Melesia and Its Main Plant Geographical Divisions. *Flora Malesiana* 1 (1): 70–75.

Vidal, J. 1979. Outline of the Ecology and Vegetation of the Indochinese Peninsula. In *Tropical Botany*, ed. O. Holm-Hensen, 109–23. New York: Academic Press.

Vinitpornsawan, S., R. Steinmetz, and B. Kanchanasakha. 2006. The Status of Bears in Thailand. In *Understanding Asian Bears to Secure Their Future*, 50–56. Ibaraki, Japan: Japan Bear Network.

Webb, E. L. 2007. Botanical Exploration of Thung Yai Naresuan Wildlife Sanctuary, Western Thailand. Final Report, National Geographic Research and Exploration Grant 6798-00.

Whitmore, T. C. 1998. *An Introduction to Tropical Rain Forests*. Oxford: Oxford University Press.

Wong, S. T., C. Servheen, and L. Ambu. 2002. Food Habits of Malayan Sun Bears in Lowland Tropical Forests of Borneo. *Ursus* 13:127–36.

———. 2004. Home Range, Movement and Activity Patterns, and Bedding Sites of Malayan Sun Bears *Helarctos malayanus* in the Rainforest of Borneo. *Biological Conservation* 119:168–81.

Woodruff, D. S. 2003. Neogene Marine Transgressions, Paleogeography and Biogeographic Transitions on the Thai-Malay Peninsula. *Journal of Biogeography* 30:551–67.

Wroughton, R. C. 1916. Mammal Survey of India, Burma, and Ceylon: Chindwin River. *Journal of the Bombay Natural History Society* 24:291–300.

Zaw, T., S. Htun, S. H. T. Po, M. Maung, A. J. Lynam, K. T. Latt, and J. W. Duckworth. 2008. Status and Distribution of Small Carnivores in Myanmar. *Small Carnivore Conservation* 38:2–28.

15

Tropical Asian Dry Forest Amphibians and Reptiles

A Regional Comparison of Ecological Communities

George R. Zug

Asian seasonally dry tropical forests extend the breadth of South Asia from the edge of the Thar desert (India, approximately lat. 72° E) to the Vietnamese coast (approximately 109° E) of Indochina (Wikramanayake et al. 2002).[1] The forests occur in a subtropical-tropical band from about 9° N to 24° N, typically at low elevations (approximately 5–300 m). Numerous physiographic and climatic factors within this large geographic area divide the dry deciduous or semi-deciduous forest into six Indian and three Indochinese forest categories.

The diversity of seasonally dry tropical forests suggests a concomitant diversity of amphibian and reptilian communities. I realized the potential difficulty of reviewing the herpetofaunas of these forests, owing to the existence of few rigorous site-specific herpetofaunal inventories of Asian dry forests. This challenge and other ones to a meaningful comparison of different dry forest herpetofaunas are detailed below. In spite of these challenges, three objectives can be realized: (1) Identify the species composition of the herpetofauna at several sites throughout the breadth of the Asian dry forests; (2) Examine each herpetofauna for a semblance of community structure; and (3) Compare species composition and community structure between sites to test for similarities/differences arising from shared/different environmental features of sites.

MATERIALS, METHODS, AND LIMITATIONS

Definition of Terms

Because my usage of ecological terms likely differs from other naturalists, I offer a few definitions to enhance communication of my results and interpretations.

A *community* is an assemblage of organisms living in the same place, interacting through mutualism, predation, competition. I accept Drury's (1998) community concept as the organisms of a locality assembled by chance and organized similarly by the stochastic interactions of physical and biotic environment, hence I use *community* and *assemblage* interchangeably. A *herpetofauna* is a community of amphibians and reptiles living together in the same area, habitat, or microhabitat. Generally, most frequently herein, herpetofauna refers to the amphibian and reptilian populations occupying the same *habitat*—another ecological term or entity used variously by me and most other naturalists.

My use of *taxon* is restricted to a species in both the biological and phylogenetic species concept sense of a population or populations of phylogenetically related and potentially interbreeding individuals. *Guilds* are groups of animals sharing similar prey and prey-capture behaviors.

Dry Forests

The dry forests referred to herein are those labeled as "dry broadleaf forests" by Wikramanayake et al. (2002, Figure 1.3, to which the ecoregion numbering scheme below refers) and specifically as "dry forests" or "dry deciduous forests" (peninsular Indian forests [16, 18–22]; Indochina forests [58, 71–72]). The primary herpetofaunas discussed below derive from four dry forest types: Khathiarbar-Gir dry deciduous forests (Gir Forest National Park), northwestern India (16); northern dry deciduous forests (central Nallamala Hills), west-central India (19); Irrawaddy dry forests (Chatthin Wildlife Sanctuary, Shwe-Settaw Wildlife Sanctuary), central Myanmar (71); and central Indochina dry forests (Sakaerat Biosphere Reserve of central Thailand and hilly eastern Cambodia; 72).

There are no sets of distributional maps of South Asian amphibians and reptiles that pinpoint the precise occurrence of individual species. I recognized this lack of a consolidated data source as the first challenge in a search for patterns of species distributions that would be concordant with the maps of the dry forest ecoregions (Wikramanayake et al. 2002). Hence, the preceding assignment of a herpetofauna to a particular forest (Figure 1) depends upon the general habitat description in the report and requires that the site lies within the mapped boundaries of a dry forest ecosystem.

Herpetofaunal Surveys of Dry Forest

Lists of herpetofaunas for dry forest sites in tropical Asia are of variable quality. Only two sites have received rigorous year-round inventorying using several sampling techniques: Sakaerat Environmental Research Station (Nakhon Ratchasima Province, Thailand, 14°30′ N 101°55′ E; Inger and Colwell 1977) and Chatthin Wildlife Sanctuary, Sagaing Division, Myanmar, 23°34.46′ N 95°44.26′ E; Zug et al. 1998, 2004; G. Zug, unpublished data). These two sites had weekly or more-frequent inventories for one year or longer. The inventory techniques included random quadrat searches, cruise collecting along transects, general-random collecting, and additionally at Chat-

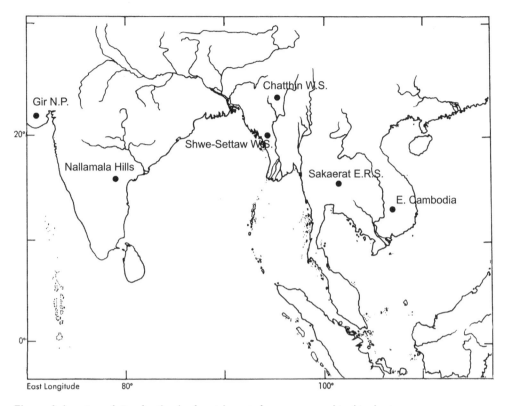

Figure 1. Location of sites for the dry forest herpetofaunas compared in this chapter.

thin, a week of drift-fence pitfall trapping each month. Thus, these herpetofaunas serve as touchstones for the sites less-thoroughly inventoried.

The herpetofaunal surveys at the other sites derived largely from general-random collecting. In India, the inventory of the Gir Forest National Park (Gujarat State, 21°08′ N 70°48′ E) derived from intermittent fieldwork over a one-year interval (Bhatt et al. 1999; Vyas 2000); the fieldwork was supplemented by interviews with local people and park personnel and a review of unpublished reports. The Nallamala Hills (Andhra Pradesh State, 15°30′ N 79°15′ E) inventory encompasses a large area (approximately 13,000 km²), and species occurrence data drew on over a century of assorted fieldwork, with some recent (1995–2004) fieldwork of more-intensive sampling of a few areas. To make the herpetofaunal listing of Nallamala more comparable in geographic scope to those of other sites, I use only the herpetofauna of the central portion of the hills (see Appendix A) for comparison.

Additional Burmese herpetofaunas are available for comparison. These herpetofaunas derived from moderately thorough herpetological inventories of variable periods, some for more than 30 days. Two such sites provide intra-Burma comparison: Shwe-Settaw Wildlife Sanctuary (Magway Division, 20°06 N 94°44 E); Min-Gon-Taung

Wildlife Sanctuary (Mandalay Division, 21°24 N 95°47 E). The Shwe-Settaw site has experienced 44 survey days in both dry and wet seasons, and its herpetofauna is enumerated in Appendix A. Min-Gon-Taung had 38 survey days, but all fieldwork occurred in the latter third of the wet season.

I have been unable to locate a site-specific inventory of a dry forest area in northeastern Cambodia or southern Vietnam. Site-specific inventories in these two countries have focused on evergreen and moist forests. A sequential survey (Stuart et al. 2006) at five sites in hilly eastern Cambodia comes closest to matching the preceding herpetofaunal surveys, and its results are listed (Appendix A) although the duration at all sites totaled 37 days and much of the fieldwork appears to have been concentrated in evergreen forests (those taxa are excluded from the site species list). The areas surveyed included Seima Biodiversity Conservation Area, Phnom Nam Lyr Wildlife Sanctuary, and Virachey National Park (three areas centered roughly at 13° N 107° E).

Other sites throughout the dry forest band have had surveys of varying intensity and duration, often concentrating on either amphibians or reptiles, or a subset of one these groups. Although their inventories are incomplete and of variable quality (e.g., inadequate confirmation of species identification), they are useful for comparative purposes but not for numerical analysis.

Taxonomic Inequalities

Comparisons of faunas require the accurate identification of component species comprising each fauna. In Asia and especially in Burma, such accuracy has become increasingly difficult, not because we are unable to recognize each component species within each site's herpetofauna and readily assign a scientific name to most specimens, but rather because the names we assign to populations in northern or southern Burma and those of central Thailand, for example, likely do not represent the same taxon. These taxonomic differences are becoming increasingly evident as geographic coverage and voucher-sample sizes increase permitting detailed examination of morphological variation.

Two examples are sufficient to demonstrate the unreliability of depending upon our current taxonomy of Asian amphibians and reptiles to determine the genetic sameness of populations in India, Burma, and Indochina. The classic example (Wüster and Thorpe 1989, 1990, 1992; Wüster 1996) of regional differentiation in Asian cobras is now nearly two decades old, and yet their model of analysis of population variation across the breadth of Asia has been little followed, and then only recently so. Until Wüster's studies, Asian cobras were considered a single, wide-ranging species, *Naja naja*, although several subspecies were recognized. By 1996, Wüster had recognized ten species in his review article for toxicologists and medical doctors. Notably, the geographic distributions of formerly recognized subspecies are not concordant with the currently delimited ranges of *Naja* species.

In frogs, we still lack a breadth-of-Asia study. The necessity of such studies is evident in widespread species, such as *Polypedates leucomystax* and its putative sister taxa (Matsui et al. 1986; Orlov et al. 2001), but the best anuran example of high regional differentiation is the *Fejervarya limnocharis* complex. The details of speciation in this group of frogs remain largely unresolved, but the taxonomic history outlined in Table 1 gives a hint at the complexity of the situation. This table contains fifteen taxa; more than twenty species had been recognized (published) as of June 2008. Our herpetological surveys in Burma have identified a minimum of six species in the north and central regions; two sympatric species were identified during my initial fieldwork at Chatthin, based on two sizes of gravid females (30–39 mm and 47–67 mm svl [snout–vent length]; Zug et al. 1998).

It is essential to remember in the subsequent text that the same species from distant locations, e.g., *Microhyla ornata* in western India and eastern Vietnam, are not the same genetic entity and possibly not even near relatives.

RESULTS

Species Inventories and Completeness

Species Accumulation Rates

The herpetofaunas of Gir, Nallamala, Chatthin, and Sakaerat (Appendix A) are assumed to be nearly fully inventoried. Confirmation of an inventory's completeness

Table 1. Chronological summary of the recognition of speciation in the paddy or rice frog *Fejervarya limnocharis* complex.[a] Inger (1954) was the first researcher to examine geographic variation of morphology within these frogs. He conclusively demonstrated that the Philippine populations (*vittigera*) differed substantially from those (*limnocharis*) elsewhere in Southeast Asia. He chose to recognize the two populational groups as subspecies. His taxonomy was followed into the 1980s when the subspecies concept was largely abandoned in herpetology.[b]

Fejervarya Species Name	General Distribution	Source
l. limnocharis	Tropical Asia, Pakistan to the Philippines	Inger 1954
l. vittigera	Philippines	Inger 1954
nepalensis	Central and Eastern Nepal	Dubois 1975
pierrei	Central and Eastern Nepal	Dubois 1975
syhadrensis	Eastern and Western India, adjacent Nepal	Dubois 1975
andamensis	Andaman Islands	Dubois 1984
nilgiris	Kerala and Tamil Nadu, India	Dubois 1984
teraiensis	Southern Nepal, adjacent India	Dubois 1984
vittigera	Philippines	Dubois 1984
mysorensis	Karnataka, India	Dutta and Singh 1996
orissaensis	Orissa, India	Dutta 1997
iskandari	Java	Veith et al. 2001
limnocharis	Java (implicit restriction of occurrence)	Veith et al. 2001
sakishimensis	Southern Ryukyu Islands	Matsui et al. 2007
mudduraja	Central Western Ghats, India	Kuramoto et al. 2007
kudremukhensis	Central Western Ghats, India	Kuramoto et al. 2007
caperata	Central Western Ghats, India	Kuramoto et al. 2007

[a]This listing of *F. limnocharis* complex species is intentionally incomplete (selective). The goal is to demonstrate the ongoing recognition of diversity (speciation) within paddy frogs. As of June 1, 2008, more than twenty species of paddy frogs had been recognized.
[b]For a broader perspective on the worldwide recognition of increasing species diversity in amphibians, see Köhler et al. 2005.

can be assessed by examination of species accumulation rates. Time-of-capture data are available for the intensely surveyed Chatthin and Sakaerat herpetofaunas, and for the less-intensively surveyed Shwe-Settaw and Min-Gon-Taung faunas. The rate of species accumulation for Chatthin and Sakaerat (Figure 2a) are similar (and typical) with an initial rapid vouchering of the sites' herpetofaunas and then a gradual slowing of the acquisition of new taxa. For these two sites, the near-total herpetofaunas were discovered in 45 and 40 weeks, respectively. Although the discovery rates were similar between these two sites, survey design and available man-power yielded different rates. The Sakaerat survey obtained 50 percent of the herpetofauna in 5 weeks and 90 percent in 19 weeks; Chatthin sampling, in contrast, obtained 50 percent in 2 weeks and 90 percent after 43 weeks. Note that collecting effort, as measured in man-hours, differed between the two sites, and this effort is impossible to quantify precisely owing to many staff and visitors capturing specimens and giving them to the survey. There was also wide variation in the abilities of collectors to see and capture animals. Additionally, the surveys started in different seasons: the middle of the dry season at Sakaerat (February), and the middle of the wet season at Chatthin (late July). The near-total species diversity is not known for either Shwe-Settaw or Min-Gon-Taung; nevertheless, a total similar to Chatthin is probable for each. The 8 and 7 week surveys, respectively, were able to inventory 50 percent of the presumed herpetofauna at each site by the fifth week. The accumulation curves of Shwe-Settaw and Min-Gon-Taung (Figure 2b) do not rise as steeply as Chatthin's, although they attain the same level by the end of two months of inventory.

Although not directly evident from the curves but extractable from the date-of-first-capture data, there was a different sequence of captures for Chatthin and Sakaerat. Both show the typical rapid rise in species discovery; however, at Chatthin, frogs dominated early captures because of an early rainy season start. Within the first week, 81 percent of the frog fauna and 46 percent of the lizards were vouchered. At Sakaerat, the survey began in the late dry season and lizards dominated; 56 percent of the lizard species were captured in the first week, contrasting to 32 percent of the frogs. At Sakaerat, all frog species were documented by the end of the twenty-first week and most lizards (90 percent) by the end of the fifteenth week. Similarly at Chatthin, most frogs (94 percent) were vouchered by the end of the third week, but to obtain most lizards (90 percent) required 43 weeks. Snake species were documented more slowly, most (94 percent) by the end of week 25 at Sakaerat, and 91 percent by the end of week 45 at Chatthin. In neither site was the herpetofauna fully inventoried at the end of the first year. Two additional snake species were discovered at Chatthin in the second year of the survey. A lizard (*Varanus bengalensis*) and a snake (*Cylindrophis ruffus*) were vouchered prior to initiation of the inventory (mid-July 1997) and were not documented during the subsequent three years of the Chatthin project.

Even though the duration of the Shwe-Settaw inventory was only eight weeks (not continuous and occurring over four years, including visits in both wet and dry seasons), the inventory attained the same species numbers (80 percent) as at Chatthin by the eighth week (Figure 2B). The rate of capture of new frog, lizard, and snake species was similar at Shwe-Settaw and Min-Gon-Taung.

A. Chatthin & Sakaerat

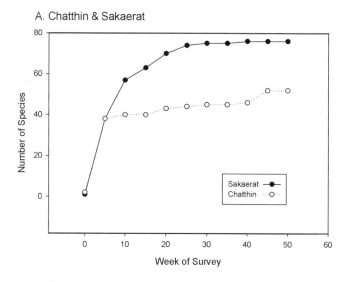

B. Within Myanmar

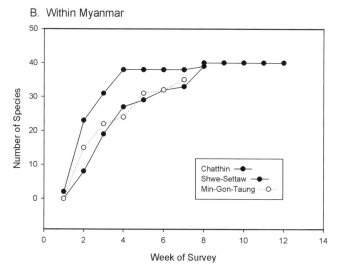

Figure 2. Rate of discovery of new species at several dry forest sites. (*A*) Comparison of weekly cumulative totals of new species inventoried at a Burmese dry forest site (Chatthin Wildlife Sanctuary) and a Thailand site (Sakaerat Environment Research Station). (*B*) Comparison of cumulative results for three Burmese dry forest surveys (Chatthin Wildlife Sanctuary, Shwe-Settaw Wildlife Sanctuary, and Min-Gon-Taung Wildlife Sanctuary).

Species Composition and Abundance

Species Composition—α Diversity

The total and component compositions of the six major dry forest herpetofaunas are summarized in Table 2. Of the two best-studied sites, Sakaerat has 22 (greater than 40 percent) more species than Chatthin. This α-diversity is nearly matched by Nallamala. The remaining three sites were less-intensely surveyed and, not unexpectedly, their species count is lower (Table 2). Shwe-Settaw at the southern end of the Burmese central dry zone with only eight survey-weeks has 81 percent of the Chatthin herpetofaunal total, and Min-Gon-Taung with six survey-weeks is 65 percent (35 species) of the Chatthin total. Cambodia has the lowest diversity (Table 2), but also the fewest survey-weeks.

Table 2. Numerical summary of the amphibian and reptilian components of some Asian dry forest herpetofaunas (Appendix A).

Taxon	Gir	Nallamala	Chatthin	Shwe-Settaw	Sakaerat	Cambodia
Caecilians	0	0	0	0	1	0
Frogs	7	19	16	11	20	13
Turtles	2	4	3	5	2	0
Lizards	14	22	13	15	19	8
Snakes	17	30	22	13	36	9
Total	40	75	54	44	78	30

In the Indian dry forest sites (Table 2), Gir's 40 species seems unrealistically low, especially considering that Vansda National Park, also in Gujarat State although mainly with moist deciduous forest habitats, has 54 species (Vyas 2004). The diversity at Nallamala is nearly 90 percent greater than that at Gir and equivalent to the diversity at Sakaerat.

Although the actual numbers of species differ among the sites, the relative diversity of each herpetofaunal component is similar (Figure 3; Table 2). At all sites, snakes have the highest diversity, and for the three best-surveyed sites, snakes comprise 40–41 percent (Nallamala and Chatthin) to 48 percent (Sakaerat) of the herpetofaunal assemblage. As detailed in the next section, this diversity associates with low abundance. Frogs and lizards share a similar diversity (approximately 25 percent each) to one another at each site and between sites (Figure 3).

The similarity in component diversity at the three well-inventoried sites across a broad latitudinal distance highlights the incompleteness of the inventory data from the other sites. The low diversity of frogs (18 percent) in the Gir forest contrasts sharply with the other sites and suggests insufficient attention to nocturnal surveys; however, in the moist deciduous forests of "nearby" Vansda National Park (Vyas 2004), frog diversity is only 22 percent. This comparison of these two Gujarat state forests is problematic because their faunal lists derive from inventories by the same researcher and presumably share the same bias or limitation for nocturnal surveys. A contrasting situation is evident for the eastern hills of Cambodia where frog diversity represents 43 percent of the herpetofauna, indicating a bias toward an amphibian inventory.

As noted previously, dry forest herpetofaunal inventories are rare. A survey of the Kalakad-Mundanthurai Tiger Reserve (Tamil Nadu State, India; Vijayakumar et al. 2006) describes the lower hill forest as a mixture of scrub, dry deciduous thicket, dry deciduous savanna, dry evergreen forest, and riparian forest. This "dry forest" Kalakad assemblage consists of 17 frogs, 2 turtles, 22 lizards, and 10 snakes (total herpetofauna: 51). An inventory of the entire reserve (Cherian et al. 2000) including higher elevations and the evergreen sholas yielded 0 caecilians, 32 frogs, 2 turtles, 13 lizards, and 15 snakes (total herpetofauna: 65). The differences in number of species and kind reflect more than the inclusion of the shola habitat but a difference in the manner of surveying and in the habitats surveyed. In both, the snakes are poorly inventoried, and in combination, the two surveys still inventoried only 22. Although the total diversity is less than for central Nallamala, the frog diversity is nearly double that

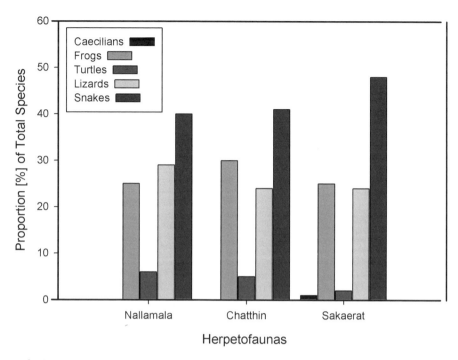

Figure 3. A portrait of relative species numbers (diversity) within each herpetofauna at three intensely surveyed sites: Nallamala Hills, India; Chatthin, Burma; and Sakaerat, Thailand. The site percentages derive from the total number of in-site species. Actual species occurrence is listed in Appendix A.

of Nallamala, and snake diversity is about two-thirds the latter (likely inadequately surveyed; a minimum of 30 snake species is predicted).

Species Composition—Taxonomy

Among the major dry forest sites, a caecilian occurs only at Sakaerat. Frogs occur at all sites (Appendix A) and represent a substantial component of the herpetofaunal community. The six sites share five anuran families. The endemic Ranixalidae is represented by a single species (*Indirana leithii*) at Nallamala. Two anurans, the black-spiny toad (*Duttaphrynus melanostictus*) and ornate narrow-mouthed frog (*Microhyla ornata*) occur at all sites. Both of these species occur geographically and ecologically beyond the dry forest zone. Neither is a human commensal, although both regularly populate anthropogenic habitats; both occur in natural forest from sea level to low montane (up to 300 m). As emphasized in Materials, Methods, and Limitations above, pan-Asian species are suspect, that is, are they the same species throughout South Asia? For *M. ornata*, the answer is no (Matsui et al. 2005) but no nomenclatural change has been proposed. The population genetics for the spiny toad is unstudied. A similar pattern is evident among other frogs where some resolution of regional and sympatric genetic differentiation among pan-Asian species has been identified and taxonomically formalized. In these taxa (e.g., the *Fejervarya limnocharis* species group [SG], or complex;

the *Hoplobatrachus tigerinus* SG; the *Polypedates leucomystax* SG), the SG has a pan-Asian distribution and its members have similar appearances, and our largely anecdotal knowledge of their ecology and behavior indicates "ecological equivalence" throughout the SG's distribution.

Another pattern among the non–pan-Asian species is dual geography of some species, that is, a species occurs in the dry forest of two of three geographic pairs: India-Burma, Burma–Southeast Asia, or India–Southeast Asia. The India-Burma taxa are *Sphaerotheca breviceps* and *Microhyla rubra*; the Burma–Southeast Asia taxa are *Glyphoglossus molossus*, *Kaloula pulchra*, *Microhyla pulchra*, *Pelophylax lateralis*, and *Chiromantis nongkhorensis*.

Among turtles, there are no pan-Asian species. This absence results from better taxonomic studies. A single dual geography exists for two geographic pairs: India-Burma, *Melanochelys trijuga*; and Burma–Southeast Asia, *Indotestudo elongata*. There are no dry forest endemics.

Lizards display the same distributional patterns and share the same taxonomic difficulties as anurans. The latter problem is highlighted by the *Calotes versicolor* SG, which until recently (Zug et al. 2006) was considered a pan-Asian species, but the recognition of two sympatric Burmese species (*C. htunwini*, and *C. irawadi*, the former potentially a dry forest endemic) revealed a multiplicity of distinct populations.

Only one lizard species, the house gecko (*Hemidactylus frenatus*), occurs in all three areas. Because of its near-total anthropogenic association, it cannot be considered a true dry forest resident. There is no lizard species occurring in India-Burma. The Burma–Southeast Asia species are *Calotes mystaceus*, *Gekko gecko*, *Eutropis multifasciata*, and *Sphenomorphus maculatus*. A number of taxa occur in one area only, such as *Calotes rouxi* and *Hemidactylus triedus* in India or *Bronchocoela smaragdina* and *Lygosoma bowringii* in Southeast Asia. None of these appears to be endemic to dry forest, although such endemicity might occur for the two *Leiolepis* species.

Snakes exhibit the same patterns and taxonomic difficulties as noted for frogs and lizards. A few snakes (*Ahaetulla nasuta*, *Chrysopelea ornata*, *Lycodon aulicus*, *Ptyas mucosus*, *Amphiesma stolatum*, *Ramphotyphlops braminus*) occur in India, Burma, and Southeast Asia dry forests. All these species also occur in a variety of habitats, including anthropogenic ones. No snake species occurs at all sites, although *Ahaetulla nasuta* and *Ptyas mucosus* likely occur in southeastern Cambodia as they occur in southern Vietnam (Campden-Main 1970).

Geographic pairs exist: India-Burma, *Xenochrophis piscator* and *Python molurus*; Burma–Southeast Asia, *Boiga multomaculata*, *Boiga ocellata*, *Coelognathus radiatus*, *Dendrelaphis subocularis*, *Sibynophis collaris*, *Rhabdophis subminiatus*, *Bungarus fasciatus*, *Crypteltropis albolabris*, and *Xenopeltis unicolor*. With the exception of *Naja mandalayensis* and potentially some of the *Oligodon* species, dry forest endemicity is lacking, and even *Oligodon* species likely overlap into drier or moister habitats.

Species Composition—Size Structure

Size influences what a species eats, where it lives, and what eats it. Thus, a comparison of amphibian and reptilian body sizes among the six dry forest sites offers

another means to examine similarities and differences of the six selected communities. To make such a comparison, I categorized body size into five classes for frogs,
three for turtles, six for lizards, and five for snakes (see Appendix B for categoriza-
tion procedure).

For frogs (Figure 4), only Chatthin possessed a miniature species (less than 20 mm
adult svl). Small species (22–44 mm svl) were generally the most frequent size class
among all sites, ranging from a proportional frequency of 29 percent (Gir) to 58 percent (Cambodia), although medium-sized frogs (46–70 mm svl) are the most frequent
class at the Gir forest (43 percent) and its small frogs have the same frequency as Chatthin (33 percent). Medium-sized species typically represent the second most frequent
size class. Large (72–96 mm svl) and big (≥ 98 mm svl) species often occur with the
same frequency (Figure 4).

The few turtle species occurring in dry forests do not permit a meaningful comparison among sites.

Lizards have about the same diversity at a site as frogs (Table 2), yet lizards parti-
tion into more size classes. The smallest (less than 42 mm svl) and largest (greater than
150 mm svl) size classes have the lowest frequency of occurrence at all sites (Figure
4): 6–8 percent, which equals one species. For the largest class, that taxon is *Varanus
bengalensis*, which is three to four times larger than the next-largest class (120–148
mm svl). For the smallest class, the lizard is either a gecko (*Cnemaspis*, *Dixonius*, or
Hemidactylus) or a lacertid (*Ophisops*). Chatthin and Cambodia have no small species. Nallamala has three small species, whereas a single species each occurs at all other
sites. The medium-small class (22–44 mm svl) dominates at Shwe-Settaw and Sakaerat,
the medium class (64–94 mm svl) dominates at Gir, and Chatthin has equal frequency
of medium-small and medium class taxa. Moderately large (98–118 mm svl) and large
(120–148 mm svl) taxa represent about a quarter of the lizard taxa at the Indian and
Burmese sites, and more than a third of the taxa at the Southeast Asian ones.

Snake sizes clustered into five discrete classes (Appendix B). The frequency of the
size classes differs strikingly among the six sites (Figure 4), although the medium-small
(300–590 mm svl) and medium (600–990 mm svl) classes comprise at least 60 percent
of the taxa present at all sites. The disparity in sampling effort is evident for Shwe-
Settaw and Cambodia compared to the other sites. Both latter sites lack the large class
(greater than 1,500 mm svl) taxa, and the small class (less than 300 mm svl) is also
absent for Cambodia. These absences skew the frequency of the other classes and make
their comparison to other sites suspect. The other four sites have near-equal frequencies of the medium-large (1,000–1,500 mm svl) class taxa. At Nallamala, medium-
small taxa are most abundant, which is striking in comparison to the other sites. Gir
and Sakaerat have roughly equal frequency of taxa in the medium and medium-small
classes. At both Burmese sites, medium-sized snakes dominate.

Behavioral Preferences

Behavioral preferences (categories) for this review include activity pattern (diurnal or
nocturnal), habitat choice or use (fossorial, arboreal, aquatic), and diet (herbivory,
omnivory, and several prey-classes of carnivory) at a gross level (Appendix B).

A. Frogs

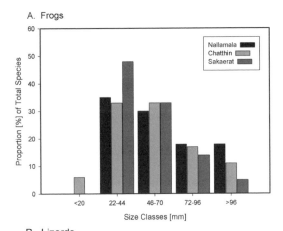

B. Lizards

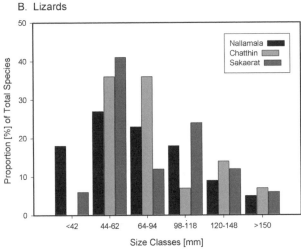

C. Snakes

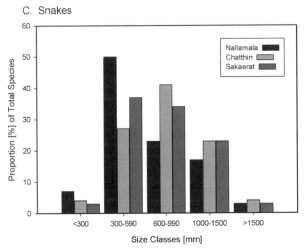

Figure 4. A portrait of frequency of size classes for frogs (A), lizards (B), and snakes (C) at three intensely surveyed sites: Nallamala Hills, India; (Chatthin, Burma; and Sakaerat, Thailand. The frequencies (%) derive from the total number of in-site species. Size classes are defined in Appendix B.

In activity patterns, frogs are largely nocturnal, although a few semiaquatic and aquatic taxa, e.g., *Euphlyctis cyanophlyctis* and *Occidozyga lima*, occur regularly at water's edge during the day; reproduction and feeding are at night. Turtles appear predominantly diurnal, and with the exception of the nocturnal geckos, lizards are diurnal. The relative homogeneity of the preceding activity patterns within a group shows that this biological aspect is "evenly" distributed among the six sites.

Snakes are the most variable in time of activity. The pythons and viperids are nocturnal, seemingly associated with the capture of sleeping birds or mammals and the ambush-capture of nocturnally active rodents. This apparent foraging strategy is shared also with the colubrids *Boiga* and *Lycodon* taxa. Most natricines are nocturnal, although *Amphiesma stolatum* and *Atretium schistosum* are diurnal species. A similar clade dichotomy occurs among the elapids with the kraits (*Bungarus*, *Calliophis*) being nocturnal and cobras (*Naja*) being diurnal/crepuscular. Most other colubrids are diurnal, and although the *Oligodon* species are commonly labeled as nocturnal, our Chatthin surveys suggest otherwise. The homalopsids and slug-eaters (*Pareas*) are nocturnal. No differences in community organization is evident among the sites.

With their nocturnal behavior, frogs have adapted to the full spectrum of subaerial habitats. Dry forest toads (Bufonidae) are terrestrial as are the microhylids with fossorial (*Glyphoglossus*, *Uperodon*) to semifossorial (some *Microhyla*) species. The other terrestrial-semifossorial frogs are *Sphaerotheca*. The arboreal frogs are *Chiromantis* and *Polypedates*. The ranids and dicroglossids are semiaquatic to aquatic, and this segregation is often equivocal, depending upon moisture level. The frog communities have three organizational patterns, and at all sites, terrestrial species are the most numerous. Gir and Shwe-Settaw lack arboreal and fossorial taxa; terrestrial species are two or more times as numerous as the aquatic and semiaquatic ones. Fossorial frogs are not documented at Cambodia; semiaquatic and aquatic species are equal in number and nearly so to the terrestrial taxa; arboreal frogs are the fewest. The pattern at Nallamala, Chatthin, and Sakaerat is similar. Terrestrial species are two to three times more numerous than any of the other behavior classes; arboreal, semiaquatic, and aquatic species are roughly equal in number, and a single fossorial species occurs at each site.

For turtles, the behavioral differences between semiaquatic (*Cyclemys*, *Melanochelys*) and aquatic (*Pangshura*, all trionychids) species are distinct. Aquatic taxa emerge from the water only to bask and lay eggs. Tortoises (*Geochelone*, *Indotestudo*) are terrestrial. Again, turtles are too few to show differences in testudine habitat use among sites.

Among lizard taxa, only terrestrial and arboreal behaviors occur. Some of the small skinks (e.g., *Lygosoma*) and lacertids (*Ophisops*) forage beneath the surface litter, but this behavior is not truly fossorial. All skinks are terrestrial, and most geckos are arboreal. The agamids are also predominantly arboreal with two terrestrial taxa (*Leiolepis*, *Psammophilus*). Monitors (*Varanus*) forage mainly on the ground but are adept climbers. With only two habitat behaviors, community organization is simple. At both Indian sites, the number of arboreal and terrestrial taxa is the same. Terrestrial lizards outnumber arboreal ones by about a third in Burma, and the converse in Southeast Asia, although the numbers of species exhibiting the two behaviors are nearly equal at Sakaerat.

Snakes occupy the full range of habitat use, although most dry forest species are either arboreal or terrestrial. The preference is often clade-related, e.g., arboreal:

Ahaetulla, *Boiga*, *Dendrelaphisor*; and terrestrial: *Coelognathus*, *Oligodon*, *Ptyas*, Sand-boas (*Eryx*), *Cylindrophis*, *Xenopeltis*, and blindsnakes (Typhlopidae), which are fosso-rial. The natricines are regularly semiaquatic, and the truly aquatic homalopsid snakes occur in dry forest only where some ponds or streams have water year-round. Compari-son of snake community habitat use is best done with the two Indian sites, Chatthin, and Sakaerat, both of which have complete snake inventories. Terrestrial snakes dominate at all four sites. The number of arboreal snakes is about one-half to two-thirds that of the terrestrial species at Chatthin and Sakaerat, and roughly a third of the terrestrial ones at Gir and Nallamala. Aquatic snakes are absent at Chatthin and exist as two to three species at the other sites. All sites have a few (1–3) fossorial snakes.

The diets of frogs and most lizards are insects and other invertebrate prey; how-ever, focused dietary studies for Asian amphibians and lizards are lacking. Indeed, dietary data for all Asian amphibians and reptiles are largely anecdotal.

Dietary diversity among lizards is herbivory in *Leiolepis* (Burma and Southeast Asia) and vertebrate carnivory in adult *Gekko gecko* and *Varanus bengalensis*. Among turtles, the testudinids are herbivores; the aquatic species are typically identified as carnivores, but most species tend toward omnivory. Because of the relative uniformity among these groups, there is no evident community organization.

Snakes display the greatest diversity of prey and prey-specialization. None is an herbivore or omnivore; all eat living or recently dead prey that the individual snake has killed. A few specialize on invertebrates: the blindsnakes (Typhlopidae) on termites and ants, and *Pareas* on terrestrial molluscans. *Pareas* occurs mainly in Southeast Asian moist evergreen forest; blindsnakes likely occur at all sites although are not yet vouchered at all. Most snakes eat vertebrates, and my prey categories (Appendix B) emphasize this dietary preference.

The diversity of prey preference in snakes permits an examination of possible community structure, and as in habitat choice, Nallamala, Chatthin, and Sakaerat are the only sites examined owing to the completeness of their inventorying. The relative frequency of the prey classes is remarkably similar (Table 3) across the three sites. The differences occur between fish-eaters (none at Chatthin) and strictly bird predators (only at Chatthin). Overall, snake community structure is the same among the three sites, i.e., much more similar than different.

Relative Abundance

Availability of data on population densities of different species in the different forest sites, and restrictions on chapter size, allow only a few observations. First and per-haps foremost is the effect of seasonality and multiyear populational fluctuation on the visibility of species. Obviously, the size of frog populations cannot be assessed in the dry season or for that matter for a few days or even a couple of weeks within a single wet season. Only Inger's Sakaerat survey (Inger and Colwell 1977; Inger 1980) provides data allowing a quantitative assessment of relative abundance, and those data are for a single calendar year, 1969. The Sakaerat data used herein derives not from Inger's published tables but from my tabulation of his specimen voucher collection at the Field Museum of Natural History (FMNH). This reexamination of abundance could not segregate the counts of species occurring in both the dry and the evergreen

Table 3. The relative frequency (percentage) of prey preference among the snake assemblages at Nallamala, Chatthin, and Sakaerat. The prey classes are defined in Appendix B.

Prey Preference	Nallamala	Chatthin	Sakaerat
Insects	7	4	6
Fish	10	0	6
Anurans	7	9	9
Reptiles	27	23	26
Ectotherms	6	8	9
Birds	0	4	0
Mammals	7	4	3
Endotherms	7	9	11
Vertebrates	30	31	31

forest, hence I might have slightly over-estimated the abundance for some species. The voucher-based data yield a total herpetofauna of 2,904 individuals (tadpole numbers are not included): 2 (less than 0.01 percent) caecilians; 1,410 (48.6 percent) frogs; 1,242 (42.8 percent) lizards; 230 (7.9 percent) snakes; and 20 (less than 0.01 percent) turtles. Within the frogs, *Microhyla heymonsi* was the most abundant taxon (235 individuals), then *Polypedates leucomystax* (200 individuals), *Fejervarya* "*limnocharis*" (157 individuals), and *M. butleri* and *M. ornata* (122 individuals each). Relative abundances of the frogs decline in a smooth curve to the least abundant, *Kaloula mediolineata* (5 individuals). Among the lizards, three species (*Dixonius siamensis* [198 individuals], *Eutropis macularia* [190 individuals], and *Calotes* "*versicolor*" [189 individuals]) have near-equal abundance. Abundance declines in a step-like fashion to a single individual each of *Eutropis longicaudata* and *Varanus bengalensis*. *Gongylosoma scripta* is the most abundant snake (27 individuals); thereafter snake abundance declines sharply, but smoothly, to seven species represented by 2 individuals and six species by 1 individual.

Our inventory vouchering at Chatthin was less intense and provides no quantitative data for comparison with Sakaerat. Frogs were definitely the most abundant component of the herpetofauna, and because of the scarcity of all lizard species, except the commensal *Hemidactylus frenatus*, I estimate that frogs constitute 80 percent or more of the total herpetofaunal abundance. *Calotes* "*versicolor*" is one of the more-abundant lizard taxa at Chatthin (where it is actually two taxa [*C. htunwini, C. irawadi*]), yet transect surveys in the forest during the late dry season yielded less than one "*versicolor*" sighting per kilometer. *Amphiesma stolatum* and *Lycodon aulicus* were the most common snakes, about one of each seen each week of surveying.

DISCUSSION

Species Inventories and Completeness

Species Accumulation Rates

How long does it take to inventory completely a herpetofauna of a well-demarcated site? The species accumulation curves from Sakaerat and Chatthin demonstrated the majority (90 percent) of a herpetofauna is vouchered within one year. Our Chatthin survey continued to find "new" species after the first year, that is, two snakes in the second year and another one in the third year. This continual discovery phenomenon is not unexpected and occurs even at the most intensely inventoried sites. The best confirmation of continual discovery derives from the Savannah River Ecological Laboratory (SREL) reserve in the Piedmont of South Carolina. This reserve of more than 800 km^2 of mixed forests and aquatic habitats hosts 95 species of amphibians and reptiles (Gibbons and Semlitsch 1991), determined by weekly monitoring for four decades. Yet, it required 21 years and the capture (and release) of over 6,000 snakes before the first pine-wood snake (*Rhadinaea flavilata*) was captured (Whiteman et al. 1995). Even more striking is the rediscovery (Luhring 2008) of Brimley's chorus frog (*Pseudacris brimleyi*) at SREL after 41 years of regular surveys and over 50 years subsequent to this species' previous vouchering in the reserve area. These discoveries highlight the difficulty of obtaining a total herpetofaunal inventory and the necessity of continuous and rigorous monitoring to obtain such an inventory.

In spite of the protracted time and high man-power requirements for a total inventory, the use of species accumulation curves remains a valuable and relatively accurate tool to assess an inventory's success. The shared similarity of the accumulation rate of new species (Figure 2) supports the probability of equal accessibility to the dry forest herpetofauna in different areas. Flattening (plateauing) of the curve at different species densities, but within a similar time-frame, indicates the robustness of the technique and the ability to obtain a reliable assessment of a dry forest site's herpetofauna with a year of intense survey.

Species accumulation curves are available for only two other tropical Asian sites. A dry forest site in the Western Ghats (Vijayakumar et al. 2006) yielded 13 amphibians and reptiles over a four-month survey, suggesting either a depauperate herpetofauna, or that the diurnal transect protocol used was a poor inventory strategy, hence the accumulative curve is uninformative. Bain and Nguyen (2004) examined two disturbed moist forests in northern Vietnam and documented 36 amphibian and 16 reptile species in 26 days of intense and varied surveying. Their accumulation curve began to flatten at the end of their third week, which might indicate about 90 percent of the local amphibian species had been inventoried; however, only 16 reptile, and particularly only 12 snake species, suggests an incomplete inventory.

Species Composition and Abundance

Species Composition—α Diversity

Of the data presented (Figure 2), the number of species for each site (Nallamala, Chatthin, Sakaerat) represents their near-total diversity. Nallamala and Sakaerat have near-equal diversity. Chatthin has significantly fewer species than these two sites, although all three sites share a similar proportion of frogs, lizards, and snakes—the main components of dry forest herpetofaunas. Diversity at the other three sites is low. Shwe-Settaw and Cambodia were knowingly incomplete faunas when presented (Appendix A), and I anticipate that the diversity of the former will match Chatthin and the latter Sakaerat when fully inventoried. Gir has an unrealistically low total herpeto-fauna of 40 species. Frog diversity (7 species; 18 percent) is a strong component of this low diversity, and snake diversity also is low. Whether Gir is depauperate or incompletely sampled remains unclear. Another Gujarat state reserve (Vansda National Park) has only 12 frog species (22 percent of total herpetofauna), nearly double the Gir frog diversity. The data from both Gujarat dry forests show low frog diversity. Perhaps the Gir herpetofauna is actually depauperate; however, I suspect that the low number of frog species indicates an incomplete inventory.

Other Indian areas have been inventoried. Some of these inventories are of dry forest, but often other forests are included, and discrimination of which species derive from the dry forest is not explicit in the reports. The Agasthyamalai Hills (Tamil Nadu State) is a dry forest mosaic, and a survey (Vijayakumar et al. 2006) at several sites during the dry season revealed 10 species of frogs, 10 species of lizards, and 7 species of snakes. As is common for many Indian inventories, the survey provides an initial faunal assessment but is inadequate for reliable comparison of dry forest herpetofaunas across tropical Asia. Aside from Sakaerat, no other Southeast Asian dry forest site has been inventoried. The similarity of Nallamala and Sakaerat diversity suggests a shared environment factor. That factor may be an adjacent evergreen forest, which serves as a population reservoir for some less arid-adapted species. In contrast, Chatthin is distant from a moist forest habitat, and this isolation yields a fauna that can persist under alternating extremes of wet and dry. This association of decreasing diversity with increasing aridity is a general phenomenon of tropical forest (Heatwole 1982), although the concept does not appear to have been linked with differential regional diversity.

Finally, in spite of differences in the number of total species and the broad geographic distances between sites, the proportional similarities among Nallamala, Chatthin, and Sakaerat are striking and unexpected. An explanation is not readily apparent.

Species Composition—Taxonomy

The proportional similarities of the herpetofaunal components might obtain from numerous shared species. I earlier noted, however, the low likelihood of shared species among India, Burma, and Southeast Asia. The sharing is mainly of membership in species groups, hence similarity in appearance (morphology), behavior, and ecology. Using the concept of ecological or niche taxon-equivalents permits interregional examination of these "taxonomic-ecological" equivalent taxa to interpret the numerous similarities in community taxonomic organization.

Frogs and their inability to prevent and tolerate dehydration are, surprisingly, a major component of the dry forest community and, herpetologically, the component that becomes numerically dominant when the monsoon begins. The most abundant paddy or pond-side frogs consist of two taxon-groups, *Fejervarya* and *Hoplobatrachus* (medium and large species, respectively). These two groups occur at all sites (occasionally a dry forest site has two differently sized species of one of them), and at all sites, *Fejervarya* is abundant throughout the first half of the monsoon, while *Hoplobatrachus* is most visible during the initial weeks and its own reproductive splurge. A waterside skittering species (*Euphlyctis* [India], *Occidozyga* [Burma, Southeast Asia]) occurs abundantly at all sites. Typically two or more *Microhyla* occur at each site; these are small semifossorial taxa. A much larger microhylid (*Glyphoglossus* [Burma, Southeast Asia], *Uperodon* [India]) is present at all sites. There is a conundrum in the dominance of terrestrial frogs in dry forest sites, considering the duration and intensity of aridity at these sites, but the smaller number of aquatic and semiaquatic frogs correlates with the presence of less-permanent water.

Similar interregional taxon-equivalents exist among lizards and snakes. Within lizards, the arboreal, and often upside-down on a tree trunk, *Calotes "versicolor"* occurs broadly in Asian dry forest. *Hemidactylus* geckos, typically two or more species, are present; often one is predominantly terrestrial in spite of the enlarged digital pads, and the other occurs on tree trunks. Several skinks are dry forest residents. *Lygosoma* is always present, and this elongate lizard lives beneath the floor litter.

Snakes are represented in Asian dry forests by a variety of taxon-equivalents, ranging from diurnal (*Ahaetulla*, *Dendrelaphis*) and nocturnal (*Boiga*) tree snakes to semifossorial hognose snakes (*Oligodon*). A *Python* species occurs throughout the dry forest zone, as also do large ratsnakes (*Ptyas*, *Coelognathus*) and a modest-sized terrestrial ambush predator (*Echis* [India], *Daboia* [Burma, Southeast Asia], *Calloselasma* [Southeast Asia]).

Even with the uncertainty of species boundaries and distributions that encompassed more than one geographic area, I identified geographic pairs of frogs, lizards, and snakes for India-Burma and Burma–Southeast Asia. Numerically, there are more Burma–Southeast Asia pairs in all three groups. No India–Southeast Asia pair exists, although a few "species" occur in all three areas.

Finally, I note the rarity of dry forest endemics. A few exist, such as *Naja mandalayensis*, but presently it is impossible to predict whether this low endemicity results from inadequate inventory of dry forest herpetofaunas, hence poor sampling and inadequate taxonomic study, or the broad ecological tolerance of dry forest species and their reproductive success in other habitats—ones less physiologically stressful although potentially with more predators and competitors. It is also noteworthy that many of the widespread taxon-equivalent species live successfully in anthropogenic habitats.

Species Composition—Size Structure

Comparison of communities by body size shows a more-variable structure. Part of this variation arises from the use of proportional representation of community components. Proportional representation is quasi-quantification, and the smaller the fauna, the greater is the effect of the presence or absence of a single species or size class in altering the depiction of organization. Again, the focus is on the three well-surveyed sites.

Small and medium-sized frogs are the dominant classes at all three sites, particularly so at Sakaerat with nearly 50 percent of the species in the small class (Figure 4). The proportion of large and big frogs declines from India to Southeast Asia. For lizards, Chatthin lacks the smallest class; the medium-small and medium classes dominate and are proportionately equal. This pattern contrasts with Nallamala and Sakaerat. In the former, the small and medium-large classes are well represented, although there are somewhat fewer medium-small and medium classes. At Sakaerat, the medium-small

classes numerically dominate, and combined with the medium-large class, these lizards comprise approximately 65 percent of the lizard fauna. The two largest classes
are similar for the three sites. Snakes at the major sites are represented by one or two
small and one large species (*Python*). The middle three size classes dominate (over 80
percent) at the three sites, although in a slightly different way at each. Medium-large
snakes have the lowest representation at all sites; medium-large snakes are about one-
half the snake species at Nallamala, medium-sized ones the majority at Chatthin, and
medium-small and medium snakes roughly equal at Sakaerat. These different frequencies might reflect different dietary cohorts at the three sites; however, diet structuring
of the snake assemblages does not match the size organization.

Behavioral Preferences

Of the three behaviors, only diets suggest community organization. There is a distinct
taxon/clade association of diurnal and nocturnal behavior. Habitat selection displays
more variety in choice, and within all three herp groups, proportional representation
among the habitat classes is similar at Nallamala, Chatthin, and Sakaerat.

Diet is relatively uniform among frogs and lizards. As in the previous two behaviors, snakes utilize a broad variety of prey and, depending upon the species, can be
highly specialized or generalists within a broad prey class. The snake communities of
the three intensively surveyed sites are similarly structured (Table 3).

Cadle and Greene (1993) examined community structure among neotropical rain
forest snakes. Although their emphasis was phylogenetic, the general structuring of
their broad community possesses organization similarities in each behavioral category
of this study. These similarities, in spite of the snake communities' occupancy of different continents and habitat types, support Cadle and Greene's advocacy of phylogenetic
relationships in community organization. The similarity also demonstrates how snake
diversity enables more species to occupy more niches and, as a group, to dominate species representation in a herpetological community but at the "cost" of low abundance.

Relative Abundance

The abundance numbers are derived from the Sakaerat voucher collection (FMNH);
some of these data are also presented in Inger's analyses (Inger and Colwell 1977, Tables
6, 7; Inger 1980, Tables 1, 2) but partitioned differently. The two data tabulations show
the same pattern for lizards and frogs (all species of each) and show similar relative abundance. In both lizards and frogs, a few species have high abundance, but most species
occur at much lower densities. Lizards numerically dominate at Sakaerat, although they
are much less abundant at Chatthin. This latter observation is not supported yet by data
analysis, as abundance data for other Asian dry forest sites are lacking.

In Other Forests

The similarity of community structure among tropical Asian dry forest herpetofaunas was unexpected, and immediately generated questions on the level of similarity

between faunas of dry forest and nearby evergreen forest, and between Asian dry forests and those of other continents. Some tentative answers are possible; a more rigorous examination will follow.

Again relying on the Sakaerat data of Inger and Colwell (1977) and examining only species diversity, we find the adjacent evergreen forest has 87 species, 32 percent each for frogs and lizards, 34 percent for snakes, and 1 percent for turtles. This moister forest has proportionately more frogs and lizards, and fewer snakes. A dry forest area (approximately 10°28′ N 85°22′ W) in Guanacaste Province, Costa Rica (Scott et al. 1983) has 78 species composed of 1 (1 percent) caecilian, 22 (28 percent) frogs, 3 (4 percent) turtles, 17 (22 percent) lizards, and 35 (45 percent) snakes. The total species matches the Sakaerat herpetofauna (Table 2), but the community has fewer lizard species and more frogs. Proportionately, the pattern is similar to that for other tropical Asian dry forests.

PRÉCIS

1. The near-total herpetofauna (greater than 90 percent) of Asian dry forest sites can be determined in less than one year by intensive weekly inventories using a variety of sampling techniques.
2. Three sites, Nallamala, Chatthin, and Sakaerat (representing India, Burma, and Southeast Asia, respectively), serve as the major herpetofaunas for examining community structure in tropical Asian dry forest habitats owing to the intensity of survey effort at these sites.
3. The total Asian dry forest herpetofauna contains more than 45 species of amphibians and 130 species of reptiles, more than one-half of which are snakes.
4. The α-diversity of Nallamala and Sakaerat are nearly equal; the Chatthin herpetofauna is roughly two-thirds of the Nallamala and Sakaerat faunas.
5. Although the total number of species differs between sites, the relative frequency of frogs, lizards, and snakes is quite similar in each dry forest community.
6. A few taxa occur throughout the breadth of Asia's dry forest, but the majority (greater than 50 percent) of the dry forest taxa in each region (India, Burma, Southeast Asia) represents different species, although they are representatives of shared species groups.
7. Size-class organization of the various communities differs slightly among the regional dry forests, especially in the frequencies of the mid-range size classes.
8. There is no apparent community organization in activity patterns or habitat use among frogs, lizards, and snakes.
9. Only snakes show community organization in diet, and the same organization occurs among the three well-surveyed forests.

ACKNOWLEDGMENTS

The Chatthin herpetological team (1997–1999: Htun Win† and Thin Thin† [team leaders], Win Zaw Lhon, Than Zaw Min, and Kyaw Kyaw†; 1999–2000: Mya Than Da Nyeine [team leader], Kyaw Kyaw†, Than Zaw Min, Win Zaw Lhon, Kyi Aung, and Kyaw Zin Tun) was responsible and essential for year-round monitoring of the Chatthin herpetofauna. I greatly appreciate and thank them for their enthusiastic assistance. Special thanks go to Alan Resetar (Field Museum of Natural History) and Jens Vindum (California Academy of Sciences), who queried their respective catalog databases to provide me with date-of-first-capture data for examining species-accumulation rates for the Sakaerat, Shwe-Settaw, and Min-Gon-Taung herpetofaunas.

The Biodiversity Survey and Inventory Program of the Smithsonian's National Museum of Natural History provided support (1997–2000) for my travel to Myanmar and for the Chatthin monitoring team's monthly monitoring and inventory work. The National Science Foundation Biodiversity Surveys and Inventories program (DEB-9971861 and DEB-0451832) has supported our (the herpetological staffs of the California Academy of Sciences, the National Museum of Natural History, and the Myanmar Nature and Wildlife Conservation Division) all-country survey of the Burmese herpetofauna. The preceding survey has provided the data on the Burmese dry forest sites reported herein. These data would have been unattainable without the logistic support and encouragement of directors of the Myanmar Nature and Wildlife Conservation Division: U Uga, U Khin Maung Zaw, and U Tin Tun. Our survey team (1999–2004) was exceptionally diligent, and I am most appreciative of the excellence of their survey and inventory work, often under difficult environmental and climatic conditions. They are Awan Kien Shain, Hla Tun, Htun Win†, Kyi Soe Lwin, Sai Wunna Kyi, San Lwin Oo, and Thin Thin†.

As always, I have received assistance from colleagues and readily thank them for their help. Raoul Bain, Sayantan Biswas, and Whit Gibbons provided prompt and detailed assistance with references on the herpetofaunas of Indochina, India, and the Savannah River Reserve, respectively. Several colleagues (Raoul Bain, Steve Busack, Ron Crombie, Al Leviton, Rom Whitaker, and Pat Zug) reviewed drafts of this chapter and eliminated a variety of errors; I am responsible for those that remain and any failure to accept their good advice.

APPENDIX A

Asian Dry Forest Herpetofaunas

The species are arranged alphabetically within general group by family, then by genus, and finally by species. Taxonomy has been updated from original sources[1] to conform to Frost's amphibian and Uetz's and Hallerman's reptile websites (June 2008).

Taxon	Gir[1]	Nallam[1]	Chatthin[1]	Shwe-S[1]	Sakaerat[1,2]	Cambodia[1,3]
CAECILIANS						
Ichthyophiidae						
Ichthyophis kohtaoensis	–	–	–	–	+	–
FROGS						
Bufonidae						
"*Bufo*" *scaber*	–	+	–	–	–	–
"*Bufo*" *stomaticus*	+	+	–	–	–	–
Duttaphrynus melanostictus	+	+	+	+	+	+
Dicroglossidae						
Euphlyctis cyanophlyctis	+	+	–	–	–	–
Euphlyctis hexadactyla	+	+	–	–	–	–
Fejervarya "*limnocharis*"/std	–	–	+	+	–	–
Fejervarya "*limnocharis*"/small	–	–	+	+	–	–
Fejervarya "*limnocharis*"/India	+	+	–	–	–	–
Fejervarya "*limnocharis*"/Thai	–	–	–	–	+	+
Hoplobatrachus crassus	–	+	–	–	–	–
Hoplobatrachus rugulosus	–	–	+	+	+	+
Hoplobatrachus tigerinus	–	+	+	–	–	–
Occidozyga lima	–	–	+	+	+	+
Occidozyga martenseii	–	–	–	–	+	+
Sphaerotheca breviceps	+	+	+	+	–	–
Sphaerotheca dobsoni	–	+	–	–	–	–
Microhylidae						
Calluella guttulata	–	–	–	–	+	–
Glyphoglossus molossus	–	–	+	–	+	–
Kalophrynus interlineatus	–	–	+	–	–	–
Kaloula mediolineata	–	–	–	–	+	–
Kaloula pulchra	–	–	+	+	+	–
Microhyla berdmorei	–	–	–	–	+	–
Microhyla butleri	–	–	–	–	+	–
Microhyla heymonsi	–	–	–	–	+	+
Microhyla ornata	+	+	+	+	+	+
Microhyla pulchra	–	–	–	+	+	+
Microhyla rubra	–	+	–	+	–	–
Microhyla sp.	–	–	–	+	–	–
Microhyla sp.-mini	–	–	+	–	–	–
Micryletta inornata	–	–	–	–	+	–
Ramanella variegata	–	+	–	–	–	–
Uperodon globulosus	–	+	–	–	–	–
Uperodon systoma	–	+	–	–	–	–
Ranidae						
Hylarana erythraea	–	–	–	–	+	–
Hylarana macrodactyla	–	–	+	–	–	–

Taxon	Gir[1]	Nallam[1]	Chatthin[1]	Shwe-S[1]	Sakaerat[1,2]	Cambodia[1,3]
Hylarana taipehensis	−	−	−	−	−	+
Hylarana sp.	−	+	−	−	−	−
Pelophylax lateralis	−	−	+	−	+	−
Rana johnsi	−	−	−	−	−	+
Ranixalidae						
Indirana leithii	−	+	−	−	−	−
Rhacophoridae						
Chiromantis nongkhorensis	−	−	+	−	+	+
Chiromantis vittatus	−	−	−	−	+	−
Polypedates leucomystax	−	−	+	−	+	+
Polypedates maculatus	−	+	−	−	−	−
TURTLES						
Geoemydidae						
Cyclemys fusca	−	−	−	+	−	−
Cyclemys oldhamii	−	−	−	−	+	−
Melanochelys trijuga	−	+	−	+	−	−
Pangshura tentoria	−	+	−	−	−	−
Testudinidae						
Geochelone elegans	+	+	−	−	−	−
Geochelone platynota	−	−	−	+	−	−
Indotestudo elongata	−	−	+	+	+	−
Trionychidae						
Amyda cartilaginea	−	−	+	−	−	−
Lissemys punctata	+	+	−	−	−	−
Lissemys scutata	−	−	+	+	−	−
LIZARDS						
Agamidae						
Bronchocoela smaragdina	−	−	−	−	−	+
Calotes htunwini	−	−	+	+	−	−
Calotes irawadi	−	−	+	+	−	−
Calotes mystaceus	−	−	+	+	+	+
Calotes rouxii	−	+	−	−	−	−
Calotes "versicolor"	+	+	−	−	+	+
Leiolepis belliana	−	−	+	+	−	−
Leiolepis reevesii	−	−	−	−	+	−
Physignathus cocincinus	−	−	−	−	+	−
Psammophilus blanfordanus	−	+	−	−	−	−
Psammophilus dorsalis	−	+	−	−	−	−
Sitana ponticeriana	+	+	−	−	−	−
Chamaeleonidae						
Chamaeleo zeylanicus	+	+	−	−	−	−
Gekkonidae						
Cnemaspis sp.	−	+	−	−	−	−
Cyrtodactylus sp.	−	−	−	+	−	−
Dixonius siamensis	−	−	−	−	+	−
Geckoella collegalensis	+	−	−	−	−	−
Gehyra lacerata	−	−	−	−	+	−

(Continued)

Taxon	Gir[1]	Nallam[1]	Chatthin[1]	Shwe-S[1]	Sakaerat[1,2]	Cambodia[1,3]
Gehyra mutilata	−	−	−	−	+	+
Gekko gecko	−	−	+	+	+	+
Hemidactylus aquilonius	−	−	+	−	−	−
Hemidactylus brookii	+	+	−	−	−	−
Hemidactylus flaviviridis	+	+	−	−	−	−
Hemidactylus frenatus	−	+	+	+	+	−
Hemidactylus giganteus	−	+	−	−	−	−
Hemidactylus karenorum	−	−	−	+	−	−
Hemidactylus leschenaultii	−	+	−	−	−	−
Hemidactylus platyurus	−	−	−	−	+	+
Hemidactylus reticulatus	−	+	−	−	−	−
Hemidactylus thayene	−	−	−	+	−	−
Hemidactylus triedrus	+	+	−	−	−	−
Lacertidae						
Ophisops jerdonii	+	+	−	−	−	−
Ophisops leschenaultii	−	+	−	−	−	−
Ophisops minor	−	+	−	−	−	−
Takydromus sexlineatus	−	−	−	−	+	−
Scincidae						
Eutropis carinata	+	+	−	−	−	−
Eutropis dissimilis	−	−	+	−	−	−
Eutropis longicaudata	−	−	−	−	+	−
Eutropis macularia	+	+	−	−	+	+
Eutropis multifasciata	−	−	−	+	+	+
Eutropis novemcarinata	−	−	+	+	−	−
Eutropis quadricarinata	−	−	+	−	−	−
Lygosoma albopunctatum	+	−	−	−	−	−
Lygosoma bowringii	−	−	−	−	+	−
Lygosoma guentheri	+	+	−	−	−	−
Lygosoma lineolatum	−	−	+	+	−	−
Lygosoma punctatum	+	+	−	−	−	−
Lygosoma quadrupes	−	−	−	−	−	+
Scincella reevesii	−	−	−	−	+	−
Sphenomorphus indicus	−	−	−	+	−	−
Sphenomorphus maculatus	−	−	+	+	+	−
Varanidae						
Varanus bengalensis	+	+	+	+	+	−
SNAKES						
Boidae						
Eryx conicus	−	+	−	−	−	−
Eryx johnii	−	+	−	−	−	−
Colubridae-Colubrinae						
Ahaetulla nasuta	+	+	+	+	+	−
Ahaetulla prasina	−	−	−	−	+	−
Ahaetulla pulverulenta	+	−	−	−	−	−
Argyrogena fasciolata	+	+	−	−	−	−
Boiga cyanea	−	−	−	−	+	−
Boiga forsteni	−	+	−	−	−	−
Boiga multomaculata	−	−	+	+	+	−
Boiga ochracea	−	−	+	−	−	−
Boiga ocellata	−	−	+	−	+	−

Taxon	Gir[1]	Nallam[1]	Chatthin[1]	Shwe-S[1]	Sakaerat[1,2]	Cambodia[1,3]
Boiga trigonata	+	+	−	−	−	−
Chrysopelea ornata	+	−	+	+	+	−
Coelognathus flavolineata	−	−	−	−	+	−
Coelognathus helena	−	+	−	−	−	−
Coelognathus radiatus	−	−	+	+	+	−
Dendrelaphis pictus	−	−	−	+	+	−
Dendrelaphis subocularis	−	−	+	−	+	−
Dendrelaphis tristis	−	+	−	−	−	−
Dryocalamus davisonii	−	−	−	−	+	−
Gonyosoma oxycephalum	−	−	−	−	+	−
Liopeltis calamaria	−	+	−	−	−	−
Liopeltis stoliczkae	−	−	−	−	−	+
Lycodon aulicus	−	+	+	+	+	−
Lycodon laoensis	−	−	−	−	+	−
Lycodon striatus	+	+	−	−	−	−
Lycodon travancoricus	−	+	−	−	−	−
Oligodon arnensis	+	+	−	−	−	−
Oligodon cinereus	−	−	−	−	+	+
Oligodon planiceps	−	−	−	+	−	−
Oligodon quadrilineatus	−	−	−	−	+	−
Oligodon splendidus	−	−	+	+	−	−
Oligodon taeniatus	−	−	−	−	+	−
Oligodon taeniolatus	+	+	−	−	−	−
Oligodon theobaldi	−	−	+	+	−	−
Oligodon travancoricus	−	+	−	−	−	−
Ptyas korros	−	−	−	−	+	+
Ptyas mucosus	+	+	+	+	+	−
Sibynophis collaris	−	−	+	−	+	−
Sibynophis subpunctatus	+	+	−	−	−	−
Colubridae-Natricinae						
Amphiesma stolatum	−	+	+	+	+	−
Atretium schistosum	−	+	−	−	−	−
Macropisthodon plumbicolor	−	+	−	−	−	−
Rhabdophis chrysargos	−	−	−	−	+	−
Rhabdophis nigrocinctus	−	−	−	−	+	−
Rhabdophis subminiatus	−	−	+	−	+	+
Xenochrophis flavipunctatus	−	−	−	−	+	−
Xenochrophis piscator	+	+	−	+	−	−
Cylindrophiidae						
Cylindrophis ruffus	−	−	+	−	−	−
Elapidae-Elapinae						
Bungarus caeruleus	−	+	−	−	−	−
Bungarus fasciatus	−	−	+	−	+	+
Calliophis maculiceps	−	−	−	−	+	−
Calliophis melanurus	+	−	−	−	−	−
Naja naja	+	+	−	−	−	−
Naja kaouthia	−	−	−	−	+	−
Naja mandalayensis	−	−	+	−	−	−
Elapidae-Psammophiinae						

(Continued)

Taxon	Gir[1]	Nallam[1]	Chatthin[1]	Shwe-S[1]	Sakaerat[1,2]	Cambodia[1,3]
Psammophis condanarus	–	–	–	+	–	–
Psammophis leithii	+	–	–	–	–	–
Homalopsidae						
Enhydris enhydris	–	+	–	–	–	–
Enhydris plumbea	–	–	–	–	+	+
Homalopsis nigroventralis	–	–	–	–	–	+
Pareatidae						
Pareas carinatus	–	–	–	–	+	–
Pareas margaritophorus	–	–	–	–	–	+
Pythonidae						
Python molurus	+	+	+	–	–	–
Python reticulatus	–	–	–	–	+	–
Typhlopidae						
Grypotyphlops acutus	–	+	–	–	–	–
Ramphotyphlops braminus	+	+	+	–	+	–
Typhlops porrectus	+	–	–	–	–	–
Viperidae						
Calloselasma rhodostoma	–	–	–	–	+	+
Crypteltropis albolabris	–	–	+	–	+	–
Daboia russelii	–	+	–	–	–	–
Daboia siamensis	–	–	+	+	–	–
Echis carinatus	+	+	–	–	–	–
Trimeresurus gramineus	–	+	–	–	–	–
Viridovipera vogeli	–	–	–	–	+	–
Xenopeltidae						
Xenopeltis unicolor	–	–	+	–	+	–

[1]Species occurrence sources: Gir—Bhatt et al. 1999, and Vyas 2000; Nallamala—Srinivasulu and Das 2008; Chatthin—Zug et al. 1998, personal data, CAS, MBM-NWCD, and USNM; Shwe-Settaw—Zug et al. 2004, CAS, and USNM; Sakaerat—Inger and Colwell 1977 (Table 1), taxonomy updated (June 2008) from FMNH database; hilly eastern Cambodia—Stuart et al. 2006.
[2]This list includes the species occurrences for both the deciduous forest and agricultural lands listed separately in Table 1 of Inger and Colwell (1977). Our sampling at Chatthin and Shwe-Settaw was not as precisely recorded as Inger's more rigorous data gathering; additionally he notes, "Though collected only in agricultural land, species almost certainly occur[s] in deciduous forest."
[3]This list of species occurrences (Stuart et al. 2006) includes records from anthropogenic habitats (Table 2 in Stuart et al.) and dry deciduous forest. Species from evergreen forest were purposefully excluded, even if deciduous forest occurs at the site, because data did not allow discrimination of precise habitat origin of voucher specimens.

APPENDIX B

Size and Ecological Coding for Asian Dry Forest Amphibians and Reptiles

Size—Each of the four groups (frogs, turtles, lizards, snakes) is individually categorized for body size (carapace length [cl] for turtles; snout-vent length [svl] for the other three groups). All size data are in millimeters (mm). Adult size data derive from numerous sources (literature; G. Zug, unpublished data) and include ranges for females, ranges for males, median/mean for females, and maximum size. These data and the subsequent ecological coding-data are available from the author, as space is not available for presentation in this book. From the size data, incomplete for many taxa, I selected a midpoint size representative for each taxon using the median or mean when available or an estimate of the midpoint when median or mean was not available. These midpoints for each herp group were plotted (bar graphs) to identify size classes by clustering of midpoints. This strategy yielded five size classes for frogs (less than 20, 22–44, 46–70, 72–96, and greater than 96 mm); three classes for turtles (less than 230, 250–300, and greater than 400 mm); six classes for lizards (less than 42, 44–62, 64–94, 98–118, 120–148, and greater than 150 mm); and five classes for snakes (less than 300, 300–590, 600–990, 1,000–1,500, and greater than 1,500 mm).

Habitat Preference—These ecological categories are at a gross level, in part because of our incomplete knowledge for many taxa but also because broader categories allow a more even comparison of community structure between dry-forest sites. The categories are fossorial-semifossorial, terrestrial, arboreal (usually found off the ground in shrubs and trees), semiaquatic (typically waterside, often feeding there), and aquatic (uncommonly found outside of water).

Activity Time—This ecological parameter identifies whether a taxon pursues most of its life-history activities during daylight (diurnal) or at night (nocturnal). Observations are insufficient to recognize any truly crepuscular taxa, and many diurnal reptile taxa shift to dawn and twilight behavior when daily temperatures soar.

Diet—Most amphibians and reptiles are carnivorous as juveniles and adults, hence herbivory is not subdivided. A few reptiles, mainly turtles, are omnivores. The carnivores' prey are partitioned into nine categories: insects and other invertebrates, fish, amphibians, reptiles, ectothermic vertebrates, birds, mammals, endothermic vertebrates, and vertebrates (amphibians to mammals). Some taxa specialize in one life-history stage, for example, reptile eggs for *Oligodon*; such specializations are not coded, and the taxon's diet is considered as one of the preceding nine categories.

NOTE

1. This study and chapter are dedicated to Robert F. Inger and the staff of the Chatthin Wildlife Sanctuary 1997–2000. Dr. Inger's studies of the Asian herpetofauna span the full spectrum of systematics and ecology, and are a hallmark of thoughtful design, intensive fieldwork, and rigorous analysis. His Sakaerat community research encouraged and guided my research at Chatthin Wildlife Sanctuary. The friendship and enthusiastic support of the Chatthin staff made my research visits productive and enjoyable.

REFERENCES

Bain, R. H., and N. Q. Truong. 2004. Herpetofaunal Diversity of Ha Giang Province in Northeastern Vietnam, with Descriptions of Two New Species. *American Museum Novitate* 3454:1–42.

Bhatt, K., R. Vyas, and M. Singh. 1999. Herpetofauna of Gir Protected Area. *Zoos' Print Journal* 14 (5): 27–30.

Cadle, J. E., and H. W. Greene. 1993. Phylogenetic Patterns, Biogeography, and the Ecological Structure of Neotropical Snake Assemblages. In *Species Diversity in Ecological Communities*:

Historical and Geographical Perspectives, ed. R. E. Ricklefs and D. Schluter, 281–93. Chicago: University of Chicago Press.

Campden-Main, S. 1970. *A Field Guide to the Snakes of South Vietnam*, i–iv, 1–114. Washington, DC: Division of Reptiles and Amphibians, National Museum of Natural History, Smithsonian Institution.

Cherian, P. T., K. Rema Devi, and M. S. Ravichandran. 2000. Ichthyo and Herpetofaunal Diversity of Kalakad Wildlife Sanctuary. *Zoos' Print Journal* 15 (2): 203–6.

Drury, W. H., Jr. 1998. *Chance and Change: Ecology for Conservationists*, i–xxiii, 1–223. Berkeley: University of California Press.

Dubois, A. 1975. Un nouveau complexe d'espèces jumelles distinguées par le chant: Les grenouilles du Népal voisines de *Rana limnocharis* Boie (amphibiens, anoures). *Comptes rendus de l'Académie des Sciences, Paris*, séries D 281, 1717–20.

———. 1984. Note préliminaire sur le groupe de *Rana limnocharis* Gravenhorst, 1829 (amphibiens, anoures). *Alytes* 3 (4): 143–59.

Dutta, S. K. 1997. A New Species of *Limnonectes* (Anura: Ranidae) from Orissa, India. *Hamadryad* 22 (1): 1–8.

Dutta, S. K., and N. Singh. 1996. Status of *Limnonectes limnocharis* (Anura: Ranidae) Species Complex in Asia. *Zoos' Print Journal* 11 (8): 15, 21.

Frost, D. R. 2007. *Amphibian Species of the World: An Online Reference*. Version 5.1, http://research.amnh.org/herpetology/amphibia/index.php (accessed October 2007).

Gibbons, J. W., and R. D. Semlitsch. 1991. *Guide to the Reptiles and Amphibians of the Savannah River Site*. Athens: University of Georgia Press.

Heatwole, H. 1982. A Review of Structuring in Herpetofaunal Assemblages. In *Herpetological Communities*, ed. N. J. Scott, Jr., 1–19. Wildlife Research Report 13. U.S. Department of the Interior Fish and Wildlife Service.

Inger, R. F. 1954. Systematics and Zoogeography of Philippine Amphibia. *Fieldiana: Zoology* 33 (4): 183–531.

———. 1980. Relative Abundance of Frogs and Lizards in Forests of Southeast Asia. *Biotropica* 12 (1): 14–22.

Inger, R. F., and R. K. Colwell. 1977. Organization of Contiguous Communities of Amphibians and Reptiles in Thailand. *Ecological Monographs* 47:229–53.

Köhler, J., D. R. Vietes, R. M. Bonett, F. H. García, F. Glaw, D. Steinke, and M. Vences. 2005. New Amphibians and Global Conservation: A Boost in Species Discoveries in a Highly Endangered Vertebrate Group. *BioScience* 55 (8): 693–96.

Kuramoto, M., S. H. Joshy, A. Kurabayashi, and M. Sumida. 2007. The Genus *Fejervarya* (Anura: Ranidae) in Central Western Ghats, India, with Descriptions of Four New Cryptic Species. *Current Herpetology* 26 (2): 81–105.

Luhring, T. M. 2008. "Problem Species" of the Savannah River Site, Such as Brimley's Chorus Frog (*Pseudacris brimleyi*), Demonstrate the Hidden Biodiversity Concept of an Intensively Studied Government Reserve. *Southeastern Naturalist* 7 (2): 371–73.

Matsui, M., H. Ito, T. Shimada, H. Ota, S. K. Saidapur, W. Khonsue, T. Tananka-Ueno, and G.-F. Wu. 2005. Taxonomic Relationships within the Pan-Oriental Narrow-Mouth Toad *Microhyla ornata* as Revealed by mtDNA Analysis (Amphibia: Anura, Microhylidae). *Zoological Science* 22:489–95.

Matsui, M., T. Seto, and T. Utsunomiya. 1986. Acoustic and Karyotypic Evidence for Specific Separation of *Polypedates megacephalus* from *P. leucomystax. Journal of Herpetology* 20 (4): 483–89.

Matsui, M., M. Toda, and H. Ota. 2007. A New Species of Frog Allied to *Fejervarya limnocharis* from the Southern Ryukyus, Japan (Amphibia: Ranidae). *Current Herpetology* 26 (2): 65–79.

Orlov, N. L., A. Lathrop, R. W. Murphy, and Ho Thu Cuc. 2001. Frogs of the Family Rhaco-phoridae (Anura: Amphibia) in the Northern Hoang Lien Mountains (Mount Fan Si Pan, Sa Pa District, Lao Cai Province), Vietnam. *Russian Journal of Herpetology* 8 (1): 17–44.

Scott, N. J., J. M. Savage, and D. C. Robinson. 1983. Checklist of Reptiles and Amphibians. In *Costa Rican Natural History*, ed. D. H. Janzen, 367–74. Chicago: University of Chicago Press.

Srinivasulu, C., and I. Das. 2008. The Herpetofauna of Nallamala Hills, Eastern Ghats, India: An Annotated Checklist, with Remarks on Nomenclature, Taxonomy, Habitat Use, Adaptive Types and Biogeography. *Asiatic Herpetological Research* 11:110–31.

Stuart, B. L., K. Sok, and T. Neaung. 2006. A Collection of Amphibians and Reptiles from Eastern Hilly Cambodia. *Raffles Bulletin of Zoology* 54 (1): 129–55.

Uetz, P., and J. Hallermann. 2008. *The Tiger Reptile Database*. www.reptile-database.org.

Veith, M., J. Kosuch, A. Ohler, and A. Dubois. 2001. Systematics of *Fejervarya limnocharis* (Gravenhorst, 1829) (Amphibia, Anura, Ranidae) and Related Species. 2. Morphological and Molecular Variation in Frogs from the Greater Sunda Islands (Sumatra, Java, Borneo) with the Definition of Two Species. *Alytes* 19 (1): 5–28.

Vijayakumar, S. P., A. Ragavendran, and B. C. Choudhury. 2006. Herpetofaunal Assemblage in a Tropical Dry Forest Mosaic of Western Ghats, India: Preliminary Analysis of Species Composition and Abundance during the Dry Season. *Hamadryad* 30 (1, 2): 41–54.

Vyas, R. 2000. Supplementary Notes on Herpetofauna of Gir Forets [sic]. *Zoos' Print Journal* 15 (5): 263–64.

———. 2004. Herpetofauna of Vansda National Park, Gujarat. *Zoos' Print Journal* 19 (6): 1512–14.

Whiteman, H. H., T. M. Mills, D. E. Scott, and J. W. Gibbons. 1995. Confirmation of Range Extension for the Pine Woods Snake (*Rhadinaea flavilata*). *Herpetological Review* 26 (3): 158.

Wikramanayake, E., E. Dinerstein, C. J. Loucks, D. M. Olson, J. Morrison, J. Lamoreux, M. McKnight, and P. Hedao. 2002. *Terrestrial Ecoregions of the Indo-Pacific: A Conservation Assessment*. Washington, DC: Island Press.

Wüster, W. 1996. Taxonomic Changes and Toxicology: Systematic Revisions of the Asiatic Cobras (*Naja naja* species complex). *Toxicon* 34 (4): 399–406.

Wüster, W., and R. S. Thorpe. 1989. Population Affinities of the Asiatic Cobra (*Naja naja*) Species Complex in South-East Asia: Reliability and Random Resampling. *Biological Journal of the Linnean Society* 36:391–409.

———. 1990. Systematics and Biogeography of the Asiatic Cobra (*Naja naja*) Species Complex in the Philippine Islands. In *Vertebrates in the Tropics*, ed. G. Peters and R. Hutterer, 333–44. Bonn: Museum Alexander Koenig.

———. 1992. Asiatic Cobra: Population Systematics of the *Naja naja* Species Complex (Serpentes: Elapidae) in India and Central Asia. *Herpetologica* 48 (1): 69–85.

Zug, G. R., H. H. K. Brown, J. A. Schulte II, and J. V. Vindum. 2006. Systematics of the Garden Lizards, *Calotes versicolor* Group (Reptilia, Squamata, Agamidae), in Myanmar: Central Dry Zone Populations. *Proceedings of the Californian Academy of Sciences* 57 (1): 1–33.

Zug, G. R., Htun Win, Thin Thin, Than Zaw Min, Win Zaw Lhon, and Kyaw Kyaw. 1998. Herpetofauna of the Chatthin Wildlife Sanctuary, North-Central Myanmar with Preliminary Observations of Their Natural History. *Hamadryad* 23:111–20.

Zug, G. R., Sai Wunna Kyi, and Htun Win. 2004. Shwesettaw Wildlife Sanctuary (Myanmar) Turtles. *Turtle & Tortoise Newsletter* 8:4.

16

The Impact of Local Communities on a Dry Tropical Forest Sanctuary

A Case Study from Upper Myanmar (Burma)

Melissa Songer, Myint Aung, Khaing Khaing Swe, Thida Oo, Chris Wemmer, Ruth DeFries, and Peter Leimgruber

Seasonally dry tropical forests are less common, more degraded, and proportionately more threatened than rain forests throughout the world (Janzen 1988; Wilson 1992). Yet these forests receive little attention from the conservation community, and most research, policy, and conservation efforts are directed toward rain forest protection (Janzen 1988; Wilson 1992; Bullock et al. 1995; Sanchez-Azofeifa et al. 2005). Though tropical dry forests have lower levels of biodiversity, they are important habitats for many endangered species, have unique ecological and physiological adaptations, and provide essential ecosystem services to the people living nearby (Stott 1990; Bullock et al. 1995; Gentry 1995; Medina 1995).

Conventional protected area strategies, strictly regulating access to the forest through an external administrative force and prohibiting resource use are not likely to succeed in conserving tropical dry forests because frequently these forests are found in areas that are densely populated, have fertile soils, and lie on low-elevation and low-relief topography allowing irrigation (Bunyavejchewin 1982; Maas 1995). Higher human population levels result in higher extraction pressures and greater threat to these dry forests (Murphy and Lugo 1986; Bullock et al. 1995). Murphy and Lugo (1986) estimate that globally as much as 80 percent of wood extraction from tropical dry forests is for the purpose of cooking and heating rather than timber (Murphy and Lugo 1986). Other important products derived from dry forests include medicines, rubber, cork, resin, cooking oils, wild game, fruit, nuts, and materials for shelter (Stott 1990; Murali et al. 1996).

Even protected dry forest areas frequently experience high levels of use (Myint Aung et al. 2004). Understanding resource requirements and use in local communities adjacent to remaining tropical dry forests is essential for developing strategies for sustainable use and long-term conservation of these ecosystems. To examine these issues we selected Chatthin Wildlife Sanctuary (CWS), Myanmar, for a case study. Like other protected areas in tropical dry forest, CWS faces intense pressures from surrounding communities. Over the past century, land cover in and around CWS was converted from an intact mixed deciduous forest into a mosaic of degraded second-growth forest (Myint Aung et al. 2004). Most large mammals and all top predators have disappeared from the area during the past twenty years, making it more accessible for grazing and extraction. Increased law enforcement, research and training activities, and transfer of many staff from headquarters located thirty-five kilometers away to inside the park, began during the past decade. However, despite increased patrolling, efforts to limit encroachment, and community outreach, the forest continues to be degraded (Myint Aung et al. 2004; Songer et al. 2009). We conducted a comprehensive study of the resource use and needs of local communities to determine why the degradation of CWS is continuing. Specifically, we addressed the following research questions:

1. How do local people use forest products at CWS?
2. Which essential household needs drive forest product use?
3. What factors characterize households with the highest levels of forest product use?

MATERIALS AND METHODS

Study Area

Chatthin Wildlife Sanctuary covers 269 km² with an additional buffer zone of 53 km² around the southern boundary (Figure 1). It is located on the northern edge of the central dry zone (95°24′ E–95°40′ E, 23°30′ N–23°42′ N), which has the highest population density in the country. Historically, the central dry zone was dominated by forest (Kurz 1877; Stamp 1925) but now has mostly been converted to agriculture. The climate is monsoonal, with three seasons including the rainy season (approximately 40 cm rainfall annually), the cool-dry season, and hot-dry season (Myint Aung et al. 2004). Crops are grown for half the year, August through January. During the dry seasons the forest floor burns from human and natural causes, and green shoots begin to appear soon after burning.

Chatthin Wildlife Sanctuary is one of the older parks in Myanmar and has a complex environmental history that has been described in detail (Myint Aung et al. 2004). Starting out as a fuel reserve in 1919 under British rule, CWS was gradually transformed into a sanctuary to conserve the endangered Eld's deer, *Cervus eldi*. Active management of the sanctuary began in 1955 and staff consisted of one person until the early 1970s. Gradually staff increased and is currently made up of about sixty people. Three villages located inside the boundaries have been "grandfathered" in; another thirty-one villages are within ten kilometers of CWS, including about 4,000 households and over 25,000 people (Figure 1).

Survey Design and Sampling

In 1999 a Smithsonian Institution research team conducted interview surveys to assess local people's perceptions of CWS and its staff (Allendorf et al. 2006). We built on this

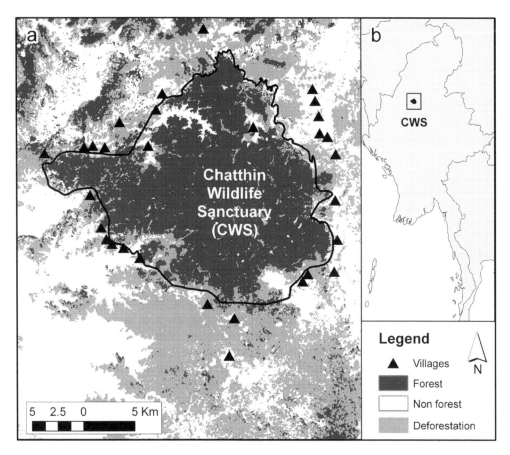

Figure 1. (*a*) Chatthin Wildlife Sanctuary (CWS) and surrounding villages (background based on land cover change analysis comparing satellite imagery from 1973 and 2005; see also Songer et al. 2009); (*b*) location of CWS in Myanmar (Burma).

work and added components to assess forest product extraction and socioeconomic conditions. In 2003, we retrained two schoolteachers from Chatthin village, Khaing Khaing Swe and Thida Oo, in survey methods. Teachers are ideal interviewers; they are well-known and respected, and local people are less inhibited to talk openly with a teacher than with CWS staff. Participants understood the teachers were working with CWS staff, and they were assured of anonymity and that they could answer without fear of repercussions.

Surveys took place during June, July, and August 2003, approximately two to four months after the end of harvest season in February. Our team conducted interviews by selecting households randomly from lists of residents in twenty-eight villages in and around CWS. Using stratified-random sampling, the team then selected two females and two males from each of seven predefined age groups for interviews, resulting in an even distribution across ages: 18–25 (n = 107), 26–33 (n = 118), 34–41 (n = 158), 42–49 (n = 148), 50–57 (n = 104), 58–65 (n = 74), and over 65 (n = 75). The average

age of respondents was 43 and ranged from 18 to 84. Interviewers asked questions and recorded answers in Burmese during one-hour interviews. Answers were subsequently translated into English, recorded in a spreadsheet, and categorized as qualitative responses for statistical analysis.

Questions were mostly open-ended, with eight fixed response questions included. Twelve questions were directed specifically toward the respondent, such as information regarding age, gender, occupation, education, and how many times they visit the forest each month. Fifty-five questions covered household information rather than information about the individual. Interviews covered general household information, agricultural practices, use of forest products by the household, and household assets.

Forest Product Surveys

Respondents were asked to list household products and food gathered by members of their household from the forest, whether the products were collected inside or outside the sanctuary, and to estimate the level of the product's importance to the household (ranging from not necessary to essential). With regard to medicinal plants, each respondent was asked about types of plants they used for medicine and where they were collected. Respondents were asked where they obtained fuelwood—whether they collected it inside the sanctuary, outside the sanctuary, or purchased it. Questions about hunting practices of the household included what animals were hunted and where they were hunted.

Each respondent was asked what foods the household had eaten the day prior to the interview and where each item was obtained, that is, from their own fields or livestock, purchased at the market, collected in the forest, or hunted. The respondents were asked about household cooking and heating habits and whether the household would be willing to use a stove rather than an open fire for cooking.

Only nineteen households (2.4 percent) did not list any use of forest products. Since forest product use is so pervasive, we relied on two other factors to indicate high-use levels. We considered households that had eaten something from the forest the day before the survey, and households that hunted inside the sanctuary, as having high-use levels.

Socioeconomic and Demographic Surveys

People in rural areas in Myanmar often rely on trade of goods and products rather than cash transactions. Therefore, to determine household wealth, we measured several household assets in addition to monetary income. These assets included amount of cash income earned, amount of land owned, amount of rice surplus or shortage, and number of livestock. To combine different livestock into one household asset, we calculated a "protein" score for each household, based on the average weight of each livestock species (Food and Agriculture Organization of the United Nations 2003). We multiplied the average weights by the number of each species owned by the household and summed the totals.

Education levels are based on the government standard levels 1–10, which may be followed by university training, either through distance learning or by attending classes. Thirty-three percent of the respondents received their education through monasteries rather than the standard government system. We categorized this form of education as 1.5 since it is the equivalent of the government's standard level 1 or 2 (Myint Aung, personal communication). The majority of the people living around the sanctuary are Buddhists, with some Muslims and Hindus, not exceeding 5 percent of the population.

Statistics

Much of the survey data are not normally distributed, so we used Kruskal-Wallis analysis of variance by ranks for comparing two groups. For example, households reporting high-use of forest product were compared to groups with low-use of forest product.

RESULTS

Villagers in the area are primarily farmers (82 percent), with some wage laborers (11 percent), and fishermen (3 percent). Households average 5.7 people and have 2.4 ha of land (median: 3 ha), primarily rice, with some maize, sugar cane, and ground nuts. The average cash income earned for goods or services is $579/year (median: $421/year), ranging from $0 to $10,316/year per household. Education levels on average are low (3.1 on the government standard system of 1–10 levels).

Household Use of Forest Products

Survey results show CWS communities depend on the sanctuary for fuelwood, food, medicine, and building materials (Figure 2). Almost all forest products are collected for direct use and represent essential subsistence items in this region. Only 13 percent of respondents sell a portion of what they collect; the rest use all products to meet household needs. Over forty-four forest products are used for food or household needs. On a scale ranging from essential to unnecessary (Janzen 1988; Stott 1990; Wilson 1992; Bullock et al. 1995; Gentry 1995), most respondents ranked at least one of these products as essential (62 percent) or very important (36 percent), and none as unnecessary.

Hunting, though rarely admitted in conversation, still appears to be an important way to acquire protein. Over 34 percent of respondents confirmed hunting and fishing in the area by household members. A considerable number (14 percent) admitted to illegal hunting and fishing inside the sanctuary. Fish and frogs are collected by over 80 percent and 60 percent of households, respectively (Figure 3a). At least 20 percent of the hunters also take medium-sized mammals such as muntjac (*Muntiacus muntjak*), pangolin (*Manis javanica*), rabbit (*Lepus peguensis*), jungle cat (*Felis chaus*), and rat (*Rattus* sp.). Other important forest uses include wood for houses, and grazing livestock (Figure 2). Most families (96 percent) manage livestock, including an average of 3.4 cows, 1.5 buffalos, 2.4 pigs, and 9.7 chickens per household. Cows and buffalo are grazed primarily in paddy fields (78 percent), though some grazing occurs inside the sanctuary (14 percent) and buffer zone (40 percent), and 31 percent of the livestock are fed purchased grain (Figure 3b).

Most households (75 percent) rely on forests to provide food to supplement their harvest, with 8 percent reporting they never go to the market and rely solely on

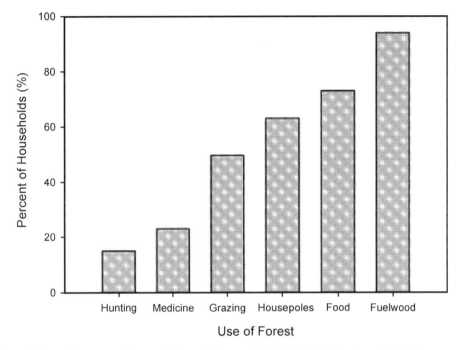

Figure 2. Overall forest product use by households near Chatthin Wildlife Sanctuary (*n* = 784).

farming and forest products (Figure 3c). Forest foods are used so regularly that 41 percent of respondents confirmed that someone in their household had eaten a forest product on the previous day. We identified over seventy medicinal plant species during the survey, and 22 percent of all households collect at least one. Plant gathering was mostly done inside the sanctuary (75 percent of households).

Socioeconomics of Forest Product Use

Socioeconomic conditions drive the use of forest resources in rural communities surrounding CWS. The villages are subsistence-based agricultural communities relying on nearby forests to supplement frequently meager rice and livestock harvests. Just over half of all surveyed households either typically did not grow enough rice to support their members or grew just enough with no surplus.

Poor households with little education depend on forest products for subsistence (Table 1). Land owned and yearly income per household member were significantly lower for households that ate forest products the day prior to the survey or that hunted inside the sanctuary than for other households (Table 1). These households also had significantly greater rice shortages. In addition, both high-use groups had significantly larger families with significantly lower education levels than other households. Not surprisingly, both hunting and forest food consuming households visited the sanctuary significantly more often than other households.

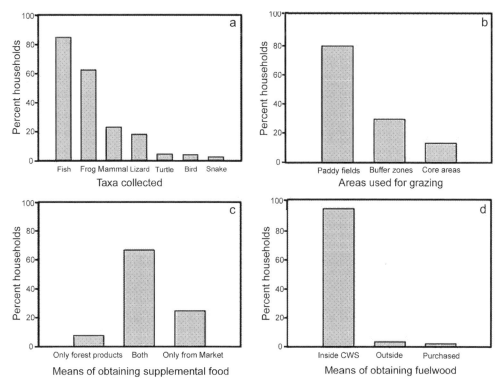

Figure 3. Summary of forest product use near Chatthin Wildlife Sanctuary (CWS): (*a*) protein sources from hunting and fishing; (*b*) locations used for grazing; (*c*) means of obtaining supplemental food; (*d*) means of obtaining fuelwood (*n* = 784).

Fuelwood

Almost all households utilize wood collected from the sanctuary for cooking fuel, and fuelwood is the most commonly extracted forest product. Usually, households (94 percent) collect fuelwood inside the sanctuary, including core areas and buffer zone. The remaining households collect outside the sanctuary (3 percent) or purchase fuel (2 percent; Figure 3d). Households primarily use branches or trees; only five households used charcoal. All cook meals twice daily, and during the cold season fires are kept up throughout the day for heat.

DISCUSSION

Resource Use Patterns and the Decline of CWS

Our research shows how the livelihood of local people is closely intertwined with the sanctuary; communities rely on the forest to meet basic needs including food, shelter, and medicine. Previous analysis of a series of satellite imagery covering the area from

Table 1. Factors affecting the livelihood of households that eat forest products or hunt inside Chatthin Wildlife Sanctuary. Statistics are based on Kruskal-Wallis one-way analysis of variance.

| Variable | Mean | | Mann-Whitney | |
	Use	Non-use	U Statistic	p Value
	Hunting inside the sanctuary			
Household size	5.6	5.7	25 487.5	0.798
Household education level	3.3	3.5	26 947.0	0.203
Yearly income/household size (US$)	$96.37	$108.23	32 651.5	0.000**
Land/household size (ha)	0.50	0.55	29 953.5	0.029*
Livestock (kg)	1,202.0	1,152.0	24 839.5	0.546
Rice shortage/household size (bins[a])	8.4	6.8	22 583.0	0.050*
Distance (km) to sanctuary	1.20	1.94	32 479.0	0.000**
Visits to park per month	4.9	1.5	15 001.0	0.000**
	Consumption of foods from inside the sanctuary			
Household size	5.84	5.52	66 938.0	0.017*
Household education level	3.32	3.64	83 636.5	0.000**
Income/household size (US$)	$93.56	$114.61	81 872.0	0.014*
Land/household size (ha)	0.51	0.58	81 386.5	0.022*
Livestock (kg)	1,172.0	1,147.0	69 501.5	0.128
Rice shortage/household size (bins[a])	8.41	6.79	22 583.0	0.050*
Distance (km) to sanctuary	1,739.57	1,963.85	78 752.5	0.147
Visits to park per month	2.46	1.32	51 990.0	0.000**

*$p \leq 0.05$;
**$p \leq 0.0001$
[a] one bin = approximately forty liters.

1973 to 2005 has shown a steady deterioration of the CWS forest from the boundary inward, in a pattern suggesting small-scale extraction rather than broad-scale conversion (Songer et al. 2009; see also Figure 1). Though CWS staff have prevented large-scale timber removal and agricultural conversions within the sanctuary boundaries, we continue to observe degradation and forest loss as well as the loss of all large predators and decline of many mammal species (Myint Aung et al. 2004). In Myanmar, a survey including two-thirds of the protected areas found 85 percent had evidence of forest product extraction, and this ranked highest among activities recognized as incompatible with protected area values (Rao et al. 2002). Unchecked and continued use of CWS by the local population will likely result in resource depletion, degradation of one of the largest remaining patches of protected dry tropical forest in Southeast Asia, and further loss of critical habitat for the last remaining viable population of Eld's deer.

The high resource use levels we demonstrate for CWS are found in tropical dry forest regions throughout the world (Singh et al. 1984; Appasamy 1993; Chopra 1993; Ganesan 1993; Hegde et al. 1996; Murali et al. 1996; Wickramasinghe et al. 1996; Abbot and Mace 1999; Mahapatra et al. 2005), contributing to a global decline that is more severe for this forest ecosystem than for others. Research shows that dry forests have the highest rates of deforestation compared to other tropical forest types throughout the world (Janzen 1988; Wikramanayake et al. 2001; DeFries et al. 2005). Recent studies show that deforestation in Myanmar demonstrated that dry dipterocarp forests were especially threatened by deforestation (Leimgruber et al. 2005) and that the dry forests that remain are less protected than other forest types (Songer 2006). Degradation of forest and loss of species used heavily by humans has been shown in many tropical dry forests (Appasamy 1993; Ganesan 1993; Murali et al. 1996; Peh et al. 2005; Karanth 2002; Davidar et al. 2007).

Fuelwood extraction is the most pressing threat to CWS; it is required on a daily basis for cooking by most households. Heavy reliance on protected dry forests for fuelwood and fodder has been reported across the tropics (V. Singh and J. Singh 1989; Appasamy 1993; Ganesan 1993; Fox et al. 1996; Wickramasinghe et al. 1996; Davidar et al. 2007). Research on fuelwood and fodder extraction in dry forests of India has shown that the amount of wood and fiber extracted from forests is well beyond what is sustainable, resulting in steady degradation of the forests and depletion of natural resources (Pandey and J. Singh 1984; V. Singh and J. Singh 1989); Ganesan 1993; Bahuguna 2000). Not surprisingly, fuelwood availability is declining at CWS. Over the past five to ten years, local people have traveled further into the forest and spent more time collecting enough wood than previously (Maung Ngwe and Taw Hla, personal communication). The same pattern has been observed in dry forests of the central Himalayan hills where women have to go further each year to collect fuelwood (J. Singh et al. 1984).

Grazing also has a negative impact on CWS forests that may continue to increase. More than 50 percent of households with livestock graze their animals inside CWS, an activity that has been shown to negatively impact forest regeneration and biodiversity (Ganesan 1993; Yonzon and Hunter 1991; Fox et al. 1996; Cabin et al. 2000; Udaya Sekhar 2003). Grazing is a subsistence strategy for low-income households that cannot afford land but have enough resources to purchase livestock. Livestock purchases require less capital, and animals can be grazed on common land or inside CWS. This conclusion is supported by our results showing that high-use households had significantly less land but on average had more livestock. In developing alternative use strategies, increasing livestock does not appear to be a means of reducing forest use, unless the livestock is contained entirely outside CWS. Curtailing grazing inside CWS needs to be a priority for enforcement as human population grows and available land outside the sanctuary diminishes.

Forest Product Use and Protected Areas

Poverty is prevalent in Myanmar's rural populations (Van Schendel 1991; Mya Maung 1998; Booth 2003) and is the driving force behind substantial extraction of food, fuelwood, fodder, medicine, and housing materials from CWS. Poverty, and limited resources, education, and work opportunities pressure rural populations to deplete their natural resources, which makes it even more difficult for people who depend on the forests to overcome poverty (Shaw 1989; Appasamy 1993; Ganesan 1993; Hegde et al. 1996; Mahapatra et al. 2005). These resources are especially helpful for the livelihood of half the people at CWS—the half that reported shortages in the amount of rice they needed and have to rely on forest products to supplement their crops. Studies show that forest products also provide safety nets or natural insurance in times of economic difficulties, disease, and natural disasters, and are particularly important for poor households (Barham et al. 1999; Pattanayak and Sills 2001; Shackleton and Shackleton 2004; Das 2005). Our results parallel those found by research into communities around dry tropical forests, which shows that people with fewer means, such as land and income (V. Singh and J. Singh 1989; Appasamy 1993; Ganesan 1993; Hegde et al. 1996; Wickramasinghe et al. 1996; Abbot and Mace 1999; Barham et al. 1999; Das 2005; Mahapatra et al. 2005), and lower education (Hegde et al. 1996), are more dependent on forest products.

In the past, efforts to mitigate extraction impacts at CWS consisted mainly of patrolling and enforcement of wildlife laws, aimed at stopping hunting and timber extraction. Our results show that ramping up protection and penalties for extraction will impact the poorest segments of CWS communities the most. Other studies of communities adjacent to protected dry forests in Nepal have reached the same conclusion (Lehmkuhl et al. 1988; Reddy and Chakravarty 1999). Poorer households are already disadvantaged due to lack of access to resources, education, and opportunities; taking away access to resources they have traditionally used puts them at further disadvantage and has the potential to negatively affect their attitudes toward protected areas and biodiversity conservation. Strict protectionist approaches that do not take people into account can be counterproductive by causing animosity between local people and forest officials and negative attitudes toward protected areas (Das 2005; Alkan et al. 2009).

In many protected areas such as CWS, significantly increasing patrolling and enforcement efforts is logistically and financially impossible due to limited budgets and insufficient staff (Myint Aung et al. 2004; Myint Aung 2007). This is a problem facing most protected areas in Myanmar (Rao et al. 2002; Myint Aung 2007) and throughout the tropics (Brunner et al. 1998). Even in highly valued protected areas such as the tiger sanctuaries of the Nilgiris Biosphere Reserve, there are not enough forest officers to stop illegal extraction of fuelwood and fodder (Ganesan 1993). Budgetary outlooks for protected areas are not likely to improve in the near future. This is not to suggest that protection and enforcement efforts are not important or cannot be effective to some extent, but rather that there are limitations to what can be accomplished solely with this approach.

Possible Mitigation Strategies

Given that our results, and studies of tropical dry forests throughout, show fuelwood is the most pervasive need for communities living near dry forest, developing strategies for reducing the impact of fuelwood extraction should be a top priority for dry forest conservation. Potential strategies range from simple and inexpensive steps that can be taken by individual households to more-complex programs that require resources and cooperation from many stakeholders.

Overall demand for fuelwood can be reduced through improved fuel efficiency and alternative fuels (Datta and Dutt 1981; Kalia 2000; Adeoti et al. 2001; Bazile 2002; Nijaguna 2002). Technological developments such as biogas stoves and solar ovens are improving and becoming more available. In Myanmar, simple, energy-efficient, inexpensive stoves are easily available, yet currently most households around CWS still rely on traditional open fires for cooking. A majority of our respondents (74 percent) said they would be willing to try stoves, and discussions with local stove owners suggest these stoves might cut fuelwood consumption by half, reducing the need for trees from the forest and decreasing the average workload for households (Khin Hla and Maung Soe, personal communication). Currently, international scientists working in CWS are encouraging increased use of the stoves through supporting outreach education and subsidies to help households purchase the stoves. Often simple energy solutions may exist, but it requires outside involvement to raise awareness or make them accessible.

Community forestry has become an important strategy for forest managers, development agencies, and government decision makers around the world (Poffenberger 2006), and has been successful in reducing extraction pressure on tropical dry forest protected areas (Nagendra 2003; Adhikari et al. 2004). Since the 1990s community forestry has been official policy in Nepal and India, and case studies have shown it can slow degradation, improve forest quality, and enhance community involvement (Saigal 2000; Gautam et al. 2002). Community forestry plots are typically governed by committees composed of community members, with varying degrees of oversight and direction from government officials, while nongovernmental organizations (NGOs) often provide financial support and capacity building. While there has been much progress, research has shown there are still shortcomings, such as conflict between and within stakeholder groups and inequitable representation of gender and classes (Saigal 2000; Adhikari et al. 2004; Poffenberger 2006). To overcome this problem, committee design should include mechanisms for representing less-powerful groups such as women and poor households. Often forestry officials have no background in community forestry, and there is need for a change in philosophy toward community forestry as well as for training in communication, conflict management, and even marketing (Saigal 2000). Community forestry involves a complex web of forestry department officials, community members, and NGO stakeholders navigating a range of policies and traditional and legal rights, and balancing the needs of people with the capacity and biodiversity values of the dry forest ecosystem. Continued oversight and support through skills training, user networks, and policy adaptations to meet site-specific requirements are needed.

Another key management strategy for conservation in rural ecosystems is the inclusion of local people as stakeholders in the decision-making process and protection of the sanctuary (Ingles 1995; Pardo 1995; Food and Agriculture Organization of the United Nations 1998; Castellanet and Jordan 2002; Poffenberger 2006; Alkan et al. 2009). In most cases, protected areas were established without input from or consultation with the people living in and around the areas. It is not surprising that the people resent losing access to what was traditionally theirs. Discussion forums and workshops bringing together staff and local people can improve people-park relationships and give people a voice (Reddy and Chakravarty 1999; Fiallo and Jacobson 1995; Allendorf et al. 2006). Without a stake in the process, local people have little incentive for curtailing detrimental activities or preserving protected areas.

CONCLUSIONS

Typically, dry forests are surrounded by dense human populations and are subject to higher extraction pressures than most other natural ecosystems. Our case study in Chatthin Wildlife Sanctuary, along with research in protected dry forest areas across the tropics, shows that socioeconomics and poverty often drive dependency on these ecosystems. As in many other dry forest protected areas, fuelwood for cooking and heating is the most important household resource and most pressing threat to the sanctuary. Conserving what remains of dry tropical forests should start with the recognition that lives and livelihoods have been connected with the forests for more than two thousand years. Strategies designed to reduce fuelwood demand through improved energy efficiency and alternative energy sources are needed, along with developing community forestry and participatory management of protected areas. Dry forest ecosystems face increasing pressures as human population increases; conserving what remains requires involving local people, NGOs, and government officials to work together to develop ways of managing protected areas and surrounding forests in ways that benefit wildlife and people in the long term.

ACKNOWLEDGMENTS

This research was funded by the Conservation and Research Center. We thank U Uga, Daw Thandar Kyi, and the staff of Chatthin Wildlife Sanctuary for logistic support in the field, and Kate Jenks, Hugh Stimson, and Melanie Delion for assistance in the lab.

REFERENCES

Abbot, J. I. O., and R. Mace. 1999. Managing Protected Woodlands: Fuelwood Collection and Law Enforcement in Lake Malawi National Park. *Conservation Biology* 13:418–21.

Adeoti, O., T. D. Adegboyega, and T. A. Ayelegun. 2001. An Assessment of Nigeria Biogas Potential from Agricultural Wastes. *Energy Sources* 23:63–68.

Adhikari, M., S. Nagata, and M. Adhikari. 2004. Rural Household and Forest: An Evaluation of Household's Dependency on Community Forest in Nepal. *Journal of Forest Research* 9:33–44.

Alkan, H., M. Korkmaz, and A. Tolunay. 2009. Assessment of Primary Factors Causing Positive or Negative Local Perceptions on Protected Areas. *Journal of Environmental Engineering and Landscape Management* 17 (1): 20–27.

Allendorf, T., Khaing Khaing Swe, Thida Oo, Ye Htut, Myint Aung, K. Allendorf, L. Hayek, P. Leimgruber, and C. Wemmer. 2006. Community Attitudes toward Three Protected Areas in Upper Myanmar (Burma). *Environmental Conservation* 33 (4): 344–52.

Appasamy, P. 1993. Role of Non-Timber Forest Products in a Subsistence Economy: The Case of a Joint Forestry Project in India. *Economic Botany* 47:258–67.

Bahuguna, V. 2000. Forests in the Economy of the Rural Poor: An Estimation of the Dependency Level. *Ambio* 29:126–29.

Barham, B., O. Coomes, and Y. Takasaki. 1999. Rain Forest Livelihoods: Income Generation, Household Wealth and Forest Use. *Unasylva* 50:34–42.

Bazile, D. 2002. Improved Stoves as a Means of Poverty Alleviation. *Boiling Point* 48:20–22.

Booth, A. 2003. The Burma Development Disaster in Comparative Historical Perspective. *South East Asia Research* 11 (2): 141–71.

Brunner, J., K. Talbott, C. Elkin, and World Resources Institute. 1998. *Logging Burma's Frontier Forests: Resources and the Regime*. Washington, DC: World Resources Institute.

Bullock, S. H., H. A. Mooney, and E. Medina. 1995. *Seasonally Dry Tropical Forests*. Cambridge: Cambridge University Press.

Bunyavejchewin, S. C. 1982. Canopy Structure of the Dry Dipterocarp Forest of Thailand. PhD diss., University of Washington, Seattle.

Cabin, R. J., S. G. Weller, D. H. Lorence, T. W. Flynn, A. K. Sakai, D. Sandquist, and L. J. Hadway. 2000. Effects of Long-Term Ungulate Exclusion and Recent Alien Species Control on the Preservation and Restoration of a Hawaiian Tropical Dry Forest. *Conservation Biology* 14:439–53.

Castellanet, C., and C. F. Jordan. 2002. *Participatory Action Research in Natural Resource Management: A Critique of the Method Based on Five Years' Experience in the Transamazonica Region of Brazil*. New York: Taylor & Francis.

Chopra, K. 1993. The Value of Non-Timber Forest Products: An Estimation for Tropical Deciduous Forests in India. *Economic Botany* 47:251–57.

Das, B. K. 2005. Role of NTFPs among Forest Villagers in a Protected Area of West Bengal. *Journal of Human Ecology* 18:129–36.

Datta, R., and G. S. Dutt. 1981. Producer Gas Engines in Villages of Less-Developed Countries. *Science* 213:731–36.

Davidar, P., M. Arjunan, P. Mammen, J. Garrigues, I. Puyravaud, and K. Roessingh. 2007. Forest Degradation in the Western Ghats Biodiversity Hotspot: Resource Collection, Livelihood Concerns and Sustainability. *Current Science* 93:1573–78.

DeFries, R., A. Hansen, A. C. Newton, and M. C. Hansen. 2005. Increasing Isolation of Protected Areas in Tropical Forests over the Past Twenty Years. *Ecological Applications* 15:19–26.

Fiallo, E. A., and S. K. Jacobson. 1995. Local Communities and Protected Areas: Attitudes of Rural Residents towards Conservation and Machalilla National Park, Ecuador. *Environmental Conservation* 223:241–49.

Food and Agriculture Organization of the United Nations. 1998. *Asia-Pacific Forestry towards 2010: Report of the Asia-Pacific Forestry Sector Outlook Study*. Rome: Food and Agriculture Organization of the United Nations.

———. 2003. *The Livestock Industries of Thailand*. Bangkok: Food and Agriculture Organization of the United Nations.

Fox, J., P. Yonzon, and N. Podger. 1996. Mapping Conflicts between Biodiversity and Human Needs in Langtang National Park, Nepal. *Conservation Biology* 10:562–69.

Ganesan, B. 1993. Extraction of Non-Timber Forest Products, including Fodder and Fuelwood, in Mudumalai, India. *Economic Botany* 47 (3): 268–74.

Gautam, A. P., E. L. Webb, and A. Eiumnoh. 2002. GIS Assessment of Land Use/Land Cover Changes Associated with Community Forestry Implementation in the Middle Hills of Nepal. *Mountain Research and Development* 22:63–69.

Gentry, A. H. 1995. Diversity and Floristic Composition of Neotropical Dry Forests. In *Seasonally Dry Tropical Forests*, ed. S. H. Bullock, H. A. Mooney, and E. Medina, 146–94. Cambridge: Cambridge University Press.

Hegde, R., S. Suryaprakash, L. Achoth, and K. Bawa. 1996. Extraction of Non-Timber Forest Products in the Forests of Biligiri Rangan Hills, India. 1. Contribution to Rural Income. *Economic Botany* 50:243–51.

Ingles, A. W. 1995. Community Forestry in Nepal: Conserving the Biological Diversity of Nepal's Forests. In *Conserving Biodiversity Outside Protected Areas: The Role of Traditional Agroecosystems*, ed. P. Halladay and D. A. Gilmour, 183–204. Geneva: International Union for Conservation of Nature and Natural Resources.

Janzen, D. H. 1988. Tropical Dry Forests: The Most Endangered Major Tropical Ecosystem. In *Biodiversity*, ed. E. O. Wilson and F. M. Peters, 130–44. Washington, DC: National Academy Press.

Kalia, A. K. 2000. Biogas as a Source of Rural Energy. *Energy Sources* 22:67–76.

Karanth, K. U. 2002. Nagarahole: Limits and Opportunities in Wildlife Conservation. In *Making Parks Work: Strategies for Preserving Tropical Nature*, ed. J. Terborgh, C. van Schiak, L. Davenport, and M. Rao, 189–202. Washington, DC: Island Press.

Kurz, S. 1877. *Forest Flora of British Burma*. Calcutta: Office of the Superintendent of Government Printing.

Lehmkuhl, J. F., R. K. Upreti, and U. R. Sharma. 1988. National Parks and Local Development: Grasses and People in Royal Chitwan National Park, Nepal. *Environmental Conservation* 15:143–48.

Leimgruber, P., D. S. Kelly, M. K. Steininger, J. Brunner, T. Müeller, and M. Songer. 2005. Forest Cover Change Patterns in Myanmar (Burma) 1990–2000. *Environmental Conservation* 32:356–64.

Maas, J. 1995. Conversion of Tropical Dry Forest to Pasture and Agriculture. In *Seasonally Dry Tropical Forests*, ed. S. H. Bullock, H. A. Mooney, and E. Medina, 399–422. Cambridge: Cambridge University Press.

Mahapatra, A., H. Albers, and E. Robinson. 2005. The Impact of NTFP Sales on Rural Households' Cash Income in India's Dry Deciduous Forest. *Environmental Management* 35:258–65.

Medina, E. 1995. Diversity of Life Forms of Higher Plants in Neotropical Dry Forests. In *Seasonally Dry Tropical Forests*, ed. S. H. Bullock, H. A. Mooney, and E. Medina, 221–42. Cambridge: Cambridge University Press.

Murali, K., U. Shankar, R. Shaanker, K. Ganeshaiah, and K. Bawa. 1996. Extraction of Non-Timber Forest Products in the Forests of Biligiri Rangan Hills, India. 2. Impact of NTFP Extraction on Regeneration, Population Structure, and Species Composition. *Economic Botany* 50:252–69.

Murphy, P. G., and A. E. Lugo. 1986. Ecology of Tropical Dry Forest. *Annual Review of Ecology and Systematics* 17:67–88.

Mya Maung. 1998. *The Burma Road to Capitalism: Economic Growth versus Democracy*. Westport, CT: Praeger Publishing.

Myint Aung, U. 2007. Policy and Practice in Myanmar's Protected Area System. *Journal of Environmental Management* 84:188–203.

Myint Aung, Khaing Khaing Swe, Thida Oo, Kyaw Kyaw Moe, P. Leimgruber, T. Allendorf, C. Duncan, and C. Wemmer. 2004. The Environmental History of Chatthin Wildlife Sanctuary, a Protected Area in Myanmar (Burma). *Journal of Environmental Management* 72 (4):

205–16.Nagendra, H. 2003. Tenure and Forest Conditions: Community Forestry in the Nepal Terai. *Environmental Conservation* 29:530–39.

Nijaguna, B. T. 2002. *Biogas Technology*. New Delhi: New Age Publishers.

Pandey, U., and J. S. Singh. 1984. Energy-Flow Relationships between Agrosystem and Forest Ecosystems in Central Himalaya. *Environmental Conservation* 11:45–53.

Pardo, R. 1995. Community Forestry Comes of Age. *Journal of Forestry* 93:20–24.

Pattanayak, S. K., and E. O. Sills. 2001. Do Tropical Forests Provide Natural Insurance? The Microeconomics of Non-Timber Forest Product Collection in the Brazilian Amazon. *Land Economics* 77:595–612.

Peh, K., J. de Jong, N. Sodhi, S. Lim, and C. Yap. 2005. Lowland Rainforest Avifauna and Human Disturbance: Persistence of Primary Forest Birds in Selectively Logged Forests and Mixed-Rural Habitats of Southern Peninsular Malaysia. *Biological Conservation* 123:489–505.

Poffenberger, M. 2006. People in the Forest: Community Forestry Experiences from Southeast Asia. *International Journal of Environment and Sustainable Development* 5:57–69.

Rao, M., A. Rabinowitz, and S. T. Khaing. 2002. Status Review of the Protected-Area System in Myanmar, with Recommendations for Conservation Planning. *Conservation Biology* 16:360–68.

Reddy, S. R. C., and S. P. Chakravarty. 1999. Forest Dependence and Income Distribution in a Subsistence Economy: Evidence from India. *World Development* 27:1141–49.

Saigal, S. 2000. Beyond Experimentation: Emerging Issues in the Institutionalization of Joint Forest Management in India. *Environmental Management* 26:269–81.

Sanchez-Azofeifa, G., M. Kalacska, M. Quesada, J. Calvo-Alvarado, J. Nassar, and J. Rodri-guez. 2005. Need for Integrated Research for a Sustainable Future in Tropical Dry Forests. *Conservation Biology* 19:285–86.

Shackleton, C., and S. Shackleton. 2004. The Importance of Non-Timber Forest Products in Rural Livelihood Security and as Safety Nets: A Review of Evidence from South Africa. *South African Journal of Science* 100:658–64.

Shaw, R. P. 1989. Rapid Population Growth and Environmental Degradation: Ultimate versus Proximate Factors. *Environmental Conservation* 16:199–208.

Singh, J. S., U. Pandey, and A. K. Tiwari. 1984. Man and Forests: A Central Himalayan Case Study. *Ambio* 13:80–87.

Singh, V. P., and J. S. Singh. 1989. Man and Forests: A Case-Study from the Dry Tropics of India. *Environmental Conservation* 16:129–36.

Songer, M. 2006. Endangered Dry Deciduous Forests of Upper Myanmar (Burma): A Multi-Scale Approach for Research and Conservation. PhD diss., University of Maryland.

Songer, M., Myint Aung, B. Senior, R. DeFries, and P. Leimgruber. 2009. Spatial and Temporal Deforestation Dynamics in Protected and Unprotected Dry Forests: A Case Study from Myanmar (Burma). *Biodiversity & Conservation* 18:1001–18.

Stamp, L. D. 1925. *The Vegetation of Burma from an Ecological Standpoint*. University of Rangoon Research Monograph, 1. Calcutta: Thacker, Spink and Company.

Stott, P. 1990. Stability and Stress in the Savanna Forests of Mainland South-East Asia. *Journal of Biogeography* 17:373–83.

Udaya Sekhar, N. 2003. Local People's Attitudes towards Conservation and Wildlife Tourism around Sariska Tiger Reserve, India. *Journal of Environmental Management* 69:339–47.

Van Schendel, W. 1991. *Three Deltas: Accumulation and Poverty in Rural Burma, Bengal and South India*. Indo-Dutch Studies on Development Alternatives, 8. Newbury Park, CA: Sage Publications.

Wickramasinghe, A., M. Pérez, and J. Blockhus. 1996. Non-Timber Forest Product Gathering in Ritigala Forest (Sri Lanka): Household Strategies and Community Differentiation. *Human Ecology* 24:493–519.

Wikramanayake, E., E. Dinerstein, C. J. Loucks, D. M. Olson, J. Morrison, J. Lamoreux, M. McKnight, and P. Hedao. 2001. *Terrestrial Ecoregions of the Indo-Pacific: A Conservation Assessment*. Washington, DC: Island Press.

Wilson, E. O. 1992. *The Diversity of Life*. Cambridge, MA: Belknap Press of Harvard University Press.

Yonzon, P. B., and M. L. Hunter, Jr. 1991. Cheese, Tourists, and Red Pandas in the Nepal Himalayas. *Conservation Biology* 5:196–202.

Subsistence Living in the Burmese Forest

Chris Wemmer

It was all happening in the cool shadows of morning. The morning market, or *mahnet zay*, offers little to the Western consumer. As a biologist however, I recognized this wonderland for what it was—a slice of rural Burmese life and a window on a subsistence economy. It is unmistakably a woman's world. Mothers, daughters, and grannies were hunkered to the ground or in makeshift stalls making change for clients, frying griddle cakes, arranging goods, or lost in neighborly banter. And there was music in the background. But there the similarity to Western markets ended. No refrigeration, no plastic-wrapped meats, and no bar codes here. This was the market as most people know it on planet Earth. In the long rays of the morning sun a vast array of roots, vegetables, and fish in varying states of preservation were spread across an acre of littered ground. I stopped before a young lady who expertly cleaved the detached head of a large carp. In the parlance of zoological dissection, it was a flawless sagittal section. The piscine brain and all its structures were laid bare as if for a lab exam in comparative anatomy. Big-eyed girls peeked and giggled at me from behind baskets, but young women averted their eyes as if I lacked some critical element of clothing. The matrons, on the other hand, eyed me with unabashed bemusement, often through a cloud of cigar smoke. I was clearly beyond the pale of tourism, and a curiosity 180 miles northwest of Mandalay, the nearest metropolis in this corner of Asia.

This is where I discovered U Chit Thein. Actually, it was his wife, Ma Thwe, who caught my eye. She was poised over a platter of rodents, deep-fried to a crispy brown. Aha! I thought, squirrels! More exact identification was impossible. The diagnostic features—feet and tail—were missing, and everything in front of the eyes had been removed. I hurried off to fetch my trusted colleague and translator, U Myint Aung, who was swapping stories in a nearby tea shop. In deference to the relaxed sense of time, I had a cup of tea, ate a ball of salted sticky rice, and then led my friend back to the market. A lively dialogue ensued in the blustery voices of Burmese market talk. My identification of the "squirrel" was incorrect. The lady was selling *chaw chewet*, or fried rats! Her husband, Chit Thein—whose name translates roughly as "lots of love"—was the rat catcher of Ponangyi village. I marveled at my good fortune. Here lived a modern-day Pied

Piper of Hamelin. Would her husband be so kind as to visit our camp in Chatthin Wildlife Sanctuary? She promised to extend our invitation.

Obviously something had been missed in translation, for that afternoon the rat catcher appeared in our camp wearing a look of total befuddlement. U Myint Aung greeted him with cheer and ordered a fresh thermos of tea. Chit Thein was a man of few words. His first and only question was why the foreigner had singled him out. Myint Aung ad-libbed: the foreigner was a kind American scientist, a world-famous expert on wild animals, and he wanted to learn about Chit Thein's work. To the rat catcher, who knew nothing of the modern world, Myint Aung's explanation seemed plausible. Chit Thein, we learned, was thirty-seven years old and had been supporting his family as a rat catcher for ten years. I eagerly jotted notes as he described his improbable methods. I was also convinced that this rat-catching business had to be seen.

That night we boarded the wildlife sanctuary's only tractor and rumbled away to visit Chit Thein at home. Our conveyance lacked headlights. The puny beam of a handheld flashlight guided us through degraded forest and across the stubble of the dry paddy fields. As our beacon faded to the brightness of a cigarette ash, the tractor decelerated to a snail's pace. This was a source of great amusement to the park rangers, and finally U Myint Aung translated: "Sorry, Chinese batteries!"

The village was dark except for a few kerosene lamps behind the split-bamboo walls of the houses. At the edge of the rice paddy stood Chit Thein and Ma Thwe's dwelling, a three-walled bamboo lean-to. From the spindly house poles hung a few plastic bags with the family's worldly belongings, and a mat of palm thatch covered the bare earth. In a corner a baby slept in a basket. Our man proudly displayed the tools of his trade. On the small of his back he wore a wicker basket and a cudgel of cane to dispatch his prey. In one gnarled hand he carried an eight-foot spear tipped with two steel points. In the other, he held a spotlight wired to a wet-cell battery on his belt. This, he explained, was not a totally satisfactory arrangement as he had difficulty concentrating on the hunt when battery acid spilled down his leg.

Chit Thein clearly had a streak of Rube Goldberg. Attached to the spotlight was a curious mechanical outgrowth: his sound decoy. He demonstrated how he used his thumb to modulate the whirring noise made by the plastic strips spinning on the little motor. To rats, he explained, it sounded like food—a grounded beetle or cicada bumbling in the dirt.

While U Myint Aung and our small entourage settled in with cups of tea, I left my friends and followed in the rat catcher's footsteps. In the beam of Chit Thein's spotlight, the paddy fields were a stark wasteland riddled with holes and the sparse leavings of the rice harvest. In three months it would be transformed into a muddy broth of cow manure, water, and rice seedlings, seething with invertebrates, fish, amorous frogs, and hungry snakes.

We had marched less than ten minutes when Chit Thein froze in his tracks. Twenty feet away a large yellow rat was scrounging obliviously in the stubble. Gripping the end of the spear, the hunter padded forward in a crouch. Wide-eyed, I followed. Then Chit Thein lowered his spear and leaned forward. The rat was three yards away. Amazingly the spear's points advanced within inches of the creature without alerting it. When the luckless rodent finally detected something amiss, it was too late. The spear struck home in a feat of remarkable eye-hand coordination, and the impaled victim was dispatched with the cudgel.

In forty-five minutes the hunter bagged eight rats and delivered me back to the lean-to. There my friends were munching deep-fried gourd and listening to U Myint Aung's lively stories. Ma Thwe found us to be eager learners, and shyly laughed as she answered our questions about rat cookery. The lessons were simple. First, gut your rat with a large knife. Leave the liver and heart in place, but save the entrails for the family dog. Second, parboil your rat for a minute so you can wipe off the hair. At this point, your rodent will look pale and really clean, but it will still look like a rat. Third, hack off the feet, tail, and nose. I watched with morbid fascination as she dipped the pale carcasses into sizzling peanut oil. Then a wonderful smell filled the air.

Ma Thwe cooked up the whole lot, and offered us the crispy, golden-brown carcasses. As a biologist—and I might add, one who prides himself as culturally sensitive—there was no choice. Then it dawned on me. I was looking at the same dry brown morsels the camp cook at the wildlife sanctuary had been serving me for lunch and dinner the past week. I may not have believed it was chicken, but for some reason, I had never given my doubts a voice; now I knew. I gingerly bit into the squirrelly meat. Yes, it was the same thing, only these pieces were hot.

I was buoyed by my encounter with Chit Thein and the prospect of undertaking my first anthropological study. The rat catcher offered me a special opportunity to explore subsistence hunting, and at the same time I could learn about the local rodent fauna. Would he be so kind, I asked, as to collect some information about his quarry? He solemnly agreed, I quickly drafted a data form, and U Myint Aung dutifully translated it into Burmese.

When we met our man the next night, we asked him to sort the rats by sex. He had no problem distinguishing the adults, for mature males are endowed with a scrotum resembling a small duffel bag. But it was obvious he hadn't taken undergraduate mammalogy. I showed him how to tell immature males from females, and explained the curious details of the rodent estrous cycle. After some coaching, he passed the beginners test with flying colors. He agreed to register the weight and sex of all kills in a notebook. I also gave him my pedometer, and he promised to reset it each night before venturing forth. Finally, I gave Myint Aung money to pay Chit Thein for the data. This particular trip to Burma was quickly coming to a close. Although I obsessed about the project on the long flight home, in the ensuing months it became dormant in the back of my mind.

In May of the following year I returned to Chatthin with my long-time Smithsonian colleague and friend Don Wilson. Our task was to train junior staff of the Wildlife Division in mammal survey methods. The colonial British were marvelous amateur naturalists, but their collections favored birds and butterflies. The mammals of Burma had never been thoroughly surveyed. Now, a half century later, the Burmese wildlife director was asking us what kinds of small mammals lived in the parks. A reflective mammalogist of repute, Don had a field catalogue that was in the thousands, testimony to years of field time skinning and stuffing specimens for museums where the world's biodiversity is catalogued. His dry wit was exactly what was needed during the muggy pre-monsoon heat of May.

Collecting and preserving scientific specimens is the first step in exploring biodiversity for species that are not readily identifiable by sight. Large- and medium-size mammals—from wild cattle to hog badgers—can be identified from hair, footprints, and photos taken by camera traps. But species of bats and rodents often look alike to the inexperienced eye. Sometimes age and sex differences are so great that a single species looks like two or three different species. What's more, the taxonomic status of many small mammals, especially in developing countries like Burma, is far from well known. A properly prepared specimen provides undeniable proof of a particular plant or animal's identity. For this reason, our goal was to train our students to scientifically inventory small mammals and create a reference collection.

Chit Thein was there the first day of class. His creased face stood out among the smooth cheeks of the youthful staff of the Wildlife Division. Don and I took turns lecturing on mammalogy. The class diligently took notes in the open-sided classroom, now and then dodging giant carpenter bees engaged in aerial combat over their nest holes in the rafters. That afternoon, Don and I led the twenty-four students into the forest and coached them in reading animal sign and setting trap lines. The next day we divided them into four teams. Each team got one hundred rat traps, a bag of unshelled peanuts as bait, and surveyor's tape to mark trap locations. Thereafter, the daily routine was to check traps in the morning, prepare specimens till early afternoon, take lunch, bathe, review lecture notes, and set traps before dark.

Chit Thein was our stand-in. In case of low trapping success, his job was to supply enough rodents so all members of the class would have a specimen to prepare. Each morning he showed up with a bucket of rats, mice, house shrews, and assorted other small mammals. He spent the first few days forlornly watching the class prepare specimens. When he showed up one morning with several skins of his own making, we realized he wanted to be a full participant. We gave him a skinning kit and found that he was a natural.

Each morning when the trap teams returned, we heaped praise on the team that turned in the biggest catch; the whole idea was to incite competition between the trapping teams. It worked on all but one group: a team of foresters by training that had a record for zero trapping success. On the fourth day, we decided a remedy was in order. We assigned the lackluster team to Chit Thein. We instructed him to spare no effort to show them how it was done, even if it meant spreading the traps over hill and dale. Our rat catcher looked none the worse for wear when his bedraggled team arrived at the camp that night. But the next morning they didn't come in empty-handed.

Day by day the mammal collection grew in number and species, but muggy weather made it hard to dry the specimens properly. Almost every day, Don tenderly arranged the specimens to dry in the sun, and when not shooing flies, he tried to get an idea of what species lived in the region by thumbing through Corbet and Hill's *Mammals* of *the Indomalayan Region*. Squirrels and tree shrews were not a challenge to identify, and the black rat and bamboo rat were distinctive. Much more of a challenge was a batch of look-alike rodents that ranged in size from a small rat to a house mouse; they could be any of several species.

A clever idea suddenly struck. We would give Chit Thein a "lab quiz." While I fetched the rat catcher and Myint Aung, Don scrambled the specimens on the planking of the porch. Chit Thein pondered the assemblage for a few moments, and then methodically sorted the specimens into fourteen groups. Nine months later—when the skulls had been impeccably cleaned by the museum's dermestid beetles—Don identified sixteen species. Unaided by taxonomic keys, diagrams, cleaned skulls, and a dissecting scope, our rat catcher came closer to assessing the real number of species than we did. He had underestimated the small rodents by two species, but insisted there were three species of bandicoot rat. Corbet and Hill's book showed only two species occurring in the area. Our class had discovered a range extension of the Burmese bandicoot rat (*Bandicota savilei*)!

Another year passed before U Myint Aung arrived in the States for a six-week study visit. He also delivered the long-awaited and stained record book of Chit Thein's nocturnal forays. Chit Thein hunted primarily four species: the cosmopolitan black rat and the three bandicoot rats. During twenty-seven weeks spread across the year, our rat catcher caught 1,259 rats. The achievement loomed even larger when we summed the carcass weights. They weighed almost 2,800 pounds—nearly a ton and a half. His average nightly catch weighed 9 pounds, but there was a clear pattern of seasonal variation. During the months of January and February, the nightly harvest topped out at a phenomenal 24 pounds, many of them juvenile rats. U Myint Aung explained, "Winter months wonderful time for rat. Farmers harvest sixteen varieties of rice and six varieties of ground nut in December." It seems the fields become a cornucopia for teenage rats born during the post-monsoon months.

Research has a way of raising more questions than it answers. I was skeptical about the pedometer data. If accurate, Chit Thein's outings, which never exceeded two miles, amounted to nothing more than strolls. There was also a glaring thirteen-week gap in the data. I wondered if some minor inconvenience, like the loss of the ballpoint pen accounted for the gaps. Why should a man who lives hand-to-mouth go out of his way to collect data for an eccentric American? I wrote to Myint Aung to get the answer. Three months later he answered that the rat catcher had packed up and moved.

Another year passed before we tracked down our friend. He was now working in Chatgyi village as a day laborer in the Forest Division's tree nursery. His family had grown to four children, the salary was predictable, and the work was year-round. With his usual economy of words, Chit Thein explained the obvious. His nightly beat was small because there were plenty of rats close to home. He could get another rat at the same burrow on consecutive nights. Ma Thwe sold the fried rats in the *manet zay* for 10–15 kyat a piece. A quick calculation showed the family earned the equivalent of $73 that year. By comparison, a government driver earns about K850 per month, or $28 per year.

As for the gaps in the data, I learned that Chit Thein didn't bother to hunt during the monsoon because the prime rat areas (paddy fields) were flooded, and rodents were hard to see elsewhere in the dense undergrowth. He also stopped hunting rats again that fall when he encountered a bumper crop of *nga-shint*, or eels, in the drying rice paddies. The eel is legendary in Burma. Not only is its flesh considered a delicacy, but its form embodies the Burmese epitome of beauty: a lithe, supple body. The best-tasting eels, I was informed, came from cemeteries where the subterranean fish feast on corpses. The eel is exquisitely adapted to seasonal extremes. On moist monsoon nights it squirms through the grass to find new pools of water, and during the dry season it burrows deep into the mud, forms a cocoon, and enters a state of physiological suspension. Cemeteries are not a challenge to hungry eels during the wet season. But after the monsoon, when the eels gather in the last drying pools, they are most vulnerable. Chit Thein did extremely well trapping eels that fall, taking 1–2 viss (about 3.6–7.2 pounds) per night, which he sold for 50–60 kyat per viss. Then villagers from Kawlin township discovered his hunting ground. The market was soon glutted with eels, and the price plummeted. He gladly switched to rat catching when the cool weather set in. That's how Chit Thein supported his family during the latter part of 1996.

At last I had heard the full story. Chit Thein had accounted for all the gaps in the data. He hadn't abandoned the project. The man exhibited the opportunism of a good predator. When the usual prey became scarce, he switched to new prey. If he encountered a more abundant and marketable species, he took advantage of it. Despite my esteem for his jungle lore, he made no claims for his curious survival skills. Nor was he ashamed of them. It was only a way to survive. But it made me wonder. Where does someone like Chit Thein come from? I needed another interview to understand the man's past.

My final appointment with Chit Thein materialized with Myint Aung's announcement "Chit Thein arrive for *Little Big Man* story." To my friend, the title of the Dustin Hoffman movie is synonymous with biography. It was the last night of my latest visit, and the entire staff of the wildlife sanctuary had gathered in the bungalow to sit quietly in the balm of burning mosquito coils. Ye Htut translated the story.

Chit Thein was the fourth of five children born to *taunggya*, or slash-and-burn cultivators, in the heart of Burma's dry zone. His early years were normal for a village boy. When he was eight his parents sent him to the monastery, where he entered the life of the *ko-yin*, or novitiate. There he learned discipline in work and study, and how to read scripture and recite prayers. The boys slept in a row on the floor, with exactly one and a half feet between them. They were forbidden to disturb each other. Horseplay was a secret activity. It was peaceful and it was good, but after a few months his parents wanted him to return to normal life. Chit Thein reflected a moment, and said he often remembered his life in the monastery.

Then his mother died. He was ten years old. His father died two years later. Life hadn't been bliss, but it quickly changed for the worse. Chit Thein and his youngest brother became a burden to a family without parents. His teenage sisters and brother were unwilling to look after the two young boys, and "paid them neglect." The two boys left the farm and looked for work. A few miles away lived an older farmer, Yhe Pak, and his wife, Ohn Khin, who took the boys in. They owned eighty acres of land and annually grew one thousand bushels of rice. For the next eight to nine years the two boys worked for their room and board in the old couple's dwelling and each year received eighty baskets of rice for their labor.

At this point in Chit Thein's narrative, as the camp generator strained, the lights in our room grew dim and then went out.

When he was twenty-two years old, Chit Thein realized he had no belongings and no money. He decided to strike out on his own. He took his brother to the monastery and said goodbye. Then he moved to the village of Khin Oo, and looked for work. There he met a carpenter named U Thaung Tin, and he became the carpenter's apprentice. He spent six years as a helper to his teacher. He also sent a little money to the monastery for his brother's needs. But he was only earning 15 kyats a day, and his savings were meager.

He befriended U Maung Oh, the rat catcher. The price of rats at the time was 50 pyas. He could catch about 150 rats a night, and earn about 70 kyats a day, much more than he was earning as a carpenter. Maung Oh agreed to become Chit Thein's *saya*, or teacher. In return for the teachings, Chit Thein became his master's servant. He hauled water, collected firewood, and helped his teacher. At night, he followed him for rat capturing. His master taught him how to find the rats at different times of the year, what methods were suitable under different conditions, how to use a torch light, and how to capture the rodents. After four months his training was complete. As a token of gratitude, Chit Thein gave his *saya* 500 kyats, and then he went to the pagoda and prayed for him.

Chit Thein began his new life as a rat catcher. During the next two years he started to discover secret techniques never taught by his master. That's when he invented the motorized sound decoy. It worked better than the bamboo-and-string device used by his master.

In Khin Oo there was a family with four daughters. The oldest was as pretty as a *nga-shint*. She had passed the second standard (fourth grade) in school. Chit Thein courted that sister, and they married when she was eighteen years old. In the next thirteen years they had four children, first a daughter, then three sons. All are still living.

I was haunted by one thing—the break up of the family. I asked how he feels about his older siblings? He said he still feels bad about what happened. Neither he nor his younger brother ever went home again. He did not want his brother and sisters "to know about his lifestyle." With his story complete, Chit Thein reflected a moment. The room was silent except for the struggling generator in the distance. Chit Thein then said that if he got rich in the future, maybe he would visit his long-separated brothers and sisters. But even if he gets rich, he's not really sure he wants to see them. Hope springs eternal, even if you are a rat catcher.

Governing the Flame

Bunong Management of Fire Regimes in Mondulkiri Province, Northeast Cambodia

Megan MacInnes

The balance of human-fire-environment dynamics in seasonally dry forest eco-systems has always been complex, and today agencies involved in conservation of dry forest biodiversity are exploring management approaches that integrate the ecological, socioeconomic, and technical aspects of fire. Based on an ethnographic case study of traditional use of fire by local indigenous communities in northeast Cambodia (MacInnes 2007), this chapter discusses the role that community-based fire regimes and their governance institutions can play within decentralized fire management strategies.

During the twentieth century, forest policies, particularly those of the colonial powers, have been dominated by what S. J. Pyne describes as "Europe's Pyrophobia" (1993, 225): the view of fire as an external disturbance to forests that threatens economically valuable timber resources (Kull 2004). Indigenous fire use in areas such as the Lower-Mekong dry forest ecoregion (LMDFE) was considered short-term resource exploitation at the expense of long-term sustainability, and was believed to be based on environmental ignorance and a lack of alternatives (Blaikie and Brookfield 1987; Kull 2004, 26; Biot et al. 1986, 16). The response was comprehensive fire prevention programs that attempted to protect people and resources from the threat of fire. Pyne (1993) and Kull (2004) provide global perspectives on these prevention and suppression policies.

Recent initiatives have to some extent reversed this understanding of the relationship between fire, humans, and the environment (Moore et al. 2002; Project Firefight South East Asia 2003); nevertheless, as this chapter shows, key drivers within human-fire-environment dynamics remain contested. In 2007 the Global Fire Initiative reported that "fire is a natural process that has played a major role in shaping our environment and maintaining biodiversity ... [and] ... fire-dependant eco-regions cover 53 percent of global terrestrial area" (Shlisky et al. 2007, 1, 7). International scientists (such as Moore et al. 2002; Kull 2004; and Myers 2006) now believe that preventing fire has had a detrimental effect on global biodiversity, furthered species loss, and increased the risk of intense and "bad" fires due to the build-up of flammable vegetation. The challenge for the twenty-first century, they state, is to develop fire management strategies that minimize the risks of "bad" fires while simultaneously maximizing the benefits from "good" fires.

The evolution of ecological theories related to fire is framed within broader changes to biodiversity conservation ideologies that increasingly recognize the importance of integrating diverse and multiple views, levels, and epistemologies within environmental management strategies (Sullivan and Homewood 2003; Brechin et al. 2002). The term *integrated fire management* (IFM) is used to describe the "integration of science and society with[in] fire management technologies" (Myers 2006, 2). Given the diverse range of situations and needs within which IFM is applied, there is a great variety in its implementation. Planning and decision making range from being controlled by governmental and external agency staff working with local community representatives who participate as locally available human resources by managing controlled fires, to local communities controlling all elements of fire management strategies, with support from governmental/external agencies as required. Examples of these can be found in Moore et al. (2002) and Project Firefight South East Asia (2003).

Community-based fire management (CBFiM) describes activities at the decentralized end of this spectrum. It has been defined as "a type of land and forest management in which a locally resident community (with or without the collaboration of other stakeholders) has substantial involvement in deciding the objectives and practices involved in preventing, controlling or utilising fires" (International Forest Fire News 2003). Epistemologically, CBFiM belongs to "community-based" natural resource management trends. It emphasizes the decision-making role of communities (Goldammer et al. 2002; Moore et al. 2002; International Forest Fire News 2003) and is based on

1. The importance of including the traditional knowledge and experience of local peoples in safe fire management of fire-dependent ecosystems;
2. The inability of governments to effectively implement centralized fire management policies due to a lack of capacity and resources, especially in remote areas; and
3. The limit to which communities can effectively manage fire on their own and the external support required in terms of capacity building, strengthened tenure security, and favorable revisions to policies and legislation.

As with any system for the management of resources, the structure governing its use plays an essential role in determining sustainability, therefore concepts relating to common-property resources (CPRs) can provide a framework for analyzing how communities govern fire regimes. Rights to access, use, and control of CPRs are managed by *governance institutions* defined as "group ownership in which behaviour of all members is subjected to rules and monitored, and there is a culture of conformity against anti-social behaviour" (Bromley and Cernea 1989, 17). CPR "rules," therefore, refer to prescriptions that create management authorities; "rights" refer to the actions that are authorized by the "rules"; and these "rights" come with "duties" to observe and monitor the "rules" (Schlager and Ostrom 1992). The theory of comparative advantage holds that small-scale, local CPRs are more effectively managed and governed by the users themselves than by external agencies (Gebremedhin et al. 2003; Agrawal 2001; Schlager and Ostrom 1992; Wade 1987; Araral 2009). Two reasons are given for this: (1) since CPRs are usually salient to the livelihoods of users, these users have the greatest incentive to govern efficiently, equitably, and sustainably, as well as being best placed to resolve conflicts; and (2) information costs for resource management are reduced as management is decentralized (Araral 2009). However, the political economy of globalization is changing the context within which CPRs are managed. What were once believed to be homogenous communities following traditional management practices are now recognized as heterogeneous groupings that negotiate daily about access to, use of, and control over natural resources (Dolsak and Ostrom 2003).

Given this changing political economy, researchers have attempted to identify factors that promote sustainable management of CPRs. Critical analysis of this literature has identified four sets of variables that provide enabling conditions for the sustainable management of CPRs by local governance institutions: (1) the characteristics of the resources, (2) the nature of the group depending on the resources, (3) the institutional regimes through which the resources are managed, and (4) the relationship between the group and external forces and authorities (Table 1). For CBFiM regimes used within CPR management, these enabling conditions provide a framework for analyzing the sustainability of the governance institutions themselves and the resource management strategies they employ.

The inductive, ethnographic case study on which this chapter is based was undertaken during 2006 and 2007 in two villages primarily populated by the Bunong ethnic minority group (indigenous to Southeast Asia) in Mondulkiri Province, northeastern Cambodia. The first village, Pu Tang, is located on the Haut Chhlong Plateau while the second, Koh Myeul Leu, is on the banks of the Srepok River (Figure 1). The research was a collaboration between an MSc student from the University of East Anglia, WWF–Cambodia, the International Institute for Environment and Development

Table 1. Enabling conditions for sustainable management of common-property resources.

Condition	Definition
Resource characteristics	Small size, clear boundaries, low mobility, possibility to store benefits from the resource, and predictability.
Group characteristics	Small size, defined membership, legitimate leadership, homogeneity of identities and interests, social capital, and low poverty.
Resource accessibility and use	User close to resource location, high value and dependence on resource, equitable allocation, low demand, and gradual changes in demand.
Internal institutional arrangements	Simple rules (devised and managed by users), easy enforcement, low-cost adjudication, and accountability of monitors to users.
External environment	Accessible and appropriate technology, low (and/or gradually increasing) interaction with markets, supportive government institutions, policy frameworks, and external assistance.

Source: Agrawal 2001; Gebremedhin et al. 2003; Araral 2009.

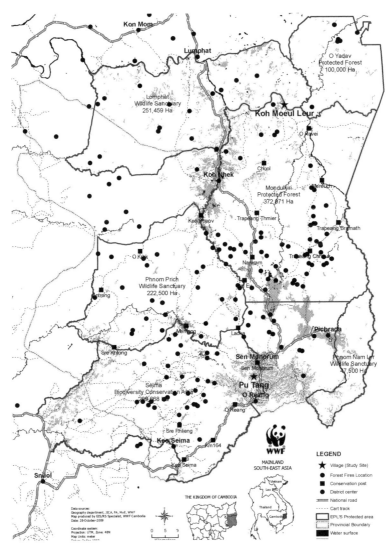

Figure 1. Map of Mondulkiri Province, northeast Cambodia, showing the two study sites, the boundaries of the areas under environmental protection, the conservation posts, and the forest fire locations monitored in February 2002 using Landsat 7 ETM+. Map courtesy of WWF Cambodia.

(IIED), and the Forestry Administration (FA) Office of Cambodia's Ministry of Agriculture, Forestry and Fisheries (MAFF). The research team consisted of a Scottish researcher and three members of the WWF–Cambodia staff (two government counterpart staff and one Bunong employee). Qualitative methodologies were used, namely, a combination of group discussions and in-depth interviews with villagers and external stakeholders. In total, three group discussions with a total of twenty-one individuals were held in each village (one discussion was composed of village elders and two were gender disaggregated). These were followed by interviews with three to four key informants with long-term experience of fire use (identified through "snowball sampling") in each village, in addition to the village chief. Nine interviews were held with government representatives, namely, local authorities, and provincial and national staff of FA, MAFF, and the Ministry of the Environment. Informal discussions were held with international staff from WWF–Cambodia throughout the research period. The analyzed results were shared with each village during data-validation meetings held at the end of the research period.

The research was limited by the initial reluctance of some villagers to discuss actual fire use with "outsiders" (defined by participants as those not originating from within the community, including nongovernmental organization [NGO] staff, government authorities, and Khmer in-migrants to the province). This was overcome through long discussions in each community and field trips undertaken without the presence of WWF–Cambodia counterpart staff. As a case study, the results presented in this chapter may not be representative of other community-based fire management of the Bunong or other indigenous minorities in the LMDFE, and replication of the research is necessary to validate and expand understanding.

Mondulkiri Province, which borders with Stung Treng, Kratie, and Ratanakiri provinces and Vietnam, is one of Cambodia's largest provinces, covering 14,288 km², and until recently was one of the most remote and inaccessible areas of Cambodia (Figure 1). The province itself consists of a mosaic of deciduous dipterocarp forest (DDF), mixed deciduous forest (MDF), and semi-evergreen forest (SEF), while the southeast of the province is known as the Haut Chhlong Plateau: a savanna grassland that is at a higher elevation (up to 700 m asl [above sea level]) than the rest of the province (Maxwell and Pinsonneault 2001). The Mondulkiri Protected Forest (MPF) covers 370,000 ha, was established in 2003, and is comanaged through the Srepok Wilderness Area Project (SWA) by the FA and WWF–Cambodia (Bauld 2007).

Until the early nineteenth century the province was entirely populated by indigenous ethnic minority groups, the largest group being the Bunong (Hickey 1982), however war and in-migration increased the numbers of Khmer (the Cambodian ethnic majority), Vietnamese, and Cham peoples. From 1998 to 2005 Mondulkiri's population increased by 53 percent; current ethnic minorities are estimated to comprise only 60 percent of the population (estimated in 2005 to be 49,612) although the Bunong still constitute the ethnic majority (Mondulkiri Provincial Department of Planning, personal communication). Traditionally the Bunong have been animist, matri-local, matri-lineal, and have lived in fortified villages of fewer than ten households that moved every five to seven years. Agriculture is based on non-irrigated shifting cultivation in which upland rice and other crops are grown. Like that of many indigenous groups, their livelihood is dependent on a diversified range of non-timber forest products (NTFPs) such as resin, honey, medicines, meat, and fish (Nikles 2006). Years of war, internal displacement, and recent economic integration into the rest of Cambodia have required traditional livelihood practices to be adaptable to external changes. Today, many Bunong living in the lowland (northern and western) areas of the province cultivate irrigated paddy-rice, whereas those in the southern highland areas continue traditional shifting cultivation (M. MacInnes, unpublished field research; personal observation). Nevertheless all remain strongly dependent on forest resources for their livelihoods.

Land and natural resources of the Bunong have traditionally been collectively owned and managed by governance institutions composed of village elders (M. Mac-Innes, unpublished field research; personal observation). These individuals, male or female, are selected for their proven ability to provide spiritual and pragmatic leadership, resolve conflicts, and make decisions beneficial to the community as a whole. Each community's assets can therefore be described as CPRs, and their sustainable management depends on the enabling conditions identified in Table 1. However, as the results of this case study reveal, within this CPR framework, some assets (such as shifting cultivation fields and sources of thatching grass) are allocated to the management of individual households, which enables them to restrict access and use by other community members, even though they remain the overall property of the community, whereas other assets (such as the grassland savanna grazing areas) are managed communally and each household has equal access and use rights.

The LMDFE Biological Assessment undertaken by WWF between 2002 and 2004 concluded that the ecoregion is "globally outstanding" in terms of species diversity, proportion of intact habitat, and downstream ecosystem services provided; within Cambodia it is a major component of the Lower Mekong watershed (Tordoff et al. 2005). The assessment described these mosaic habitat landscapes dominated by DDF as having low elevation (10–500 m asl), a tropical and strongly monsoonal climate, high fire frequency, high herbivore biomass (due to extensive grassland and grassy understory), and relatively low human population density. The assessment identified five focal processes that are integral to the LMDFE, all of which have significant current or potential threats: the fire regime; climatic seasonality; habitat modification by Asian elephants and ungulates, including potential moderation of fire regimes; predation of large ungulates by tiger, leopard, and dhole; and the annual flood cycle. According to WWF, a periodic fire regime is considered essential, especially in DDF communities, because of

> its role in maintaining the structure, composition and function of plant communities and, thus, the habitats of many focal species. The high frequency of fires throughout most of the eco-region prevents build-up of dead plant materials to levels that could support high-intensity crown fires . . . [c]onsequently, although fires occur every year in the DDF, they do not generally kill trees or invade adjacent areas of dense forests. Thus the fire regime in the LMDFE probably causes no significant degradation to dense forests, but maintains the structure of open forests (Tordoff et al. 2005, 19).

WWF considers fire to have particular importance for vegetation control given the low population density of ungulates, which have proven in other areas and at higher densities to have controlled build-up of dead plant material, thus moderating the fire regime. These annual fire regimes are primarily anthropogenic, of high frequency, and therefore of low intensity, whereas natural fires are predicted to take place later in the season and less frequently (Tordoff et al. 2005; Maxwell 2004; WWF–Cambodia, personal communication). However the role of fire in defining boundaries between DDF and SEF remains controversial; it is not clear what impact a change to "natural" fire regimes would have. Existing literature for Southeast Asia regarding the role of fire within habitat mosaics varies from those who perceive DDF to be degraded SEF, to those who believe DDF and SEF have stable and specific edaphic boundaries (Tordoff et al. 2005; Maxwell 2004; WWF–Cambodia, personal communication).

Two threats to the fire regime in the DDF have been identified by WWF: reduced ungulate populations and changing patterns of anthropogenic regimes due to suppression efforts. The first requires the restoration of ungulate population densities to natural levels. The second is of short-term significance due to the lack of capacity for fire management over extensive DDF areas, and of long-term significance unless a "better understanding of the ecological role of anthropogenic fires is gained and incorporated into fire management regimes" (Tordoff et al. 2005).

The perspective recognizing the importance of fire for the LMDFE is not reflected in Cambodia's legal framework. Although fire is not considered to be a primary threat to Cambodia's forests (in comparison to illegal logging or forest conversion for agricultural purposes), the government has adopted a strong anti-fire policy that criminalizes the setting of fires in forested areas by nonstate agencies (government stakeholders, personal communication). Specific legislation prohibiting fire use includes the Forestry Law (2002), the Protected Area Law (2007), the Environmental Law (revised in 2001), and provincial regulations for the MPF (2006).

STUDY FINDINGS: BACKGROUND

The results of this study revealed that Bunong people describe water and fire as the essential foundations of life. Fire is used by them in a range of domestic activities: cooking and boiling water; decorating handicrafts; maintaining soil fertility; enabling resin collection; and as a treatment for sicknesses. Smoke is used to prevent pests from destroying stored rice, rice seed, and other seeds; for curing bottle gourds; and for protecting thatching from insects. The study found that as a tool for natural resource management, Bunong use of fires is based on the understanding that it can "clean" the environment: fields are cleaned in preparation for planting, forest paths are cleaned for easier walking, and savanna grasslands are cleaned to control vegetation and therefore prevent build-up of fuel loads that cause uncontrollable fires.

Study participants believed their use of fire to be the same historically as during their ancestor's time, with an overall increase recently due to population increase. The exception to this was during a period of regional and national conflict (ca. 1968–98) when many fire regimes were discontinued because of mass internal population displacement. Because Bunong people are mainly animist, they believe that the health and well-being of themselves, their family, and the wider community depends on maintaining reciprocal relationships with ancestral spirits and natural spirits in the environment. The effect of these beliefs on how Bunong people use fire is that traditionally many religious ceremonies were required before and after using fire to clear new shifting cultivation fields; if these were not followed, then problems would follow such as bad harvests and disease (Condominas 1977). However, this research concluded that today in both case-study villages, fire appears to have a decreasing religious, social, and cultural importance. Informants said that young people are less interested in continuing traditional natural resource management practices, including showing a respect for its animist importance. During group discussions in Koh Myeul Leu village, informants stated they could not afford the financial burden of continuing expensive animal sacrifices (traditionally required before using fire), which they no longer believe in.

Both villages stated their experience was that almost all fires in Mondulkiri are anthropogenic; very few people could recall fires starting "naturally." Human-set fires center around villages and are abundant in the region (Figure 1). The study inquired as to the Bunong understanding of the relationship between fire and the environment and found that the central belief is that fire is not regarded as the most important factor in ecological formation: the ecology itself determines if an area will burn or not, differentiated by the presence of existing fuel-loads; that is, dry forest, bamboo forest, savanna grassland, and thatching-grass areas will burn, but semi-evergreen and evergreen areas are un-burnable because (1) they have no ground-level fuel, and (2) they have not been cut and dried so their leaves are "cold" and therefore nonflammable. Therefore the driver of ecological formation is instead believed to be edaphic qualities.

The Bunong have a detailed knowledge of fire behavior. They understand that flame temperature depends on the strength of wind, presence of a flammable fuel-load, moisture levels, atmospheric temperature, and fire-line speed. Flora are more susceptible to high-temperature fires, but flora vulnerability and resistance are dependant on species, age, size, and presence of a protective and undamaged bark. The Bunong do not believe that high-intensity fires will negatively impact soil biodiversity. Nor do they believe fire affects insect or pest populations since these insects are not present in crop residues when fires are set; however, the health of wild and domestic animals is known to depend on nutritional intake from minerals left in ash following vegetation burns.

ANALYSIS AND DISCUSSION

Table 2 summarizes the four main fire regimes analyzed during the study: agricultural burning of specific fields during the shifting cultivation cycle (used mainly in Pu Tang); thatching-grass fires (burning of small 2–3 m plots of *Imperata cylindrical*, used in both villages); grassland fires (burns of unlimited size in the savanna grassland areas of the southern Haut Chhlong Plateau and therefore not used in Koh Myeul Leu); and dry forest fires (used mainly in Koh Myeul Leu and again of unlimited size). From the management strategies summarized in the third column of this table, it is evident that users have clear, customary laws for fire management that range from strict rules regarding agricultural burning to more relaxed, opportunistic management of dry forest fires. Agricultural and thatching-grass fires are controlled and target specific areas within a wider landscape that is not burned. Customary law described by the participants dictates that within the broader communal property management framework, these resource types are individually allocated and managed. This means that only specific and recognized individuals can access and benefit from this fire regime, whereas the costs of these fire regimes being mismanaged will impact the larger community. Strict rules and preventative measures are therefore employed to collaborate with neighboring property owners. Grassland and dry forest fires, on the other hand, target broad landscapes of unlimited size and resources that are collectively owned. This means the

Table 2. Synopsis of four Bunong fire regimes.

Fire Regime	Objective of Fire Regime	Impact of Removing Fire Regime from Landscape	Management Strategy
Shifting cultivation fields	Improve harvests (clear area, maintain soil fertility, and minimize weed regrowth)	Crop failure (loss of soil fertility, difficult planting, fast weed regrowth) Increased labor clearing vegetation	Control with boundary Set headfire[a] in average strength wind to burn "uphill" as "fire rises" Burn fire breaks to prevent uncontrolled fires, in cooperation with neighbors Let men lead decision making
Thatching grass	Thatch Improve hunting Clear vegetation Provide greenshoot[b]	Thatching-quality loss Difficult walking	Control with boundary Don't consider climatic/ topographical conditions Burn fire breaks to prevent uncontrolled fires, in cooperation with neighbors Let men lead decision making
Savanna grassland	Provide greenshoot[b] Clear paths and improve visibility Improve hunting Provide traditional custom and fun	Lack of greenshoot[b] Dangerous vegetation build-up (hazardous fuel load, difficult walking, hiding places for dangerous animals)	Don't consider climatic/ topographical conditions; don't control with boundary Don't take preventive measures; owner has responsibility to protect property in high-risk areas Allow gender-equal decision making
Dry forest	Clear vegetation Maintain soil fertility Provide greenshoot[b] Improve forest quality (ease of walking, visibility) Improve hunting	Dangerous vegetation build-up (hazardous fuel load, difficult walking, hiding places for dangerous animals, low visibility) Loss of soil fertility Poor greenshoot quality	Don't consider climatic/ topographical conditions; don't control with boundary Don't take preventive measures; don't monitor impacts (opportunistic) Allow gender-equal decision making

[a]A fire-line that burns in the same direction as the prevailing wind.
[b]The regrowth of vegetation following a burn, which is considered the most nutritious for grazing animals (domestic and wild) such as cattle and buffalo.

whole community can access and benefit from this fire regime, and equally the costs of mismanagement are felt throughout the community. Both villages explained that they had no management rules for these fire regimes because there is no property in need of protection. Protection of privately owned resources located within these collectively owned areas is the responsibility of the owner, not the fire setter. In both villages there was a clearly articulated difference between *collectively owned* resources and *open access* regimes; resources that are collectively owned are traditionally considered to belong to specific communities, whose ownership is respected by their neighbors. This concept of individual, collective, and open access resources within a broader system of CPR management therefore plays an important role in both villages in terms of the institutional rights and obligations of different fire regimes.

The majority of participants could recall examples of uncontrollable fires destroying property, and both villages provided examples of conflict resolution. These mechanisms focus on fires started by individuals that destroy the property of others by accident, sometimes through mismanagement and sometimes through lack of experience. In line with the management strategies outlined in Table 2, resolution processes for agricultural and thatching-grass fires are more precise than for grassland savanna and dry forest fires. The overall goal described was to resolve the conflict between the two parties and thereby realign community harmony; mediation is frequently provided by village elders if the parties cannot resolve the case alone. Traditionally, compensation was paid in an agreed period of manual labor by an adult from the perpetrator's household to the victim's household. Today, financial compensation is more commonly used.

Five of the nine government representatives interviewed recognized the importance of fire use for local livelihoods and for maintaining dry forest ecology, whereas two strongly supported its legislative criminalization. However, because government agencies lack the technical and financial resources to implement the law and establish prescribed fire regimes, those responsible for implementing the Forestry Law described an informal policy of focusing limited resources on public awareness instead of enforcement. In addition, all those interviewed recognized the limited understanding of government and conservation organizations regarding the relationship between fire, humans, and the environment; for example, how divergent current fire regimes are from optimal levels (Tordoff et al. 2005). From a policy perspective, because key elements of the legal framework related to fire management have not been completed, this research identified the potential for decriminalizing specific CBFiM strategies and decentralizing components of fire monitoring and management to local communities. WWF–Cambodia is developing options to address this knowledge gap (such as the research on which this chapter was based) and to identify policy options for comanagement of fire to enable all stakeholders to benefit from "good" fire while minimizing the risks of "bad" fires (WWF–Cambodia, personal communication).

These management strategies of the Bunong for fire were compared with enabling conditions (based on the four sets of variables noted above and as described in Table 1) for sustainable use of CPRs. Conditions 1 (resource characteristics) and 2 (group characteristics) define the resources and the community using them, while the relationship between these forms condition 3 (resource accessibility and use), which indicates the changing supply and demand of resources. Condition 4 (internal institutional arrangements) outlines the governance institution that monitors and enforces the rules, rights, and duties of CPR use. The fifth condition (external environment) describes the way in which CPR management adapts to external pressures. As Dolsak and Ostrom (2003) noted, this is becoming an increasingly important factor as demand for land and natural resources increases globally.

Condition 1: Resource Characteristics—small size, clear boundaries, low mobility, possibility to store benefits from the resource, and predictability (Agrawal 2001).

At a local scale, shifting cultivation and thatching-grass fire regimes fit this condition: the fires are small in size and used within clearly demarcated areas, and the majority of resources managed by these fires have storage potential. The other two fire regimes (dry forest and savanna-grassland fires) do not fit this condition at the local scale as their ecologies represent proportionally significant areas of the LMDFE.

Condition 2: Group Characteristics—small size, defined membership, legitimate leadership, homogeneity of identities and interests, social capital, and low poverty (Agrawal 2001).

Until the 1960s both villages would have met all the characteristics of condition 2 except low poverty. However, years of conflict combined with recent socioeconomic and infrastructural developments have increased social heterogeneity and weakened social capital and the legitimacy of traditional leadership of the elders, particularly within younger generations who are often now more interested in adopting the social norms of the dominant Khmer culture. Despite these sociocultural changes, indigenous villages in Mondulkiri are still considered some of the nation's poorest (Beresford et al. 2004). From a broader perspective, increasing external demand for land and resources from actors who do not respect or comply with Bunong customary laws has the potential to delegitimize these structures to the point that collectively owned regimes become open access.

Condition 3: Resource Accessibility and Use—user close to resource location, high value and dependence on resource, equitable allocation, low demand, and gradual changes in demand (Agrawal 2001).

Bunong management of all four fire regimes would traditionally have met this condition, however the inclusion of "gradual changes" is a key caveat for the ability of traditional CPR management approaches to deal with changing patterns of supply and demand. Within the context of weakening group homogeneity or solidarity described under condition 2, Bunong communities are less able to respond effectively to the rapidly increasing external demand on land and natural resources they have traditionally owned.

Condition 4: Internal Institutional Arrangements—simple rules (devised and managed by users), easy enforcement, low-cost adjudication, and accountability of monitors to users (Agrawal 2001).

The internal institutional arrangements used by the Bunong to manage all four fire regimes meet condition 4, although the increasing use of financial rather than labor-based compensation payments may increase the cost of adjudication in the future.

Condition 5: External Environment—accessible and appropriate technology, low (and/ or gradually increasing) interaction with markets, supportive government institutions, policy frameworks, and external assistance (Agrawal 2001).

Technology is not adequately accessible or appropriate to the villages studied. Access to markets is slowly improving, however the Bunong frequently find themselves at a disadvantage in market interactions with Khmer traders (M. MacInnes, personal observation). The criminalization by the Forestry Law and other legislation of local-level fire use is further disempowering. It exacerbates the delegitimization of traditional governance institutions while preventing government agencies from benefiting from traditional fire management knowledge. Nevertheless, current gaps in the policy framework provide an opportunity for increasing decentralized decision making in fire management and empowering indigenous communities to actively engage with the external environment on their own terms.

The results of the comparison of these enabling conditions within Bunong fire regime governance reveal that different fire regimes are managed under different types of common-property regulations. Prescribed management strategies will therefore only be successful if they are appropriate to each individual fire regime. The ability of Bunong fire management strategies to adapt to changes in the external environment is highly dependent on, and will directly influence, the other four enabling conditions. Therefore, the extent to which the Bunong will be able to take advantage of new opportunities provided by the government's decentralization program for natural resource management depends on whether the enabling framework is empowering, or disempowering, and this is determined by the level of support provided by external agencies and policies. Should this not be provided, then traditional fire regimes and the resource governance structures on which they depend risk fragmenting to open-access regimes. In summary, these results indicate the need for WWF–Cambodia, the FA, and other stakeholders to prioritize two areas for future CBFiM intervention. First, strategies for strengthening social capital and the internal legitimacy of governance institutions are essential for Bunong communities to sustainably manage resources with fire (condition 2). Second, external support in the form of technical and financial assistance is required, in addition to the formation of favorable policy and legislative frameworks (condition 5). The provision of both of these will strengthen the Bunong ability to respond to increasing external pressures and maintain sustainable control over fire regimes and natural resources (condition 3).

Available literature indicates that policy and practice pertaining to CBFiM in mainland Southeast Asia is not as developed as in Africa or Central and Latin America (Moore et al. 2002; Mistry and Berardi 2005). Nevertheless, there are some lessons that can be learned to help policy makers and practitioners identify the most appropriate CBFiM approach for the LMDFE.

THE PA-TONG NETWORK IN NAN PROVINCE IN NORTHERN THAILAND

The forests of Nan Province, northern Thailand are a mix of evergreen and DDF, and the Nan river supplies 56 percent of northern Thailand's water.[1] The area is populated by a number of indigenous ethnic minority groups who traditionally depend on upland rice and tree crops in upland areas and irrigated crops in the lowland areas (rotations of maize, rice, peanuts, and tobacco). Traditionally, in the Pa-tong project area, fire regimes of the indigenous groups focused on clearing fields and maintaining soil fertility within shifting cultivation cycles. Only limited fire prevention practices were used, such as suppression through "back-burning," nor were there any collaborative efforts between villages to manage fire. Forest fires (blamed on uncontrolled spread from local agricultural practices, NTFP collection, and hunting) are considered by the Thai government to be one of the major causes of deforestation, leading to soil erosion, loss of fertility, and a reduction in water supplies for crop irrigation and local fisheries. Government fire policies, under the responsibility of the Department of National Parks, Wildlife and Plant Conservation, have until recently controlled fires through suppression and law enforcement. Local communities have been involved in government initiatives to conserve and restore the forest since 1979. However, these activities included only forest fire control activities without any inter-village collaboration, and as a result government initiatives saw no reduction in fire frequency or forest degradation.

Since the mid-1980s, villagers in the Pa-tong area have self-mobilized to coordinate reforestation and environmental enrichment activities. These activities received limited external support from NGOs, but after representatives of villages and local authorities joined a government-sponsored study tour to neighboring Chiang Mai Province in 1993, these local representatives decided to formalize natural resource governance networks. During the 1990s, government policies for forest management had simultaneously decentralized. The Upper Nan Project (coordinated by the Department of National Parks, Wildlife and Plant Conservation) was implemented from 1996 to 2003 and considered the establishment of these community-based networks one of its key achievements. The networks aimed to reduce further natural resource degradation (including forest fire control), resolve inter-village disputes, and strengthen communication with local government authorities. Department staff believe that the success of these networks was dependent on two factors: the inclusion and leadership of respected village headmen, and the support and involvement of government policies and authorities. Network rules and regulations were developed from village-level assessments of problems and needs. Once developed, the village rules and regulations were combined at the subdistrict level and then used by department staff to develop overall management plans. Despite government support, compliance with these rules was not enforceable by law, only social sanction; this lack of legal recognition continues to be reported as a concern by the Pa-tong communities.

The CBFiM activities of the network were based on rules and regulations that prohibited encroachment into forests, outlined a system of fines for allowing the spread of fire, and described penalties if villagers didn't participate in fire management activities set by the village committee. Fire management issues addressed at the local level covered when to begin early burning; when and where to prepare fire-breaks; fire patrolling; equipment required; and availability of funding for these activities. External training in fire management was provided to community members by NGOs and the Forest Fire Control Unit in Chiang Mai, and included forest fire protection and early burning techniques. In addition, villagers improved their technical capacity through employment as laborers in the Government Forest Fire Control programs. Network activities resulted in a reduction in fire frequency and intensity and an increased capacity to manage fire. External funding through the Upper Nan Project ended in 2003, and although by 2005 activities had remained the same, the Regional Community Forestry Training Centre case study noted that it was not clear how long this would continue.

In terms of lessons applicable to the Mondulkiri context, the results of the Pa-tong Network suggest that CBFiM will only be effective if integrated within broader natural resource management strategies, implemented across landscapes, and built upon village-to-village networks and support groups. The combination of internal, village-level commitment and leadership legitimacy, external support from local administrative authorities, and favorable policies proved essential for the project's success. In addition, generation of rules and regulations from village-level needs and priorities was critical to ensuring local-level ownership of the networks' objectives. However, the lack of legal enforcement for compliance with these rules and regulations continues to be a threat to local management strategies. Finally, this case study reveals the positive role that NGOs can have supporting CBFiM initiatives by supplementing areas where government capacity is limited, and strengthening village-to-village networks.

IUCN NON-TIMBER FOREST PRODUCT PROJECT, SALAVAN PROVINCE, LAO PEOPLE'S DEMOCRATIC REPUBLIC

The DDF and pine forests of the Lao People's Democratic Republic (PDR) are believed to be actively maintained by fire.[2] The government's forest fire control initiatives aim to increase forest cover and ensure that shifting cultivation fires do not escape and cause forest destruction, while also protecting plantations and reforestation efforts. The Forest Law (1996), prime minister's decrees 164 and 169 (1993), and Order 54/ Ministry of Agriculture and Forestry (1996) give local people rights to use the forest according to their customary beliefs but forbid burning in protected areas. In additional to national-level legislation, Order 2094/MAF "Fighting Forest Fires during the Dry Season" (1999) gives authority to provincial and district authorities to implement forest fire management activities, and it supports the involvement of local communities in forest fire management.

The IUCN Non-Timber Forest Product Project, which is the subject of this case study and operated from 1995 to 2001, was located on the edge of the Xe Bang Nouan National Biodiversity Conservation Area (NBCA). Local fire regimes here are used for shifting cultivation, clearing vegetation, improving movement and hunting, providing cattle forage, cultivating mushrooms, and stimulating grass for thatching and fodder. The project aimed to develop models for sustainable use of NTFPs within the NBCA, and at its inception community members identified two threats to sustainable livelihoods: recognition of user rights by other villages, and fire.

In early 1999, devastating fires destroyed a large portion of the NBCA, leading to a significant decline in NTFP harvest rates that negatively impacted livelihoods.

Attempts to fight the fires were hampered by lack of inter-village coordination. Following the fires, Salavan provincial authorities asked the IUCN project to support the establishment of fire control committees in each village (based on Order 2094) in order to protect the forest inside the NBCA through coordination of fire protection and control. Project staff worked with existing village-level natural resource management committees and district authorities to strengthen villagers' understanding of the threats of fires by using problem analyses of "good" and "bad" fires. Fire control committees were established in each village; multi-village agreements were developed on fire preparedness strategies, management rules, and regulations; and the operations of both were supported by monthly meetings at the district to consolidate information on fire management. Twelve months after the fire, the provincial and district authorities endorsed the village comanagement plans, which included securing local government support for the fire control committees.

In terms of lessons applicable to the Mondulkiri context, the case study concludes that the success of the project was due to its having strong local level impetus from shared experiences of uncontrolled fires and being grounded within broader community-based natural resource management activities that valued local knowledge and encouraged local people to make their own decisions. External support from sympathetic local government staff and structures, policies promoting decentralization of fire management, and a favorable legal framework were also essential. Although Order 2094 would have been implemented by district officials even without the IUCN project's support, the case study suggests that the project's support in terms of community organization, planning, training, and funding significantly increased its effectiveness.

CONCLUSIONS

Although understanding of human-fire-environment dynamics is improving, my review reveals that key interactions within dry forest ecologies and management remain contested, especially in countries such as Cambodia, which have traditionally adopted anti-fire forest management frameworks. In Mondulkiri, the Bunong people view fire as an essential tool for natural resource management that directly impacts their livelihood. Within the four fire regimes studied, fire was used to clear vegetation and reduce weed regrowth, "clean" the environment of hazardous fuel loads and dangerous animals, maintain soil fertility, ensure good harvests, stimulate greenshoot for domestic animals and thatching, and improve mobility, hunting, and visibility. The research participants believed that if the Cambodian government enforced its ban on fires, then harvests would fail, there would not be sufficient fodder for animals, hunting and traveling through the forest would be difficult and dangerous, and fuel loads would increase, risking "uncontrollably destructive" fires. Fire is not recognized as a driver of environmental change; rather, the environment itself is believed to determine if a fire will burn or not. From this perspective, it appears that local understanding of the relationship between humans, fire, and the environment share much with the "pro-fire" scientific understanding of the relationship between fire and the environment in the LMDFE (Tordoff et al. 2005), in particular the role of edaphic characteristics.

Governance strategies of the four fire regimes included in this study fall into two categories and are regulated by customary rules dependant on (1) the expected outcome, (2) the resource tenure system, and (3) the perceived risk to property. When these governance strategies were analyzed within a framework of conditions suspected to enable sustainable management of common-property resources, two recommended focal areas for CBFiM support were identified for WWF–Cambodia and the Forestry Administration: first, the importance of strengthening governance institutions; and second, the importance of developing external support from local authorities and an enabling policy and legislative environment. This conclusion is in line with theoretical approaches of IFM and CBFiM frameworks and is also supported by experiences from establishing indigenous CBFiM strategies in northern Thailand and Laos PDR. Unfortunately, current Cambodian laws criminalize the setting of fires in forested areas, even though many government officials admit that they do not have the capacity or resources to manage fire effectively in remote areas like Mondulkiri. Nevertheless, while fire legislation and management guidelines remain incomplete, there is an opportunity for policy makers to learn from IFM and CBFiM models implemented in similar contexts. This may enable the Cambodian government to constructively benefit from indigenous traditional fire regimes; to decentralize management of fire through including local peoples; and to support CBFiM through legal frameworks and local government structures.

NOTES

1. This section has been adapted from Regional Community Forestry Training Centre for Asia and the Pacific (2005).
2. This section has been adapted from London (2003).

REFERENCES

Agrawal, A. 2001. Common Property Institutions and Sustainable Governance of Resources. *World Development* 29 (10): 1649–72.

Araral, E., Jr. 2009. What Explains Collective Action in the Commons? Theory and Evidence from the Philippines. *World Development* 37 (3): 687–97.

Bauld, S. 2007. WWF Greater Mekong Cambodia Country Programme: Eco-Tourism Feasibility Study. Srepok Wilderness Area Project Technical Paper Series 3, www.panda.org/what_we_do/where_we_work/project/projects_in_depth/dry_forests_ecoregion/publications/?149282/Ecotourism-feasibility-study.

Beresford, M., S. Nguon, R. Rathin, S. Sau, and C. Namazie. 2004. *The Macro-Economics of Poverty Reduction in Cambodia*. Phnom Penh: Asia-Pacific Regional Programme on the Macroeconomics of Poverty Reduction, United Nations Development Programme, www.un.org.kh/undp/knowledge/publications/the-macro-economics-of-poverty-reduction-in-cambodia.

Biot, Y., P. M. Blaikie, C. Jackson, and R. Palmer-Jones, eds. 1989. *Rethinking Research on Land Degradation in Developing Countries*. World Bank Discussion Paper 289. Washington, DC: World Bank.

Blaikie, P., and H. Brookfield, eds. 1987. *Land Degradation and Society*. London: Methuen.

Brechin, S. R., P. R. Wilshusen, C. L. Fortwangler, and P. C. West. 2002. Beyond the Square Wheel: Toward a More Comprehensive Understanding of Biodiversity Conservation as a Social and Political Process. *Society and Natural Resources* 15:41–64.

Bromley, D. W., and M. M. Cernea. 1989. *The Management of Common Property Natural Resources: Some Conceptual and Operational Fallacies*. Washington, DC: World Bank.

Condominas, G. 1977. *We Have Eaten the Forest: The Story of a Montagnard Village in the Central Highlands of Vietnam*. New York: Farrar, Straus and Giroux.

Dolsak, N., and E. Ostrom, eds. 2003. *The Commons in the New Millennium: Challenges and Adaptations*. Cambridge, MA: The MIT Press.

Food and Agriculture Organization of the United Nations Project Firefight South East Asia. 2003. *Community-Based Fire Management: Case Studies from China, The Gambia, Honduras, India, the Lao People's Democratic Republic and Turkey*. Forest Resources Development Service Working Paper FFM/2. Bangkok: Food and Agriculture Organization of the United Nations / Forest Resources Development Service, www.fao.org/DOCREP/006/AD348E/AD348E00.HTM.

Gebremedhin, B., J. Pender, and G. Tesfay. 2003. Community Natural Resource Management: The Case of Woodlots in Northern Ethiopia. *Environment and Development Economics* 8:129–48.

Goldammer, J. G., P. Frost, M. Jurvelius, E. M. Kamminga, T. Kruger, S. I. Moody, and M. Pogeyed. 2002. Community Participation in Integrated Forest Fire Management: Experiences from Africa, Asia and Europe. In *Communities in Flames: Proceedings of an International Conference on Community Involvement in Fire Management*, ed. P. F. Moore, D. Ganz, L. Cheng Tan, T. Enters, and P. B. Durst, 33–52. Bangkok: Food and Agriculture Organization of the United Nations, www.fao.org/docrep/005/ac798e/ac798e00.htm.

Hickey, G. C. 1982. *Sons of the Mountains: Ethnohistory of the Vietnamese Central Highlands to 1954*. New Haven, CT: Yale University Press.

International Forest Fire News. 2003. Strategic Paper: Community-Based Fire Management. In *Outcomes of the International Wildland Fire Summit, Sydney, Australia, 8 October 2003, Part V*. International Forest Fire News 29 (July–December 2003), www.fire.uni-freiburg.de/iffn/iffn_29/IWFS-4-Paper-5.doc.

Kull, C. A. 2004. *Isle of Fire: The Political Ecology of Landscape Burning in Madagascar*. Chicago: University of Chicago Press.

London, S. 2003. Community-Based Fire Management in the Lao People's Democratic Republic: Past, Present and Future. In *Community-Based Fire Management: Case Studies from China, The Gambia, Honduras, India, the Lao People's Democratic Republic and Turkey*, 97–120. Forest Resources Development Service Working Paper FFM/2. Bangkok: Food and Agriculture Organization of the United Nations / Forest Resources Development Service, www.fao.org/DOCREP/006/AD348E/AD348E00.HTM.

MacInnes, M. 2007. Bunong Use of Anthropogenic Fire as a Natural Resource Management Tool in Mondulkiri Province, Northeast Cambodia. Master's thesis, University of East Anglia, UK.

Maxwell, A. M. 2004. Fire Regimes in North-Eastern Cambodian Monsoonal Forests, with a 9300-year Sediment Charcoal Record. *Journal of Biogeography* 31:225–39.

Maxwell, A. M., and Y. Pinsonneault. 2001. *Proceedings of the Conservation Strategy Workshop: Dry Forest Landscapes of Northern and Northeastern Cambodia, 27–29th July 2001*. Phnom Penh: WWF Ecoregion-Based Conservation Programme / WWF–Cambodia.

Ministry of Agriculture, Forestry and Fisheries. 2002. *Forestry Law*. Phnom Penh: Ministry of Agriculture, Forestry and Fisheries, Royal Government of Cambodia, www.forestry.gov.kh/Documents/Forestry%20Law_Eng.pdf.

Mistry, J., and A. Berardi. 2005. Assessing Fire Potential in a Brazilian Savanna Nature Reserve. *Biotropica* 37 (3): 439–51.

Moore, P. F., D. Ganz, L. Cheng Tan, T. Enters, and P. B. Durst. 2002. *Communities in Flames*: *Proceedings of an International Conference on Community Involvement in Fire Management*. Bangkok: Food and Agriculture Organization of the United Nations, www.fao.org/ docrep/005/ac798e/ac798e00.htm.

Myers, R. L. 2006. *Living with Fire*: *Sustaining Ecosystems and Livelihoods through Integrated Fire Management*. Arlington, VA: Nature Conservancy Global Fire Initiative.

Nikles, B. 2006. Use and Management of Forest Resources in Phnong Villages, Mondulkiri. Master's thesis, University of Zurich.

Pyne, S. J. 1993. Keeper of the Flame: A Survey of Anthropogenic Fire. In *Fire in the Environment*: *The Ecological, Atmospheric and Climatic Importance of Vegetation Fire*, ed. P. J. Crutzen, and J. G. Goldammer, 245–66. Chichester, UK: John Wiley & Sons.

Regional Community Forestry Training Centre for Asia and the Pacific. 2005. *Case Study*: *Community-Based Fire Management of the Pa-tong Network*. Bangkok: Regional Community Forestry Training Center for Asia and the Pacific / Kasetsart University, Bangkok.

Schlager, E., and E. Ostrom. 1992. Property-Rights Regimes and Natural Resources: A Conceptual Analysis. *Land Economics* 68 (3): 249–62.

Shlisky, A., J. Waugh, P. Gonzalez, M. Gonzalez, M. Manta, H. Santoso, E. Alvarado, A. Ainuddin Nuruddin, D. A. Rodríguez-Trejo, R. Swaty, D. Schmidt, M. Kaufmann, R. Myers, A. Alencar, F. Kearns, D. Johnson, J. Smith, D. Zollner, and W. Fulks. 2007. *Fire, Ecosystems and People*: *Threats and Strategies for Global Biodiversity Conservation*. GFI Technical Report 2007-2. Arlington, VA: Nature Conservancy.

Sullivan, S., and K. Homewood. 2003. On Non-equilibrium and Nomadism: Knowledge, Diversity and Global Modernity in Drylands (and Beyond . . .) CSGR Working Paper 122/03, www2.warwick.ac.uk/fac/soc/csgr/research/workingpapers/2003/wp12203.pdf/.

Tordoff, A. W., R. J. Timmins, A. Maxwell, K. Huy, V. Lic, and E. H. Khou, eds. 2005. *Biological Assessment of the Lower Mekong Dry Forests Ecoregion*. Phnom Penh: WWF–Cambodia.

Wade, R. 1987. The Management of Common Property Resources: Collective Action as an Alternative to Privatization or State Regulation. *Cambridge Journal of Economics* 11:95–106.

18

Local Residents' Attitudes toward Seasonally Dry Forests at Selected Sites in Nepal and Myanmar

Teri D. Allendorf

Humans have a long history of impacting and modifying seasonally dry tropical forests, and some of these forests are far from a "natural" condition (Janzen 1988). Their conservation and management requires understanding the ecological processes, as well as the social processes and contexts, associated with them (Lockwood et al. 2006). Many communities living near dry forests are heavily dependent on them for resources. Engaging these communities must be a central piece of conservation, and incorporating their perspectives into conservation is critical (Dasmann 1984; Machlis and Tichnell 1985; Zube 1986; Brandon and Wells 1992; Newmark et al. 1993; Fiallo and Jacobson 1995; Furze et al. 1996).

An important step in engaging residents is to understand how they perceive their relationship with dry forest. This entails understanding not only their dependence on forest resources, but also other ways that they relate to the forest as a part of their life and their sense of place (Raval 1994). People who depend on these ecosystems for survival are spending much of their time in these forests, whether cutting fuelwood, grazing livestock, extracting non-timber forest products such as mushrooms and berries, taking a picnic, or sitting in the shade of a tree. Understanding the meaning that the environment has to people, not just in terms of how it helps them maintain their livelihoods, but how they experience it every day, is an important piece in understanding how we can better conserve this threatened ecosystem.

Struhsaker et al. (2005) found that the attitudes of neighboring communities toward protected areas is one of the top two factors that predict protected area success in sixteen protected areas in Africa. Understanding residents' attitudes can help find solutions to conflicts and unsustainable extraction and is essential to good governance of these areas (Lockwood et al. 2006). Engaging with the attitudes of residents around protected areas will enhance discourse on parks (Hanna et al. 2008) and will facilitate the recognition and understanding of the multiple values and meanings of protected areas within communities (Neufeld 2008). Recognition of these multiple values and meanings within communities can be helpful for conservation, rather than an obstacle to be overcome.

In this chapter, I describe the attitudes and perceptions of people living within, or near, six dry forest protected areas in Nepal and Myanmar (Allendorf et al. 2006, 2007; Table 1). This work is based on survey data and supplemented by in-depth interviews in one of the areas, Bardia National Park (Allendorf et al. 2007).

NEPAL AND MYANMAR

Nepal is one of the most biologically diverse regions in Asia and has populations of large mammal species such as tigers, elephants, and rhinoceros (Chaudhary et al. 2009). It also has one of most progressive systems of protected areas and community forestry in a developing country, including buffer zone legislation that distributes revenues to local communities for development (Heinen and Shrestha 2006). Over 19 percent of Nepal is in a protected area system that includes nine national parks, three wildlife reserves, one hunting reserve, three conservation areas, and eleven buffer zones covering an area of 28,999 km² (Chaudhary et al. 2009). Conservation of the protected areas includes a progressive system of sharing tourism revenues with local communities (Heinen and Mehta 1999).

Unlike many other nations in Southeast Asia, Myanmar has maintained large tracts of its natural habitats (Leimgruber et al. 2003). It is also one of the most biologically diverse regions in Asia (Myers et al. 2000; Wikramanayake et al. 2001). In 1990, more than half of the remaining forests in mainland Southeast Asia could be found in Myanmar (Dinerstein and Wikramanayake 1993), and they support relatively large populations of mammals such as tigers and elephants (Leimgruber et al. 2003; Lynam 2003). However, Myanmar's growing human populations have increased pressure on its natural resources and protected areas, and resources are increasingly strained by demands from Myanmar's neighbors, China, India, Thailand, and Bangladesh (Myint Aung et al. 2004). Most of the threats to protected areas in Myanmar come from small-scale activities of local communities (Rao et al. 2002), activities such as deforestation (Leimgruber et al. 2005), hunting (Lynam 2003; Rao et al. 2002), and agricultural practices (Myint Aung et al. 2004; Leimgruber et al. 2005), which have caused significant declines in wildlife populations and loss of natural habitats. Environmental education and conservation and development activities have been scarce around Myanmar's protected areas and, as in many countries, protected area managers have neither the human capacity nor technical and financial resources to manage the areas without the cooperation of local communities (Myint Aung 2007; Rao et al. 2002).

In Nepal, the study areas were Kaakri Bihaar, a "natural park"; Bardia National Park (NP); and the northern section of the Lumbini Development Project, a wildlife sanctuary (Figure 1). These areas, one national park, one local park, and one development zone that incorporated a conservation area, differ more in management than in human population pressure and have different histories and management strategies, including legal access for local residents. All the areas are surrounded by agricultural land with little to no forest. Bardia NP is relatively well protected by the military, Kaakri Bihaar has forest department guards who are stationed there but do little to stop illegal extraction of fuelwood, and Lumbini was newly established area with no official system of protection at the time of the surveys (Allendorf 2007).

Table 1. Summary description of protected areas studied in southwestern Nepal and upper Myanmar.

	Nepal			Myanmar		
	Kaakri Bihaar	Royal Bardia National Park	Lumbini	Alaungdaw Kathapa National Park	Chatthin Wildlife Sanctuary	Htamanthi Wildlife Sanctuary
Management objective	"Natural park"	National park	Wildlife sanctuary	National Heritage Site; biodiversity conservation	Conservation of thamin (Eld's deer)	Conservation of Sumatran rhinoceros
Management authority	District Forestry Office	Department of National Parks and Wildlife Conservation	Lumbini Development Project, and NGOs	Nature & Wildlife Conservation Division, Forest Department	Nature & Wildlife Conservation Division, Forest Department	Nature & Wildlife Conservation Division, Forest Department
Size (km^2)	1.76	968	1.20	1,606	268	2,151
Entry	Freely	With permit	Freely	Limited	Limited	Limited
Extraction	Informally, deadwood and fodder	Thatch once per year	Thatch once per year; informally, fodder	Illegal	Illegal	Illegal
Habitat	Sal (*Shorea robusta*) and pine (*Pinus roxburghii*) forest.	Sal (*Shorea robusta*) forest, with grassland, savanna, and riverine forest	Grassland	Semi-monsoon dipterocarp; mixed deciduous, evergreen, and pine forest	Monsoon dipterocarp forest	Mixed evergreen forest
Year established	1974	1969 (reserve); 1989 (national park)	Early 1970s (development area); 1995 (wildlife sanctuary)	1984	1941	1974

Figure 1. Location of protected areas in Nepal (Allendorf 2007).

In Myanmar, the study areas were Alaungdaw Kathapa National Park (AKNP), Chatthin Wildlife Sanctuary (CWS), and Htamanthi Wildlife Sanctuary (HWS; Figure 2). They represent a range of human population pressure, from relatively low in Htamanthi WS, which is surrounded by extensive intact forest, to high in Chatthin WS, which is surrounded by agricultural land and severely degraded forest (Allendorf et al. 2006). Alaungdaw Kathapa NP is intermediate in population pressure because it is surrounded by a mix of relatively intact buffer zone forests and agricultural fields.

ATTITUDES TOWARD DRY FOREST PROTECTED AREAS

Attitude is defined as a human psychological tendency expressed by evaluating a particular object with favor or disfavor (Ajzen and Fishbein 1980), such as liking or disliking a national park. Attitudes consist of beliefs, which are associations people establish between an attitude object and various attributes. In the phrase "a national park is part of a country's wealth," *national park* is the attitude object, *country's wealth* is an attribute, and *is part* is a relational term. These beliefs, or perceptions, can be positive or negative. Because the term *beliefs* can also refer to a person's core values (Dietz et al. 2005) rather than attributes of a particular attitude object, I will use the term *perceptions* to refer to people's beliefs about protected areas.

Using this definition of attitude, I conducted a survey in the three areas in Nepal, and with colleagues, in the three areas in Myanmar, to determine people's attitudes and perceptions toward the protected areas. A list of perceptions about each protected area was generated by asking people why they liked or disliked the area and the benefits and problems the area caused for them. Verbatim responses were recorded to the following questions: "What are the problems the protected area causes you?"

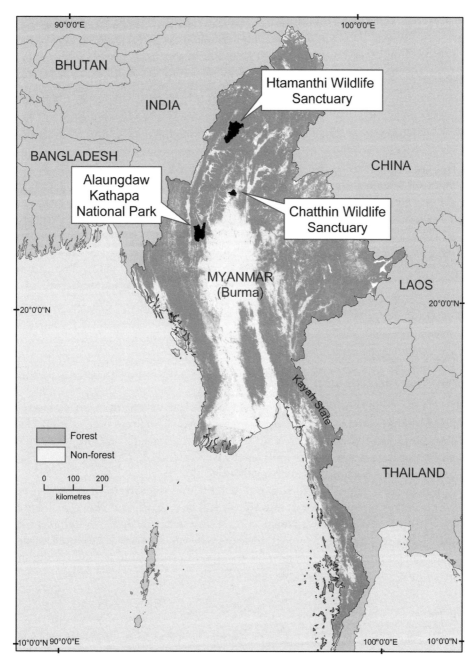

Figure 2. Location of protected areas in Myanmar (Allendorf et al. 2006).

"What are the benefits of the protected area?" "Do you like or dislike the protected area?" and "Why?" To facilitate analysis of the data, people's perceptions of benefits and problems and reasons for liking or disliking the areas were combined and sorted into major categories. The categories were created inductively after consideration of the responses gathered in the survey. Positive perception categories are conservation and ecosystem service benefits, recreational and aesthetic benefits, extraction benefits, management benefits, and country benefits. Negative perception categories are no extraction, problems with management, and wildlife conflict. In addition, we collected five socioeconomic variables for each respondent: gender, age, education, occupation, and landholding.

The following section describes the categories of perceptions that people hold toward the protected areas. As the data demonstrate, each area is different in which perceptions are mentioned most and least frequently, while some perceptions played roles in some areas and not others (Table 2). It is important to remember that people generated their responses on their own initiative, so the numbers reflect the number of people that generated the response on their own, not the number who might say a particular aspect was important if given a predetermined list. Tables 3 and 4 provide a detailed list of the responses that were categorized for each protected area.

Positive Perceptions

Conservation and Ecosystem Services

In all the areas, people appreciate the areas for the clean and healthy environment and good climate they provide. People appreciate that protected areas make the air cleaner and the soil moister because more rain falls. They appreciate that the forest is conserved. In Nepal, they say forests are Nepal's wealth and often use the phrase *Hariyo ban Nepalko dhan* [Green forest is Nepal's wealth].

People feel it is important to conserve wildlife and the forest: "It is good that the government takes care of the forest and later it will be available for children and grandchildren." Some respondents in Nepal said the forest and animals would all be gone because of the pressure from people if the areas were not protected: "Wildlife are good. We would all go and cut the forest. In Nepal, the forest is not in a good condition for wildlife, so the government made this place for them."

Extraction

Extraction, legal and illegal, is occurring in all the areas because people rely on the areas' resources for common and daily needs. People extract, for example, fuelwood, fodder, thatch, wood for building houses and making furniture, and leaves for fertilizer. As one respondent in Nepal said, "They will not let us take grass and wood, but if we really need it we can steal it." Interestingly, some people recognize that the illegality of extraction is protecting resources for extraction. As one respondent said in Nepal, "If the area wasn't protected, it would all be gone in one day."

Table 2. Characteristics of survey respondents living adjacent to protected areas in southwestern Nepal and upper Myanmar.

	Nepal			Myanmar		
Socioeconomic Variables[a]	Kaakri Bihaar	Royal Bardia National Park	Lumbini	Alaungdaw Kathapa National Park	Chatthin Wildlife Sanctuary	Htamanthi Wildlife Sanctuary
Mean age ±SD	35.2 ± 12.3	32.8 ± 13.6	38.0 ± 14.4	44.83 ± 15.97	43.76 ± 16.16	41.12 ± 14.33
Female (%)	47	51	25	37	44	29
Education (years) ±SD	2.93 ± 4.2	2.61 ± 4.0	3.51 ± 4.2	3.67 ± 2.7	3.32 ± 2.39	3.93 ± 2.7
No education (%)	60	64	54	13	13	7
Occupation (%)						
Farming	55	70	80	86	87	64
Other (labor, business, teaching)	45	30	20	10	11	30
Land (ha) ±SD	0.64 ± 1.25	1.10 ± 1.74	1.51 ± 1.72	3.99 ± 4.16	6.34 ± 5.75	1.99 ± 2.08
Landless (%)	32	21	12	13	18	36

[a]SD = standard deviation.

Table 3. Residents' positive and negative perceptions of three protected areas in Nepal. Subcategories do not sum to 100% as some respondents had multiple answers.

Positive	Kaakri Bihaar %	Royal Bardia National Park %	Lumbini %
Conservation and ecosystem services	29	32	14
Forest conservation	11	13	7
Better climate	6	8	6
More rain/water	5	9	2
Wildlife conservation	4	7	1
Forest is country's wealth	2	4	2
For the future	9	2	1
Healthy	1	1	1
Would be gone if not protected	2	4	
Good for environment	1		
Clean		2	
Extraction	58	36	11
Fodder	49	5	5
Fuelwood	24	12	2
Fruits/vegetables	15	3	1
Wood for lumber	1	1	
Leaves for fertilizer	4		
Mud/clay	3		
Flowers	2		
Fish			1
Can extract thatch	2	27	8
Can steal	1	4	
Grazing	1		
Recreation/aesthetics	64	30	55
Good to see/beautiful	19	14	27
Green	12	12	18
Take walks	29	7	19
Shade	10	2	8
See new people	2	4	9
Jungle all around	6	2	1
Sit/rest	4	1	1
For show	3	2	
See wildlife	7	3	
For people to visit	4	1	
See things	1	1	
Temple conservation	12		10
Bathe		1	
Management activities		2	11
Employment		2	8
Development		1	5
Road			4
Other			
Area is nearby	24	1	
Tourism	2	2	
Good for area	1	1	

Negative	Kaakri Bihaar %	Royal Bardia National Park %	Lumbini %
Extraction (no extraction of the following)	34	67	32
Fuelwood	29	53	18
Fodder	13	29	7
Fruits/vegetables	1	1	
Water		2	1
Mud/clay	4		
Leaves for fertilizer	2		
Lumber		6	
Fish		3	
No extraction in general	7	26	13
No grazing	2	18	5
Thatch limited/expensive		5	2
Conflicts with management	6	30	50
No access	4	18	2
Grazing fines	2	6	
Benefit for government or foreigners		4	4
Negative interactions with guards		5	1
Government took land			38
No work available			19
Problems with wildlife	2	35	28
Wildlife eat crops	2	35	28
Livestock depredation		4	18
Fear of wildlife		1	5

Recreation and Aesthetics

Interestingly, in Myanmar, recreational and aesthetics were not mentioned as benefits of the protected area. However, in Nepal, many people appreciate that protected areas are nice to look at, or beautiful, and they like the "greenness" of the areas. People like walking in the area, sitting in the shade of the trees, and enjoying the cool breezes: "It is for people to have a cool, shady place to visit." Others mentioned enjoying the opportunity to see and talk with people who visit the area from other places in Nepal and from other countries: "You came. Conversation is available with people like you."

In Nepal, residents also enjoy the opportunity to see wildlife in the areas. In Nepal, people use the word *ramaailo*, a word that means "entertaining," "enjoyable," or "fun." A few people mentioned it was generally good to have the protected area for the benefit of the surrounding area; for example, one man said that Kaakri Bihaar is the "superhit" of Surkhet because it makes the area famous all over the country. In Kaakri Bihaar and Lumbini, residents enjoy visiting the temples and, in Lumbini, they enjoy watching the progress of the hotels and monasteries as they are being built. The areas also provide a place for private ablutions: one respondent said that Bardia NP provided a place to bathe; one respondent said a benefit of Lumbini was having a place to go to the toilet.

Table 4. Positive and negative perceptions of protected areas of villagers living around three parks in Myanmar. Subcategories do not sum to 100% as some respondents had multiple answers.

Positive	Alaungdaw Kathapa National Park %	Chatthin Wildlife Sanctuary %	Htamanthi Wildlife Sanctuary %
Conservation	63	37	52
Protect deer		6	
Protect elephant/tiger			4
Conserve forest	33	25	21
Improve climate	27	14	4
Wildlife conservation	28	4	27
Wildlife live in freedom		3	
Conservation of natural resources	2		2
Park or sanctuary authority obligation to conserve	1		
Joy in greenness	19		
Extraction	14	7	7
Fuelwood		3	
Fertilizer	1	1	
Furniture	10	1	4
Leaves for packing		1	
House poles		1	
Food	1	2	1
Thatch		1	
Rice		2	
Bamboo		1	3
Palm leaf			3
Cane			3
Deodar (perfume)			1
Fishing materials	3		
Management	16	6	
"Clean and Green Chatthin" program		1	
Increased conservation education	4	1	
Youth programs		1	
Education signs		1	
Development	3	3	
Transportation	5	1	
Employment	1	1	
No robbers	1	1	
Grazing in return for reserve work		1	
Conservation of shrine	1		
Ecotourism	2		
Country	4	4	31

Negative	Alaungdaw Kathapa National Park %	Chatthin Wildlife Sanctuary %	Htamanthi Wildlife Sanctuary %
Extraction (no extraction of the following)	22	50	46
Fuelwood	1	26	
House poles	7	19	
Fodder	1	15	
Food		8	
Furniture wood	6	5	17
Thatch		5	18
Bamboo		2	18
Leaves for packing		1	
Fertilizer		<1	
Palm leaf			30
Cane			21
Resin	8		7
Gold			6
Fish	4		5
Other	1		2
Orchids	4		
Hemp	6		
Pasture	2		
No extraction in general		3	
No extraction for profit		1	
Charcoal/fuelwood are expensive		<1	
Management	4	39	9
Cropland reclaimed by park or sanctuary authority		21	<1
Restricted access for through travel		13	
Park or sanctuary authority staff punish people	2	9	8
No cropland in protected area	2	5	1
Resettlement issues		1	
No timber allowance		<1	
Wildlife damage		14	1

Protected Area Management Activities

In some areas, people recognized that the protected area management provided some benefits to local communities. People felt that work opportunities provided by the protected areas and development activities, such as road-building, were benefits of the protected areas. For example, in Lumbini, the building of monasteries and hotels had the potential to create manual-labor job opportunities for local residents. In Chatthin WS, they mentioned environmental awareness programs being conducted in one area and grazing privileges in return for help in capturing deer for a research project. In Alaungdaw Kathapa NP they mentioned the conservation of a shrine and tourism associated with the shrine.

The protected area can also increase the security in an area. In Chatthin WS, people mentioned the lack of robbers in villages because of protected area patrol activities, and in Bardia NP, people mentioned the guards allowing residents to use guard vehicles to take people to the hospital.

Negative Perceptions

Lack of Extraction

Although people view extraction as a benefit of the protected areas, as described earlier, they also consider a lack of legal extraction to be a problem. The same items they recognize as resource benefits of the area they also resent not being able to extract freely or legally. In some areas, people also resent that they cannot graze their livestock and that they are fined when their animals are caught illegally grazing.

Protected Area Management Activities

Management activities cause many of the problems mentioned by residents. People resent that management restricts their access to the areas. In some of the areas, such as Kaakri Bihaar, where they can freely enter the area, this response implied frustration with not being able to extract. However, in other areas, such as Bardia NP and Chatthin WS, they are frustrated that they cannot legally enter, even for recreational purposes, without permission. As one respondent in Bardia NP said, "The army tells us we can't even go to the toilet." People resent being fined and harassed by guards. They resent not being allowed access to the timber in the protected areas for houses and public buildings, although in some of the areas at the time, such as Bardia NP, management did allow some timber extraction for public buildings, such as schools, if prior approval was solicited. Residents resent that park staff can access timber for building staff quarters and offices, and in the case of Bardia NP, the tour operators are allowed timber to build their hotels inside the protected area. Additional conflicts with management in Chatthin WS include cropland that was reclaimed by the protected area and resettlement that is planned for villages inside one of the protected areas.

Problems with Wildlife

In some of the areas, crop damage and livestock depredation by wildlife are problems. People in some villages, such as around Bardia NP where there are deer, rhinos, and elephant, keep watch all night to protect their crops. In Lumbini and Chatthin WS, bluebuck (*Boselaphus tragocamelus*) and Eld's deer (*Rucervus eldi*), respectively, eat crops. As one respondent said, "All animals are good to us, but we fear them and they eat the crops in the fields."

Wildlife also threaten people's safety and lives. In Bardia NP, respondents mentioned rhinoceros and elephants attacking people. These fears are justified, as between 1988 and 1994, tigers, rhinoceros, and elephants injured eight and killed ten people around this park (Bhatta 1994).

ATTRIBUTES OF PERCEPTIONS

While I have presented the perception categories here as discrete and distinct, there are some larger themes that are important to understand people's perceptions and place them in a broader context.

Interconnectedness of Perceptions

Although people's perceptions of protected areas can be considered as discrete categories for analysis, perceptions can be interconnected in lived experiences and may have multiple meanings to people. For example, extraction and recreation can occur simultaneously. In Nepal, groups of women often go to the forest together to cut wood or fodder. It can be an opportunity to get out of the house and gives them the opportunity to enjoy one another's company. They may take a break from collecting and sit in the shade of a tree to rest, eat, and chat. At Lumbini, people's resentment at being resettled can be considered a conflict with management. However, resettlement also can include for people a feeling of loss and nostalgia.

Another example of the multiple meanings of perceptions is residents' resentment at being denied access to the areas. A lack of access can be a conflict with management because it is a policy that management sets and enforces. It can also be considered an extraction problem if people place it in the context of not being allowed to extract. However, people may also place it in the context of not even being able to recreate in the area. People in Bardia NP and Chatthin WS mentioned not being able to even take walks in the area.

Scale of Perceptions

People recognize the value of the protected areas at different scales, not just the benefits the areas can provide them. They can recognize that the areas benefit their communities as well as their country. For example, wildlife may be appreciated not only for personal enjoyment, but also because it is the country's wealth. As one resident explained, "Tomorrow's generation will not be able to see wildlife that we can see now. How many animals are already finished? Let us talk about rhinos. We know how rhinos look, we have seen them, but in our children's time, how will they know how rhinos are? They will not know except from books."

Relationship of People's Perceptions of Protected Area to Other Entities

People's perceptions of management can play an important role in their attitudes toward protected areas (Adams and Hulme 2001; Alexander 2000; Fiallo and Jacobson 1995; Holmes 2003b; Infield and Namara 2001; Ite 1996; McClanahan et al. 2005; Parry and Campbell 1992; Picard 2003). In a review of factors influencing attitudes toward protected areas, a poor relationship with park staff was the only variable that was associated with negative perceptions of protected areas in all six areas reviewed (Fiallo and Jacobson 1995). Some studies have found that people's attitudes toward management can be even more negative than their attitudes toward the protected area (Infield 1988; Newmark et al. 1993).

People's perceptions of the areas described in this chapter demonstrate that management can play both a positive and negative role simultaneously in people's attitudes toward protected areas. As with the protected areas themselves, people can be ambivalent toward protected area management. They resent some aspects of management, such as limited access to resources, yet they may appreciate that management conserves resources that could otherwise be consumed by them or others. People may dislike not having free access to resources in the park, such as fuelwood and fodder, but they also recognize that if it were an open access area, people would be extracting a greater quantity of resources and people from greater distances away would be extracting. As one man said in Bardia NP, "Open forest would be finished in one day. Now it is closed. People steal, but, at the same time, they are afraid [to steal]."

In Nepal, it is not only people's perceptions of park management that affect their attitude toward protected areas, it is also their perceptions of the government. While people may understand the reasons why the government has strictly protected these areas, such as for the role that these areas play in protecting Nepal's national heritage, their judgment of the government's reasons can be positive or negative depending on how they perceive the government. For example, if they respect the government, they are likely to respect the government's reasons for protecting these areas. However, if they view the government as corrupt, they may perceive the protected area as another form of selfishness on the part of the politicians. In the case of Bardia NP, people's memories of the king and his support of protected areas is important to their positive perceptions of the area, while their perception of the current government—particularly local officials—as corrupt has a negative effect on their attitude toward the area. In Kaakri Bihaar, people talk positively about the king and queen visiting during the establishment of the park to plant saplings.

Surprisingly, in Nepal, although international non-governmental organizations (NGOs) have been working in two of the areas, one for many years, residents did not mention NGOs or their work as a benefit or as a problem in the survey. In contrast, in Chatthin WS, where management has implemented small, local projects conducted by staff and community members, such as garbage collection in the largest town near the sanctuary headquarters, environmental essay contests in the schools, and road maintenance programs, people mentioned these benefits more than people in Bardia NP mentioned the big NGO projects. Additionally, although relatively small numbers of people mentioned positive management activities in Myanmar, these activities were significant predictors of positive attitudes in two of the areas, Chatthin WS and Alaungdaw Kathapa NP (Allendorf et al. 2006). Negative management activities in all three areas, such as reclamation of protected area land from local communities and punishment for illegal extraction, were associated with a negative attitude toward the protected areas.

Ambivalence of Perceptions

An important and often overlooked aspect of people's attitudes is that most people are neither completely in favor of nor opposed to a protected area. People are often ambivalent toward protected areas, meaning they simultaneously maintain strong positive *and* negative evaluations of the protected area (de Liver et al. 2007). For example, on one hand, the wildlife eats residents' crops and residents' access to resources is restricted because of the protected area. On the other hand, the area provides resources (legally and illegally), environmental services, and aesthetic benefits.

Understanding residents' ambivalence is a key point in starting to create collaborative and constructive relationships with local communities to conserve protected areas. People's positive perceptions, once identified, can provide common ground for managers, conservationists, and local communities to begin working together. Understanding people's ambivalence can help managers develop different strategies for communication with people and also for dealing with practical, day-to-day issues of management.

CONCLUSIONS

Local communities near seasonally dry forest protected areas in Nepal and Myanmar have complex and ambivalent perceptions of the protected areas, with the majority of people reporting liking each area, except one area in Myanmar. Two types of benefits were recognized in all the areas: conservation and ecosystem services, and extraction benefits. Two types of problems were recognized in all the areas: people's inability to legally extract forest resources, and people's conflicts with management. Wildlife conflicts play a key role in areas with wildlife that enters fields and harms people, for example, elephants and rhinos in Bardia National Park; Eld's deer in Chatthin Wildlife Sanctuary; and *nilgai*, or blue bull, in Lumbini. Additionally, in Nepal, recreational and esthetic benefits are an important component of people's perceptions. In some areas, people also recognize benefits that are related to management of the area, such as road-building and education activities.

As people's perceptions demonstrate in these six areas, seasonally dry forests, even in the more-strictly protected context of protected areas, are important to people for a variety of reasons, and people value them for more than the resources they provide. While people may resent not having access to freely extract from these areas, their perceptions are not limited to subsistence issues but encompass a broader range of values. People appreciate the multiple benefits of dry forests, in addition to the natural resources they can extract, such as conservation and ecosystem service benefits and recreational and aesthetic benefits.

An important benefit to understanding the perceptions of local people is that it can help us test the assumptions that sometimes affect our approach to managing people around reserves. Sometimes we assume that people with subsistence livelihoods do not recognize non-economic values of the environment, such as ecosystem services, or we assume we need to teach these values to local communities. Another assumption is the idea that park-people relationships consist primarily of the threats that local people pose to biodiversity and that these threats can be mitigated by livelihood solutions, such as integrated conservation and development projects (ICDPs). While these assumptions may be true in specific cases, they are not universal, and may only represent one aspect of the resident's relationship toward a protected area. These assumptions prevent us from finding positive aspects of the relationship that people themselves may recognize, and limits our ability to recognize creative and innovative strategies for improving the park-people relationship. Additionally, as these biases tend to be economically or livelihood oriented, they lead toward considering development strategies as the sole way to improve park-people relationships (Infield 2001).

The complexity of people's perceptions highlights the need for a more-nuanced approach to working with the communities surrounding parks and reserves. More often than not, conservationists emphasize the threats people pose to biodiversity and posit meeting people's "needs" as a way to conserve biodiversity (Brandon and Wells 1992). Meanwhile, there is little to no discussion of the role that people's positive perceptions may be able to play in conservation (Mugisha and Infield 2009). Integration of the full set of values people hold into conservation strategies, in conjunction with approaches such as sustainable extraction and alternative livelihood strategies, can provide an important complement, or even alternative, to economic benefits in gaining people's support for protected areas (Norton 1989; Kaltenborg et al. 1999; Infield 2001).

REFERENCES

Adams, W. M., and D. Hulme. 2001. If Community Conservation Is the Answer in Africa, What Is the Question? *Oryx* 35 (3): 193–200.

Ajzen, I., and M. Fishbein. 1980. *Understanding Attitudes and Predicting Social Behavior.* Englewood Cliffs, NJ: Prentice Hall.

Alexander, S. E. 2000. Resident Attitudes towards Conservation and Black Howler Monkeys in Belize: The Community Baboon Sanctuary. *Environmental Conservation* 27 (4): 341–50.

Allendorf, T. D. 2007. Residents' Attitudes toward Three Protected Areas in Southwestern Nepal. *Biodiversity and Conservation* 16 (7): 2087–2102.

Allendorf, T., Khaing Khaing Swe, Thida Oo, Ye Htut, Myint Aung, K. Allendorf, L. Hayek, P. Leimgruber, and C. Wemmer. 2006. Community Attitudes toward Three Protected Areas in Upper Myanmar (Burma). *Environmental Conservation* 33 (4): 344–52.

Allendorf, T. D., J. L. D. Smith, and D. Anderson. 2007. Understanding Park-People Relationships: Residents' Perceptions of Royal Bardia National Park, Nepal. *Landscape and Urban Planning* 82:33–40.

Bhatta, S. R. 1994. Beginning with Buffer Zone Management: A Case Study from Royal Bardia National Park, Nepal. Master's thesis, Agricultural University of Norway.

Brandon, K. E., and M. Wells. 1992. Planning for People and Parks—Design Dilemmas. *World Development* 20:557–70.

Chaudhary, R. P., K. C. Paudel, and S. K. Koirala. 2009. *Nepal Fourth National Report to the Convention on Biological Diversity.* Kathmandu, Nepal: Ministry of Forests and Soil Conservation.

Dasmann, R. F. 1984. The Relationship between Protected Areas and Indigenous People. In *National Parks, Conservation, and Development: The Role of Protected Areas in Sustaining Society,* ed. J. A. McNeely and K. R. Miller. Washington, DC: Smithsonian Institute Press.

de Liver, Y., J. van der Pligt, and D. Wigboldus. 2007. Positive and Negative Associations Underlying Ambivalent Attitudes. *Journal of Experimental Social Psychology* 43:319–26.

Dietz, T., A. Fitzgerald, and R. Shwom. 2005. Environmental Values. *Annual Review of Environmental Resources* 30:335–72.

Dinerstein, E., and E. Wikramanayake. 1993. Beyond "Hotspots": How to Prioritize Investments to Conserve Biodiversity in the Indo-Pacific Region. *Conservation Biology* 7:53–65.

Fiallo, E. A., and S. K. Jacobson. 1995. Local Communities and Protected Areas: Attitudes of Rural Residents towards Conservation and Machalilla National Park, Ecuador. *Environmental Conservation* 22 (3): 241–49.

Furze, B., T. de Lacy, and J. Birckhead. 1996. *Culture, Conservation, and Biodiversity.* New York: John Wiley & Sons.

Hanna, K. S., D. A. Clark, and D. S. Slocombe. 2008. *Transforming Parks and Protected Areas: Policy and Governance in a Changing World.* New York: Routledge.

Heinen, J. T., and J. N. Mehta. 1999. Conceptual and Legal Issues in the Designation and Management of Conservation Areas in Nepal. *Environmental Conservation* 26:21–29.

Heinen, J. T., and S. K. Shrestha. 2006. Evolving Policies for Conservation: An Historical Profile of the Protected Area System of Nepal. *Journal of Environmental Planning and Management* 49 (1): 41–58.

Holmes, C. M. 2003a. Assessing the Perceived Utility of Wood Resources in a Protected Area of Western Tanzania. *Biological Conservation* 111 (2): 179–89.

———. 2003b. The Influence of Protected Area Outreach on Conservation Attitudes and Resource Use Patterns: A Case Study from Western Tanzania. *Oryx* 37 (3): 305–15.

Infield, M. 1988. Attitudes of a Rural Community towards Conservation and a Local Conservation Area in Natal, South Africa. *Biological Conservation* 45 (1): 21–46.

———. 2001. Cultural Values: A Forgotten Strategy for Building Community Support for Protected Areas in Africa. *Conservation Biology* 15 (3): 800–802.

Infield, M., and A. Namara. 2001. Community Attitudes and Behaviour towards Conservation: An Assessment of a Community Conservation Programme around Lake Mburo National Park, Uganda. *Oryx* 35 (1): 48–60.

Ite, U. E. 1996. Community Perceptions of the Cross River National Park, Nigeria. *Environmental Conservation* 23 (4): 351–57.

Janzen, D. H. 1988. Tropical Dry Forests: The Most Endangered Major Tropical Ecosystem. In *Biodiversity*, ed. E. O. Wilson and F. M. Peters. Washington, DC: National Academy Press.

Kaltenborg, B., H. Riese, and M. Hundeide. 1999. National Park Planning and Local Participation: Some Reflections from a Mountain Region in Southern Norway. *Mountain Research and Development* 19:51–61.

Leimgruber, P., J. B. Gagnon, C. Wemmer, D. S. Kelly, M. A. Songer, and E. R. Selig. 2003. Fragmentation of Asia's Remaining Wildlands: Implications for Asian Elephant Conservation. *Animal Conservation* 6:347–59.

Leimgruber, P., D. S. Kelly, M. K. Steininger, J. Brunner, T. Muller, and M. Songer. 2005. Forest Cover Change Patterns in Myanmar (Burma) 1990–2000. *Environmental Conservation* 32 (4): 356–64.

Lockwood, M., G. L. Worboys, and A. Kothari. 2006. *Managing Protected Areas: A Global Guide.* Sterling, VA: Earthscan.

Lynam, A. 2003. *A National Tiger Action Plan for the Union Of Myanmar.* Myanmar Forest Department, Ministry of Forestry.

Machlis, G. E., and D. L. Tichnell. 1985. *The State of the World's Parks: An International Assessment for Resource Management, Policy and Research.* Boulder, CO: Westview.

McClanahan, T., J. Davies, and J. Maina. 2005. Factors Influencing Resource Users and Managers' Perceptions towards Marine Protected Area Management in Kenya. *Environmental Conservation* 32 (1): 42–49.

Mugisha, A., and M. Infield. 2009. People-Oriented Conservation: Using Cultural Values in Uganda. *Oryx* 43 (1): 13–16.

Myers, N., R. A. Mittermeier, C. G. Mittermeier, G. A. B. da Fonseca, and J. Kent. 2000. Biodiversity Hotspots for Conservation Priorities. *Nature* 403 (6772): 853–58.

Myint Aung. 2007. Policy and Practice in Myanmar's Protected Area System. *Journal of Environmental Management* 84 (2): 188–203.

Myint Aung, Khaing Khaing Swe, Thida Oo, Kyaw Kyaw Moe, P. Leimgruber, T. Allendorf, C. Duncan, and C. Wemmer. 2004. The Environmental History of Chatthin Wildlife Sanctuary, a Protected Area in Myanmar (Burma). *Journal of Environmental Management* 72 (4): 205–16.

Neufeld, D. 2008. Indigenous Peoples and Protected Heritage Areas: Acknowledging Cultural Pluralism. In *Transforming Parks and Protected Areas: Policy and Governance in a Changing World*, ed. K. S. Hanna, D. A. Clark, and D. S. Slocombe, 181–99. New York: Routledge.

Newmark, W. D., N. L. Leonard, H. I. Sariko, and D. G. M. Gamassa. 1993. Conservation Attitudes of Local People Living Adjacent to Five Protected Areas in Tanzania. *Biological Conservation* 63 (2): 177–83.

Norton, B. G. 1989. The Cultural Approach to Conservation Biology. In *Conservation for the Twenty-first Century*, ed. D. Western and M. C. Pearl, 241–46. New York: Oxford University Press.

Parry, D., and B. Campbell. 1992. Attitudes of Rural Communities to Animal Wildlife and Its Utilization in Chobe Enclave and Mababe Depression, Botswana. *Environmental Conservation* 19 (3): 245–52.

Picard, C. H. 2003. Post-apartheid Perceptions of the Greater St Lucia Wetland Park, South Africa. *Environmental Conservation* 30 (2): 182–91.

Rao, M., M. Rabinowitz, and S. T. Khaing. 2002. Status Review of the Protected-Area System in Myanmar, with Recommendations for Conservation Planning. *Conservation Biology* 16 (2): 360–68.

Raval, S. R. 1994. Wheel of Life—Perceptions and Concerns of the Resident Peoples for Gir National Park in India. *Society and Natural Resources* 7:305–20.

Struhsaker, T. T., P. J. Struhsaker, and K. S. Siex. 2005. Conserving Africa's Rainforests: Problems in Protected Areas and Possible Solutions. *Biological Conservation* 123:45–54.

Wikramanayake, E., E. Dinerstein, C. J. Loucks, D. M. Olson, J. Morrison, J. Lamoreux, M. McKnight, and P. Hedao. 2002. *Terrestrial Ecoregions of the Indo-Pacific: A Conservation Assessment*. Washington, DC: Island Press.

Zube, E. H. 1986. Local and Extra-local Perceptions of National Parks and Protected Areas. *Landscape and Urban Planning* 13:11–17.

Conservation Planning and Management in Southeast Asia's Seasonally Dry Forest Landscapes

Experiences from Cambodia

Andrew L. Maxwell and Nicholas J. Cox

This chapter summarizes experiences within the Cambodia program of the World Wide Fund for Nature (WWF) in conservation management of seasonally dry tropical forests. The summary encompasses about twelve years of experience, 1998–2009, both describing the process of regional management planning and prioritization, and discussing important thematic issues in direct management of a landscape in eastern Cambodia. The authors were involved in the sequence of planning and management described here, but the views expressed in this chapter do not necessarily reflect those of WWF.

Of the three global zones generally recognized as supporting large expanses of relatively natural tropical deciduous forest (i.e., cerrado, miombo, and deciduous dipterocarp forest), this chapter discusses planning and management issues in the Southeast Asian deciduous dipterocarp forest (DDF) and associated plant communities. The vegetation mosaic dominated by DDF occurs almost exclusively in the mainland section of Southeast Asia, with its pre-1960s extent covering parts of central Myanmar (Burma), much of northeastern Thailand and Southern Laos, the central highlands of Vietnam, and most of northern and eastern Cambodia. For reasons discussed below, this summary focuses on dry forest management in Cambodia, based on assessment, prioritization, and planning for a region of analysis including most of Laos, Vietnam, Cambodia, and Thailand.

This chapter can be read as a case study, as some of the important management issues are reflected in other landscapes on other continents. As with almost all case studies, however, there is the caveat that some aspects of the situation in Cambodia are unique, primarily because formal conservation efforts have been building only very recently in Cambodia due to war, political and economic isolation, and social instability during the period 1968–98.

ECOREGIONAL PLANNING AND PRIORITIZATION

The ecoregional approach to conservation was developed within WWF during the 1990s with the key documents including articles that describe the "Global 200" (Olson and Dinerstein 1998, 2002). The approach was developed to focus biological conservation efforts on relatively large areas, or *ecoregions*, that are defined and prioritized by ecological characteristics. The approach for this project is outlined in a flowchart (Figure 1). The ecoregional approach was intended to improve upon previous conservation efforts by (1) focusing on the most globally or regionally important representative ecosystems, (2) expanding the scales of planning and management both temporally and geographically, (3) conserving ecosystem processes as well as key species, and (4) integrating biodiversity conservation priorities within regional land-use planning and policies. Similar broad-scale approaches to conservation prioritization and management were also being developed by other international organizations at about the same time (Maltby et al. 1996; Mittermeier et al. 1998; Stattersfield 1998).

Among the Global 200, four ecoregions were considered priorities within mainland Southeast Asia: the Annamite Range Moist Forests, Central Indochina Dry Forests, and the Cardamom Mountains Moist Forests in the terrestrial realm; and the Mekong River in the freshwater realm (Olson and Dinerstein 2002; Wikramanayake et al. 2002).

As the original identification of the Global 200 was essentially a desk study, WWF and its conservation partners convened the first of several ecoregional planning workshops in Phnom Penh, Cambodia, in March 2000, with the participation of eighty scientists and conservation staff who had experience working in mainland Southeast Asia (Baltzer et al. 2001). This "visioning" workshop had as its goal the establishment of a biological vision and broad targets for the region. The region of analysis covered much of mainland Southeast Asia, i.e., all of Cambodia and most of Vietnam and Laos, except for a far northern section with transitional subtropical characteristics. Parts of Thailand also were considered, especially the dry forest zone in the northeast; and Thailand-based scientists participated in the workshop, but the workshop organizers observed that most of the Mekong watershed in Thailand had been converted to agriculture, so its small natural areas were not included in the analysis.

Through a series of group discussions on distribution of key mammals, birds, vegetation, reptiles, and butterflies, the visioning workshop's primary output was a four-class prioritization of conservation landscapes throughout the region of analysis (Baltzer et al. 2001) and encompassing the three terrestrial ecoregions. Three of the classes were hierarchical in importance, and the fourth class consisted of areas suspected to be important but that needed more research data to be comparable to other priority areas in the analysis. Of five priority landscapes within the Central Indochina Dry Forests Ecoregion (name later changed to Lower Mekong Dry Forests Ecoregion), four were classed as critical, or of the highest priority. Three of these, the Northern Plains, Eastern Plains, and Central Cambodian Semi-evergreen Lowland Forests, occurred almost entirely within Cambodia, while the fourth, the Mekong River and major tributaries, was centered in Cambodia but included significant territory in Laos.

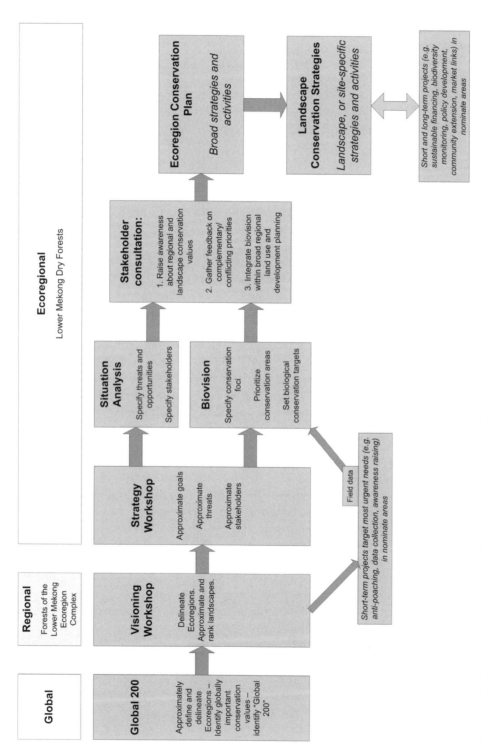

Figure I. A flowchart of the sequence of inputs and analysis involved in management planning for the Lower Mekong Dry Forest Ecoregion.

Results of the visioning workshop in 2000 included at least two important perspectives on dry forests that were carried into subsequent planning and management. The first was that candidate dry forest landscapes were large in area because most of the known priority species, large-bodied mammals and birds, require large home ranges; if management were to focus on small sites, it would not be successful in conserving these key species in their natural habitats. The second perspective was illustrated by including the Mekong River and tributaries (i.e., riparian zones), as well as the central Cambodian semi-evergreen forest (SEF) landscape, within the dry forests ecoregion. Their inclusion emphasizes that an ecoregion is not just a forest or vegetation type, but a geographical area that may include diverse habitats, especially when that habitat diversity is a defining characteristic, as in the case of dry forests. This point is discussed more fully in the section below on the biological assessment.

With a regional biological vision covering the Lower Mekong Basin, the next step was to refine and adapt general ecoregional guidelines to each of the proposed focus ecoregions, including the central Indochina dry forests. Following a flurry of survey work conducted during the 2000–2001 dry season, a three-day workshop was held in Phnom Penh in June 2001, to develop a guiding conservation strategy for the dry forests ecoregion (Maxwell and Pinsonneault 2001). Again this workshop involved mostly conservationists among the fifty-eight participants, but included more representation for local natural resource users throughout the ecoregion.

Workshop participants first reached a consensus on long-term goals (condition in twenty years) that reflected the ecoregional approach to conservation, namely (1) conserving key species, (2) conserving the ecological processes that support key species and communities, (3) integrating local grassroots development within the conservation perspective, and (4) working to ensure sustainability through ongoing multi-stakeholder support. With these goals in mind, workshop participants reviewed, revised, and prioritized landscapes for conservation investment, which resulted in a large area in eastern Cambodia's Mondulkiri Province being chosen as a top priority.

Workshop participants agreed that, up to 2001, hunting was the most important threat to the conservation values of dry forests. This threat was reflected in the situation that relatively natural habitat was extensive, but key species populations were seriously depleted (see Duckworth and Hedges 1998). The workshop then developed a broad strategy to address the threat of hunting, which was significant for several reasons. Addressing the threat at the ecoregional scale required more than just strict law enforcement by well-armed rangers. In a situation possibly unique to Cambodia (see thematic discussion below), basic laws and supporting legislation needed to be developed to guide protected area (PA) management, forest management, and conservation law enforcement. The workshop identified needs for building the capacity of government staff at all levels and both inside and outside the conservation sector. This capacity-building would involve education and training, including new legislation, plus equipment and ongoing routine material support. Local communities also needed awareness-raising on the importance of wildlife conservation and sustainable resource management for their livelihoods, especially after Cambodia's long period of civil instability and the social and political isolation of indigenous communities living throughout the dry forests. Communities should be encouraged and supported to actively participate in land-use and conservation planning and management, but government at all levels needed to recognize that local participation would probably be dependent on guaranteed rights of access for local resource users. Finally, conservation efforts needed monitoring of key species populations, of the effectiveness of law enforcement, and of project and program management. This broad range of activities was needed to address the threat of hunting effectively.

One other important outcome of the strategy workshop was an acknowledgment that hunting, despite currently being the primary threat, also represented an important opportunity in the long run if conservation efforts were successful. Cases of hunting and non-hunting tourism were presented by participants from South Africa, India, Nepal, and Sri Lanka, indicating that the dry forests had relatively good potential for eventually generating revenues to cover conservation costs, as long as key species populations could be restored to viable levels, such as those described by Wharton (1957, 1966) in the 1950s.

The next step in the management planning process was the identification of specific biological priorities for the renamed Lower Mekong Dry Forests Ecoregion, and of measurable conservation targets to guide management actions. This biological assessment, or "Dry Forests Biovision" (Tordoff et al. 2005), was carried out as a systematic series of consultations among biological experts, conservation managers, and field staff from October 2002 through 2004. Although much of the information supporting a prioritization had been introduced in previous meetings, the biological assessment covered the available data in a more systematic way, and the analysis aimed for more durable and defensible conclusions. Important parts of this assessment were establishing a clear definition for the ecoregion, specifying conservation foci (key biodiversity "components" to be conserved), using the foci for area prioritization, and developing measurable conservation targets to guide management. Foci were included in or excluded from the analysis based on their global significance and the significance of threat to their existence. All of the foci were ecological or biological; this assessment did not evaluate socioeconomic factors, other than the controversial role of anthropogenic fire.

According to the biovision, "The Lower Mekong Dry Forests Ecoregion (LMDFE) comprises the landscapes characterized by habitat mosaics (at the large scale) dominated by deciduous dipterocarp forest (DDF) in Cambodia, Lao PDR, Thailand and Vietnam" (Tordoff et al. 2005). As mentioned earlier, this definition is important for highlighting that the ecoregion consists of landscapes of diverse habitat, and not just a single deciduous forest type, even though DDF is the foundation forest type throughout the mosaic. This mosaic pattern is recognized in the assessment as a conservation focus, the only pattern recognized as a focus.

Most of the conservation foci used in the assessment and area prioritization were faunal species, mostly birds and mammals, for which conservation success could be measured and monitored by population numbers. As the ecoregional approach includes conservation of ecosystem functions, five of the foci were ecological processes:

- fire regime
- climatic seasonality
- habitat modification by elephants and other large ungulates (including influence on fire regime)
- predation of large ungulates by tiger, leopard, and dhole
- annual flood cycle

From a management perspective, these process foci may be difficult to set targets for, considering the lack of data and capacity for ecological analysis. However, they need to be considered in planning and management as essential ecological components so that management actions will not overlook or compromise them. Fire regime and habitat modification by ungulates are two processes for which management targets could eventually be set, depending on adequate, good quality data to support analysis.

The biovision identified five large multi-foci priority areas (Figure 2), at the scale of landscapes, where conservation investment should be focused to preserve the most important components of the ecoregion's biodiversity. This map did not contain any surprises compared to previous prioritization efforts, but the priority areas reflect clearer, more defensible criteria and values used in the assessment.

The biological assessment represented the culmination of planning at the expert technical level, establishing biological and ecological targets that needed to be achieved through conservation action, and monitored. The ecoregional planning called for the further step of preparing an overall conservation plan that addressed the remaining general objectives, integrating conservation within regional land-use planning and policies, and exploring options for sustaining the conservation efforts over the long term.

A formal situation analysis was carried out in 2004, which broadly analyzed threats and opportunities across the ecoregion, and also identified major stakeholders both within and outside the conservation sector. In a series of stakeholder consultations across all four countries (Cambodia, Laos, Thailand, and Vietnam), the objectives and targets of the biovision were reviewed and discussed, and accepted or revised as needed. The stakeholders, most from relevant government agencies, provided input on their needs to reflect and support national policies, which generally held economic development as a higher priority than biodiversity conservation. Other conservation and natural resource management partners also contributed ideas or additions to the plan.

The consultation process included, in Cambodia, a short-lived Dry Forests Coalition with major representation from national and provincial government partners, and other international and national conservation nongovernmental organizations (NGOs). Although the coalition could not be sustained in the long term (discussed below under governance), the occasional meetings did help with direct communication among a wide variety of stakeholders during development of the conservation plan. Through all these consultations, the possibilities of integrating conservation within land-use planning and economic development were broached, and useful refinements of the conservation plan were developed.

These region-wide consultations resulted in the Dry Forests Ecoregion Conservation Plan (WWF Greater Mekong Programme 2005). The conservation plan contains a twenty-year vision, revised broad objectives, targets, and milestones that fit within a collaborative stakeholder plan.

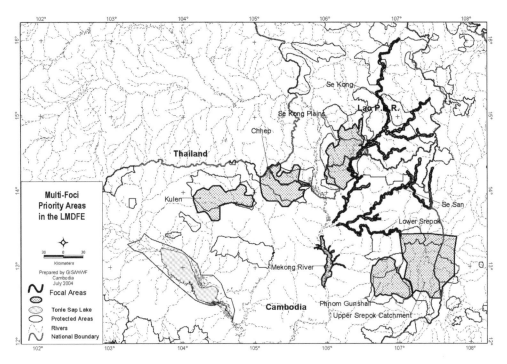

Figure 2. From the Biological Assessment of the Lower Mekong Dry Forest Ecoregion (adapted from Tordoff et al. 2005).

The objectives in the Dry Forests Ecoregion Conservation Plan were

1. *Conserve species, habitats, and their interactions*—Protect and restore populations of priority species and wildlife communities, and the key habitats and ecological processes of which they are an integral component.
2. *Improve the sustainability of livelihoods*—Protect priority landscapes and corridors through sustainable use of forest resources, sustainable agriculture, and forest restoration.
3. *Address and mitigate the most urgent, broad-scale pressures on biodiversity*—Approaches are developed and implemented to minimize impacts of key threats such as agriculture and wildlife trade on biological diversity and ecological processes.
4. *Raise awareness, increase participation, and build capacity*—Local, national, and international stakeholders are aware of, actively participate in, and have the capacity to plan long-term biodiversity conservation in the Dry Forests Ecoregion.

The conservation plan was compiled to cover all important issues related to conservation in the ecoregion, regardless of the specific agents of implementation for any particular action, i.e., it was intended as an ecoregional plan, not merely a WWF plan. The WWF Greater Mekong Programme held subsequent internal consultations to formulate its own Dry Forests Action Plan, a subset of targets, milestones, and activities for which WWF chose to take responsibility for achievement.

The management planning process described so far may seem to some readers to have involved too much discussion and planning, and not enough action. Actually, the agencies and organizations involved in conservation in Cambodia recognized that urgent action was needed to prevent local or global extinction of some key species, and therefore the most basic steps, e.g., ranger training and support for law enforcement together with community awareness–raising, were being taken even as the assessment and planning process moved ahead. Collaborative projects among relevant government agencies and international partners were initiated in the Northern Plains (Wildlife Conservation Society [WCS] partnering with the Forestry Administration in a Global Environment Facility–funded project) and Eastern Plains by 2000. However, a complex and detailed planning process was needed to propagate conservation effectiveness beyond the site level and beyond a time scale of one to two years. Conservationists and supporters needed to know where the efforts were going over the long term, including whether the goals and objectives were realistic, how conservation effectiveness could be evaluated and improved, and how Cambodia or the other national government agencies might be able eventually to take the lead on conservation management. It was deemed necessary to plan conservation interventions based on long-term goals, and not just on short-term objectives or based around activities needed to address problems as they arise. Such planning takes time, especially covering a twenty- to fifty-year time horizon and an area representing more than 20,000 km² of ecologically and socially diverse landscapes. The integrated plan was accepted by most stakeholders as more credible, more relevant, and more durable than a series of short, crisis-oriented projects, and the plan is adaptable based on periodic review and ongoing stakeholder consultations (Figure 3).

MANAGEMENT EXPERIENCES

Since WWF and other international organizations (e.g., WCS, FFI, Conservation International, WildAid–Wildlife Alliance) began site-specific conservation support in Cambodia in the late 1990s, most projects focused on establishing or improving PA management, in direct partnership with relevant government agencies. This contrasts somewhat with other conservation approaches, in which organizations work more independently from government, concentrating on scientific research, environmental education and awareness raising, or on developing input to policy development. The reasons for the more direct approach in Cambodia were

- The Protected Area and Protected Forest (PF) systems were formalized only recently (1993 and 2002, respectively), and there was no ongoing government conservation system. So the government requested financial and technical assistance with conservation planning and management in protected areas and protected forests, as well as with community-based management of natural resources in buffer zones.
- Key species of wildlife were seriously threatened with extinction by uncontrolled hunting, so improvements in law enforcement were urgently needed, and relevant agencies did not have experience in conservation law enforcement.

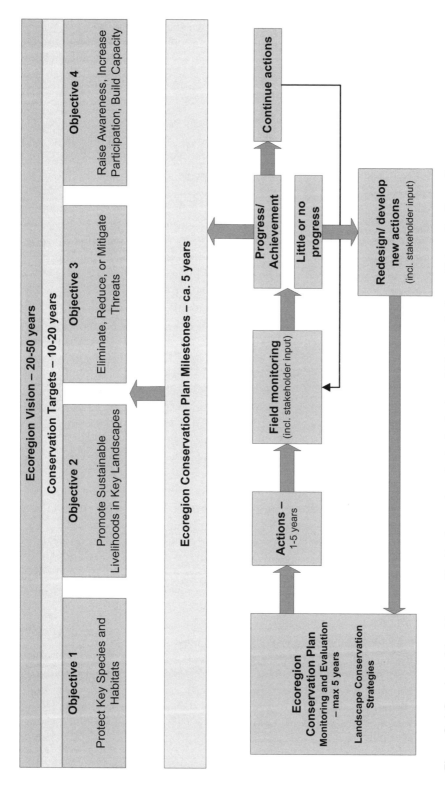

Figure 3. A flowchart of the vision, specific objectives, and monitoring feedback established for the Lower Mekong Dry Forest Ecoregion program.

The first, most immediate need was to support rangers on patrol, not only to arrest violators, but also to wean the rangers themselves from unsustainable hunting (most rangers are former hunters from villages near the conservation core zones), and to disseminate grassroots conservation awareness, which was not evident before about 1995.

In Cambodia, WWF began to move its focus toward the Eastern Plains landscape, particularly Mondulkiri Province, in 2000 (see Figure 4). This area had been subject to early (for Cambodia) wildlife surveys (Desai and Lic 1996; Olivier and Woodford 1994), and showed potential for conservation of representative Southeast Asian dry forest biota and habitats. This focus became sharper as the planning process described above proceeded.

One biological survey carried out in the northern parts of the province in May–June 2000 (Timmins and Ou 2001) covered eastern parts of what was already designated as Phnom Prich Wildlife Sanctuary (PPWS) and survey blocks farther east near the Srepok River in what would become the Mondulkiri Protected Forest (MPF). Although the survey could not confirm tiger presence inside PPWS (one of the main objectives), it did confirm relatively high numbers of banteng (*Bos javanicus*) and Asian elephant (*Elephas maximus*), while also confirming the presence of dry forest indicator species, e.g., Green Peafowl (*Pavo muticus*) and jungle cat (*Felis chaus*; see Duckworth et al. 2005). In the eastern survey section, signs of banteng and gaur (*Bos gaurus*) were abundant, and wild water buffalo (*Bubalus arnee*) and Eld's deer (*Cervus eldii*) populations were confirmed. Because of the abundance of water holes, together with the Srepok River, water birds were both more diverse and more abundant in the eastern section than in most other parts of the Eastern Plains, including White-shouldered Ibis (*Pseudibis davisoni*) and Black-necked Stork (*Ephippiorhynchus asiaticus*).

Another survey about the same time also confirmed the conservation value of the upper Srepok area (Long et al. 2000), and tigers were later confirmed as present in both PPWS and MPF. Subsequent international reviews continue to confirm the importance of northern Mondulkiri for populations of key mammal and bird species (Seng et al. 2003; Sanderson et al. 2006).

Based on the results of the surveys, and continuing on a MOU with the Ministry of Environment, WWF initiated support to PPWS in 2002, with preliminary community and ecological surveys carried out during 2001. PPWS covers 2,250 km^2 of rich habitat mosaic, the majority being DDF, but with significant expanses of semi-evergreen and evergreen forests, particularly on hills in the southeastern section that rise over 500 m in elevation.

The first management steps were similar to those taken in previous WWF involvement in Cambodia's Virachey National Park, namely, consultation with the ministry to identify a competent director, competitive selection of rangers, initial training of rangers, and establishment of reporting procedures. Five ranger stations were initially rented in local villages, but were built new starting in 2003. Throughout the course of PPWS management, significant support came from the U.S. Fish and Wildlife Service, especially through the Asian Elephant Conservation Fund (USFWS–AsECF).

Locator Map for Eastern Cambodia

- Major rivers
- Protected Areas and Conservation Areas
- Provincial boundaries
- National boundaries
- ⊙ PPWS-MPF Communities
- ● Provincial towns

30   0   30   60
Kilometers

Figure 4. Eastern Cambodia with nationally recognized conservation areas described in the chapter, including Phnom Prich Wildlife Sanctuary (PPWS) and Mondulkiri Protected Forest (MPF).

Initial ranger training in 2002 covered basic principles of conservation and PA management, plus basic field procedures in orienteering, use of GPS, and data recording and reporting. A focused training on wildlife monitoring procedures, based on line transects, was conducted as permanent transects were established in May 2003. Law enforcement training for the full staff of twenty rangers was first conducted in November 2003, subcontracted to WildAid, which had an MOU with the Ministry of Environment for a national-level ranger training program. The program was designed to reflect regional standard terms of reference, developed within the ASEAN Regional Centre for Biodiversity Conservation (ARCBC).

Because of transportation problems and complications in provision of firearms to rangers, initially most enforcement actions involved simple on-site actions, i.e., confiscation of illegal materials, a lecture on the importance of conservation and on wildlife protection laws, and a contract signed or thumbprinted by the violator to stop conducting illegal activities. Live wildlife were returned to the forest, dead wildlife and large materials (timber) were burned on site, and small materials (e.g., chainsaws, shock-fishing rigs) were stored in the ranger station as court evidence. This relatively "soft" approach was necessary at the beginning because most local people were not aware of any legal controls on hunting, fishing, or logging, and most did not know what a PA was, or that they were living near one. Also, rangers did not have the resources to transport violators or large quantities of evidence to the provincial court. The PPWS management team felt that strict enforcement, aside from being logistically impossible, would turn local people away from long-term conservation efforts, and this was true throughout most of Cambodia. This initial phase of awareness-raising through confiscation and contracts did appear to reduce the number of incidents of opportunistic hunting. The situation changed over time, as more local people became aware of the importance of conservation and the risks of confiscation, and a smaller number of better-organized violators would take the risk of confiscation for more valuable forest products.

By late 2003, WWF developed a project to support the recently designated Mondulkiri Protected Forest (MPF), in cooperation with the Forestry Administration (FA) of the Ministry of Agriculture, Forestry and Fisheries, and with diverse sources of funding, including significant private funding. MPF covers 3,560 km^2 of generally flat terrain with scattered hills, and with about 80 percent of its area covered in DDF. MPF holds more seasonal waterholes than PPWS, and there is a short section of the Srepok River flowing through the northeastern corner, representing both a significant regional fishery and also good tourism potential. The MPF area also had been surveyed and subject to awareness-raising by the Community Wildlife Ranger Project (CWRP, with support from Cat Action Treasury and USFWS), which transferred many trained staff to the new WWF project.

The MPF project attracted a higher level of financial support than PPWS for reasons including better tourism potential (Goodman et al. 2003), more diverse and abundant key wildlife species, and a novel (for Cambodia) approach to management. The project reflected an opportunity to apply principles of conservation through sustainable use, along the lines of a southern African model (Prins et al. 2000; Pierce et al. 2002). Although sustainable hunting was not an immediate option because of depressed key species populations, northeastern Mondulkiri showed the best potential in the ecoregion for a long-term, self-sustaining project based on wildlife viewing and possibly trophy hunting. In 2006, a small group of Cambodian government and community project staff visited Namibia to learn about how communities monitor wildlife for trophy hunting through a simple daily record system (MOMS—management oriented monitoring system). On returning to Cambodia, MOMS was redesigned to fit the local situation, focusing on community monitoring of biodiversity and natural resources rather than wildlife inventory for hunting. The nature of the system means it could easily be adapted in the future should sustainable hunting become feasible.

Another novel management approach for Cambodia was to have as many staff as possible, including foreign technical advisors, stay full time on site at the PF headquarters, which was not near any village. This called for a higher level of initial investment in infrastructure because there were no buildings standing at the site chosen for the headquarters (station and living quarters), roads had to be at least minimally maintained for truck traffic, and trucks were needed for regular transportation of people and materials. All of this represented a quantum leap in investment compared to PPWS, but seemed justifiable in light of the good potential for the project to restore wildlife populations and return some tourism-based revenues in the long term, which interested even some non-conservation government stakeholders.

The MPF support project (initially called the Srepok Wilderness Area) was started with a staff structure different from PPWS, because the FA had never defined a role for rangers. The FA staff served as managers or section chiefs, with local community people hired as rangers or guides, and police seconded for patrolling and law enforcement. Initial training focused on data recording and reporting using a standard monitoring program (MIST-GIS—monitoring information system), while most of the basics of law enforcement procedures were already understood by the seconded police. Because the MPF project started later than PPWS, and was located in an area both already subject to conservation awareness-raising (by CWRP) and adjacent to an international border, patrolling rangers and police enforced conservation laws strictly from the beginning without significantly alienating local communities. There were still, however, logistical and political problems with some types of enforcement, especially related to logging.

By 2007 it was becoming increasingly unwieldy for WWF to manage PPWS and MPF as separate projects. In line with the ecoregional approach, WWF project management was integrated within a single landscape program, although government administration continued under the two separate agencies. The Eastern Plains Landscape Programme (EPLP) was set up with three major management components, i.e., PA/PF management, biological research and monitoring, and community extension, each managed by single individuals covering both designated areas, who were in turn supervised by a single landscape manager. Although the integration increased the administrative burden for some of the managers (the landscape manager and community extension team leader had previously been working only in MPF), there were many of the expected benefits of the ecoregional approach, generally resulting in broader understanding of seasonal wildlife movements and habitat use, community resource use patterns, and threats from outside the PAs.

As of the middle of 2009, each of the PAs employs about 30 field staff for patrolling and law enforcement, 5 staff for biological research and monitoring, and 5 staff working with local community networks on extension, i.e., grassroots community development and natural resource management. These numbers are approximate because roles change or there is cross-sector cooperation according to seasonal or short-term project work, e.g., three patrolling rangers joined the research team for collection of elephant fecal DNA samples in early 2009; and in the past some research rangers have joined resin tree survey teams in the community extension component. Patrolling rangers are required to be on patrol for a minimum of 16 days per month in a combination of 5-day and 3-day patrols, and with the times staggered to avoid predictability.

There are 8 large ranger stations and 5 small stations distributed across the two areas, with plans for adding at least 2 more small stations. In 2009 the first tourism facility was built in MPF, a community guest house, and plans were completed for a well-outfitted tented camp modeled on African examples. Roads inside the core zones are only dirt or rock tracks, and there are no bridges in either area.

Both PPWS and MPF have draft management plans, but it has been difficult to finalize approval for either of them, primarily because of problems with official recognition of PA zoning, discussed below under economic issues. PA management is monitored through the WWF/World Bank Protected Area Management Effectiveness Tracking Tool (METT; Stolton et al. 2005), applied every two years. Biological monitoring systems for tiger, tiger prey, and elephant populations are being implemented and refined, in collaboration with other international conservation partners.

Regarding budget, not including the national level administrative overhead that is extremely difficult to calculate, the Eastern Plains Landscape Program requires at least US$500,000 per year for routine operations and minimal growth. For an upper end of the range, the program could effectively use US$1.5–2 million per year, or more depending on the nature of inputs to tourism development, or expansion of activities to Lomphat Wildlife Sanctuary. Although MPF has had reasonable success as a target for fundraising, notably in terms of private donor funding for tourism development and some species-specific interventions, PPWS has not seen the same effort or results. So for the conservation landscape as a whole, the funding is still subject to vagaries in focus and timing, which make planning, implementation of plans, and monitoring difficult.

MAJOR MANAGEMENT ISSUES, CONSTRAINTS, AND OPPORTUNITIES

There are salient management issues that directly affect the achievement of conservation objectives and goals in Cambodia's Eastern Plains. This section discusses selected issues and experiences in management of the Eastern Plains landscape. This is not a comprehensive assessment; it focuses on significant problems, as they may provide useful lessons-learned. However, a few successful or promising interventions are also described as they reflect more enlightened adaptations of a global conservation approach to the local situation. This section also is not an inventory of threats, i.e., hunting, habitat conversion, or root causes, although those are interwoven throughout most of the management issues.

Planning

The advantages of the ecoregional approach, as compared to more traditional conservation approaches, are mentioned above, and are touted in various policy documents (e.g., Olson and Dinerstein 1998). There are also inherent disadvantages to the approach that are not well documented, and mostly pertaining to a tension between broad-scale planning and policy versus local-scale action.

Ecoregional conservation is an iterative process of prioritization, a series of assessments that gradually narrow the focus to the most important places for conservation. As conservation funding is restricted, the priority places become fewer with each stage of assessment. This means not only that the most important places are identified and targeted for investment (the keystone advantage of the approach), but also that slightly less important places lose international support. This can include places that possibly support significant *representative* biodiversity (e.g., unique communities or ecosystems with low species richness) or communities suspected as being species-rich but that have not yet been investigated. As the number of focal ecoregions decreases, even though the chosen ones are relatively species-rich, the broad objective of global representation of biodiversity is weakened.

The Lower Mekong dry forests are currently included among top *global* priorities for conservation investment, but probably mostly because they are part of the Mekong River watershed rather than for their own representative biodiversity values. Data were insufficient during the assessment phase to confirm any but the coarsest conservation values, i.e., the occurrence of large-bodied mammals and birds, while suspected botanical values are still uninvestigated (e.g., possible canopy tree species richness in central Cambodia dry evergreen forests, seasonal wetland herb species richness, and possible orchid endemism). Although it has been encouraging to receive support from the international level, the situation on the ground sometimes feels tenuous, particularly when it seems the prioritization process itself may be inconsistent.

Despite the strong advantages of an ecoregional approach focusing on globally important places, the transition periods involving new prioritization schemes were difficult to navigate at the national and local scales. WWF, between 1998 and 2006, was setting priorities based on at least three distinct schemes: target driven programs (TDPs), ecoregions, and drivers of biodiversity loss (DBLs). And recently another management approach, the Network Initiative (NI), has reared up, designed partly as a "new" approach to identifying the key strategies required to transform conservation, and also for refocusing fundraising efforts. Prioritization criteria in some cases have overlapped, and in such cases the choice of an area for investment is conveniently clear and justifiable. But in some cases a project of top global priority in one scheme will be irrelevant in another.

In the case of the dry forests, for example, the Eastern Plains landscape was a high priority for tiger and Asian elephant conservation (species TDP), and of medium-high importance for its ecoregion, but not necessarily because of its tiger and elephant populations. The landscape is of low importance for DBLs for a variety of reasons, e.g., it is primarily on the supply side of wildlife trade, and most of the supply flow is undocumented. Furthermore, if a relict, "old school" project of support for PA management is effective, the landscape becomes less important for DBLs because the supply of tiger bone from here to international markets is decreased, which is a significant objective for both TDPs and the ecoregional approach. On top of this, while a small, poor country like Cambodia harbors globally significant conservation values, the national economy will never be significant on a global scale, so macroeconomic criteria as adopted in the case of DBLs will not likely set any landscape in Cambodia as a priority. This is just one example, but one that is illustrative of how planning complexities at the global scale can confuse project managers trying to make things work on the ground.

Add to this a trend of increasing emphasis on stakeholder involvement in conservation, and the issue of relationships with local partners quickly becomes even more complex. Governments at all levels need to know what to expect over at least the period of national development plans, e.g., usually five years. Rangers and community supporters also need to know. Depending on the communications infrastructure in a given area, it may be that staff and counterparts of the international organization are the individuals who must explain conservation priorities and investments to local stakeholders. In the Eastern Plains, it was difficult to explain clearly WWF's international criteria for support during the period 2002–2006, and that period of uncertainty may or may not have negatively affected stakeholder cooperation at the local level.

Over the past few years, with the completion of the stakeholder conservation plan, and with a prioritization frenzy at the international level seeming to settle down, the advantages of the ecoregional approach are becoming clearer, and more clearly expressed in the conservation landscape. More and more, the focal area with the highest concentration of representative values is being managed as a landscape, with the government agencies, as principle stakeholders, coming to a better understanding of the need for coordinated work across the territories of several agencies. The ecoregional conservation plan has enough focus to guide the preservation of recognized conservation values, while also leaving enough scope for research into less-recognized but equally important conservation values (including ecosystem processes) that may be essential for guiding future conservation efforts.

One more tension in the planning process should be mentioned. Because the approach expands the geographical scale of conservation, it necessitates more extensive and complex engagement in social, economic, and political issues. This approach therefore requires more investment, at least of time, in stakeholders with political influence. Investment in political influence, in the best cases, is necessary for broad-scale improvements in policy that can ensure long-term success of conservation interventions. However, politicians sometimes exploit international supporters by rhetorically adopting attractive policies that the government is either unwilling or incapable of implementing and monitoring. For conservationists and local counterparts, this can often be frustrating in the short term and demoralizing in the long term, especially when seen as an opportunity cost that depletes the level of financial and staff support available for direct, local interventions. Conservation planners at the regional and global scales must consider how investment in policies will realistically yield positive outcomes, and should consider that some investment in policy may be relatively wasteful compared to direct support to local partners.

Physical

Extent of Ecosystem

The broad extent of the dry forests is both an opportunity and constraint. During the planning process, Cambodia had an incomparable extent of relatively natural dry forest mosaic, with eastern, northern, and southwestern Cambodia holding as much as 25,000 km² of relatively intact dry forest habitat, albeit with depleted populations of key faunal species. Even with recent habitat conversion (see below) there is still at least 17,500 km² (Eastern Plains landscape: 11,500 km²; Northern Plains landscape: 6,000 km²) of dry forest mosaic designated national PAs or PFs, and also some areas of excellent habitat that are subject to local protection measures (western Siem Pang district, Stung Treng; the Kratie hog deer area). This extent is itself a primary conservation value because many of the key species have huge ranges (Asian elephant, Eastern Sarus crane, vultures), probably linked to the open, grassy habitats and scattered seasonal waterholes that can support abundant mega-herbivores, their carnivore predators, and scavengers.

However, the huge extent of relatively flat terrain means that access is very difficult to control. Cart tracks or motorbike roads can cross PA boundaries almost anywhere, and truck roads can go in quickly with an economic incentive like targeted logging or mining. Large mammals, most being key species, have been, and in some places still are, relatively easy to hunt in the large tracts of open vegetation. In particular Eld's deer and hog deer (*Axis porcinus*) are noted for behavior that makes them easy hunting targets. Other quarry species may be harder to find, but the extent of habitat means that hundreds of indiscriminate snares can be set in the wilderness without being detected for more than a week. Rangers generally work hard under difficult conditions, but the current level of about 60 patrolling rangers to cover over 6,000 km² in MPF and PPWS is well below the level needed considering the status of the wide-ranging key species. Ranger patrol coverage is even more problematic in adjacent PAs.

Seasonality

Strong seasonal variation in precipitation and related natural elements is intrinsic to the ecoregion (Tordoff et al. 2005), which makes management difficult, particularly in terms of infrastructure and access. Generally, until recently the government had not invested in year-round roads and bridges to remote villages. During the rainy season, small streams are raging torrents and dangerous for rangers patrolling or local travel. Malaria is ubiquitous in the rainy season, affecting PA rangers, their families, and local communities. At the same time, access to health services is very difficult and expensive.

These problems can be found also throughout perhumid conservation areas, but the dry forests also have another suite of problems associated with the dry season. Proceeding from the lack of rain for at least five months, agriculture or horticulture is impossible and grain stores may be inadequate, so people traditionally have relied on hunting. With wildlife numbers depleted, many of the people now can find only muntjac and turtles, and the current urgent need for turtle and tortoise conservation is very difficult for local people to understand and accept. The lack of water also causes health problems, particularly for eyes and skin, probably because of dust and often unclean water for washing.

These conditions affect management either through significant ranger health and safety issues, or through cooperation and support from local communities. Some of the core sites for conservation are valuable because they are dry season refuges with year-round water. People who are foraging for their livelihoods need that water and associated forest resources also, but effective conservation requires some areas of strict exclusion from human use. So negotiations can be complex and time-consuming for both conservationists and local people.

Fire

Fire is an intrinsic characteristic of the ecoregion, related directly to seasonality. It also is generally acknowledged that most or all fires in these dry forests are set by humans (Stott 1988; Kanjanavanit 1992; Maxwell 2004; MacInnes 2007). Beyond those basic observations, there is little consensus in Southeast Asia on the role of fire in conservation, and therefore how to manage it. Entrenched, non-local prejudices put forward a view that all fire destroys forests and wildlife, and so must be suppressed. There is also the perspective that dry forests are degraded forms of natural SEF, and usually the degradation is attributed to a history of human abuse through burning (Wharton 1966). Even the more thoughtful and specific descriptions of fire's impacts (Desai et al. 2002) lack supporting data. These perspectives are already institutionalized in Cambodia at the national level, with the national Forest Law and Protected Areas Law containing clauses that flatly prohibit, and set punishments for, burning in all forests.

If it were actually possible to enforce anti-burning laws, it could be disastrous for the dry forest mosaic as we know it. The burning regime of the recent past is based on very frequent, even annual fires that are small in size and patchy in distribution, and burn at low intensity mostly in the ground layer, and so seem to be of limited threat to the forest canopy (Maxwell 2004). There is no evidence that recent anthropogenic burning has allowed DDF to invade semi-evergreen or evergreen forest blocks, at least not in the northeastern belt of dry forest in Cambodia. Some research suggests that fire suppression would encourage build-up of fuel, leading eventually to more intense, extensive fires (Stott 1986, 1988), which by killing canopy trees could actually reduce the total extent of forest.

Obviously, there is a wide variation in perspectives on the role of fire. In the short term, the most urgent management question is whether restricting local people's access is beneficial by reducing the impacts of seemingly uncontrolled fires, or whether exclusion of local people threatens the whole landscape by disrupting the long-term fire regime, and risking an eventually huge and hot canopy fire. This question is urgent because there is not enough technical knowledge or enough staff to implement controlled burns by PA staff, so abrupt restrictions on customary access may be harmful for habitat management by changing the fire regime over the core conservation zones.

It is clear only that more research is needed to provide guidance in formulation of a fire management strategy and procedures that effectively support conservation. Research should contribute to a better understanding of the role of humans in landscape evolution, and the ways in which human inputs are changing. Also needed is modeling of herbivory in the past and present, projected in a scenario that allows for restored populations of mega-herbivores. As of 2009, formulation of fire management strategy and procedures is stalled by a lack of data at the landscape scale, despite supportive efforts at the local level (MacInnes 2007).

Technical

Data Availability

In many important topics, scientific data are lacking to support management decisions. Gaps in data are directly related to the problems of capacity, and historical conflict and isolation, discussed below. From before the conflict period (ca. 1968–98), there are no scientifically derived population estimates for key wildlife species, and the few ecological observations are part of forestry studies (Rollet 1962, 1972; Schmid 1974). Field research was not possible over most of Cambodia during the years of conflict, and with constraints in national capacity, population monitoring could not begin immediately when the major conflicts were resolved by 1998. So in many ways, the data to support planning and management are only recent baseline data, making it very difficult or impossible to identify and respond to trends in populations or habitat conditions.

Wildlife Populations

There are no pre-conflict population estimates, but careful observations from before the Indochina war period indicate relatively high densities of large-bodied mammal and bird species (Wharton 1957). During the early conflict period, Khmer Rouge controlled the Northern Plains and Eastern Plains from 1970–79 and controlled hunting strictly, but not to maintain wildlife populations. Some rural communes were assigned the responsibility of hunting in specialized groups to provide meat to the combat front (Loucks et al. 2009), or to provide wildlife products for trade to China (Kiernan 1996, 376–85; Loucks et al. 2009). After 1979, hunting was not controlled, during a period through the 1980s when guns were abundant throughout Cambodia and international market demand for wildlife was increasing. By the mid-1990s, many parts of the Cambodian plains were depleted of all large mammal species, with the possible exceptions of muntjac and wild pig. From a management standpoint, low wildlife population densities make both monitoring and protection very difficult. Low density of tigers also makes conservation fundraising difficult, even though relatively natural habitat is still extensive, prey populations seem adequate, and the tiger population probably could recover quickly with effective protection.

However, with only the recent few years of direct support to PA management, the signs of recovery of certain species are very encouraging. In PPWS, camera trapping has consistently shown numerous calves and juveniles of Asian elephants, with only three confirmed cases of poaching since 2002. Formal monitoring based on fecal

DNA analysis is being carried out to determine the total elephant population that ranges across the landscape. For top carnivores, in 1996, 223 km of walking transects inside what is now MPF revealed just one leopard sign and six apparent tiger signs (Desai and Lic 1996). This compares to recent surveys in the same general area where leopard signs have been recorded every 1–2 km (WWF, unpublished data; T. Gray, personal communication), and tigers are still being confirmed by signs and camera-trap photos. These improvements are largely due to implementation of actions identified in the conservation plan, i.e., investment in international standards of protection and monitoring (in partnership with WCS and other technical stakeholders) through the engagement of international experts working in the field with local rangers and community counterparts.

Dynamic Social Setting, Shifting Threats

A changing social setting is not unique to the dry forests, but the timing of some changes, coupled with gaps in national management capacity, has made adaptive management difficult. As the planning process began, hunting was clearly the primary threat, which, although demanding a complex response, seemed manageable because the most threatening activities were either illegal or becoming illegal, and the number of local stakeholders benefiting from hunting for trade seemed relatively small.

With an improved security situation in Cambodia's hinterlands after 1998, national development strategies allowed or encouraged migration to seemingly empty, degraded forests for conversion to agriculture, both small-scale and commercial. From a management standpoint, this threat has been, and still is, much more difficult to resolve than hunting, because the habitat conversion activities may not be illegal (e.g., converting "degraded" land to productive land), and at the same time the government has evolved to a perspective that the total area under protection is too large and Cambodia cannot afford to have large areas "locked up" in conservation. This means that conservation organizations have shifted gradually from being perceived as supporting sustainable community development and preserving natural heritage to denying people's access to agriculture and food security. This perception may be a common problem regionally and globally, but it came late to Cambodia, after initial steps in the planning and management process had been initiated. Societal changes that led to rampant habitat conversion came up quickly and proceeded faster than the rate of adaptation of the conservation management process.

In one case, however, the broad landscape perspective led to valuable collaboration in preventing a potential problem. In 2004, with planning and upgrading of a major road through MPF, uncontrolled land encroachment for agriculture along the road accelerated rapidly, particularly by people from other provinces. Knowing that new settlement and agriculture would significantly increase the risk of conflict between seasonally migrating elephants and new settlers, WWF worked with the FA and provincial government agencies to delineate an area along the upgraded road that would be off-limits for new settlement, while also calling for participatory land-use planning with indigenous communities already in the area. That "corridor" was approved by various agencies in early 2005. Although there have been problems controlling illegal logging in the corridor, land encroachment has been almost completely stopped, and there has been no report of human-elephant conflict or of disruptions in seasonal elephant migration. Cambodia remains the only Asian elephant range state where human-elephant conflict is not a serious problem.

Suitability for Conversion to Rice

Dry forest soils generally are poor in nutrients and structure, being mostly derived from old (Triassic) acidic sandstones (Fontaine and Workman 1978). The terrain is low-relief, large areas are relatively flat, and the soils typically form a hardpan near the surface (Crocker 1962). These conditions allow for formation of seasonal water-holes in shallow depressions of varying size scattered throughout the landscape and some lasting until late in the dry season. These waterholes are excellent habitat for a diverse array of wildlife species, including populations of some large waterbirds that presently are difficult or impossible to find in neighboring countries, e.g., Giant Ibis (*Thaumatibis gigantea*), White-shouldered Ibis (*Pseudibis davisoni*), and Black-necked Stork (*Ephippiorhynchus asiaticus*).

These same physical characteristics also make the terrain suitable for conversion to rice cultivation (Crocker 1962). This, coupled with a current national perception that the north and east are huge expanses of unused frontier waiting for pioneer development, makes paddy-rice production tempting for migrants, leading to habitat conversion and encroachment on or near PAs. With water scarce in the dry season, the land can support only one rice crop per year, and yields typically do not last through the year. This may lead farmers to resort to hunting or other forms of foraging to make up the annual gap in food, as mentioned above in relation to seasonality. As a result, habitat conversion is compounded by opportunistic hunting, including competition between newcomers and indigenous residents for resources, all of which complicate conservation management.

Capacity—Material and Human Resources

A range of issues related to national capacity affect conservation management on a daily basis. It would be overly simplistic to attribute all these issues to the history of conflict during the past forty years, but many of the links between recent history and current capacity gaps are obvious. Some parts of this situation are unique to Cambodia, especially the actions and impacts of Khmer Rouge rule. But some of the residual impacts may reflect situations in other places, and there may be important lessons from the experience in Cambodia.

To briefly summarize the history of conflict, when Charles Wharton conducted his ecological survey of the Kouprey, beginning in 1951, some parts of Cambodia already were off-limits because of security concerns, e.g., Viet Minh activities east of the Mekong River (Wharton 1957). Cambodia was drawn into the American-Vietnam War when American forces began bombing in Cambodia in the late 1960s. In 1970 almost all the Northern Plains and Eastern Plains were abandoned by the Cambodian government and became subject to Khmer Rouge rule even as civil war continued. After the Khmer Rouge formally took over the country in 1975, dry forest communities in Mondulkiri were forcibly moved, and usually concentrated in large multi-village cooperatives far from traditional village sites. Vietnamese forces invaded in 1979 and established a Cambodian regime to replace the Khmer Rouge, but open civil war continued in the countryside through the 1980s, with lingering conflict in the 1990s. During the 1980s, land mines were used extensively, especially along the Cambodia-Thailand border, which remain a problem to this day in parts of the Northern Plains. Khmer Rouge forces became entrenched during the 1980s and 1990s in remote areas of Kratie, Mondulkiri, and Preah Vihear provinces, until final surrender and "re-integration" in 1998–99.

During this period, 1968–98, in the remote dry forest areas, governance by Khmer Rouge or subsequent national governments focused on winning armed conflict and, after 1993, reinforcing the political base of the ruling party. Nature conservation and sustainable resource management were not government priorities, and often the natural resource base was exploited for trade to maintain military activities. Aside from the impacts of uncontrolled hunting described above, there were at least three major areas of impact in terms of national and landscape capacity for conservation: infrastructure, scientific research capacity, and governance.

Infrastructure

During this thirty-year period, funding was generally not allocated by the government for road development, communications, or social services in conflict zones. Local people remained poor, with negligible formal education and almost no access to health services. With no mainstream economic development, hunting for external trade was either the easiest or the only option for income, especially as international trade networks and demand for wildlife products were increasing during the 1980s (Loucks et al. 2009). Although large roadless areas can be good for conservation, in Cambodia's case the poor infrastructure was the result of neglect and not of management for preservation of wilderness. Hunters had adequate access, but rangers, beginning their work about 1998, could not find or pursue violators. This problem was compounded by better-developed road networks in adjacent areas of Laos and Vietnam, so that access to conservation core areas has actually been easier from outside Cambodia.

The threat is perhaps greatest from the upgrading of existing roads rather than the widespread development of new and interconnecting roads, though this is sure to follow. The only real strategy available is to assist local government agencies with integrated planning to ensure that any new roads are planned carefully, and that upgrading of existing roads includes a plan to mitigate and control associated threats such as in-migration, particularly where roads pass through or close to PAs or biodiversity corridors. This was the approach taken successfully in the elephant migration corridor mentioned above.

Scientific Research Capacity

Especially during 1975–79, the Khmer Rouge obliterated basic education for a generation of young Cambodians. There were no schools, and intellectuals were killed, exiled, or forced into hiding. During the 1979–89 period, Cambodia was still subject to civil war and economic isolation (because of the Vietnamese occupation), with higher education going to a very select few sent out to study either in Soviet bloc countries or Vietnam. Up to 1993, many high-level national staff in natural resource management agencies had pursued any technical educational opportunities they could find, earning bachelor's degrees in forestry (with an emphasis on production and trade) or in engineering, and with the instruction conducted in various languages: Vietnamese, Russian and other Slavic languages, or in a few cases, French or Spanish.

By the time of the UN-sponsored election and initial post-conflict opening up of Cambodia in 1993, the most qualified agency staff and officers were gearing up to learn English for new positions in new, externally funded projects incorporating the newest international trends in conservation and PA management. Even for these champions of resilience, there had been very limited primary or secondary science education. Despite their university degrees, they had minimal understanding of natural systems as systems, so biodiversity generally was understood and portrayed as an abundance of big mammals or birds, and PA management was often understood primarily as a law enforcement effort to hold an agency's territory. The problem of relatively weak national capacity in conservation science was compounded by gaps in international technical advice, as most foreign advisors were also inexperienced in conservation management and incapable of functioning in Khmer or any other language besides English.

In recent years though, we have started to see improvements in scientific capacity, initially through overseas training and study opportunities for a number of up-and-coming young scientists and natural resource managers. Additionally, several internationally funded conservation projects have facilitated in-country training and learning opportunities both through practical experience and academic opportunities. The first biodiversity-focused master's degree course, initiated in 2005 with support from the Darwin Initiative, has recently seen some of its graduates apply their new knowledge in a range of field-level conservation and research projects, including surveys and population monitoring for tiger and yellow-cheeked crested gibbon.

Though not a strictly scientific approach to biodiversity monitoring, adaptation of the Namibian MOMS to Cambodia through a pilot project in two of Mondulkiri's PAs, as noted above, represents a new tool for building capacity in government and communities to manage natural resources in the short- and long-term. The process began in 2006 with a visit to Namibia and was piloted first in MPF in 2007 and expanded to PPWS in 2008. Though originally based on specific needs in community conservancies in parts of Namibia, the concept has been fully adapted by Cambodian government and community project staff in Mondulkiri based on local needs and the local context. Data is collected on a range of biodiversity and natural resource indicators including wildlife numbers, water availability, and poaching and other illegal incidents.

The Namibian experience, as well as other technical study visits to South Africa, India, and Nepal between 2003 and 2008, has shown the value of learning from other landscapes where conservation challenges are similar. It is clear that in effectively building local capacity, a common measure of success is the degree to which local needs are fully accounted for when designing conservation management tools.

Eventually, any successful adaptation strategies must be flexible and owned by communities and government agencies who have the most at stake. There is nothing particularly surprising about that, but nevertheless it serves as an important reminder not to overlook local needs when setting out to develop capacity building strategies to achieve conservation goals.

Governance

One of the more pervasive and intractable of the constraints limiting conservation in Cambodia is that all major governmental decisions follow a chain of command, with currently only one person, not a conservationist, at the top. In terms of management capacity, this pattern of decision making discourages or actively suppresses the critical thinking that is essential for adaptive conservation management. Working only within the government, senior technical staff, project officers, and project managers may never get a chance to see the outcomes of rational, innovative thinking because management options are limited to directives from high-level government officials who, with the exception of a few enlightened individuals, show almost no understanding of ecosystem function. To some extent, this is a problem in any country where conservation is perceived as threatening economic development. But in Cambodia, it has been especially difficult because national mid-level staff or agency junior officers have had difficulty presenting scientific arguments to guide management, often due to the educational gaps described above.

Patronage politics also make it difficult to manage threats because most major threats (logging, land encroachment, some poaching) are directed from more powerful, more dangerous levels of society. Trying to contain these threats, even through legal action, can be life threatening, as even the courts are not independent of patronage. This situation stifles national capacity because conservation best practices can only be encountered on foreign study tours, and local conservationists can feel the situation is hopeless in Cambodia.

This attitude affected the Dry Forests Coalition, which was conceived as an ongoing national forum for regular stakeholder consultation during and after the planning phase, eventually to be led cooperatively by relevant government agencies. Most government participants, however, saw it as NGO-driven, therefore participants would generally support the ideas presented because travel costs for participation were covered by the NGO. Independent NGOs like WWF could function as interlocutors for cooperation, but in a patronage system where government agencies function only hierarchically, most government officials seem unwilling to develop lateral, cooperative efforts; they consider it not their role, or they don't know how.

However, this is not to say that the situation cannot change. As a country like Cambodia develops and its society becomes more complex, more government decisions will require professional input to reduce the risks of failure. This will call for a certain degree of experimentation when it comes to governance structures and mechanisms. The first efforts may be made not at the national level, but more locally. There have been some recent small successes from the Eastern Plains landscape, one example being the establishment of a working group in Mondulkiri Province designed to improve planning coordination and information sharing between key provincial government line agencies, especially on issues related to conservation, land-use planning, tourism, and other development activities that concern natural resource management. Steps like these may be incremental, but eventually the cumulative success of cooperative actions should be apparent.

Economic

Competing National–International Investments

In the 2001 strategy workshop, an observation was made that dry forests have few economic alternatives other than conservation. This was believable at the time because the dry forests are known to have poor soils, frequent fires, six months of water shortage annually, and limited access to major markets. However, it now appears that these few economic alternatives are capable of generating significant impacts for conservation management.

Mineral exploration for eventual exploitation has become a national priority in Cambodia, with an explicit government position guaranteeing that no part of any PA is excluded from mineral exploration. Related to that, PA management planning is hobbled because no PA zoning plans will be approved at the national level until each area has been assessed for mineral potential, which effectively guts national policies of biodiversity protection. The situation on the ground is not so dire, as actual mining activities in Cambodia are still fairly limited in area, and up to 2009 all "legally approved" activities were technically exploratory, i.e., not fully-fledged mining operations. But the total concession area is huge, and all of it potentially available for habitat degradation by mining and related activities (Global Witness 2009). Any criticism of mining expansion is viewed by many government agencies as opposition to economic development.

One approach to resolve this threat has been to identify and establish communications with those mining companies that have a more public face, and those that have already shown a willingness to work toward sharing benefits with local communities and conservationists. Such international companies are increasingly being held accountable at home for their impacts overseas. These companies are more likely to have a voice in recommending best practices for a country like Cambodia that is just now developing new policies. Improving the awareness of communities of possible impacts is also important so that they might participate more meaningfully in consultations and planning.

In addition to mining, agricultural concessions for rubber, cassava, jatropha, teak, and other crops have exploded across the Eastern and Northern Plains landscapes since 2004, converting most areas that were not legally designated for protection. Generally, DDF is considered "degraded forest" (not productive in timber), and native plant conservation or ecosystem services are not factored into a cost-benefit analysis, partly because of a lack of data. In 2006, MPF lost about 56,000 ha from its southern part to a complex of concessions, and management was not informed until after the concession contracts were signed.

As with mining there are, however, some tools being used to assist government agencies with their planning and to reduce the threat from haphazard conversion to agriculture. A pilot land-suitability assessment was undertaken in 2008–2009 in one district to the south of MPF with the intention of highlighting the most appropriate forms of agriculture in areas designated for development. There is a risk that this also requires careful planning to ensure that rapid agricultural development in areas adjacent to PAs does not lead to encroachment in areas where boundaries are unclear or there is insufficient capacity for effective monitoring. But the key is to anticipate and to clarify to the government the significant financial and social risks from unplanned development.

Aside from direct loss of land area from protection, commercial concessions also have created indirect impacts, and not only in Cambodia, because many of the concessions cover indigenous village lands used traditionally for rotational agriculture and forest product collection. Even aside from the obvious question of justice, in cases where this conversion does not directly encroach on PA habitat, it pushes indigenous people into internal exile (having no land or resources of their own), and sets up conflicts with territorial PA managers, although the perpetrators are actually neither local people nor PA managers. In such cases, PA management is seen as either opposing development (if it advocates for local resource tenure) or as stifling indigenous livelihoods (if it maintains the PA exclusion zones).

Conservation Marketing

Conservation projects are relatively expensive, especially where only a small fraction of the running costs (e.g., of PAs) comes from government sources. Unfortunately, competition for scarce funding necessarily arises within the conservation field in general, and within individual organizations. Much funding is dependent on public recognition of conservation values, and in this regard tropical dry forests fail miserably. Within the tropics, rain forests clearly are recognized, with corals and some major rivers falling in behind. The Mekong River is recognized, and its conservation program has done relatively well in terms of funding.

Even within WWF, one program manager, perhaps reflecting a network consensus, said that the dry forests lack an *identity* for effective communications. Despite the suggested conservation values associated with seasonality, the more photographed face of the forests, in the dry season, looks like hell, and shows about as much biological richness as the moon. At one point, a WWF webpage included the Cambodian plains as one of ten places or species around the globe where conservation is "Too Late . . . 10 things we'll probably never get to see." That entry is gone now, but running a message like that on the Internet for more than two years says something about both public and institutional recognition of dry forests' conservation values.

Although key species populations are recovering, they are still depleted relative to observations from the 1950s, it can be misleading to make comparisons between Asian dry forests and the well-known savannas of eastern and southern Africa. There is currently no existing tourism infrastructure of international standards, and indigenous material culture has suffered from the years of conflict and displacement. These factors in poor public recognition are compounded by a lack of scientific data that could support expert recognition of the ecoregion's value for globally representative biodiversity.

A related problem in marketing the Eastern Plains as a landscape project hinges on the broad extent of conservation areas. Cambodia's PA and PF systems were inaugurated relatively late in international terms, 1993 and 2002, respectively. Based on old maps, the system planners designated relatively large PAs, particularly in what were (misleadingly) mapped as less-populated dry forests. With all new management and staff, limited budgets, poor public awareness, and increasing skepticism from competing government agencies, Cambodian conservation management and staff necessarily have restricted their focus to the establishment of sound PA management, while including innovative approaches to participatory land-use planning. As the ecoregional approach to conservation has developed internationally, the Eastern Plains Program may look like old site-based PA management to some in the WWF network, despite the fact that just PPWS and MPF cover 6,000 km², including large multi-village indigenous settlements and, in the case of MPF, a major road and river. This would be a large landscape in many other countries, and is managed as a landscape project by the WWF–Cambodia program. But it has taken some effort to justify adequate support for the two adjacent PAs. This continues to be a challenge, and is part of an ongoing effort to market what is in actual fact a large and complex landscape conservation program.

Despite all these constraints to recognition, the Eastern Plains landscape has attracted enough funding to get a foothold on a promising conservation program. In the course of several study tours and site visits involving stakeholders from Cambodia, South Africa, and India, there has been a consensus that the Eastern Plains represent a phenomenal extent of diverse and relatively intact natural habitat, which can support the recovery of its wildlife populations to carrying capacity as envisioned in the conservation plan. With incremental progress in wildlife protection and recovery, tourism development, and communications, it should be possible for more and more people to see and recognize the values of the landscape, and hopefully join in support of its conservation for generations to come.

CONCLUSIONS—LOOKING FORWARD

This chapter has offered a brief overview of salient management issues in Asian seasonally dry forests, from planning through the initial years of project implementation in the field. Discussion of issues has focused on problems because they are often weakly documented in project progress reporting, and because they can offer valuable lessons, perhaps more valuable than successes.

The ecoregional process, despite some intrinsic tensions in the initial adoption phase, has provided a strong rationale and framework for conservation planning over the dry forests, and it continues to support the rationale and strategies for specific management interventions in the highest priority sites in the Eastern Plains and Northern Plains of Cambodia. Although a formal, scientific monitoring system is still under development, many informal but useful measures show progress in attaining the basic conservation objectives. Key species populations, identified through a systematic process, appear to be recovering, based on increasing frequency of visual observation by rangers; ranger observations are significant because most rangers, as noted above, were previously hunters, so they are aware of the effort that was involved in finding quarry species as late as 2001.

Regarding the extent of habitat, there are still more than 17,000 km^2 of dry forest mosaic designated at the national level for protection in Cambodia, and most of this is actually relatively intact in terms of natural processes (depending on one's perspective on fire). Habitat conversion outside the PAs is proceeding at a rapid rate, but with an increasing recognition of the economic potential for tourism in the open forest landscapes, there is a real possibility of maintaining essential ecological processes throughout these highest priority landscapes.

Regarding threats, the concerted effort in establishment of basic PA management has apparently reduced the threat of poaching across the Eastern Plains, and there is excellent potential for most key species populations to recover to pre-conflict densities. Poaching is still a very serious threat at least for tigers because of drastically low-density populations, and for chelonians and fish because of the lack of public awareness of their conservation values. But even for these high-risk taxa, their habitats are extensive and relatively intact, and the prey base for tigers is adequate, so it is possible to be optimistic about recovery as long as protection is maintained.

The threat of habitat conversion has been a moving target with a diverse set of actors, and although there have been a few serious setbacks, there is no doubt that the rate of habitat conversion over both the Northern Plains and Eastern Plains has been slower than if there had been no landscape-scale conservation efforts and no supported PAs. Cambodia still harbors a vast extent of naturally functioning seasonally dry forest mosaic, and there is potential for maintaining these large areas as viable conservation landscapes, as long as their values are recognized by a diverse group of stakeholders.

The major objective of awareness-raising to support conservation has seen limited progress among some experts and national technical staff, although many other stakeholders seem unaware of the dry forests' representative conservation values other than that of large mammals. Among conservationists and their institutions, there is an understandably practical emphasis on large-bodied key species, i.e., large waterbirds, mega-herbivores, and top carnivores, and these foci will probably continue to serve as catalysts for support.

However, there remains a serious gap in understanding among both experts and the general public of what makes the Asian seasonally dry forests globally significant. Understanding conservation values in tropical rain forests also is not perfect, but in the case of these dry forests, even some of the basic elements of their identity are not understood well enough to support rational management. To address this, the region needs scientific field research to support a more precise, accurate, and complex understanding of the identity, i.e., the representative values, of this ecoregion and other closely related tropical dry forests (for the neotropics, see Sanchez-Azofeifa et al. 2005). This characterization must continue the identification of key wildlife species, but go beyond the emphasis on species. The duality of seasonal processes, starting with climatic trends past and projected, and also including water availability and fire, must be studied, analyzed, and described as inherent landscape elements and trends, rather than as threats. This can enhance investigation of biotic interrelationships with non-biotic elements, e.g., adaptations to seasonal drought and fire, and implications for future adaptations to global and regional climate change. These research questions apply directly to the most basic management questions: Where are we going? What are the long-term vision and goals for this landscape?

With regard to future impacts of climate change and the need for appropriate mitigation and adaptation strategies, there is currently a dearth of information on what may or may not be effective, or indeed on what the most likely impacts will be across an ecosystem. However, a timely development of baselines against which climate change impacts can be monitored is clearly important for management. In the seasonally dry forests system, this could include not only population monitoring but also monitoring of nonbiological foci such as natural water sources, or the timing and distribution of fires. Protected area systems may also need to be reassessed again to consider whether habitats are effectively protected to ensure ecosystem integrity, and whether wildlife populations can be maintained based on the credible projections of future impacts.

Financing of conservation is another long-standing challenge we must face. Most conservation organizations have been struggling with securing the required funds to do conservation, including in the dry forests, and many have talked often of the need for sustainable financing and innovative funding sources (Sanchez-Azofeifa et al. 2005). The concept of payments for ecological/environmental services (PES) has been around for some time now, but there are still relatively few good examples where significant funds have flowed to where they are most needed. There has also been much talk of the potential for generating payments to cover the costs of managing and conserving forests through reducing emissions from deforestation and degradation (REDD) mechanisms, and we wait to see if this becomes a significant reality or not. There are a multitude of other ideas, such as biodiversity offsets from mining and hydropower companies. Certainly we are well overdue to really put some of these ideas to the test.

Scientific data are needed to rationalize broader-scale management decisions, including decisions handed down from skeptical politicians. Conservationists cannot expect politicians to ponder and accept only emotional pleas for protection of big animals, much less the vague term *biodiversity*, so arguments to slow the rate of ecosystem degradation will need to be more quantitatively defensible and to cover physical as well as biological components of the ecosystem. Wherever possible, these arguments need to be put into economic terms. Wildlife conservation probably cannot compete economically in the short term with development initiatives like mining and hydropower, but conservationists need to be able to represent a broad environmental point of view covering total ecosystem function, including the long-term societal costs of converting this natural ecosystem to a landscape of overlapping and conflicting engineering projects. No one can expect that rational arguments will always be heeded, but without robust data, there are few rational arguments to present. Currently, gaps in a systems perspective are found at national and international levels, and can only be filled through careful, appropriate research.

An improved systems perspective would be the fulfillment of some basic tenets of the ecoregional approach to conservation. Communications from relevant international conservation institutions could propagate a more complex system-oriented perspective, helping to raise awareness among the general public, donors, and national decision makers. Better recognition of globally representative biodiversity occurring under the stresses of strong seasonality could hopefully translate into improved support for conservation management in the seasonally dry tropical forests, not only financially but also in terms of an improved expert analysis and understanding of the role of representative seasonal ecosystem processes within the changing biosphere.

ACRONYMS AND ABBREVIATIONS

ASEAN—Association of Southeast Asian Nations
AsECF—Asian Elephant Conservation Fund (USFWS)
CWRP—Community Wildlife Ranger Project (funding from Cat Action Treasury and USFWS)
DBL—driver of biodiversity loss
DDF—deciduous dipterocarp forest
EPLP—Eastern Plains Landscape Project (WWF, partnering with FA and MoE)
FA—Forestry Administration (MAFF)
FFI—Fauna and Flora International
GEF—Global Environment Facility
GMP—Greater Mekong Programme (WWF regional programme for Cambodia, Laos, Vietnam and Thailand)
KR—Khmer Rouge, common name for the Democratic Kampuchea regime
LMDFE—Lower Mekong Dry Forests Ecoregion
MAFF—Ministry of Agriculture, Forestry and Fisheries
MoE—Ministry of Environment
MOU—memorandum of understanding
MPF—Mondulkiri Protected Forest (MAFF/FA management)
NGO—nongovernmental organization
NTFP—non-timber forest product
PA—protected area (in Cambodia, under MoE jurisdiction)
PES—payment for ecological/environmental services
PF—protected forest (under MAFF jurisdiction)
PLUP—participatory land-use planning
PPWS—Phnom Prich Wildlife Sanctuary (MoE/DNCP management)
SEF—semi-evergreen forest

USFWS—U.S. Fish and Wildlife Service
WCS—Wildlife Conservation Society
WS—wildlife sanctuary (a subset of PAs, under MoE jurisdiction)
WWF—World Wide Fund for Nature (International); World Wildlife Fund (United States)

REFERENCES

Baltzer, M. C., T. D. Nguyen, and R. G. Shore, eds. 2001. *Towards a Vision for Biodiversity Conservation in the Forests of the Lower Mekong Ecoregion Complex*. Hanoi: WWF Indochina Programme.

Crocker, C. D. 1962. *Exploratory Survey of the Soils of Cambodia*. Phnom Penh: Royal Cambodian Government Soil Commission / USAID.

Desai, A. A., S. Chak, R. Ou, V. Lic, and A. Maxwell. 2002. *Initial Surveys to Determine the Distribution of Wild Asian Elephant Populations in Selected Areas of Northeastern Cambodia, 2000–2001*. Phnom Penh: WWF Cambodia Programme.

Desai, A., and V. Lic. 1996. *Status and Distribution of Large Mammals in Eastern Cambodia*: *Results of the First Foot Surveys in Mondulkiri and Ratanakiri Provinces*. Phnom Penh: IUCN / FFI / WWF Large Mammal Conservation Project.

Duckworth, J. W., and S. Hedges. 1998. *Tracking Tigers*: *A Review of the Status of Tiger, Asian Elephant, Gaur and Banteng in Vietnam, Lao, Cambodia and Yunnan (China), with Recommendations for Future Conservation Action*. Hanoi: WWF Indochina Programme.

Duckworth, J. W., C. M. Poole, R. J. Tizard, J. L. Walston, and R. J. Timmins. 2005. The Jungle Cat *Felis chaus* in Indochina: A Threatened Population of a Widespread and Adaptable Species. *Biodiversity and Conservation* 14:1263–80.

Fontaine, H., and D. R. Workman. 1978. The Geology and Mineral Resources of Kampuchea, Laos, and Vietnam. In *Geology and Mineral Resources of Southeast Asia*, ed. P. Nutalaya, 539–603. Bangkok: Asian Institute of Technology.

Global Witness. 2009. Country for Sale: How Cambodia's Elite Has Captured the Country's Extractive Industries. Global Witness, April 2, 2009, www.globalwitness.org/media_library _detail.php/713/km/country_for_sale (accessed June 2009).

Goodman, P. S., A. J. Conway, and R. J. Timmins. 2003. Selection and Planning of an Intensively Managed Wildlife Recovery Zone in the Dry Forest Ecoregion of North Eastern Cambodia. Internal report to WWF–Cambodia.

Kanjanavit, S. 1992. Aspects of the Temporal Pattern of Dry Season Fires in the Dry Dipterocarp Forests of Thailand. PhD diss., School of Oriental and African Studies, University of London.

Kiernan, B. 1996. *The Pol Pot Regime*: *Race, Power and Genocide in Cambodia under the Khmer Rouge, 1975–79*. New Haven, CT: Yale University Press.

Long, B., S. R. Swan, and M. Kry. 2000. *Biological Surveys in Northeast Mondulkiri, Cambodia*. Hanoi and Phnom Penh: FFI Indochina Programme / Wildlife Protection Office.

Loucks, C., M. B. Mascia, A. Maxwell, K. Huy, K. Duong, N. Chea, B. Long, N. Cox, and T. Seng. 2009. Wildlife Decline in Cambodia, 1953–2005: Exploring the Legacy of Armed Conflict. *Conservation Letters* 2 (2): 82–92.

MacInnes, M. 2007. Bunong Use of Anthropogenic Fire as a Natural Resource Management Tool in Mondulkiri Province, Northeast Cambodia. Master's thesis, University of East Anglia, UK.

Maltby, E., M. A. Holdgate, M. Acreman, and A. Weir, eds. 1996. *The Scientific Basis of Ecosystem Management towards the Third Millennium*. Proceedings and Syntheses of the First Sibthorp Seminar held at the Royal Holloway Institute for Environmental Research, University of London, 21–22 June 1996.

Maxwell, A. L. 2004. Fire Regimes in North-Eastern Cambodian Monsoonal Forests, with a 9300-year Sediment Charcoal Record. *Journal of Biogeography* 31:225–39.

Maxwell, A., and Y. Pinsonneault, eds. 2001. *Proceedings of the Conservation Strategy Workshop: Dry Forest Landscapes of Northern and North-Eastern Cambodia*. Phnom Penh: WWF Cambodia Program.

Mittermeier, R. A., N. Myers, J. Thomsen, G. A. B. da Fonseca, and S. Olivieri. 1998. Biodiversity Hotspots and Major Tropical Wilderness Areas: Approaches to Setting Conservation Priorities. *Conservation Biology* 12:516–20.

Olivier, R., and M. Woodford. 1994. *Aerial Surveys for Kouprey in Cambodia March 1994*. Gland, Switzerland, and Cambridge, UK: IUCN Species Survival Commission.

Olson, D., and E. Dinerstein. 1998. The Global 200: A Representation Approach to Conserving the Earth's Most Biologically Valuable Ecoregions. *Conservation Biology* 12:502–15.

———. 2002. The Global 200: Priority Ecoregions for Global Conservation. *Annals of the Missouri Botanical Garden* 89:199–224.

Pierce, S. M., R. M. Cowling, T. Sandwith, and K. MacKinnon. 2002. *Mainstreaming Biodiversity in Development: Case Studies from South Africa*. Washington, DC: World Bank.

Prins H. H. T., J. G. Grootenhuis, and T. T. Dolan. 2000. *Wildlife Conservation by Sustainable Use*. Dordrecht: Kluwer Academic Press.

Rollet, B. 1962. *Inventaire forestier de l'Est Mekong*. Rome: United Nations Food and Agriculture Organization.

———. 1972. La végétation du Cambodge. *Bois et Forêts des Tropiques* 144:3–15; 145:23–38; 146:3–20.

Sanchez-Azofeifa, G. A., M. Kalacska, M. Quesada, J. C. Calvo-Alvarado, J. M. Nassar, and J. P. Rodriguez. 2005. Need for Integrated Research for a Sustainable Future in Tropical Dry Forests. *Conservation Biology* 19 (2): 285–86.

Sanderson, E., J. Forrest, C. Loucks, J. Ginsberg, E. Dinerstein, J. Seidensticker, P. Leimgruber, M. Songer, A. Heydlauff, T. O'Brien, G. Bryja, S. Klenzendorf, and E. Wikramanayake. 2006. *Setting Priorities for the Conservation and Recovery of Wild Tigers: 2005–2015. The Technical Assessment*. New York and Washington, DC: WCS / WWF / Smithsonian / NFWF-STF.

Schmid, M. 1974. *Végétation du Vietnam: Le massif Sud-Annamitique et les régions limitrophes*. Paris: Office de la Recherche Scientifique et Technique d'Outre-Mer.

Seng, K. H., B. Pech, C. M. Poole, A. W. Tordoff, P. Davidson, and E. Delattre. 2003. *Directory of Important Bird Areas in Cambodia: Key Sites for Conservation*. Phnom Penh: Department of Forestry and Wildlife / Department of Nature Conservation and Protection / BirdLife International in Indochina / WCS Cambodia Programme.

Stattersfield, A. J., M. J. Crosby, A. J. Long, and D. C. Wege. 1998. *Endemic Bird Areas of the World: Priorities for Biodiversity Conservation*. Cambridge: BirdLife International.

Stolton, S., M. Hockings, N. Dudley, K. MacKinnon, and T. Whitten. 2005. *Reporting Progress at Protected Area Sites*. Gland, Switzerland: WWF.

Stott, P. 1986. The Spatial Pattern of Dry Season Fires in the Savanna Forests of Thailand. *Journal of Biogeography* 13 (4): 345–58.

———. 1988. The Forest as Phoenix: Towards a Biogeography of Fire in Mainland Southeast Asia. *Geographical Journal* 154 (3): 337–50.

Timmins, R., and R. Ou. 2001. *The Importance of Phnom Prich Wildlife Sanctuary and Adjacent Areas for the Conservation of Tigers and Other Key Species: A Summary*. Phnom Penh: WWF Cambodia Conservation Program.

Tordoff, A. W., R. J. Timmins, A. Maxwell, K. Huy, V. Lic, and E. H. Khou. 2005. Biological Assessment of the Lower Mekong Dry Forests Ecoregion: Final Draft Report. Phnom Penh: WWF Indochina Programme.

Wharton, C. H. 1957. An Ecological Study of the Kouprey *Novibos sauveli* Urbain. *Monograph of the Institute of Science and Technology*, no. 5, Manila.

———. 1966. Man, Fire and Wild Cattle in North Cambodia. In *Proceedings of the 5th Annual Tall Timbers Fire Ecology Conference*, 23–65.

Wikramanayake, E., E. Dinerstein, C. J. Loucks, D. M. Olson, J. Morrison, J. Lamoreux, M. McKnight, and P. Hedao. 2002. *Terrestrial Ecoregions of the Indo-Pacific: A Conservation Assessment*. Washington, DC: Island Press.

WWF Greater Mekong Programme. 2005. *Conservation Plan for the Lower Mekong Dry Forest Ecoregion, 2006–2010*. Vientiane, Laos: WWF Greater Mekong Programme.

Index

forest products, human use of, 308–10, *310*;
    from Chatthin Wildlife Sanctuary, 308–10,
    310, 311; factors affecting, *312*; mitigation
    of, 314–16; in protected areas, 313–16. *See
    also* non-timber forest products
forestry: community, 315; fire and, 25. *See
    also* forest management
frogs: behavior of, 287; diet of, 288;
    distribution of, 283–84; diversity of, *283*;
    in herpetofaunas, *282*; size classes for,
    286; speciation of, *279*; study on, 278
frugivores: Asiatic black bear as, 253; diet
    seasonality of, 209; sun bear as, 253;
    threat to, 211
fruit: in Asiatic black bear diet, *259*, 260–61;
    availability-use relationship with, 206–8;
    in gibbons' diet, 199, 201, *202, 205*;
    seasonal availability of, 199, 201; in sun
    bear diet, *259*, 260–61
fruiting, interannual variation in, 201, 203–4,
    210
fuelwood, 4–5; from Chatthin Wildlife
    Sanctuary, *310*, 311, 313; management
    of, 88
Fulvous Treeduck, *141*

Garcinia benthamii, 202
Gardenia, 48
gaur (*Bos gaurus*): decline in, 180; ecological
    traits of, 179–80; fire and, 189; forage
    eaten by, 182, *183–84*; geographic
    distribution of, 180–81, *181*, 182, 185;
    habitat protection and manipulation for,
    189; habitat use of, 182; herd size of,
    185; home range of, 182, 185; population
    of, 185; predators of, 187–88; protected
    status of, 188; in Thailand, *181*; ungulate
    interactions with, 186–88
GDTDF. *See* global distribution of tropical
    dry forest
geology, of Mudumalai, 39–40
Giant panda (*Ailuropoda melanoleuca*), 218
gibbons (Hylobatidae): diet variation of,
    204, *205*, 206; feeding behavior of, 199,
    201, *203*, 206–8; forest types with, 208;
    fruits eaten by, 199, 201, *202, 205*;
    limits on, 208–9; population density of,
    196; ranging behavior of, 201, 209–10;
    research on, 196
gingers (Zingiberaceae), 115, *123, 127*;
    classifications of, 116, *119*; conservation

of, 128; dispersal-vicariance analysis of,
    118; dormancy capability of, 121–22,
    124, 126; ecology of, 119, 121–22;
    evergreen, *125*; family lineages of, 119;
    geographic distribution of, *121*; habitats
    of, 119, 121–22; molecular phylogeny of,
    116–17; morphological characteristics of,
    120; phylogenetic relationships of, *117–
    18*; pollinators of, 122–24; seasonality of,
    120, 125; taxonomic diversity of, 115–19;
    underground water storage organs of, *122*
Gir Forest National Park, 277
Gironniera nervosa, 202
GLC2000. *See* Global Land Cover 2000
Global 200, 366
global distribution of tropical dry forest
    (GDTDF), 62; accuracy of, 67;
    performance of, 69–70; SDTF area
    estimates with, *64*; SDTF spatial
    distribution estimates from, *65*
Global Fire Initiative, 329
Global Land Cover 2000 (GLC2000), 61,
    62; accuracy of, 67; SDTF area estimates
    from, *64*; SDTF spatial distribution
    estimates from, *65*
Globba wengeri, 123
Globbeae, *118*
Gluta usitata, 13
Golden-fronted Leafbird, *137*
Gomphandra tetrandra, 202
governance, in Cambodia, 388–89
grasses: ecotone, 152; fire regimes involving,
    337; as forage, 183
grazing: in Chatthin Wildlife Sanctuary, 309,
    310, 313; deferral of, 88; by elephants,
    151
Greater-necklaced Laughingthrush, *136*
Great Tit, *136*
Green Bee-eater, *134*
Grey-backed Shrike, *141*
Grey Bush Chat, *136, 141*
Grey-capped Pygmy Woodpecker, *134*
Grey-headed Canary-Flycatcher, *141*
Grey Wagtail, *137*
growth rings, in tropical tree species, 28

habitat: of ants, 82; Asiatic black bear use
    of, 257–58; of banteng, 185–86, 189; in
    Chatthin Wildlife Sanctuary, 132; of deer,
    169–70; definition of, 276; of Eld's deer,
    170–71; of elephants, 153–54; of gaur,

List of Contributors

Teri D. Allendorf
University of Wisconsin
Madison, Wisconsin, United States
teriallendorf@yahoo.com

Myint Aung
Friends of Wildlife
Yangon, Myanmar
myintaungwildlife@gmail.com

Megan C. Baker
Smithsonian Conservation Biology Institute
Front Royal, Virginia, United States
bakerm@si.edu

Patrick Baker
Australian Center for Biodiversity, School of Biological Sciences
Monash University
Melborne, Australia
Patrick.baker@sci.monash.edu.all

Naris Bhumpakphan
Department of Forest Biology
Faculty of Forestry
Kasetsart University
Bangkok, Thailand
ffornrb@ku.ac.th

Warren Y. Brockelman
Ecology Laboratory, Biotech Central Research Unit
National Science and Technology Development Agency, Thailand;
and Department of Biology, Faculty of Science, Mahidol University, Salaya, Thailand
wybrock@cscoms.com

Sarayudh Bunyavejchewin
Wildlife and Plant Conservation Department, Research Office
National Parks
Bangkok, Thailand
sarayudh_b@yahoo.com

Raghunandan Chundawat
Baavan, S17 Panchsheel aprt.
A1 block Panchsheel Enclave
New Delhi, 110017 India
ragu.baavan@gmail.com

Nicholas J. Cox
Dry Forests Ecoregion Program Coordinator
World Wildlife Fund Greater Mekong
Vientiane, Lao
nick.cox@wwfgreatermekong.org

Peter Cutter
World Wildlife Fund Thailand
Bangkok, Thailand

Virginia Dale
Environmental Sciences Division
Oak Ridge National Laboratory
Oak Ridge, Tennessee, United States
dalevh@ornl.gov

Handanakere Shivaramaiah Dattaraja
Center for Ecological Sciences
India Institute of Science
Bangalore, India
dattaraj@ces.iisc.ernet.in

Stuart J. Davies
Center for Tropical Forest Science
Smithsonian Tropical Research Institute and Arnold Arboretum
Harvard University
22 Divinity Avenue
Cambridge, Massachusetts, United States

Ruth DeFries
Ecology, Evolution, and Environmental Biology
Columbia University
New York, New York
Red2402@columbia.edu

Melanie Delion
Smithsonian Conservation Biology Institute
Front Royal, Virginia, United States

Prithiviraj Fernando
Research Associate, Smithsonian Conservation Biology Institute and Eco
    Health Alliance
Rajagiriya, Sri Lanka
pruthu62@gmail.com

Johann Georg Goldammer
Global Fire Monitoring Center / Fire Ecology Research Group, Max Planck
    Institute for Chemistry
c/o University of Freiburg / United Nations University
Freiburg, Germany
johann.goldammer@fire.uni-freiburg.de

Bhim Gurung
University of Minnesota
St. Paul, Minnesota, United States
guru0023@umn.edu

A. J. T. Johnsingh
Wildlife Institute of India
Dehradun, India
Present address: Nature Conservation Foundation, Mysore, India
ajt.johnsing@gmail.com

W. John Kress
Department of Botany
National Museum of National History, Smithsonian Institution
Washington, DC, United States
kressj@si.edu

Peter Leimgruber
Smithsonian Conservation Biology Institute
Front Royal, Virginia, United States
leimgruberp@si.edu

Megan MacInnes
meganmacinnes@yahoo.co.uk

Andrew L. Maxwell
Consultant Conservation Advisor
Stuart, Virginia, United States
amaxwell52@yahoo.com

Chuck McDougal
International Trust for Nature Conservation
Kathmandu, Nepal

William J. McShea
Smithsonian Conservation Biology Institute
Front Royal, Virginia, United States
mcsheaw@si.edu

Nandita Mondal
Center for Ecological Sciences
India Institute of Science
Bangalore, India
nandita@ces.iisc.ernet.in

Thida Oo
Friends of Wildlife
Yangon, Myanmar

R. Raghunath
Nature Conservation Foundation
Mysore, India

John H. Rappole
Smithsonian Conservation Biology Institute
Front Royal, Virginia, United States
rappolej@si.edu

John Seidensticker
Smithsonian Conservation Biology Institute
Washington, DC, United States
seidenstickerj@si.edu

Nay Myo Shwe
Nature and Wildlife Conservation Division
Myanmar Forestry Department
Naypajidaw, Myanmar

Achara Simchareon
Research Division, Department of National Parks
Wildlife and Plant Conservation
Bangkok, Thailand

Saksit Simchareon
Research Division, Department of National Parks
Wildlife and Plant Conservation
Bangkok, Thailand

James L. David Smith
University of Minnesota
St. Paul, Minnesota, United States
smith017@umn.edu

Melissa Songer
Smithsonian Conservation Biology Institute
Front Royal, Virginia, United States
songerm@si.edu

Robert Steinmetz
World Wildlife Fund Greater Mekong
Roberts@wwfgreatermekong.org

Raman Sukumar
Professor and Chairman, Center for Ecological Sciences
India Institute of Science
Bangalore, India
rsuku@ces.iisc.ernet.in

Hebbalalu Satyanarayana Suresh
Center for Ecological Sciences
India Institute of Science
Bangalore, India
suresh@ces.iisc.ernet.in

Khaing Khaing Swe
Friends of Wildlife
Yangon, Myanmar

Kobsak Wanthongchai
Department of Silviculture, Faculty of Forestry
Kasetsart University
Bangkok, Thailand
kobsak.w@ku.ac.th

Chris Wemmer
California Academy of Sciences
San Francisco, California, United States
chindwin@sbcglobal.net

K. Yoganand
Wildlife Institute of India
Dehradun, India
Present address: Center for African Ecology, School of Animal and Plant
    Environmental Sciences
University of Witwatersrand
PL Bag 3, Wits 2050, South Africa
k.yoganand@gmail.com

George R. Zug
Department of Vertebrate Zoology
Smithsonian Institution, National Museum of Natural History
Washington, DC, United States
zugg@si.edu